Terotechnology XIV

14th International Conference on Terotechnology,
Kielce, Poland, 16-17 October 2025

Editors

Norbert Radek[1]
Agnieszka Szczotok[2]

[1]Kielce University of Technology
[2]Silesian University of Technology

Peer review statement

All papers published in this volume of “Materials Research Proceedings” have been peer reviewed. The process of peer review was initiated and overseen by the above proceedings editors. All reviews were conducted by expert referees in accordance to Materials Research Forum LLC high standards.

Published under License by **Materials Research Forum LLC**
Millersville, PA 17551, USA

Published as part of the proceedings series
Materials Research Proceedings
Volume 62 (2026)

ISSN 2474-3941 (Print)
ISSN 2474-395X (Online)

ISBN 978-1-64490-400-8 (Print)
ISBN 978-1-64490-401-5 (eBook)

Distributed worldwide by

Materials Research Forum LLC
105 Springdale Lane
Millersville, PA 17551
USA
https://mrforum.com

Manufactured in the United State of America
10 9 8 7 6 5 4 3 2 1

Table of Contents

Preface

Terotechnology is the technology of installation, commissioning, maintenance, replacement, and removal of plant machinery and equipment, feedback on operation and design, and related subjects and practices.

It has been over twenty years since the first Conference on Terotechnology was held. The conference is still a venture of Kielce University of Technology and the Polish Maintenance Society. The Terotechnology conference began in 2001, and subsequent meetings were held from 2005 to 2009 in the annual cycle. In 2011, a biannual conference cycle was established.

The 14th International Conference on Terotechnology was held at the Kielce University of Technology, Kielce, Poland, on 16-17 October 2025, situated halfway between Kraków and Warsaw. The conference was held under the patronage of the Minister of National Defense, Deputy Prime Minister of Poland, Władysław Kosiniak-Kamysz.

The ongoing Russian aggression against Ukraine, now in its fourth year, the growing military threat from Russia, and its military needs were reflected in the content of most articles. Issues of artificial intelligence, social impact, environmental protection, and related legal regulations are also gaining importance and strongly influencing the technologies used. Circular economy and technology design adapted to the use of recycled materials are becoming increasingly prominent. Product customization and material savings are leading to the increasingly widespread use of processing techniques using concentrated energy beams, particularly lasers.

The selected papers from the 14th Conference have been reviewed by at least two experts and prepared for this volume of *Materials Research Proceedings*. The articles were chosen based on their quality, scientific merit, and relevance to the conference. The volume represents the recent advances in materials, technologies, and methods.

The Organizing Committee is grateful to the deans and workers of the Centre for Laser Technologies of Metals of the Kielce University of Technology, Department of Applied Computer Science of the Cracow University of Technology, who helped with the organization of the 14th Conference and the edition of this volume. The Committee would also like to thank all who have updated and reviewed the papers submitted to the conference, the scientific secretary for her editing work, and Materials Research Forum LLC for producing the volume.

Conference Organizers

Kielce University of Technology
Polish Maintenance Society
F.H. Barwa
Military Institute of Engineering Technology

Scientific Committee

Organizing Committee

Multiphysics Modeling of Vibrations in an Asynchronous Spindle Motor Driven by a Magnetic Eccentric

Andrii GOROSHKO[1,a *], Maryna ZEMBYTSKA[1,b] and Renata DWORNICKA[2,c]

[1]Khmelnytskyi National University, 11 Institutska st, 29016 Khmelnytskyi, Ukraine

[2]Cracow University of Technology, 37 Jana Pawla II, 31-864 Krakow, Poland

[a]horoshkoan@khmnu.edu.ua, [b]marynazembytska@gmail.com, [c]renata.dwornicka@pk.edu.pl

Keywords: Motor-Spindle, Vibrations, UMP, Eccentricity, Simscape

Abstract. One of the main reasons for the failure of asynchronous motor-spindles is the rotor vibration caused by its eccentric position. To study the dynamics of an eccentric motor-spindle, a mathematical model was created in this work, which differs from known models in that the analytical expressions for UMP are based on the law of change in inductance magnetization. This made it possible to obtain analytical dependencies not only for unbalanced magnetic forces, but also for the remaining electrical and magnetic parameters of the motor. The simulation results for the case of dynamic eccentricity are confirmed by the experimental results for a motor-spindle with a power of 2.2 kW. This confirms the accuracy of the model. In general, the theoretical and practical results obtained in the work can be used for convenient modeling of vibration processes in engines and for analyzing problems and diagnosing motor-spindle malfunctions.

Introduction

For induction motors, bearings fixed in special stands are used as rotor supports. For induction motors, bearing failures account for about 40% of the total number of failures of induction machines [1, 2]. Bearing failures are mainly caused by increased wear. The main wear factors are mechanical rotor imbalance due to the eccentricity of the rotor mass and unbalanced magnetic tension (UMP) due to magnetic eccentricity. Since induction motors have a relatively small air gap, they are more prone to UMP. Dynamic forces and moments caused by rotor eccentricity are additional internal excitation for the motor subsystem, which includes centrifugal force (CF), friction-impact forces (RIF), UMP, unbalanced force torque, and friction torque [3]. Today, the issue of induction motor dynamics is actively studied [4-6]. Special requirements for low vibration levels apply to motor spindles, since exceeding the permissible vibrations not only negatively affects the service life of the motor, but also worsens the accuracy of machining on the machine tool. Therefore, to combat vibrations, it is important to have accurate and effective models of motor spindle dynamics. Due to the high requirements for low vibration levels of motor spindles, the problem of reducing and controlling vibrations is relevant. In [7], the numerical solution of the control equation for a two-rotor system with unbalanced and defective coupling was considered. In [8], the dynamics of an eccentric motor spindle were studied; for this purpose, a model was obtained that takes into account magnetic eccentricity and mass eccentricity. In [9], the authors investigated the relationship between thermal expansion and the size of the air gap, which causes the appearance of UMP and frequency-dependent nonlinear changes in stiffness. The authors of [10] proposed a finite element model of the rotor system of the motor-spindle using the Timoshenko beam theory, taking into account the UMP and centrifugal force.

The purpose of this article is to create a mathematical model of the rotor oscillations of an asynchronous spindle motor, which are caused by the magnetic eccentricity of the rotor. Unlike traditional models for calculating UMP, which use the energy method or the integration of the Maxwell stress tensor in the air gap between the stator and the rotor, expanding the magnetic permeability of the gap into a Fourier series, this work proposes to model UMP using the variable mutual inductance of the motor. It is the change in inductance due to the uneven gap that causes the appearance of UMP. In addition, the change in inductance has an impact on the electrical and mechanical characteristics of the motor-spindle.

The issues discussed here are important not only for purely mechanical problems [11]. Industrial users of engines also encounter similar difficulties, for example in foundry [12], welding [13], but also with various coating techniques [14,15], including such specific ones as DLC coatings [16], ESD method [17,18], or laser micromachining [19]. Instability in engine operation significantly affects the deterioration of precision manufacturing processes such as laser texturing [20,21] and reduces product quality [22,23]. Induced vibrations, transferred to other structural elements, can lead to increased wear, including, among others, reduced corrosion resistance [24,25] due to the formation of vibration-induced scratches. Such quality-degrading phenomena also reduce the suitability of both the process and the products for potential military applications [26], high-precision biotechnological applications [27,28], or specific consumer applications [29,30]. In addition to the theoretical-analytical approach proposed in this article, one could also consider DOE methods widely used in industry [31,32], including those using uncertainty estimation [33], resampling [34,35], and the fuzzy approach [36-38]. This would provide a significant impetus for the development of similar analytical [39] and environmental-prognostic [40] methods.

Methodology

In the general case, in an induction motor, any rotor with mass m is characterized by a mass eccentricity e and a magnetic eccentricity e_m. When the rotor rotates at a speed ω, the eccentricity of the mass causes the appearance of an unbalanced radial force $e\omega^2 m$ (Fig. 1). The uneven air gap caused by the magnetic eccentricity leads to the appearance of an unbalanced magnetic tension, characterized by a radial force F_{UMP} (Fig. 2).

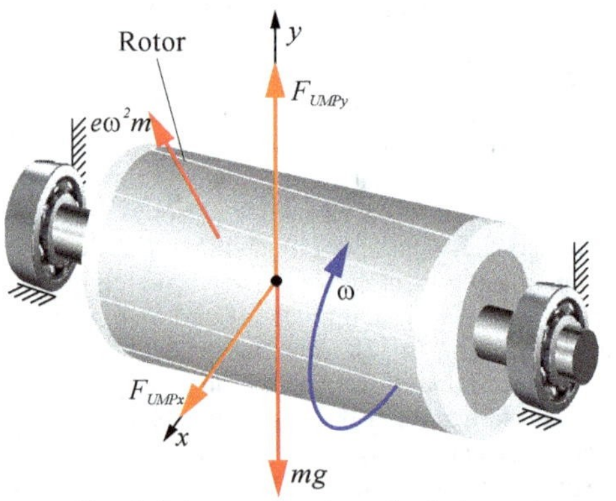

Fig. 1. Scheme of action of unbalanced forces at rotor eccentricity

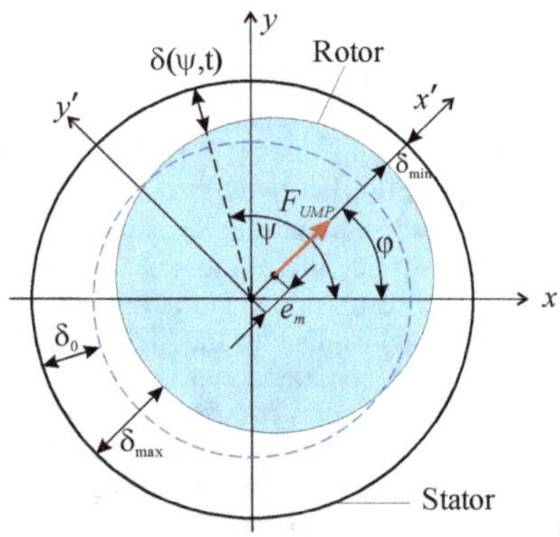

Fig. 2. Scheme of the magnetic eccentricity of the rotor

With static magnetic eccentricity, the air gap is a function of the rotor rotation angle ψ

$$\delta(\psi,t) = \delta_0 - e_m \cos(\psi-\varphi) = \delta_0\left(1-\varepsilon\cos(\psi-\varphi)\right), \tag{1}$$

where δ_0 – average air gap value when centering the rotor,

ψ - angular coordinate on the rotor circle,
e_m – magnetic eccentricity value,
$\varepsilon = \frac{e_m}{\delta_0}$ – relative eccentricity value,
φ - eccentricity position angle (static eccentricity angle of the rotor).

The magnetic permeability of the gap is a function of the rotation angle and the relative eccentricity:

$$\Lambda(\psi,\varepsilon) = \frac{\mu_0}{\delta(\psi,t)} = \frac{\mu_0}{\delta_0\left(1-\varepsilon\cos(\psi-\varphi)\right)}. \tag{2}$$

The inductance of the magnetizing circuit $L_m(\psi,\varepsilon) \sim \Lambda(\psi,\varepsilon)$ depends on the number of turns and the cross-sectional area of the magnetic core. Then, taking into account (2), we have

$$L_m(\psi,\varepsilon) = L_{m0}\left(1-\varepsilon\cos(\psi-\varphi)\right)^{-1} \tag{3}$$

where L_{m0} is the gap inductance in the absence of eccentricity.

Dependence (3) is approximate, since it does not consider the phenomenon of saturation of the magnetic core, the influence of teeth and grooves, considering the stator and rotor as absolutely smooth cylinders. The scattering of the magnetic field outside the magnetic core is not considered. The magnetic permeability is considered infinitely large, and all the magnetic resistance is concentrated in the air gap.

With dynamic eccentricity and constant rotor rotation speed ω = const, the eccentricity position angle is a function of time

$$\varphi(t) = \omega t + \varphi_0, \tag{4}$$

where φ_0 – initial angle of the dynamic eccentricity position,
ω - rotor angular velocity.

In this case, the mutual inductance $L_m(\varepsilon,t)$ for a given eccentricity is a function of time and the angle of rotation of the rotor (4). The. The instantaneous value of the eccentricity projections is defined as $e_x = e_m\cos(\omega t+\varphi_0)$, $e_y = e_m\sin(\omega t+\varphi_0)$.

The UMP force is the derivative of the accumulated magnetic energy along the displacement coordinate e_m. To account for the effect of all the energy at each point in the gap, we apply integration over the entire circle:

$$F_{UMP} = -\frac{\partial W}{\partial e_m} = -\frac{1}{2}i_m^2(t)\int_0^{2\pi}\frac{\partial L(\psi,t)}{\partial e}d\psi, \tag{5}$$

where $i_m(t)$ is the instantaneous value of magnetizing current, and

$$\frac{\partial L(\psi,t)}{\partial e_m} = \frac{L_{m0}}{\delta_0}\cdot\frac{\cos(\psi-\omega t-\varphi_0)}{\left(1-\varepsilon\cos(\psi-\omega t-\varphi_0)\right)^2}. \tag{6}$$

The total resultant magnetic tension force is defined as

$$F_{UMP} = -\pi L_{m0}i_m^2(t)e_m\delta_0^{-2}\left(1-\varepsilon^2\right)^{-\frac{3}{2}}. \tag{7}$$

The projections of the force UMP on the coordinate axis, along which the displacement of the dynamic eccentricity e_m occurs during rotor rotation, is calculated as

$$F_{UMPx} = F_{UMP}\cos\varphi(t) = -\pi L_{m0} i_m^2(t)\varepsilon\delta_0^{-1}\left(1-(\varepsilon)^2\right)^{-\frac{3}{2}}\cos(\omega t+\varphi_0),$$
$$F_{UMPy} = F_{UMP}\sin\varphi(t) = -\pi L_{m0} i_m^2(t)\varepsilon\delta_0^{-1}\left(1-(\varepsilon)^2\right)^{-\frac{3}{2}}\sin(\omega t+\varphi_0). \quad (8)$$

The instantaneous value of the magnetizing current can be given as

$$i_m = I_m \cos\omega_e t, \quad (9)$$

where ω_e - cyclic frequency of the stator current.

Considering (9), we have

$$F_{UMPx} = -\pi\varepsilon\delta_0^{-1}\left(1-\varepsilon^2\right)^{-\frac{3}{2}} L_{m0} I_m^{\,2}(1+\cos 2\omega_e t)\cdot\cos(\omega t+\varphi_0),$$
$$F_{UMPy} = -\pi\varepsilon\delta_0^{-1}\left(1-\varepsilon^2\right)^{-\frac{3}{2}} L_{m0} I_m^{\,2}(1+\cos 2\omega_e t)\cdot\sin(\omega t+\varphi_0). \quad (10)$$

Taking $K = -\pi\varepsilon\delta_0^{-1}\left(1-\varepsilon^2\right)^{-\frac{3}{2}} L_{m0} I_m^{\,2}$, we have

$$F_{UMPx} = 2K\cos(\omega t+\varphi_0) + K\cos\left((2\omega_e-\omega)t-\varphi_0\right) + K\cos\left((2\omega_e+\omega)t+\varphi_0\right),$$
$$F_{UMPy} = 2K\sin(\omega t+\varphi_0) + K\sin\left((2\omega_e+\omega)t+\varphi_0\right) - K\sin\left((2\omega_e-\omega)t-\varphi_0\right). \quad (11)$$

Here, $2K\cos(\omega t+\varphi_0)$ is the UMP power storage is based on the rotor winding frequency. The other two combined terms represent two wrapping force vectors: one that wraps prolongually, and one that extends to the wrapper.

Having looked at the partial deviation from the static eccentricity at φ=const, we can:

$$F_{UMPx} = 2K\cos\varphi + K\cos(2\omega_e t+\varphi) + K\cos(2\omega_e t-\varphi),$$
$$F_{UMPy} = 2K\sin\varphi + K\sin(2\omega_e t+\varphi) - K\sin(2\omega_e t-\varphi). \quad (12)$$

Also, because of the static magnetic eccentricity, the UMP force has a constant reserve, directed towards the smallest value of the gap $\delta_{\min}$ and reserve, which pulsates at a constant electrical frequency. With dynamic eccentricity, the stationary warehouse transforms into a harmonic at the rotor rotation frequency.

To integrate the UMP into the mechanical vibration of the rotor, it is necessary to consider the dynamic rotor-stator model, which would account for the spring-inertial and dissipative powers of the bearing supports. In the proposed model, a rigid rotor of mass m of the motor has three degrees of freedom and, rotating with an angular velocity ω, can make small translational movements in the direction of the x and y axes. In the model shown in Fig. 3, the following assumptions are made: the elastic characteristics of all supports are linear; the rotor is assumed to be rigid and its deformations are neglected; the position of the center of mass point S of the rotor is known; the stator is considered to be absolutely rigid and rigidly fixed to a rigid foundation.

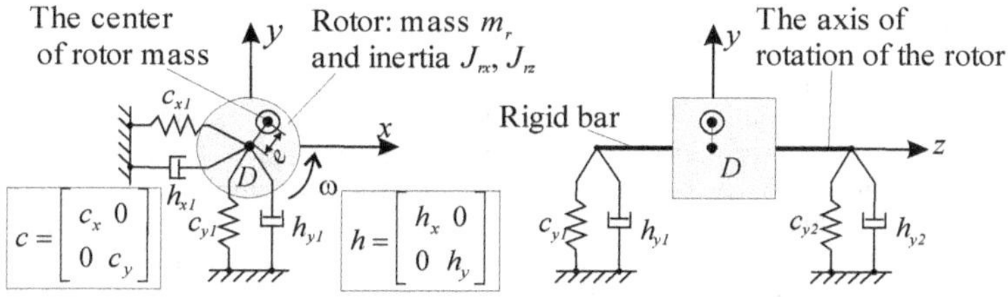

Fig. 3. Dynamic model of a rotor with mass eccentricity

In the diagram, the mass eccentricity is equal to *e*, where point *D* is obtained as a result of the intersection of the rotor axis of rotation with a plane drawn through the center of mass point *S* perpendicular to the axis of rotation of the rotor. For generalized coordinates, we take the *x*, *y* coordinates of point *D*. The differential equations of rotor oscillations are obtained from the Lagrange equation of the second kind, considering the energy dissipation during Rayleigh damping, and taking into account F_{UMP}:

$$\begin{aligned} m\ddot{x} + \left(h_{x1} + h_{x2}\right)\dot{x} + \left(c_{x1} + c_{x2}\right)x = m\omega^2 e \sin\omega t + F_{UMPx}, \\ m\ddot{y} + \left(h_{y1} + h_{y2}\right)\dot{z} + \left(c_{y1} + c_{y2}\right)z = m\omega^2 e \cos\omega t + F_{UMPy}. \end{aligned} \tag{13}$$

where $\dot{x}_s, \dot{z}_s$ are projections of the full velocities of the point *S*;

$h_{x1}, h_{x2}, h_{z1}, h_{z2}$ are the projections of the viscous friction vectors of supports 1 and 2;

$c_{x1}, c_{x2}, c_{z1}, c_{z2}$ are projections of the stiffness vectors of the supports 1 i 2.

The developed model was implemented using the standard Asynchronous Machine block from the Simulink/Simscape library. To describe the dynamic processes in an asynchronous machine, a flat orthogonal coordinate system *dq* is used here, which rotates at a speed of ω. The electrical part of the engine is described by a fourth-order system, and the mechanical part by a second-order system. All electrical variables and parameters refer to the stator. The system of differential equations describing the dynamics of an asynchronous machine when changing voltage, currents, torques, and speeds has the form [41]:

$$\begin{aligned} U_{qs} &= R_s i_{qs} + \frac{d\psi_{qs}}{dt} + \omega_e \psi_{ds}, & M_e &= 1.5p\left(\psi_{ds} i_{qs} - \psi_{qs} i_{ds}\right), \\ U_{ds} &= R_s i_{ds} + \frac{d\psi_{ds}}{dt} - \omega_e \psi_{qs}, & \psi_{qs} &= L_s i_{qs} + L_m i'_{qr}, \\ & & \psi_{ds} &= L_s i_{ds} + L_m i'_{dr}, \\ U'_{qr} &= R'_r i'_{qr} + \frac{d\psi'_{qr}}{dt} + \left(\omega_e - \omega_r\right)\psi'_{dr}, & \psi'_{qr} &= L'_r i'_{qr} + L_m i_{qs}, \\ & & \psi'_{dr} &= L'_r i'_{dr} + L_m i_{ds}, \\ U'_{dr} &= R'_r i'_{dr} + \frac{d\psi'_{dr}}{dt} - \left(\omega_e - \omega_r\right)\psi'_{qr}, & L_s &= L_{ls} + L_m, L'_r = L'_{lr} + L_m. \end{aligned} \tag{14}$$

where $U_{qs} = U_m \cos\omega t$, $U_{qs} = U_m \sin\omega t$ - voltages on the stator windings along the *q* and *d* axes, respectively; U'_{qr}, U'_{dr} - stator voltage; i_{qs}, i_{ds} - stator currents; i'_{qr}, i'_{dr} - rotor currents; ψ_{qs}, ψ_{ds} - stator *q*-axis and *d*-axis fluxes; $d\psi'_{qr}, d\psi'_{dr}$ - rotor *q*-axis and *d*-axis fluxes; L_s – total

Materials Research Proceedings 62 (2026) 1-10 https://doi.org/10.21741/9781644904015 1

stator inductance, L_{ls} stator leakage inductance; L_r' - total rotor inductance, L_m – inductance of the magnetizing circuit; ω_r – electrical power of the rotor, M_e - electromagnetic torque of the motor.

In the basic model, the *Mutual inductance* L_m*=const* parameter is entered once at the beginning of the simulation and is not expected to change during the simulation. In order to implement a variable magnetization inductance (3) in the model, the basic Asynchronous Machine block was modified (Fig. 4). For this purpose, a mask adjustment was performed, which made it possible to replace the fixed parameter L_m with a variable one $L_m(\psi,t)$, which is supplied from the outside. This provided dynamic control of the parameter during the simulation and calculation of all electrical and mechanical parameters when the inductance of the magnetization circuit changes, and more accurate calculations of the electrical and mechanical parameters of the motor.

Simulation Results and Discussion

The dynamic model of a motor-spindle with an eccentric mass and unbalanced magnetic tension, described by equations (6) and (7), was implemented as a Simscape simulation model and solved numerically using MATLAB. The STF80-2.2-ER20 spindle with an average air gap at rotor centering of 0.2 mm and a rotor eccentricity of 20 μm, which corresponds to 10% relative eccentricity, was used as the object of study.

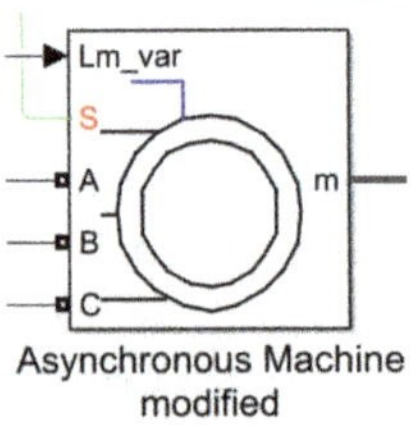

Fig. 4. Modified Simscape block

Fig. 5. Time diagram of UMP forces

Analysis of the obtained results showed that at a nominal rotation speed of 24000 rpm and a corresponding electric current frequency of 400 Hz, the dynamic magnetic eccentricity causes insignificant fluctuations in the speed, electromagnetic torque, and rotor power with the rotor rotation frequency. The time dependence of UMP, which has the form of a modulated signal, is presented in Fig. 5.

Experimental verification

For experimental verification of the model, the investigated installation with the motor-spindle STF80-2.2-ER20 with a rotor eccentricity of 20 μm, which corresponds to 10% relative eccentricity (Fig. 6), was used. To record and measure the vibration of the motor spindle, a Rigol digital oscilloscope and PI-19 amplification equipment for vibration measurements were used.

Experimental data processing was carried out in MATLAB. The vibration amplitude was measured at a nominal speed of 24000 rpm using vibration sensors installed on a support next to the spindle flange. Fig. 7 presents comparative spectral dependences obtained during simulation and during the experiment. The results are in good agreement, which indicates the high accuracy and reliability of the proposed model.

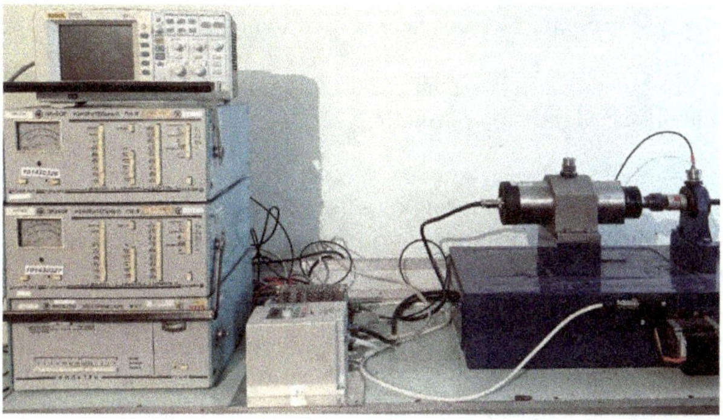

Fig. 6. Test platform for studying motor-spindle vibrations

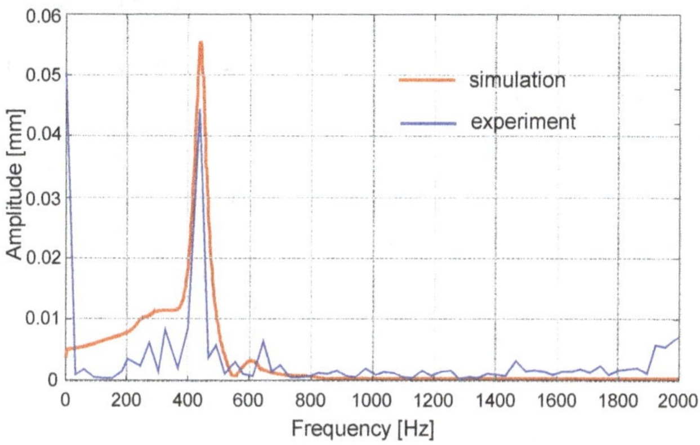

Fig. 7. Vibration spectrum obtained experimentally and in simulation

Conclusion

In contrast to the generally accepted approach to modeling UMP using Maxwell's equations and the Fourier series expansion of the magnetic permeability of the gap, this work proposes to model UMP using the variable mutual inductance of the motor. For this purpose, in the case of dynamic magnetic eccentricity, analytical dependences of magnetization inductance, dependence of UMP on magnetization inductance were obtained, and the standard Asynchronous block in Simulink/Simscape was modified. This approach makes it possible to build multiphysics models of arbitrary complexity based on the standard Simscape library. The proposed approach to modeling vibrations takes into account not only the influence of UMP, but also the influence of variable magnetization inductance on the electrical and mechanical parameters of the motor.

The simulation results confirm the insignificant influence of the permissible eccentricity on the stator current, shaft torque, and rotor speed. The obtained dependences for describing the UMP forces on the position of the rotor eccentricity and magnetizing inductance are consistent with the results obtained by other authors using the energy method and the method of integrating Maxwell's equations. Comparison of the results of full-scale experiments is in good agreement with the results of numerical simulation in Simulink/Simscape, which indicates the high accuracy and reliability of the proposed model.

The results obtained in this article are useful for diagnosing faults and predicting the dynamic eccentricity and unbalance of the motor-spindle.

References

[1] L. M. Popa et al., Condition monitoring of wind generators, 38th IAS Annual Meeting on Conf. Record of the Industry Applications Conf., Salt Lake City, Utah, USA, 3 (2003) 1839-1846. https://doi.org/10.1109/IAS.2003.1257819

[2] W. Thomson, Vibration Monitoring of Induction Motors and Case Histories on Shaft Misalignment and Soft Foot, In Vibration Monitoring of Induction Motors: Practical Diagnosis of Faults via Industrial Case Studies Cambridge: Cambridge University Press, 2020, pp. 1-46. https://doi.org/10.1017/9781108784887.002

[3] Y. Liu et al., Effect of rotor eccentricity on the dynamic performance of a traction motor and its support bearings in a locomotive, Proc. Inst. Mech. Eng., Part F: Journal of Rail and Rapid Transit. 236(9) (2022) 1080-1090. https://doi.org/10.1177/09544097211072335

[4] A. Goroshko, R. Dwornicka and M. Zembytska, Multiphysics Modeling of Direct-Start Induction Motor Considering Rotor Unbalance, Materials Research Proceedings 15 (2024) 267-276. https://doi.org/10.21741/9781644903315-31

[5] U. Werner, Influence of electromagnetic field damping on forced vibrations of induction rotors caused by dynamic rotor eccentricity, ZAMM-Journal of Applied Mathematics and Mechanics 97(1) (2017) 38-59. https://doi.org/10.1002/zamm.201500285

[6] A.V. Goroshko, M.V. Zembytska and V.P. Paiuk, Induction motor vibrations caused by mechanical and magnetic rotor eccentricity. 11(1) (2024) D66–D77. https://doi.org/10.21272/jes.2024.11(1).d8

[7] N. Wang, D. Jiang, Vibration response characteristics of a dual-rotor with unbalance-misalignment coupling faults: theoretical analysis and experimental study, Mechanism and Machine Theory 125 (2018) 207-219. https://doi.org/10.1016/j.mechmachtheory.2018.03.009

[8] Z. Wang et al., Study on the unbalanced fault dynamic characteristics of eccentric motorized spindle considering the effect of magnetic pull, Shock and Vibration, 2021(19) (2021) 5536853. https://doi.org/10.1155/2021/5536853

[9] L. Jin, Y. Li and X. Zeng, Effect of Thermal-Expansion-Induced Dynamic Air Gap Displacement on Vibration Characteristics in High-Speed Motorized Spindle Rotors, Journal of Engineering Research (2025). https://doi.org/10.1016/j.jer.2025.08.009

[10] C. Xiaoan, Z. Peng and L. Yongya, Dynamic characteristics of high-speed motorized spindles affected by rotor's eccentric mass, Journal of Vibration and Shock, 33(2) (2014) 37-40. https://doi.org/10.1088/2631-8695/ada668

[11] M. Krynke, R. Ulewicz, Analysis of the influence of slewing bearing mounting on their static load capacity, Transportation Research Procedia 40 (2019) 745-750. https://doi.org/10.1016/j.trpro.2019.07.105

[12] R. Ulewicz, K. Czerwińska and A. Pacana, A Rank Model of Casting Non-Conformity Detection Methods in the Context of Industry 4.0, Materials 16 (2023) art. 723. https://doi.org/10.3390/ma16020723

[13] N. Radek et al., Properties of Steel Welded with CO2 Laser, Lecture Notes in Mechanical Engineering 0 (2020) 571-580. https://doi.org/10.1007/978-3-030-33146-7_65

[14] L. Adamczyk, K. Giza and A. Dudek, Electrochemical preparation of composite coatings of 3,4-etylenodioxythiophene (EDOT) and 4-(pyrrole-1-yl) benzoic acid (PyBA) with heteropolyanions, Materials Chemistry and Physics 144 (2014) 418-424. https://doi.org/10.1016/j.matchemphys.2014.01.012

[15] N. Radek et al., Formation of coatings with technologies using concentrated energy stream, Production Engineering Archives 28 (2022) 117-122., https://doi.org/10.30657/pea.2022.28.13

[16] N. Radek et al., Operational properties of dlc coatings and their potential application, METAL 2022 – 31st Int. Conf. Metall. Mater. (2022) 531-536. https://doi.org/10.37904/metal.2022.4491

[17] N. Radek et al., The effect of laser treatment on operational properties of ESD coatings, METAL 2021 – 30th Anniv. Int. Conf. Metall. Mater. (2021) 876-882. https://doi.org/10.37904/metal.2021.4212

[18] N. Radek et al., Technology and applications of esd coatings before and after laser processing, METAL 2023 – Int. Conf. Metall. Mater. (2024) 500-505. https://doi.org/10.37904/metal.2023.4727

[19] A. Kalinowski et al., Laser surface texturing: characteristics and applications, System Safety: Human - Technical Facility - Environment 5 (2023) 240-248. https://doi.org/10.2478/czoto-2023-0026

[20] Ł.J. Orman et al., Laser Treatment of Surfaces for Pool Boiling Heat Transfer Enhancement, Materials 16 (2023) art. 1365. https://doi.org/10.3390/ma16041365

[21] Ł.J. Orman et al., Application of laser treatment technology for boiling heat transfer augmentation, Production Engineering Archives 30 (2024) 259-265. https://doi.org/10.30657/pea.2024.30.25

[22] R. Ulewicz, Quality management system operation in the woodworking industry, The Path Forward for Wood Products: A Global Perspective - Proceedings of Scientific Papers 0 (2016) 51-56.

[23] R. Ulewicz, B. Krstić and M. Ingaldi, Mining Industry 4.0 – Opportunities and Barriers, Acta Montanistica Slovaca 27 (2022) 291-305. https://doi.org/10.46544/AMS.v2i2.02

[24] O. Kierat, A. Dudek and L. Adamczyk, The effect of the corrosion medium on silane coatings deposited on titanium grade 2 and titanium alloy Ti13Nb13Zr, Materials 14 (2021) art. 6350. https://doi.org/10.3390/ma14216350

[25] T. Lipiński, J. Pietraszek, Corrosion of the S235JR Carbon Steel after Normalizing and Overheating Annealing in 2.5% Sulphuric Acid at Room Temperature, Materials Research Proceedings 24 (2022) 102-108. https://doi.org/10.21741/9781644902059-16

[26] W. Przybył et al., Microwave absorption properties of carbonyl iron-based paint coatings for military applications, Defence Technology 22 (2023) 1-9. https://doi.org/10.1016/j.dt.2022.06.013

[27] E. Skrzypczak-Pietraszek, J. Pietraszek, Chemical profile and seasonal variation of phenolic acid content in bastard balm (Melittis melissophyllum L., Lamiaceae), Journal of Pharmaceutical and Biomedical Analysis 66 (2012) 154-161. https://doi.org/10.1016/j.jpba.2012.03.037

[28] E. Skrzypczak-Pietraszek, J. Pietraszek, Seasonal changes of flavonoid content in Melittis melissophyllum L. (Lamiaceae), Chemistry and Biodiversity 11 (2014) 562-570. https://doi.org/10.1002/cbdv.201300148

[29] J.M. Djoković et al., Selection of the Optimal Window Type and Orientation for the Two Cities in Serbia and One in Slovakia, Energies 15 (2022) art. 323. https://doi.org/10.3390/en15010323

[30] Ł.J. Orman et al., Comparative Analysis of Indoor Environmental Quality and Self-Reported Productivity in Intelligent and Traditional Buildings, Energies 16 (2023) art. 6663. https://doi.org/10.3390/en16186663

[31] J. Pietraszek et al., The parametric RSM model with higher order terms for the meat tumbler machine process, Solid State Phenomena 235 (2015) 37-44. https://doi.org/10.4028/www.scientific.net/SSP.235.37

[32] R. Dwornicka, J. Pietraszek, The outline of the expert system for the design of experiment, Production Engineering Archives 20 (2018) 43-48. https://doi.org/10.30657/pea.2018.20.09

[33] J. Pietraszek, E. Skrzypczak-Pietraszek, The uncertainty and robustness of the principal component analysis as a tool for the dimensionality reduction, Solid State Phenomena 235 (2015) 1-8. https://doi.org/10.4028/www.scientific.net/SSP.235.1

[34] J. Pietraszek, A. Gadek-Moszczak, The smooth bootstrap approach to the distribution of a shape in the ferritic stainless steel aisi 434l powders, Solid State Phenomena 197 (2013) 162-167. https://doi.org/10.4028/www.scientific.net/SSP.197.162

[35] A. Szczotok et al., The bootstrap analysis of one-way ANOVA stability in the case of the ceramic shell mould of airfoil blade casting, Solid State Phenomena 235 (2015) 24-30. https://doi.org/10.4028/www.scientific.net/SSP.235.24

[36] J. Pietraszek, Fuzzy regression compared to classical experimental design in the case of flywheel assembly, Lecture Notes in Computer Science 7267 LNAI (2012) 310-317. https://doi.org/10.1007/978-3-642-29347-4_36

[37] J. Pietraszek, The modified sequential-binary approach for fuzzy operations on correlated assessments, Lecture Notes in Computer Science 7894 LNAI (2013) 353-364. https://doi.org/10.1007/978-3-642-38658-9_32

[38] J. Pietraszek, E. Skrzypczak-Pietraszek, The optimization of the technological process with the fuzzy regression, Advanced Materials Research 874 (2014) 151-155. https://doi.org/10.4028/www.scientific.net/AMR.874.151

[39] A. Gądek-Moszczak et al., Nano X-ray Tomography Application for Quantitative Surface Layer Geometry Analysis after Laser Beam Modification, Materials 15 (2022) art. 5935. https://doi.org/10.3390/ma15175935

[40] E. Skrzypczak-Pietraszek, A.G. Pietraszek, Phytoremediation and Allelopathy as an Element of Sustainable Circular Economy to Prevent Environmental Pollution, Materials Research Proceedings 45 (2024) 140-145. https://doi.org/10.21741/9781644903315-17

[41] P.C. Krause, O. Wasynczuk and S. D. Sudhoff. Analysis of Electric Machinery, IEEE Press, (2002) 1-663. https://doi.org/10.1002/9781118524336

Economic and Legal Perspective of Safe and Secure E-Commerce Development

Ewa KOZIEŃ[1,a*], Adam KOZIEŃ[2,b]

[1]Krakow University of Economics, College of Management and Quality Sciences, Department of International Management and Logistics, ul. Rakowicka 27, 31-510 Cracow, Poland

[2]Jagiellonian University in Cracow, Faculty of Law and Administration, Department of Administrative Procedure, ul. Gołębia 24, 31-007 Cracow, Poland

[a]koziene@uek.krakow.pl, [b]adam.kozien@uj.edu.pl

Keywords: Development, E-Commerce, Economy, Law, Safe, Secure

Abstract. Currently, the e-commerce industry is one of the fastest-growing branches of trade. Even the Ukrainian army utilizes e-commerce in logistic gamification, where armament supplies are granted according to the number of enemies killed. Enterprises that develop their own e-commerce models aim to enhance their economic potential. New trends in e-commerce have a significant impact on the dynamics of online product and service sales, as well as on strengthening their competitive position. Improving a sales process, aiming to attract the largest possible number of clients, involves leveraging key e-commerce trends such as omnichannel, personalization, m-commerce, social selling, and content marketing informed by Big Data. The dynamics of e-commerce development also pertain to legal regulations, including norms of common international law, EU law, and national law. The objective of this paper is to indicate economic and legal conditions from the perspective of e-commerce development.

Introduction

Currently, the development of enterprises is largely determined by utilizing a communicative platform, such as the Internet. The development of global communication infrastructure has enabled the dynamization of the relationship between an organization and its clients. The global dimension of e-commerce has a significant impact on the developmental expansion of both large and small enterprises, which effectively reach e-clients, thereby increasing their market share. Activities of enterprises in the scope of e-commerce are supported by the improvement of the selling process, aiming at attracting the possibly largest number of clients, which relates to the use of main e-commerce trends, such as omnichannel, personalization, m-commerce, social selling, and content marketing based on information coming from Big Data. E-commerce development requires adherence to legal regulations in respect of norms of common international law, EU law, as well as national laws. The objective of this paper is to indicate economic and legal conditions from the perspective of e-commerce development. The research hypothesis assumed that the functioning of e-commerce is influenced by various economic and legal factors, while the dynamics of change require their identification and an interdisciplinary approach.

Methodology

The article is interdisciplinary in nature, addressing the issue of e-commerce from the perspectives of economics and law. Therefore, the research methods employed are primarily based on an analysis of relevant literature and legal acts – sources of law commonly applicable in Poland – and utilize an abductive approach.

The primary method employed was an analysis of the literature on the subject, along with a critique of the literature, which was utilized in both the economic and legal sections. It was based on a review of the literature on the subject, as well as a critical analysis of it, and, on this basis, an identification of new contemporary legal and economic conditions relating to e-commerce. The method of literature review was based on a search of the literature on the subject, the preparation of a database of specialist literature, the analysis of individual assumptions, and the preparation of a report forming the basis for theoretical analyses in the article. Additionally, methods characteristic of the legal sciences were employed. The first method employed was the theoretical-legal method, which enabled conceptual, semantic, and theoretical analyses within the scope of this article, as well as considerations based on legal logic, particularly in identifying a set of concept designations and analyzing the relationships and scope of concepts. In addition, a formal-dogmatic method was used to analyze legal acts – sources of law generally applicable in Poland. This method was used to analyze normative material based on the interpretation of legal norms at three levels: linguistic (semantic), systemic, and purposive. This method also assumes the need to consider the hierarchy of sources of generally applicable law, as well as the level of detail and generality of individual norms and their interrelations, and the chronology of issued legal acts. The methodology of the article was based on the following levels:

1. Identification of the research problem.
2. Formulation of the research objective and hypothesis.
3. Identification of the state of knowledge based on an analysis of the literature on the subject and the normative material contained in legal acts – sources of generally applicable law in Poland.
4. Collection of content, text, and normative data
5. Analysis, coding, and interpretation of data
6. Compilation of research results.
7. Formulation of conclusions.

E-commerce characteristics and trends

E-commerce (e-trade) is an integral element of e-business, whose all operations take place using an electronic platform. Ph. Kotler narrows down these actions to the processes of purchase and sales, performed with electronic devices [1]. Whereas the World Trade Organization identifies e-commerce with production, advertising, and distribution of products via teleinformation networks [2]. In the scope of main e-commerce trends, i.e., omnichannel, personalization, mobile electronic trading (m-commerce), social selling, and content marketing, based on the information coming from Big Data, the development is noticeable, which relates to the use of modern technologies in building effective systems of communication with a client. Due to the omnichannel strategy that connects multiple sales channels, clients can choose any channel, including mobile, which they are using more frequently. They can also view a product prior to purchase in a convenient location and select a preferred method of delivery. Clients may also compare product offers from online shops worldwide and express their opinions, for example, on social media. Moreover, clients are enabled to personalize a product they wish to purchase and also design and modify it in any way they need. Development and changes in forming e-commerce trends are possible as a result of utilizing technological infrastructure and understanding clients' expectations within the context of the relationship between an enterprise and its clients.

Economic perspective of e-commerce development

Currently, the e-economy, defined as a new economy, is juxtaposed with the traditional economy. However, from the perspective of economic development, both the traditional economy and the e-economy based on knowledge, as well as the advancement of modern technologies, are of vital importance. Considering the value of the e-commerce market in Poland since 2011, as well as estimates for 2028, a gradual growth trend is noticeable, with forecasts indicating an average annual growth rate of 8% [3]. A growing trend suggests that economic entities in the e-commerce sector should focus their activities on developing integrated management systems to enhance client relations. These actions concern the establishment of infrastructure using high-speed Internet, which enables enterprises to implement marketing and sales strategies that allow them to reach their e-clients more effectively. Through the internet, enterprises can establish business partnerships (e.g., internet associations of enterprises) with other enterprises to meet the diverse expectations of clients for seamless, transparent sales services delivered using modern technologies. A significant role in e-commerce is played by programming enterprises that specialize in establishing internet shops, optimizing their positioning in browsers, and collecting data on client behavior and preferences [4]. Making purchases via the Internet also influences the change of behavior of „new consumers", who are characterized by, among others, freedom of choice of any offer and adjusting it to one's own needs, as well as searching, prior to making a purchase decision, for information from many sources and channels [5]. Currently, e-consumers have a strong impact on the financial benefits of e-shops. Due to the high mobility of e-clients among e-shops, the comparison of available product offers on the Internet, and the verification of product opinions, it is challenging to maintain client loyalty. A challenge for contemporary enterprises is mastering principles that function in the environment of mobile technologies, through the automation and optimization of business processes, which enables them to derive benefits from productivity growth, increase client satisfaction, reduce operational costs, and make more effective decisions [6].

New legal challenges in the e-commerce business

The appearance of e-commerce necessitated the need for an appropriate adjustment of law to the functioning of legal relations on the Internet, which does not recognize state borders. At the international level, a response to e-commerce issues was the UNCITRAL document developed in 1996 (United Nations Commission on International Trade Law) – UNCITRAL Model Law on Electronic Commerce with a Guide to Enactment (1996) [7]. The issue of e-commerce found its regulation at the level of the European Union law within the framework of Directive 2000/31/EC of the European Parliament and of the Council of 8 June 2000 on certain legal aspects of information society services, in particular electronic commerce, in the Internal Market [8] (OJ L 178, 17.7.2000). Among the legal challenges of a global character, one could mention issues of personal data protection (including Big Data problems), protection of rights and consumer interests, market monopolization by large international corporations, as well as problems related to the position of international corporations within states and their respective legal systems.

Problems related to personal data protection affect all entrepreneurs operating in e-commerce, although they are especially visible in international corporations, which process a vast amount of personal data. At the level of the European Union law an issue of personal data protection is regulated in the Regulation (EU) 2016/679 of the European Parliament and of the Council of 27 April 2016 on the protection of natural persons with regard to the processing of personal data and on the free movement of such data, and repealing Directive 95/46/EC (General Data Protection Regulation) (Text with EEA relevance) (OJ L119, 4.5.2016) [9], which as a regulation in the system of sources of the European Union law is of binding character for EU member states [10] for all EU member states and introduced more effective personal data protection also in e-commerce. However, a problem arises in global relations among the EU and its member states, as well as among non-EU states, where this type of regulation is lacking. Enterprises, and especially large international corporations active on e-commerce markets, manage, gather, and process immense amounts of personal data, which have become contemporary merchandise, also with material dimensions. Average consumer does not realize that through his/her activity in the Internet, and especially through the activity of service using character, „he/she is selling for free" sensitive data on himself/ herself and e.g. on his/her consumer's preferences, which become data on also material value, by their consequent use by enterprises aimed at adjustment of suitable individualized offers. Issues related to Big Data are becoming a huge legal challenge, not only at the national and EU levels, but especially at the international level.

The issue of protecting competition and consumers [11] is developed at the European Union level, both in primary and secondary law. However, this problem occurs at the international level, where consumers may have difficulties in pursuing their rights with global corporations (an issue of the court with proper jurisdiction, the corporate registered office, and the legal background of large international corporations), as well as consumers may be subjected to dishonest market practices in e-commerce. In the realm of e-commerce, a potential issue may arise from the monopolization of the virtual market by large global enterprises and internet portals. In the primary and secondary law of the European Union, there are regulations aimed at countering market monopolization [12]. In contrast, at the international law level, there are no effective legal regulations to counteract e-commerce monopolization. Contemporary legal challenges in the e-commerce sector can include the strong position of international corporations, which may have even greater capital than states. Moreover, especially in the Internet sphere, which is unrestricted, they become entities that do not respect the legal systems of states.

Social and Environmental Impact

The development of e-commerce has a significant impact on many industries, including industrial ones, through the local implementation of transactional and IT mechanisms specific to e-commerce. This primarily means the widespread implementation of network security mechanisms, including transmission security and authorization and authentication, which is very important both in the military area [13,14] and in the consumer area such as intelligent construction [15], but also in industrial control, including the automotive industry [16], the woodworking industry [17], foundry [18,19], and energy management [20, 21]. Economic mechanisms operating in e-commerce are also being implemented in the field of waste management [22-24], including heavy metals [25], increasing their operational efficiency. E-commerce processes have an inspiring influence on process management methodologies [26,27], especially in the area of the circular economy [28-30] requiring the implementation of a wide range of natural products [31] and remediation of degraded areas [32]. For such an approach to be effective and economically viable, it is necessary to use materials with increased corrosion resistance [33], wear resistance [34-36] and the introduction of material substitutes obtained from industrial waste [37]. This approach requires the use of process optimization methods [38,39], nonparametric methods [41], and fuzzy evaluations [Pietraszek-2012; Pietraszek-2013], which are essential in addressing the significant material and environmental disturbances encountered here. Knowledge engineering tools, including expert systems [43], which are characterized by greater precision and repeatability than currently available AI systems, prove to be of great help.

Summary

In conclusion, it can be stated that e-commerce enables higher flexibility in business operations. Flexibility of e-commerce means the lack of restrictions on conducting business activity with a view to time (real-time) and e-client (freedom of reaching a client in various locations). E-commerce functions based on a model of virtual community, where enterprises offer their products and services. The efficiency of managing the purchasing process refers to both relations with clients and the parallel connection of traditional and online trading forms. The present challenge for companies includes building various systems to gather and process data on a client, as well as connecting all channels to establish a single customer view. Enterprises selling products in both traditional and online trade should build a cohesive system for managing clients, encompassing a holistic assembly of all channels. Joint treatment of sales channels facilitates the gathering and exchange of information on clients and their preferences. E-commerce must be regulated by the generally applicable law. E-commerce currently is of global character, and because of that new

Materials Research Proceedings 62 (2026) 11-17 https://doi.org/10.21741/9781644904015-2

problems and legal challenges appear involving e-commerce, such as: an issue of personal data protection (including Big Data issues), protection of rights and a consumer's situation, monopolization of market by large corporations and internet portals with international coverage, and also an issue of international corporations' position in a context of their relations with particular states, especially in a scope of functioning such corporations in legal systems of various states. Therefore, it appears that, at the level of international law, a binding act is necessary to regulate fundamental issues concerning e-commerce.

References

[1] Ph. Kotler, G. Amstrong, J. Saunders, V. Wong, Principles of Marketing: 3rd European Edition, Pitman Publishing, 2001.

[2] M. Bacchetta, P. Low, A. Mattoo, L. Schuknecht, H. Wagerand, M. Wehrens, Special Studies – Electronic Commerce and the role of the WTO, WTO Publications, Geneva, 1998.

[3] E-commerce w Polsce, Raport Gemius, Izba Gospodarki Elektronicznej, 2025.

[4] W.M. Grudzewski, I. Hejduk, Przedsiębiorstwo wirtualne, Difin, Warszawa, 2002.

[5] D. Tapscott, Grown Up Digital: How the Net Generation is Changing Your World, McGraw Hill, 2008.

[6] U. Varshney, A. Mallow, R. Jain, P. Ahluwalia, Wireless in the Enterprise: Requirements and Possible Solutions, Proceedings of the Workshop on Wireless Strategy in the Enterprise: An International Research Perspective, University of California-Berkeley, Berkeley, 2002.

[7] UNCITRAL Model Law on Electronic Commerce with Guide to Enactment 1996 with additional article 5 bis as adopted in 1998. https://www.uncitral.org/pdf/english/texts/electcom/05-89450_Ebook.pdf.

[8] Directive 2000/31/EC of the European Parliament and of the Council of 8 June 2000 on certain legal aspects of information society services, in particular electronic commerce, in the Internal Market (Directive on electronic commerce), OJ L 178, 17.7.2000.

[9] Regulation (EU) 2016/679 of the European Parliament and of the Council of 27 April 2016 on the protection of natural persons with regard to the processing of personal data and on the free movement of such data, and repealing Directive 95/46/EC (General Data Protection Regulation) (Text with EEA relevance), OJ L119, 4.5.2016.

[10] P. Ruczkowski, 2020, [in:] Zdyb, M., Stelmasiak, J. (Eds.), 2016. Prawo administracyjne. Wolters Kluwer, Warszawa.

[11] J. Barcz, M. Górka, A. Wyrozumska, Instytucje i prawo Unii Europejskiej, Wolters Kluwer, Warszawa, 2023.

[12] C. Banasiński, 2015, [in:] H. Gronkiewicz-Waltz, M. Wierzbowski, (Eds.), Prawo gospodarcze. Zagadnienia administracyjnoprawne. Wolters Kluwer, Warszawa, 2015.

[13] M. Michalski et al., The Influence of Operational Exposure on Changes in Parameters of Effective Camouflage of Coatings Used in Military Technology, Advances in Science and Technology Research Journal 17 (2023) 182-196. https://doi.org/10.12913/22998624/156940

[14] W. Przybył et al., Microwave absorption properties of carbonyl iron-based paint coatings for military applications, Defence Technology 22 (2023) 1-9. https://doi.org/10.1016/j.dt.2022.06.013

[15] Ł.J. Orman et al., Comparative Analysis of Indoor Environmental Quality and Self-Reported Productivity in Intelligent and Traditional Buildings, Energies 16 (2023) art. 6663. https://doi.org/10.3390/en16186663

[16] A. Pacana et al., Analysis of a practical approach to the concept of sustainable development in a manufacturing company in the automotive sector, Waste Forum (2020) 151-161.

[17] R. Ulewicz, Quality management system operation in the woodworking industry, The Path Forward for Wood Products: A Global Perspective - Proceedings of Scientific Papers (2016) 51-56.

[18] R. Ulewicz, F. Novy and R. Dwornicka, Quality and work safety in metal foundry, METAL 2020 – 29th Int. Conf. Metall. Mater. (2020) 1287-1293. https://doi.org/10.37904/metal.2020.3649

[19] R. Ulewicz, D. Siwiec and A. Pacana, Sustainable Vehicle Design Considering Quality Level and Life Cycle Environmental Assessment (LCA), Energies 16 (2023) art. 8122. https://doi.org/10.3390/en16248122

[20] O. Gumen et al., Research of thermal processes in industrial premises with energy-saving technologies of heating, Diagnostyka 18 (2017) 43-49.

[21] S. Maleczek et al., Tests of Acid Batteries for Hybrid Energy Storage and Buffering System—A Technical Approach, Energies 15 (2022) art. 3514. https://doi.org/10.3390/en15103514

[22] M. Ingaldi, S. Jursová, Economy and possibilities of waste utilization in Poland, METAL 2013 – 22nd Int. Conf. Metall. Mater. (2013) 1779-1784.

[23] S. Brožová, M. Ingaldi and I. Šperlín, Economical aspects of high-temperature heating utilization for industrial waste treatment, METAL 2013 - 22nd Int. Conf. Metall. Mater. (2013) 1735-1739.

[24] A. Pietrzak, M. Ulewicz, Influence of Post-Consumer Waste Thermoplastic Elastomers Obtained from Used Car Floor Mats on Concrete Properties, Materials 16 (2023) art. 2231. https://doi.org/10.3390/ma16062231

[25] E. Radzyminska-Lenarcik, R. Ulewicz and M. Ulewicz, Zinc recovery from model and waste solutions using polymer inclusion membranes (PIMs) with 1-octyl-4-methylimidazole, Desalination and Water Treatment 102 (2018) 211-219. https://doi.org/10.5004/dwt.2018.21826

[26] A. Pacana, D. Siwiec and R. Dwornicka, Method Supporting Improving Products in Terms of Qualitative-Environmental, Materials Research Proceedings 34 (2023) 226-236. https://doi.org/10.21741/9781644902691-27

[27] A. Pacana, D. Siwiec, Method of Fuzzy Analysis of Qualitative-Environmental Threat in Improving Products and Processes (Fuzzy QE-FMEA), Materials 16 (2023) art. 1651. https://doi.org/10.3390/ma16041651

[28] D. Siwiec, A. Pacana and A. Gazda, A New QFD-CE Method for Considering the Concept of Sustainable Development and Circular Economy, Energies 16 (2023) art. 2474. https://doi.org/10.3390/en16052474

[29] A. Pacana, D. Siwiec and R. Dwornicka, Pro-quality method improving industrial products toward sustainable development with criteria of circular economy, METAL 2024 - Int. Conf. Metall. Mater. 2023-May (2024) 697-702. https://doi.org/10.37904/metal.2023.4743

[30] B. Gajdzik et al., Approaching open innovation in customization frameworks for product prototypes with emphasis on quality and life cycle assessment (QLCA), Journal of Open Innovation: Technology, Market, and Complexity 10 (2024) art. 100268. https://doi.org/10.1016/j.joitmc.2024.100268

[31] E. Skrzypczak-Pietraszek, J. Pietraszek, Chemical profile and seasonal variation of phenolic acid content in bastard balm (Melittis melissophyllum L., Lamiaceae), Journal of Pharmaceutical and Biomedical Analysis 66 (2012) 154-161. https://doi.org/10.1016/j.jpba.2012.03.037

[32] E. Skrzypczak-Pietraszek, A.G. Pietraszek, Phytoremediation and Allelopathy as an Element of Sustainable Circular Economy to Prevent Environmental Pollution, Materials Research Proceedings 45 (2024) 140-145. https://doi.org/10.21741/9781644903315-17

[33] T. Lipiński, J. Pietraszek, Corrosion of the S235JR Carbon Steel after Normalizing and Overheating Annealing in 2.5% Sulphuric Acid at Room Temperature, Materials Research Proceedings 24 (2022) 102-108. https://doi.org/10.21741/9781644902059-16

[34] N. Radek et al., Formation of coatings with technologies using concentrated energy stream, Production Engineering Archives 28 (2022) 117-122. https://doi.org/10.30657/pea.2022.28.13

[35] N. Radek et al., Operational properties of DLC coatings and their potential application, METAL 2022 – 31st Int. Conf. Metall. Mater., (2022), Ostrava, Tanger, 531-536. https://doi.org/10.37904/metal.2022.4491

[36] A. Kalinowski et al., Laser surface texturing: characteristics and applications, System Safety: Human-Technical Facility-Environment 5 (2023) 240-248. https://doi.org/10.2478/czoto-2023-0026

[37] M. Ulewicz, A. Pietrzak, Properties and structure of concretes doped with production waste of thermoplastic elastomers from the production of car floor mats, Materials 14 (2021) art. 872. https://doi.org/10.3390/ma14040872

[38] J. Pietraszek, A. Goroshko, The heuristic approach to the selection of experimental design, model and valid pre-processing transformation of DoE outcome, Advanced Materials Research 874 (2014) 145-149. https://doi.org/10.4028/www.scientific.net/AMR.874.145

[39] J. Pietraszek, N. Radek and A.V. Goroshko, Challenges for the DOE methodology related to the introduction of Industry 4.0, Production Engineering Archives 26 (2020) 190-194. https://doi.org/10.30657/pea.2020.26.33

[40] J. Pietraszek, A. Gadek-Moszczak, The smooth bootstrap approach to the distribution of a shape in the ferritic stainless steel AISI 434l powders, Solid State Phenomena 197 (2013) 162-167. https://doi.org/10.4028/www.scientific.net/SSP.197.162

[41] J. Pietraszek, Fuzzy regression compared to classical experimental design in the case of flywheel assembly, 7267 LNAI (2012) 310-317. https://doi.org/10.1007/978-3-642-29347-4_36

[42] J. Pietraszek, The modified sequential-binary approach for fuzzy operations on correlated assessments, Lecture Notes in Computer Science 7894 LNAI (2013) 353-364. https://doi.org/10.1007/978-3-642-38658-9_32

[43] R. Dwornicka, J. Pietraszek, The outline of the expert system for the design of experiment, Production Engineering Archives 20 (2018) 43-48. https://doi.org/10.30657/pea.2018.20.09

Materials Research Proceedings 62 (2026) 18-25
https://doi.org/10.21741/9781644904015-3

Fe-Cr-Co Filler Metal for Hardfacing Industrial Fittings – An Alternative to Stellite Bonds

JÓŹWIK Tadeusz[1,a*] and SZYDZIAK Paweł[1,b]

[1]TEDSPAW Sp. z o.o. ul. Bukowa 2, 26-026 Bilcza, Poland

[a]tad.jwik@post.pl, [b]tedspaw@tedspaw.pl

Keywords: Surfacing, Hardfacing, Industrial Fittings, Sealing Surfaces

Abstract. The article presents the implementation of a welding material – Fe-Cr-Co (iron-chromium-cobalt) weld metal, which is a cheaper and technically advantageous alternative to traditional cobalt alloys used for surfacing and sealing surfaces of industrial fittings. The aim was to find a material with similar tribological wear resistance to commonly used cobalt-based welds, but one that is technologically easier to use, more resistant to cracking, and easier to use during repairs. The authors proposed an Fe-Cr-Co alloy (XHD 6804 electrode) – an iron-based filler metal with the addition of chromium and cobalt. It is characterized by high heat resistance (up to 900°C), a hardness of 45–51 HRC, heat resistance, thermal shock resistance, and ease of surfacing.

Introduction

Ensuring the required durability of industrial fittings in power plants depends, among other things, on the durability of the sealing surfaces of the fittings. The durability of these elements is affected by the operational wear process, including tribological wear. In fittings used in power plants, the sealing surfaces – i.e., the surfaces of the plug, wedge, and seat – must be extremely resistant to:

- corrosion and cavitation erosion,
- abrasion during opening/closing cycles - an abrasion mechanism based on all tribological wear mechanisms, at elevated temperatures,
- wear during mutual "pressure/pressure" interaction - an abrasion mechanism based mainly on adhesive mechanisms, at elevated temperatures.

Materials and Parameters

The samples were made by welding [1] with a non-consumable electrode in a gas shield (TIG), [2] on a steel disc made of 10CrMo910 steel with a diameter of 150 mm, on which a ϕ3.25 XHD 6804 electrode was used to apply weld metal in three passes and two layers, with a weld width of approximately 25 mm and a total weld thickness of 6 mm. Preheating before welding 150^{0}C, temperatures were controlled at the surfacing site – the temperature in this process did not exceed 300^{0}C. After welding, the test disc was cooled to ambient temperature in a thermal insulation wool blanket. Then, the hardness of the surface layers of the weld was measured, and post-weld heat treatment was performed. The heat treatment was carried out in a SOLO laboratory furnace, with the following parameters: heating to the maximum tempering (aging) temperature of 200^{0}C/h, tempering at this temperature for 1.5 hours, cooling in the furnace to a temperature of 400^{0}C. After that, cooling took place outside the furnace in still air. After the above-described steps, samples were taken from the test disc for testing, which mainly consisted of examining the chemical composition of the surface layers of the hardfacing, assessing the structural composition of the hardfacing in cross-section, and measuring hardness at ambient temperatures and elevated temperatures of 550°C. The above tests were carried out for comparison on a similar test disc, on which a stellite weld was made using a buffer of 18%Cr + 8%Ni steel [3], stellite E20-UM-40CTZ (stellite 6), and an Eutec 9060N Castolin electrode (Table 1).

Table 1. Examples of chemical compositions of Stellite 6 and XHD6804 binder, and the hardness of welds at ambient temperatures and 550°C

Electrode and steel grade X-15 stellite	**Chemical composition, %**								**HRC**	
	C	Si	Mn	Cr	Ni	W	Fe	Co	20°C	550°C
EutecDur 9060N (stellite 6) E 20-UM-40 CTZ	1.04	0.08	0.79	30.0	2.33	5.05	1.79	rest	40	31
XHD6804	0.13	0.56	0.10	12.2	0.25	0.3	rest	12.5	45	36

Test Results

Hardness measurements. The basic parameter tested to determine the resistance of welds to destructive factors was hardness. The hardness of "XHD 6804 and Eutec 9060N" welds at ambient temperatures was measured using the Vickers method at a load of 10 kN on a NOVATEST 330 hardness tester. Similar hardness measurements were performed at elevated temperatures on a stand consisting of a Brinell HP 3000 hardness tester equipped with a penetrator, a 10 mm carbide ball, and a SOLO type 111-10/9/20 laboratory furnace. The operating times associated with removing the sample from the oven after heating and loading it with the indenter were practically the same. The diameters of all indentations were measured at ambient temperature. The results of the tests at ambient temperature (approximately 25°C) for HV10 after conversion were converted to HRC. At elevated temperatures, the results from HB after conversion were also converted to HRC. Results of approximate hardness measurements at elevated temperatures, performed using hardfacing with coated electrodes whose compositions are listed in Table 2.

Table 2. Chemical composition of electrodes

Electrode designation	**Hardness measurement temperature, °C**					
	25	200	30	400	500	550
XHD 6804	516 HV10	470 HB	450 HB	410 HB	370 HB	350 HB
	approx. 50 HRC	approx. 48 HRC	approx. 47 HRC	approx. 44 HRC	approx. 40 HRC	approx. 37 HRC
Eutec 9060N	452 HV10	410 HB	390 HB	350 HB	320 HB	310 HB
	approx. 46 HRC	approx. 43 HRC	approx. 42 HRC	approx. 38 HRC	approx. 34 HRC	approx. 33 HRC

The hardness distribution on the weld cross-section is shown in Figures 1-3. The hardness measurement results of the padding welds are presented in Table 3.

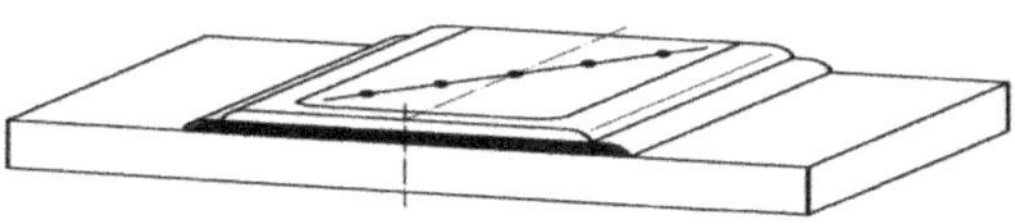

Fig. 1. Diagram showing the location of hardness measurement points - outer surface of the weld.

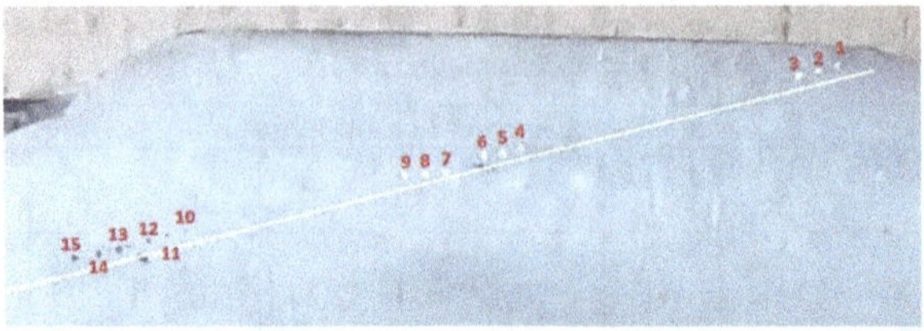

Fig. 2. Diagram showing the location of hardness measurement points – hardness distribution – cross-section of a weld made with XHD 6804 electrodes and 10CrMo910 base material.

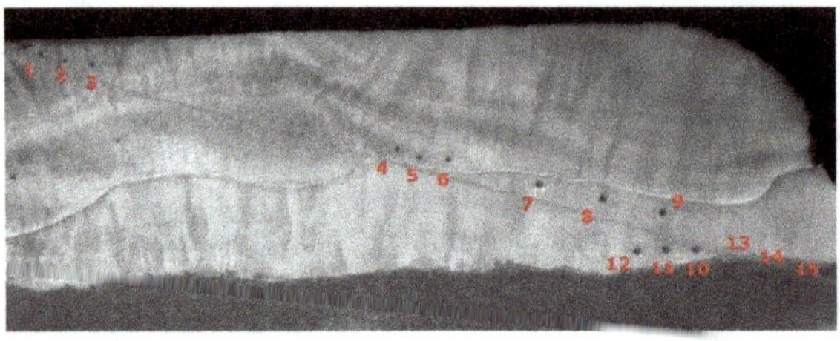

Fig. 3. Diagram showing the location of hardness measurement points – hardness distribution cross-section of a weld made with Eutec 9060N and 18/8 buffer electrodes and parent material.

Table 3. Results of hardness measurements of welds (hardfacing) made with coated electrodes whose chemical compositions are given in Table 1

Electrode designation	Hardness measurement results – HV10/measurement points													
	1	2	3	4	5	6	7	8	9	10	11	12	13	14
XHD 6804 results in the top row measurements on the cross-section in Fig. 2, in the bottom row on the outer surface in Fig. 1	566	533	544	610	598	597	566	534	546	518	526	490	211	220
	535	525	528	485	509	---	---	---	---	---	---	---	---	---
Eutec 9060N results in the top row measurements on the cross-section Fig. 2, in the bottom row on the outer surface Fig. 4	477	424	451	418	404	422	198	207	208	204	195	212	453	466
	443	459	432	462	441	---	---	---	---	---	---	---	---	---

Microscopic Examinations

Stationary grinding machines with replaceable sandpaper, featuring grit sizes ranging from 120 to 1,000, were used to prepare the surface of the samples. 3μm and 1μm diamond pastes were used for polishing. Etching was performed chemically using Nital reagent on the material on which hardfacing was performed, and Vilella and Adler II reagents on the joints, obtained from XHD

6804 and Eutec 9060N electrodes. The initial observation of the samples was conducted using a Leica DMI 3000M light microscope at magnifications of 200–500x, and the final observation was performed using Philips XL 40 and Phenom XL G2 scanning electron microscopes.

Weld made with XHD 6804 electrodes – the weld contains the chemical elements listed in Table 1. The main phase of the structure is alloy martensite. After heat treatment (approximately 620°C/2 hours), the material has a tempered martensite structure, with a martensite grain shape similar to "lamellar" and characteristic of castings. During the tests, using the etching preparation, no residual austenite was observed adjacent to the basic martensitic phase; however, carbide precipitates, presumably $Cr_{23}C_6$ or $(Fe,Cr)_7C_3$, were revealed. An example of the revealed structures of "XHD 6804" is shown in Figure 5.

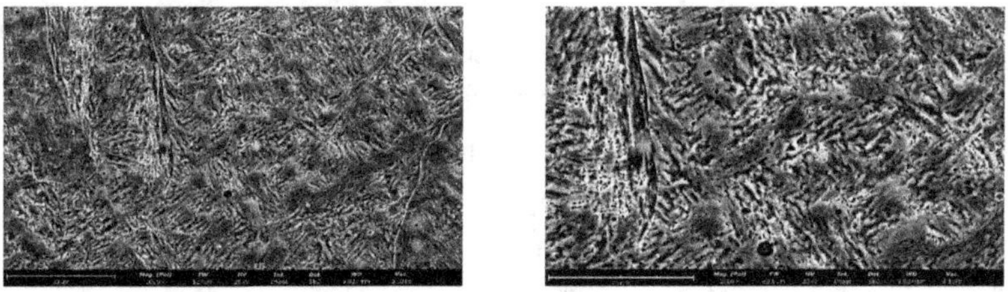

Fig. 5. Example image of the structure of the XHD 6804 weld - tempered martensite, between the martensite strips, presumably carbide precipitates – SEM observation – magnification 100 x 2000x – hardness 535 HV10.

The hardfacing was made with Eutec 9060N stellite electrodes – the revealed structure of the surface layer of the stellite hardfacing was a matrix (grains) with a regular flat-centered crystallographic structure (RCS) – dendritic structure composed of a solid solution "gamma" rich in cobalt and chromium and tungsten carbides [4]. Observation of the cross-section confirmed that during crystallization, the main eutectic carbide phases (primary carbides) were formed, distributed in the interdendritic spaces. Secondary carbides were also observed, which formed during the cooling of the weld and during heat treatment after surfacing. These carbides are dispersed and take the form of spheroidal particles [5]. In cobalt alloys, carbides $M_{23}C_6$, M_6C, and $M_7C_{(3)}$ are most commonly found. An example of the revealed structures of stellite hardfacing is shown in Figure 6.

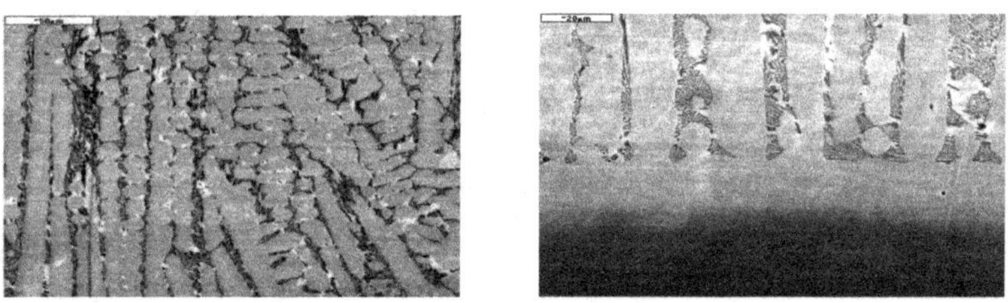

Fig. 6. Left image - observation near the surface layer, right image - boundary between stellite and "buffer" material – SEM - magnification 1000x and 500x.

Technological, Economic, Environmental, and Social Impact

The implementation of the novel Fe-Cr-Co filler metal for hardfacing industrial fittings represents a substantial shift in material and process technology, generating significant impacts across multiple dimensions.

The primary technological advantage lies in the superior performance and ease of application compared to traditional cobalt-based alloys (stellites). The Fe-Cr-Co alloy exhibits a robust tempered martensite structure with fine carbide precipitates, providing exceptional hardness and wear resistance, comparable to or exceeding stellites, especially at elevated temperatures (up to 900°C). This high-temperature stability is crucial for components operating in power plants, such as superheater pipe coils [6].

The alloy's structure and properties are fundamentally dependent on the processing method, similar to the challenges observed in controlling the structure of sintered stainless steel through various alloying techniques [7] and the microstructure refinement in alloys processed by welding [8]. Furthermore, the resulting weld metal must ensure the integrity of the overall component, particularly in welded joints, to mitigate risks such as hydrogen embrittlement in high-stress environments [9]. The achievement of a highly durable, crack-resistant surface is a key technological advance, contrasting with the inherent brittleness and tendency for cracking in stellites. The resulting material structure exhibits excellent fatigue properties [10] and resistance to external factors, critical for the operational reliability of industrial valves [11, 12].

Understanding and controlling the microstructure and mechanical properties are paramount [13]. Detailed research is continually focused on improving process parameters to achieve the required microstructural outcomes in specialized steels [14] and optimizing fatigue performance based on material factors, such as internal inclusions [15] or microstructure [16]. Advanced surface engineering techniques, such as laser alloying to obtain hard coatings [17] or the use of electrospark alloying [18], provide a broader context for developing high-performance wear layers. Furthermore, new methods, such as laser treatment for heat transfer augmentation, highlight the growing importance of surface modification in high-efficiency industrial applications [19]. For high-temperature applications, controlling fracture behavior in thermal barrier coatings provides essential context for the required durability of the Fe-Cr-Co hardfacing layer [20]. Similarly, the tribological performance of the resulting layer is confirmed by studies focused on the wear behavior of advanced composite layers [21]. The quality of structural components, such as castings used in aerospace or energy applications, also depends on rigorous control of manufacturing processes and the resulting material properties [22].

The economic benefits of using the Fe-Cr-Co alloy are significant, primarily driven by a 20–40% reduction in material and labor costs compared to stellites, as well as simplified repair procedures due to its reduced propensity for cracking and the elimination of the need for buffer layers or time-consuming heat treatments. Efficient process management is fundamental to maximizing these economic gains. The use of advanced computational and statistical techniques, such as design of experiments (DoE), is essential for process optimization, minimizing research time and resources [23]. Furthermore, the development of expert systems facilitates the application of DoE in various manufacturing environments [24]. The stability and reliability of the overall manufacturing process, including assembly and quality control of complex metal products [25], as well as the modeling of errors in manufacturing systems [26], directly translate into reduced costs and fewer errors.

In the context of modern industrial trends (Industry 4.0), process optimization faces new challenges [27], making advanced statistical methods, like fuzzy regression [28] or modified sequential-binary approaches [29], increasingly relevant for handling uncertainty and complex assessments. The maturity of process monitoring is a metric that correlates directly with efficiency and cost reduction in manufacturing [30].

The environmental and social benefits are tied to the reduction in cobalt usage, a strategic material with volatile pricing and often ethically challenging sourcing. Replacing cobalt-based alloys with an iron-based alternative (Fe-Cr-Co) improves supply chain security and reduces reliance on critical raw materials. While the primary focus is on material performance, the adoption of modern statistical tools – such as the smooth bootstrap approach for analyzing material properties [31] or non-parametric analysis for complex assessments [32] – indicates a commitment to rigorous, data-driven methodology. This ensures the reliability and repeatability of the hardfacing process, which in turn contributes to the societal benefit of increased safety and reduced maintenance downtime in critical power infrastructure. The improved corrosion resistance of components, a field of ongoing material research [33], also reduces the frequency of replacement, resulting in less waste and a lower environmental footprint. Overall, the transition to the Fe-Cr-Co alloy signifies a move toward more sustainable material choices and a more robust, cost-effective maintenance strategy in the energy sector.

Summary

The article presents the implementation of a new welding material – Fe-Cr-Co (iron-chromium-cobalt) weld metal, which is a cheaper and technically advantageous alternative to traditional cobalt alloys used for surfacing sealing surfaces of industrial fittings. The aim was to find a material with similar wear and tribological resistance to stellites, but technologically easier, more resistant to cracking, and easier to use during repairs. Industrial fittings in power plants (valves, gate valves, etc.) operate in harsh conditions: high temperatures, erosion, cavitation, abrasion. Traditional stellites, despite their high resistance, are not very ductile and are prone to cracking, which complicates repairs and requires adherence to technological regimes. The proposed Fe-Cr-Co alloy is characterized by high heat resistance (up to 900°C), hardness of 45–54HRC, heat resistance, thermal shock resistance, and ease of surfacing. Its use does not require a buffer layer or time-consuming heat treatment, and the cost of the material and work is 20–40% lower than that of stellites. The tests were carried out on 10CrMo910 steel. The hardfacing was performed with an XHD 6804 electrode and compared with Eutec 9060N stellite hardfacing. The results showed that the new material has a tempered martensite structure with Cr23C6 and (Fe,Cr)7C3 carbide precipitates, retains higher hardness at high temperatures (37 HRC at 550°C compared to 33 HRC for stellite), and has a uniform hardness distribution. Microscopic SEM examinations confirmed the absence of cracks and the homogeneity of the welds. The authors implemented the Fe-Cr-Co (XHD 6804) filler metal in 2020 in safety, relief, and reduction-cooling valves. Its high durability, resistance to cavitation and erosion, good adhesion, and resistance to thermal shocks have been confirmed in operation.

The Fe-Cr-Co alloy is a fully-fledged alternative to stellites. It provides high hardness, thermal stability, ease of welding, and significant cost reduction. Its martensitic structure guarantees durability and resistance to cracking. Operational results confirm the practical suitability of the alloy for local surface hardening of valve components that must be resistant to the tribological wear required in the energy industry.

References

[1] R.R. Garbadea, N.B. Dhokey, Overview on Hardfacing Processes, Materials and Applications, IOP Conference Series 2021: Materials Science and Engineering 1 (2021) art.012033. https://doi.org/10.1088/1757-899x/1017/1/012033

[2] E. Tasak, Metalurgia spawania, Wydawnictwo JAK, Kraków, 2008.

[3] PN-EN 14700:2022 – Welding consumables. Consumables for hardfacing – PKN 2023.

[4] M. Gajewski, A. Skrzypczyk, J. Kasińska, Testing of stellite pad welds on sealing surface on fittings, Przegląd Spawalnictwa 79(8) (2007) 43-48.

[5] Z. Bojar, Analiza wpływu struktury na odporność korozyjną i mechanizm pękania stopów kobaltu typu Vitalium, Military University of Technology, Warszawa, 1992.

[6] K. Trzewiczek et al., Evaluation of the state for the material of the live steam superheater pipe coils of V degree, Advanced Materials Research 874 (2014) 35-42. https://doi.org/10.4028/www.scientific.net/AMR.874.35

[7] A. Dudek et al., The effect of alloying method on the structure and properties of sintered stainless steel, Arch. Metall. Mater. 62 (2017) 281-287. https://doi.org/10.1515/amm-2017-0042

[8] T. Lipiński, Influence of surface refinement on microstructure of Al-Si cast alloys processed by welding method, Manufacturing Technology 15 (2015) 576-581.

[9] P. Jonšta et al., Hydrogen embrittlement of welded joint made of supermartensitic stainless steel in environment containing sulfane, Arch. Metall. Mater. 61 (2016) 709-711. https://doi.org/10.1515/amm-2016-0121

[10] T. Lipinski et al., Influence of oxygen content in medium carbon steel on bending fatigue strength, Eng. Rural Develop. 21 (2022) 351-356. https://doi.org/10.22616/ERDev.2022.21.TF116

[11] P.J. Romanowicz et al., Using the Effect of Compression Stress in Fatigue Analysis of the Roller Bearing for Bimodal Stress Histories, Materials 15 (2022) art. 196. https://doi.org/10.3390/ma15010196

[12] T. Lipiński, A. Wach, Influence of inclusions on bending fatigue strength coefficient the medium carbon steel melted in an electric furnace, Prod. Eng. Arch. 26 (2020) 88-91. https://doi.org/10.30657/pea.2020.26.18

[13] R. Ulewicz et al., Structure and mechanical properties of fine-grained steels, Periodica Polytechnica Transportation Engineering 41 (2013) 111-115. https://doi.org/10.3311/PPtr.7110

[14] D. Siwiec et al., Improving the process of achieving required microstructure and mechanical properties of 38mnvs6 steel, METAL 2020 – 29th Int. Conf. Metall. Mater. (2020) 591-596. https://doi.org/10.37904/metal.2020.3525

[15] R. Ulewicz et al., Fatigue properties of wear resistant martensitic steel, METAL 2014 – 23rd Int. Conf. Metall. Mater. (2014) 784-789.

[16] T. Lipinski, J. Pietraszek, Influence of animal slurry on carbon C35 steel with different microstructure at room temperature, Eng. Rural Develop. 21 (2022) 344-350. https://doi.org/10.22616/ERDev.2022.21.TF115

[17] M. Kuklinski et al., Laser alloying monel 400 with amorphous boron to obtain hard coatings, Materials 12 (2019) art. 3494. https://doi.org/10.3390/ma12213494

[18] N. Radek et al., The WC-Co electrospark alloying coatings modified by laser treatment, Powder Metall. Metal Ceramics 47 (2008) 197-201. https://doi.org/10.1007/s11106-008-9005-7

[19] Ł.J. Orman et al., Application of laser treatment technology for boiling heat transfer augmentation, Prod. Eng. Arch. 30 (2024) 259-265. https://doi.org/10.30657/pea.2024.30.25

[20] G. Guidoni et al., Fracture behaviour of thermal barrier coatings after high temperature exposure in air, Materials Science and Engineering: A 397 (2005) 209-214. https://doi.org/10.1016/j.msea.2005.02.022

[21] J. Korzekwa et al., Tribological behaviour of Al2O3/inorganic fullerene-like WS2 composite layer sliding against plastic, International Journal of Surface Science and Engineering 10 (2016) 570-584. https://doi.org/10.1504/IJSURFSE.2016.081035

[22] A. Szczotok et al., The Impact of the Thickness of the Ceramic Shell Mould on the ($\gamma + \gamma'$) Eutectic in the IN713C Superalloy Airfoil Blade Casting, Arch. Metall. Mater. 62 (2017) 587-593. https://doi.org/10.1515/amm-2017-0087

[23] J. Pietraszek et al., Factorial approach to assessment of GPU computational efficiency in surrogate models, Advanced Materials Research 874 (2014) 157-162. https://doi.org/10.4028/www.scientific.net/AMR.874.157

[24] R. Dwornicka, J. Pietraszek, The outline of the expert system for the design of experiment, Prod. Eng. Arch. 20 (2018) 43-48. https://doi.org/10.30657/pea.2018.20.09

[25] D. Klimecka-Tatar, R. Dwornicka, The assembly process stability assessment based on the strength parameters statistical control of complex metal products, METAL 2019 – 28th Int. Conf. Metall. Mater. (2019) 709-714.

[26] J. Pietraszek et al., Modeling of errors counting system for PCB soldered in the wave soldering technology, Advanced Materials Research 874 (2014) 139-143. https://doi.org/10.4028/www.scientific.net/AMR.874.139

[27] J. Pietraszek et al., Challenges for the DOE methodology related to the introduction of Industry 4.0, Prod. Eng. Arch. 26 (2020) 190-194. https://doi.org/10.30657/pea.2020.26.33

[28] J. Pietraszek, Fuzzy regression compared to classical experimental design in the case of flywheel assembly, Lecture Notes in Computer Science 7267 LNAI (2012) 310-317. https://doi.org/10.1007/978-3-642-29347-4_36

[29] J. Pietraszek, The modified sequential-binary approach for fuzzy operations on correlated assessments, Lecture Notes in Computer Science 7894 LNAI (2013) 353-364. https://doi.org/10.1007/978-3-642-38658-9_32

[30] K. Czerwińska, A. Pacana, Analysis of the maturity of process monitoring in manufacturing companies, Prod. Eng. Arch. 28 (2022) 246-251. https://doi.org/10.30657/pea.2022.28.30

[31] J. Pietraszek, A. Gadek-Moszczak, The smooth bootstrap approach to the distribution of a shape in the ferritic stainless steel AISI 434L powders, Solid State Phenomena 197 (2013) 162-167. https://doi.org/10.4028/www.scientific.net/SSP.197.162

[32] E. Skrzypczak-Pietraszek, J. Pietraszek, Chemical profile and seasonal variation of phenolic acid content in bastard balm (Melittis melissophyllum L., Lamiaceae), Journal of Pharmaceutical and Biomedical Analysis 66 (2012) 154-161. https://doi.org/10.1016/j.jpba.2012.03.037

[33] T. Lipiński, J. Pietraszek, Corrosion of the S235JR Carbon Steel after Normalizing and Overheating Annealing in 2.5% Sulphuric Acid at Room Temperature, Materials Research Proceedings 24 (2022) 102-108. https://doi.org/10.21741/9781644902059-16

Materials Research Proceedings 62 (2026) 26-33 https://doi.org/10.21741/9781644904015-4

Laser Micromachining of Stainless Steels Used in Military Applications

Artur KALINOWSKI[1,a*] and Krystian MULCZYK[1,b]

[1]Kielce University of Technology, Faculty of Mechatronics and Mechanical Engineering, Al. 1000-lecia P.P. 7, 25-314 Kielce, Poland

[a]akalinowski@tu.kielce.pl, [b]kmulczyk@tu.kielce.pl

Keywords: Laser Micromachining, High Energy Systems, Military Applications

Abstract. This study presents a systematic investigation into laser micromachining of stainless steel used in military applications. Using a single-mode fiber laser with controlled process parameters such as pulse repetition frequency, scanning speed, and F-theta lens focal length, the laser effect on surface microstructure was examined. Understanding the influence of key laser parameters enables optimized and reproducible processes, contributing to improved performance and lifespan of military hardware under extreme conditions.

Introduction

In the past decade, laser micromachining has established itself as a leading method for achieving ultra-precise surface modifications across diverse materials. Among various laser types, fiber laser micromachining is particularly valued for its flexibility, high efficiency, and precision. Moreover, laser-textured stainless steel produced through such micromachining methods has gained significant attention in military applications. Its enhanced durability and controlled surface properties make it particularly suitable for defense components, including weapon housing, machine components, and lightweight armor systems. The microstructuring of stainless-steel surfaces can also improve frictional characteristics, tribological performance under lubricated conditions and overall reliability in extreme operational environments [1-3]. Furthermore, laser-textured stainless-steel surfaces enhance adhesion and coating performance in protective layers, further extending the service life of critical military hardware [4-5].

The term “laser” is an acronym for “Light Amplification by Stimulated Emission of Radiation.” A laser system fundamentally comprises three critical components: a gain medium, which may be a solid, liquid, or gaseous substance capable of light amplification, an external energy source or pump to excite the gain medium and an optical resonator consisting of reflective elements that provide feedback to sustain and amplify coherent light emission [6].

Experimental

The aim of this experiment was to investigate the influence of individual laser micromachining parameters, such as pulse repetition frequency, scanning speed and focal length of the F-theta lenses. To ensure the precision of the results, a constant laser pulse energy of 0.31 mJ was maintained throughout the experiments. The specimens were made from 1.4034 stainless steel. Three types of laser scanning speeds, three levels of pulse frequency and two focal lengths of F-theta lenses were selected. After creating grooves on the sample surfaces, they were analyzed using an optical microscope to compare the produced geometries and identify the influence of the selected parameters on the texture depth. The selected laser for the study was a single-mode fiber laser SPI G3.1 with an average output power of 20 W. Laser specification is presented in Table 1. Fig.1 presents laser with galvo system scan heads (Fig.1a) and the single optical pulse shape generated by the laser operating at a pulse repetition frequency of 25 kHz (Fig.1b). The shape is typical for pulse fiber lasers.

Table 1. SPI G 3.1 20 W fiber laser specification

Name/model	SPI G 3.1 20 W
Type	Single-mode fiber laser
Wavelength, nm	1064
Average power, W	20
Pulse duration, ns	9-500
Pulse energy, mJ	≤ 0.8
Frequency, kHz	25-500
Beam quality M^2	<1.3
Beam expansion ratio	7.8
Galvo system scan head max scanning speed	10,000 mm/s
Lens	F-theta 160 mm/ F-theta 290 mm

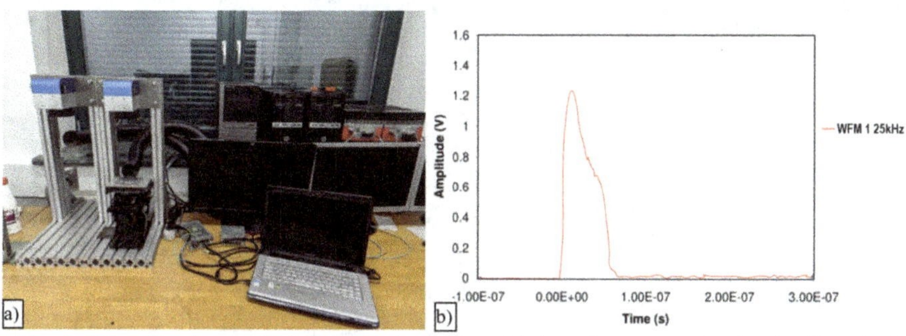

Fig. 1. Pulse fiber lasers G3.1 20W: a) the device with laser beam scanning heads, b) optical pulse shape at 25 kHz.

Table 2. Selected laser micromachining parameters

F-theta focal length	Frequency, kHz	Scanning speed, mm/s		
160 mm	10	500	1000	2000
	25	500	1000	2000
	50	500	1000	2000
290 mm	10	500	1000	2000
	25	500	1000	2000
	50	500	1000	2000

The selected laser micromachining parameters are presented in Table 2. To calculate the size of laser spot diameter with different focal lengths, equation (1) was used:

$$2w_0 = \frac{4M^2\lambda f}{\pi D} \tag{1}$$

where:

$2w_0$ – laser beam spot diameter, mm
M^2 – beam quality

λ – laser wavelength, nm
f – lens focal length, mm
D – beam diameter at lens, mm

The laser beam diameters at the focal point for the 160 mm and 290 mm F-theta lenses were calculated to be 28.3 μm and 51.4 μm, respectively. Figures 2-4 present cross-sectional profiles of microtexture produced at different laser beam scanning speeds and pulse repetition frequencies. The analysis of the presented graphs demonstrates that increasing the pulse repetition frequency from 10 kHz to 50 kHz leads to a notable increase in both the maximum and minimum depths of the produced microtexture profiles. This effect is observed for all tested combinations of scanning speed and focal length, with the greater texture depths occurring at the highest frequency of 50 kHz. Additionally, the use of a longer focal length (290 mm) consistently results in deeper surface features compared to a shorter focal length (160 mm). Conversely, higher scanning speeds generally produce shallower microtexture profiles, particularly evident in the case of the 160 mm focal length, where increasing the speed from 500 mm/s to 2000 mm/s significantly reduces the profile depth. These results indicate that both the pulse repetition frequency and focal length are dominant factors in determining the depth and sharpness of laser-induced microtextures, whereas the scanning speed predominantly affects the overall magnitude of these surface features.

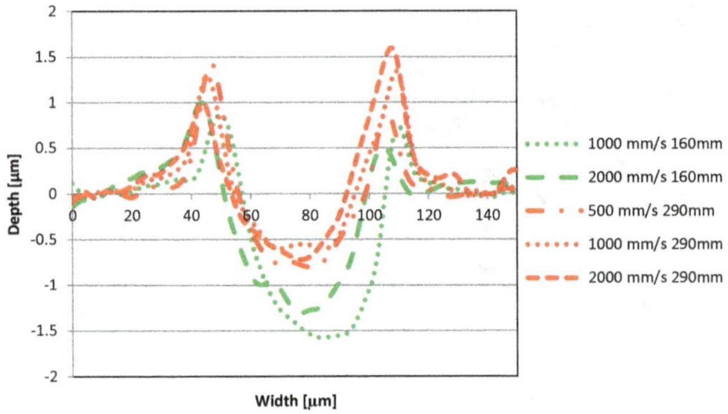

Fig. 2. Cross-sectional profile of microtexture produced at different laser beam scanning speeds and a pulse repetition frequency of 10 kHz.

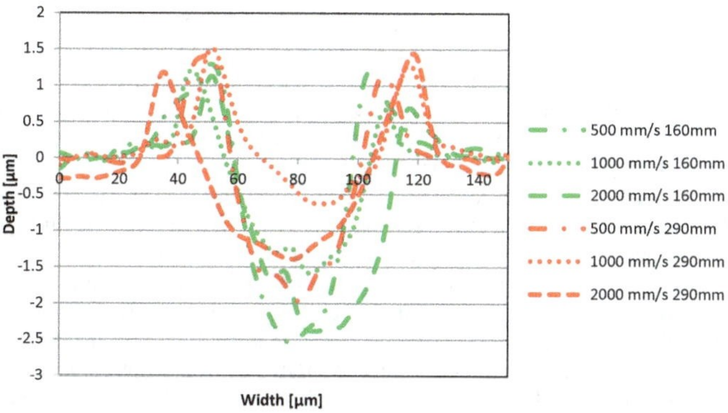

Fig. 3. Cross-sectional profile of microtexture produced at different laser beam scanning speeds and a pulse repetition frequency of 25 kHz.

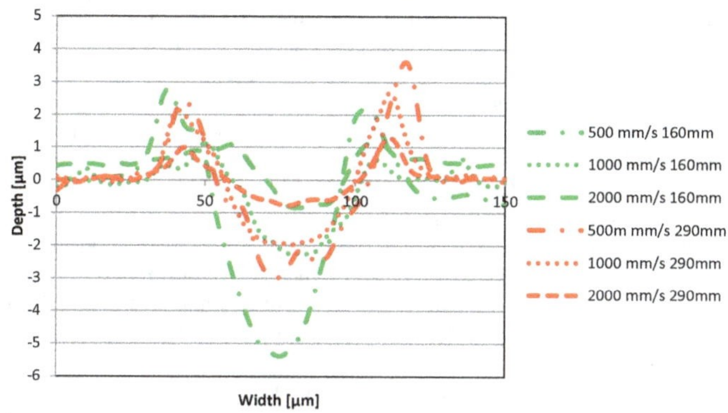

Fig. 4. Cross-sectional profile of microtexture produced at different laser beam scanning speeds and a pulse repetition frequency of 50 kHz.

Technological and Economic Impact

Laser micromachining of stainless steel, used in critical defense components, is having a significant, multifaceted impact on industry and society.

Technological advances in precision materials processing, such as laser micromachining, are crucial for achieving high performance and reliability in military components. The impact on surface properties is comparable to that achieved by other advanced material modification processes, such as electrospark deposition, which allows for the determination of the operational properties of steel impactors [7]. Precision manufacturing techniques enable precise microstructural characterization, including gamma-phase precipitation in single-crystal nickel superalloys [8], and detailed analysis of the structure and mechanical properties of alloys, such as the modifier-treated Al-7% SiMg alloy [9]. New machining methods open the door to applications in advanced control systems, such as intelligent positioning systems based on electropneumatic servo drives, essential for the stabilization of optical systems [10]. Material structure control is continually being improved through the modification of hypoeutectic Al-Si alloys with an exothermic modifier [11] and through advanced statistical and nonparametric material analysis, for example, in assessing particle shape distribution in powdered ferritic stainless steel [12] using methods such as smooth bootstrap. In the context of reliability and longevity, assessing the material condition of critical components, such as steam superheater coil tubes [13], is crucial, which directly translates into system safety and uptime. Laser texturing can increase the fatigue strength of materials, such as ductile iron in the ultra-high-cycle regime [14], which is crucial for lightweight armor systems. Laser technology also offers an alternative or complement to conventional techniques. Comparing cutting techniques for different steels, including HSLA steel [15], helps in rationally selecting the manufacturing process. Furthermore, innovations in machining are supported by the evaluation of the use of equipment for the production of metal components in civil engineering [16], which reflects the continuous development of industrial machinery. Improving the mechanical properties of alloys, such as sodium-modified AlSi9Mg alloy [17], provides a rationale for precise surface modifications. Control of surface quality is crucial, as confirmed by research on the causes of surface staining after anodized coatings [18].

In the context of quality and repeatability, metallographic study and repeatability analysis for thin-walled castings [19] are important, ensuring consistency of critical properties. Laser micromachining also has direct applications in defense systems, such as methods for controlling a scanning and tracking gyroscopic system mounted on a combat vehicle [20]. Laser micromachining, as a highly precise and repeatable method, translates directly into economic benefits. Reducing defects and increasing reliability is crucial for military components. The development of expert systems for DoE methodology [21] optimizes research time and costs, which is beneficial for the implementation of new laser processes. Analysis of the impact of rotary bearing assembly on their static load capacity [22] demonstrates the importance of durability and reliability, which micromachining can improve, reducing maintenance costs. The introduction of Industry 4.0 [23] poses new challenges for DoE methodology, but at the same time enables higher efficiency of laser processes through their full automation and digitalization. Striving to improve manufacturing processes using selected key performance indicators (KPIs) [24] is fundamental to the economic rationalization of laser machining. Furthermore, rapid prototyping [25], for example, of pneumatic control valves, allows for a shorter design cycle for critical components, reducing costs and time to market.

Although laser micromachining is generally a clean process that minimizes waste compared to conventional machining, it has broader implications for environmental impact in materials science. Chemical and statistical analyses [26] can be used to control polluting processes or in the context of sustainable chemistry. Contemporary approaches to materials science must address the problem of deposit buildup on steel samples during biomass and coal co-combustion [27], which poses a threat to energy infrastructure and the environment. Laser micromachining can be used to create coatings resistant to such phenomena. Another issue is the corrosion of stainless steel in nitric acid environments [28], which requires the use of materials with superior resistance, which can be improved with laser texturing. Laser fabrication and material modification in the context of recycling are supported by research on electrostatic separation of biodegradable polymer mixtures [29]. This demonstrates a broader engineering trend toward minimizing environmental impact. Furthermore, sustainability issues are represented by research on natural aggregate crushing [30], emphasizing the importance of durability in civil engineering. Energy optimization of buildings, for example, window type selection [31], points to the need for global energy reduction, which laser micromachining can contribute to by creating lightweight components.

Improving the reliability and durability of military equipment has a direct impact on national security and the lives of personnel. The application of precision machining in the context of materials science also touches on broader societal issues, including medicine. An example is the simulation of the impact of cancer metastasis on the biomechanics of the lumbar spine [32], which demonstrates the interdisciplinary nature of precision engineering. The development of advanced control systems, such as servo drives with pneumatic muscle actuators in 6-degree-of-freedom parallel manipulators [33], is leading to innovations in robotics and automation, which are important for work ergonomics, automation, and safety. Finally, the use of fuzzy regression for process optimization [34] and sequential-binary modifications to fuzzy operations [35] enable better uncertainty management in complex technical systems, ultimately improving safety and reliability, which have a critical social dimension in the case of military equipment.

Conclusions

Laser micromachining of stainless steels offers a highly effective and precise method for surface modification, particularly in military applications where durability, reliability, and controlled surface properties are crucial. The experimental results demonstrate that the depth and sharpness of laser-induced microtextures are primarily influenced by the pulse repetition frequency and the focal length of the F-theta lens, while the scanning speed mainly affects the overall magnitude of the surface features. Optimizing these parameters allows for improved process reproducibility and enhanced performance of critical defense components, such as weapon housings, optical mounts, and lightweight armor systems. The findings contribute to the advancement of precision laser micromachining techniques, supporting the development of more robust and reliable military hardware.

References

[1] A. Woźniak et al., The influence of laser texturing on the tribological behavior of titanium alloy Ti6Al4V in medical applications, Arch. Civ. Mech. Eng. 24 (2024) 1-19. https://doi.org/10.1007/S43452-024-00960-3

[2] A. Kalinowski et al., Selection of Process Parameters for Laser Texturing of DLC Coatings, Materials Research Proceedings 24 (2022) 46-51. https://doi.org/10.21741/9781644902059-8

[3] I. Etsion, Improving tribological performance of mechanical components by laser surface texturing, Tribology Letters 17 (2004) 733-737. https://doi.org/10.1007/S11249-004-8081-1

[4] E. Hegr et al., Evaluation of DLC coating for parts of weapons and military equipment, METAL 2020 – 29th Int. Conf. Metall. Mater., 1-10. https://doi.org/10.37904/metal.2020.3537

[5] Information on https://www.militaryaerospace.com/power/article/55140201/laser-weapons-high-power-microwaves-high-energy-weapons

[6] J. Kusiński, Lasery i ich zastosowanie w inżynierii materiałowej, Wydawnictwo Naukowe Akapit, Kraków, 2000.

[7] N. Radek, Determining the operational properties of steel beaters after electrospark deposition, Eksploatacja i Niezawodnosc 44 (2009) 10-16.

[8] A. Szczotok et al., Stereological characterization of γ' phase precipitation in CMSX-6 monocrystalline nickel-base superalloy, Materials Characterization 60 (2009) 1114-1119. https://doi.org/10.1016/j.matchar.2009.01.023

[9] T. Lipiński, The structure and mechanical properties of Al-7%SiMg alloy treated with a homogeneous modifier, Solid State Phenomena 163 (2010) 183-186. https://doi.org/10.4028/www.scientific.net/SSP.163.183

[10] J.E. Takosoglu, P.A. Laski, Intelligent positioning system of electro-pneumatic servo-drive, Annals of DAAAM and Proc. International DAAAM Symposium (2011) 1641-1642.

[11] T. Lipiński, P. Szabracki, Modification of the hypo-eutectic Al-Si alloys with an exothermic modifier, Arch. Metall. Mater. 58 (2013) 453-458. https://doi.org/10.2478/amm-2013-0017

[12] J. Pietraszek, A. Gadek-Moszczak, The smooth bootstrap approach to the distribution of a shape in the ferritic stainless steel AISI 434L powders, Solid State Phenomena 197 (2013) 162-167. https://doi.org/10.4028/www.scientific.net/SSP.197.162

[13] K. Trzewiczek et al., Evaluation of the state for the material of the live steam superheater pipe coils of V degree, Advanced Materials Research 874 (2014) 35-42. https://doi.org/10.4028/www.scientific.net/AMR.874.35

[14] R. Ulewicz et al., Fatigue strength of ductile iron in ultra-high cycle regime, Advanced Materials Research 874 (2014) 43-48. https://doi.org/10.4028/www.scientific.net/AMR.874.43

[15] P. Szataniak et al., HSLA steels - Comparison of cutting techniques, METAL 2014 – 23rd Int. Conf. Metall. Mater. (2014) 778-783.

[16] J. Selejdak et al., The evaluation of the use of a device for producing metal elements applied in civil engineering, METAL 2014 – 23rd Int. Conf. Metall. Mater. (2014) 1882-1888.

[17] T. Lipiński, P. Szabracki, Mechanical properties of AlSi9Mg alloy with a sodium modifier, Solid State Phenomena 223 (2015) 78-86. https://doi.org/10.4028/www.scientific.net/SSP.223.78

[18] S. Michna et al., Research the causes of surface stains after eloxal coating for the profile from the AlMgSi alloy using substructural analysis, Manufacturing Technology 15 (2015) 620-624.

[19] A. Szczotok et al., Metallographic Study and Repeatability Analysis of γ' Phase Precipitates in Cored, Thin-Walled Castings Made from IN713C Superalloy, Arch. Metall. Mater. 62 (2017) 595-601. https://doi.org/10.1515/amm-2017-0088

[20] I. Krzysztofik et al., Selected methods of control of the scanning and tracking gyroscope system mounted on a combat vehicle, Annual Reviews in Control 44 (2017) 173-182. https://doi.org/10.1016/j.arcontrol.2016.10.003

[21] R. Dwornicka, J. Pietraszek, The outline of the expert system for the design of experiment, Prod. Eng. Arch. 20 (2018) 43-48. https://doi.org/10.30657/pea.2018.20.09

[22] M. Krynke, R. Ulewicz, Analysis of the influence of slewing bearing mounting on their static load capacity, Transp. Res. Proc. 40 (2019) 745-750. https://doi.org/10.1016/j.trpro.2019.07.105

[23] J. Pietraszek et al., Challenges for the DOE methodology related to the introduction of Industry 4.0, Prod. Eng. Arch. 26 (2020) 190-194. https://doi.org/10.30657/pea.2020.26.33

[24] C. Karolina et al., Improvement of the production process with the use of selected KPIs, System Safety: Human - Technical Facility - Environment 2 (2020) 307-315. https://doi.org/10.2478/czoto-2020-0038

[25] S. Blasiak et al., Rapid prototyping of pneumatic directional control valves, Polymers 13 (2021) art. 1458. https://doi.org/10.3390/polym13091458

[26] E. Skrzypczak-Pietraszek, J. Pietraszek, Chemical profile and seasonal variation of phenolic acid content in bastard balm (Melittis melissophyllum L., Lamiaceae), Journal of Pharmaceutical and Biomedical Analysis 66 (2012) 154-161. https://doi.org/10.1016/j.jpba.2012.03.037

[27] M. Opydo et al., Characteristics of solids accumulation on steel samples during co-combustion of biomass and coal in a CFB boiler, Biomass and Bioenergy 120 (2019) 291-300. https://doi.org/10.1016/j.biombioe.2018.11.027

[28] T. Lipiński, D. Karpisz, Corrosion rate of 1.4152 stainless steel in a hot nitrate acid, METAL 2019 – 28th Int. Conf. Metall. Mater. (2019) 1086-1091.

[29] M. Zenkiewicz et al., Electrostatic separation of binary mixtures of some biodegradable polymers and poly(vinyl chloride) or poly(ethylene terephthalate), Polimery/Polymers 61 (2016) 835-843. https://doi.org/10.14314/polimery.2016.835

[30] A. Pacana et al., Influence of natural aggregate crushing process on crushing strength index, Sustainability (Switzerland) 13 (2021) art. 8353. https://doi.org/10.3390/su13158353

[31] J.M. Djoković et al., Selection of the Optimal Window Type and Orientation for the Two Cities in Serbia and One in Slovakia, Energies 15 (2022) art. 323. https://doi.org/10.3390/en15010323

[32] J. Lorkowski et al., Effects of breast and prostate cancer metastases on lumbar spine biomechanics: Rapid in silico evaluation, Advances in Experimental Medicine and Biology 1096 (2018) 31-39. https://doi.org/10.1007/5584_2018_192

[33] D.S. Pietrala, P.A. Laski, Design and Control of a Pneumatic Muscle Servo Drive Containing Its Own Pneumatic Muscles, Applied Sciences 12 (2022) art. 11024. https://doi.org/10.3390/app122111024

[34] J. Pietraszek, Fuzzy regression compared to classical experimental design in the case of flywheel assembly, Lecture Notes in Computer Science 7267 LNAI (2012) 310-317. https://doi.org/10.1007/978-3-642-29347-4_36

[35] J. Pietraszek, The modified sequential-binary approach for fuzzy operations on correlated assessments, Lecture Notes in Computer Science 7894 LNAI (2013) 353-364. https://doi.org/10.1007/978-3-642-38658-9_32

Permanent Casting Marking Methods and Optical Identification Methods

Szczepan KOSTECKI[a], Krzysztof BORKOWSKI[b], Gabriel BRACHA[c], Dawid S. PIETRALA[d], Hubert WIŚNIEWSKI[e] and Paweł A. ŁASKI[f] *

Kielce University of Technology, Faculty of Mechatronics and Mechanical Engineering, Department of Automation and Robotics, Al. 1000-lecia Państwa Polskiego 7, 25-314 Kielce, Poland

[a]skostecki@tu.kielce.pl, [b]kborkowski@tu.kielce.pl, [c]gbracha@tu.kielce.pl, [d]dpietrala@tu.kielce.pl, [e]hubert.w@tu.kielce.pl, [f]pawell@tu.kielce.pl

Keywords: Marked Castings, Casting Identification, Image Recognition

Abstract. In recent years, the manufacturing industry has undertaken several initiatives to enhance the monitoring of product life cycles by incorporating special markings directly during production. This is particularly important in today's global market, where unambiguous identification of genuine products is essential. Such monitoring is becoming increasingly relevant, particularly in ensuring the reliable traceability of products and related services that manufacturers, retailers, and end-users can access. As a result, databases maintained by component manufacturers, which enable the storage of information about an individual product throughout its entire life cycle, now play a crucial role. In this context, the growing popularity, advantages, and ease of use of permanent product marking encourage its application in manufacturing technologies, including foundry processes, as well as in other industrial areas. Introducing a permanent imprint into the casting, without the need for laser processing or the use of non-durable self-adhesive labels, enables the unambiguous identification of each manufactured part, as highlighted in this article. The paper analyses the types and methods of producing permanent impressions in the mold, resulting in a negative imprint on the casting, as well as the techniques for reading the markings obtained in this way.

Introduction

Currently, a significant number of components are manufactured worldwide using metal casting methods. Marking them is particularly important in addressing the growing number of spare parts, copies, and counterfeits. Marking products and uniquely identifying castings enables the monitoring of the manufacturing process, and in the event of a complaint, allows for identifying any technological errors. It also allows for distinguishing between original products and those that are copied or counterfeit. The oldest methods of marking and marking castings and forgings include alphanumeric characters and graphic symbols, such as simplified company logos. Figure 1 shows typical casting markings. Such markings are widely used, but they do not allow for the identification of individual products; only the batches for which the models and casting molds were made can be identified. Furthermore, with the current development of reverse engineering, copying and unauthorized production of non-original products are possible.

Currently, one of the most common methods of marking products is the application of graphic symbols, such as QR codes (Fig. 2). Marking foundry products using QR codes enables fast and error-free identification of parts at every stage of the production process and operation. QR codes can contain information such as the batch number, chemical composition of the alloy, heat treatment parameters, production date, or mold number. The code is usually applied by laser engraving, stamping, or by using special inserts in the mold that reproduce the code already during casting. Compared to traditional alphanumeric marking, QR codes enable the storage of a

Materials Research Proceedings 62 (2026) 34-42 https://doi.org/10.21741/9781644904015-5

significantly larger amount of data on a small surface and facilitate integration with ERP systems or quality tracking systems. As a result, full tracking of the product's history (traceability) is possible, which is particularly important in industries with high quality requirements, such as the automotive and aviation sectors. However, such markings can be removed or worn off during operation.

Fig. 1. Typical casting stamps with logos and alphanumeric markings

Fig. 2. Typical graphic symbols used in marking and identification (source: [1,3-5])

Fig. 3. Handheld code reader – QR code reading. [source: 6-11]

Currently, one of the most common methods of marking products is the application of graphic symbols, such as QR codes (Fig. 2). Marking foundry products using QR codes enables fast and error-free identification of parts at every stage of the production process and operation. QR codes can contain information such as the batch number, chemical composition of the alloy, heat treatment parameters, production date, or mold number. The code is usually applied by laser engraving, stamping, or by using special inserts in the mold that reproduce the code already during casting. Compared to traditional alphanumeric marking, QR codes enable the storage of a significantly larger amount of data on a small surface and facilitate integration with ERP systems or quality tracking systems. As a result, full tracking of the product's history (traceability) is possible, which is particularly important in industries with high quality requirements, such as the automotive and aviation sectors. However, such markings can be removed or worn off during operation.

Reading such markings from devices in the digital world is straightforward; ready-made solutions are available in the form of service readers with databases, as shown in Fig. 3, for code-reading devices.

As already mentioned, marking castings using QR codes can be problematic under certain conditions. Additional difficulties arise wherever a QR code is to be applied on a raw, unmachined surface. This is the case for foundry products manufactured in sand molds, where the surface of the products is not machined and is highly rough. Therefore, engineers and researchers are seeking alternative methods of marking cast products, particularly where products are manufactured in sand molds. A solution known from patent EP 3 736 064 A1 uses a set of three rotary keys as the marker. Fig. 4 shows a view of the marking assembly placed in the mold in a manner that allows it to be imprinted on the flat surface of the casting mold, producing an unambiguous imprint on the final product.

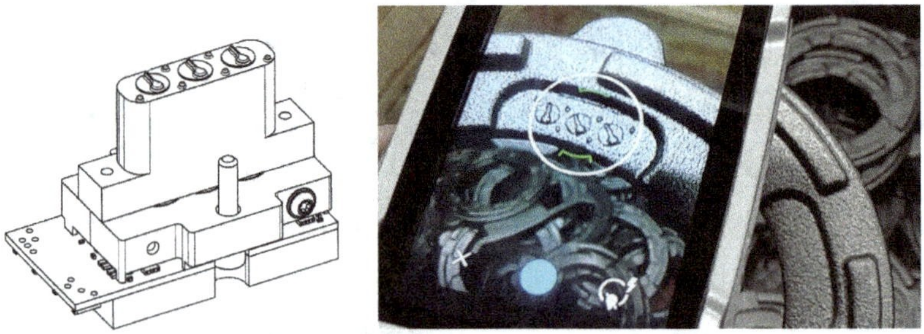

Fig. 4. Casting marking process - patent EP 3 736 064 A1 [12], advertising materials DIAS company

The number of possible combinations resulting from the angular rotation of key-shaped markers reaches tens of thousands and, in most cases, is sufficient for the complete identification of each product manufactured in series. Figure 5 shows the imprint of a marker produced in a sand mold.

Fig. 5. Casting marking process – advertising materials DIAS company

Full identification of a single product enables the linking of a casting to a specific cause of defects or the identification of process parameters that caused them during production. Furthermore, as mentioned, this method of full casting traceability allows for tracing when and how it was produced, optimizing waste, and determining the total production cost. It also enables full traceability of the casting throughout its entire life cycle. Figure 5 shows the process of producing negative, frameless casting molds. The process also illustrates the subsequent stages of sand mold production.

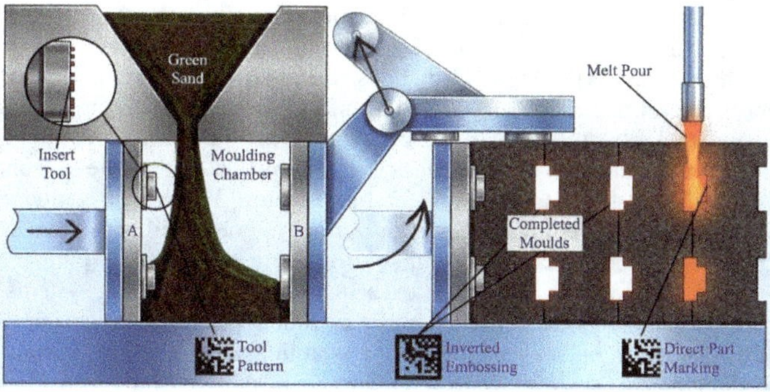

Fig. 5. The process of marking castings – with full identification of a single product. [2]

Digital Methods of Identifying Foundry Products

A proprietary module solution for creating impressions in the molding sand during pen molding was proposed. The solution is based on a similar approach to QR codes, but in a significantly simplified form. This simplification is necessary for two reasons: first, the impression created must be high enough to be read using vision systems. Moreover, it is durable enough to be significantly more difficult to remove. Furthermore, an additional feature remains that, with the right combination of both concave and convex surfaces, will significantly reduce the complete removal of the permanent marking. This is particularly important when exporting parts and components, or finished devices, that are covered by international bans. The developed single module has three possible states for each pin. The mechanism for extending or retracting individual pins is based on a cam mechanism activated by a mini motor with a reduction gear. Therefore, the total number of possible combinations is close to 20,000. When combined in a modular manner, this creates multiple possible combinations. This is particularly important in serial and large-scale production, as it allows for complete and unambiguous identification of a single product.

Fig. 6. Module for creating impressions in sand molds

Economic and Technological Impact

The introduction of permanent casting marking and tracking systems is a key element of the transformation towards Industry 4.0, posing new challenges and opportunities for the design of experiments (DOE) methodology [13]. Technological evolution requires the continuous optimization of production processes, which can be achieved through advanced statistical methods, such as fuzzy regression in process optimization [14, 15], and the use of nonparametric statistical tools, including the bootstrap approach, for model stability analysis [16, 17]. The development of material and process technologies has a direct impact on quality and efficiency. Research focuses on improving performance, for example, by determining the performance characteristics after electro-spark deposition [18] or modifying alloys such as Al-11% Si [19]. Modern processes also require precise control and modeling, including modeling the hot forging process on hydraulic presses [20] and analyzing the effect of longitudinal stresses in continuous rolling [21]. Even in civil engineering, the evaluation of metal fabrication equipment plays a role in economic efficiency [22], as analyzes the crushing strength of natural aggregates [23]. Precise components and control systems are essential in modern machines. This applies to both intelligent servo drive positioning systems [24] and proportional valves with piezoelectric actuators [25]. In mass production, modeling and error mitigation are crucial, for example, in error counting systems for PCBs [26]. Further automation and robotics, such as masonry robots [27] and the design of low-vibration machines [28], also rely on advanced technical solutions. Research into rapid prototyping of automation components, such as control valves [29], shortens time-to-market, which has a significant economic impact. The need for reliability also applies to specialized applications, such as the analysis of load capacity in rotating bearings [30] and joining processes, including friction welding [31]. Research on the production of composite coatings [32] and the evaluation of the shrinkage of bioceramic sinters [33] also contributes to the development of advanced materials technologies. Optimization of processes, such as those in meat processing machines [34], contributes to cost reduction. Fuzzy logic-based methods are employed to more accurately describe complex systems [35].

Environmental technologies and the pursuit of sustainable development are closely linked to innovation. New methods for marking castings, promoting full traceability, supporting scrap reduction, and optimizing resource consumption. Research into new energy sources, such as vertical-axis wind turbines [36], and studies on energy efficiency in construction, including optimal window selection [37], have a direct impact on the environment. Recycling is a significant area where electrostatic separation of polymer blends plays a key role in achieving a circular economy [38]. Innovative purification approaches, such as ion flotation of heavy metals, including zinc and cadmium [39], and the study of activated carbon materials for electrochemical or filtration applications [40], also contribute to environmental improvement. The impact of advanced technologies on society is primarily visible in improved safety, health, and quality of life. In the context of health, biomedical engineering plays a crucial role, for example, in assessing the biomechanics of the lumbar spine in relation to cancer metastases [41] and in predicting the compressive strength of vertebrae [42]. Furthermore, data analysis methods, including advanced statistical and nonparametric analyses, originally used in chemical composition analysis (e.g., in plant studies, where the phenolic acid profile was studied [43]), are now widely used to increase the reliability and precision of medical and engineering research, which directly translates into product safety and quality.

Summary

The article discusses contemporary methods of permanent marking of castings and optical methods for their identification, with particular emphasis on the possibility of tracking the entire life cycle of a product (traceability). The authors stress that, in the conditions of the global market, the need for unambiguous, non-removable marking of parts is increasing – both for quality control and complaint handling, as well as for combating counterfeits. Traditional marking (alphanumeric symbols, foundry logos) allows for the recognition of the production batch at most, but not an individual part. With today's reverse-engineering technologies, it can be easily copied. The authors then describe the currently widespread graphical identification codes (QR, DataMatrix, etc.), which enable the storage of a large amount of data on a small surface and facilitate easy integration with ERP systems or quality-tracking systems. However, they point out their limitations in foundry applications – especially in the case of castings from sand molds with a rough, non-machined surface and the fact that codes applied by laser or on labels can be removed or worn out during service. As an alternative, the paper discusses solutions that utilize permanent imprinting of a marker in the mold, e.g., the system of three rotary "keys" known from patent EP 3 736 064 A1, which, due to the numerous combinations, allows for the unambiguous identification of each casting. Such mechanical imprints, created directly in the mold, are much more difficult to remove, yet they allow a given casting to be linked with process parameters, which promotes production optimization, scrap reduction, and full traceability throughout the product's life cycle. In the original part of the article, the authors present their own modular system for producing convex–concave imprints in a sand mold, which, in concept, resemble a QR code but are deliberately simplified. Each module "pin" can assume three states (raised, recessed, neutral), and a cam mechanism driven by a geared motor enables their individual adjustment. This yields approximately 20,000 possible combinations for a single module and, when modules are combined, many times more. The resulting marking is clearly visible to vision systems, resistant to wear, and very difficult to remove completely, which makes the proposed method particularly suitable for series and mass production of castings that require full, unambiguous identification of every single piece.

References

[1] R. Kikuchi et al., Embedding QR codes onto B-spline surfaces for 3D printing, CAD Computer Aided Design 102 (2018) 215–223. https://doi.org/10.1016/j.cad.2018.04.025

[2] N.K. Vedel-Smith, T.A. Lenau, Casting traceability with direct part marking using reconfigurable pin-type tooling based on paraffin–graphite actuators, Journal of Manufacturing Systems 31 (2012) 113-120. https://doi.org/10.1016/j.jmsy.2011.12.001

[3] C.E.H. Ventura et al., Towards Part Lifetime Traceability Using Machined Quick Response Codes, Procedia Technology 26 (2016) 89-96. https://doi.org/10.1016/j.protcy.2016.08.013

[4] F. Chen et al., Embedded product authentication codes in additive manufactured parts: Imaging and image processing for improved scan ability, Additive Manufacturing 35 (2020) art.101319. https://doi.org/10.1016/j.addma.2020.101319

[5] S.I. Okada, E.M.S. de Souza, Digital marketing strategies in the age of search, REMark – Revista Brasileira de Marketing 10 (2011) 46-72. https://doi.org/10.5585/remark.v10i1.2199

[6] L. Liao et al., Data Extraction Method for Industrial Data Matrix Codes Based on Local Adjacent Modules Structure, Applied Sciences 12 (2022) art.2291. https://doi.org/10.3390/app12052291

[7] L. Karrach, E. Pivarčiová, Comparative Study of Data Matrix Codes Localization and Recognition Methods, Journal of Imaging 7 (2021) art.163. https://doi.org/10.3390/jimaging7090163

[8] L. Song et al., Research on a traceability process of sand core information by printing QR code on sand core surface in the casting production process, International Journal of Metalcasting 15(4) (2021) 1476-1482. https://doi.org/10.1007/s40962-021-00572-0

[9] F. Deng et al., Comparative evaluation of marking methods on cast parts of Al–Si alloy with image processing, International Journal of Metalcasting 16(3) (2022) 1122-1139. https://doi.org/10.1007/s40962-021-00661-0

[10] T. Uyan et al., Sand casting implementation of two-dimensional digital code direct-part-marking using additively manufactured tags, International Journal of Metalcasting 16(3) (2022) 1140-1151. https://doi.org/10.1007/s40962-021-00680-x

[11] J. Batista et al., QR code micro-certified gemstones: femtosecond writing and Raman characterization in diamond, ruby and sapphire, Scientific Reports 9 (2019) art.8927. https://doi.org/10.1038/s41598-019-45405-7

[12] P. Larsen, H. Wegge, Foundry production line and method of operating such foundry production line, European Patent EP 3 736 064 A1, Disa Industries A/S, 2020.

[13] J. Pietraszek et al., Challenges for the DOE methodology related to the introduction of Industry 4.0, Prod. Eng. Arch. 26 (2020) 190-194. https://doi.org/10.30657/pea.2020.26.33

[14] J. Pietraszek, Fuzzy regression compared to classical experimental design in the case of flywheel assembly, Lecture Notes in Computer Science 7267 LNAI (2012) 310-317. https://doi.org/10.1007/978-3-642-29347-4_36

[15] J. Pietraszek, E. Skrzypczak-Pietraszek, The optimization of the technological process with the fuzzy regression, Advanced Materials Research 874 (2014) 151-155. https://doi.org/10.4028/www.scientific.net/AMR.874.151

[16] J. Pietraszek, A. Gadek-Moszczak, The smooth bootstrap approach to the distribution of a shape in the ferritic stainless steel AISI 434L powders, Solid State Phenomena 197 (2013) 162-167. https://doi.org/10.4028/www.scientific.net/SSP.197.162

[17] A. Szczotok et al., A. Szczotok et al., The bootstrap analysis of one-way ANOVA stability in the case of the ceramic shell mould of airfoil blade casting, Solid State Phenomena 235 (2015) 24-30. https://doi.org/10.4028/www.scientific.net/SSP.235.24

[18] N. Radek, Determining the operational properties of steel beaters after electrospark deposition, Eksploatacja i Niezawodnosc 44 (2009) 10-16.

[19] T. Lipinski, Modification of Al-11% Si alloy with Cl-based modifier, Manufacturing Technology 15 (2015) 581-587.

[20] R. Dindorf et al., Prediction of the parameters and the hot open die elongation forging process on an 80 MN hydraulic press, Open Engineering 11 (2021) 528-534. https://doi.org/10.1515/eng-2021-0056

[21] P. Sygut et al., Theoretical analysis of the influence of longitudinal stress changes on band dimensions during continuous rolling process, Archives of Metallurgy and Materials 61 (2016) 183-188. https://doi.org/10.1515/amm-2016-0032

[22] J. Selejdak et al., The evaluation of the use of a device for producing metal elements applied in civil engineering, METAL 2014 – 23rd Int. Conf. Metall. Mater. (2014) 1882-1888.

[23] A. Pacana et al., Influence of natural aggregate crushing process on crushing strength index, Sustainability 13 (2021) art. 8353. https://doi.org/10.3390/su13158353

[24] J.E. Takosoglu, P.A. Laski, Intelligent positioning system of electro-pneumatic servo-drive, Annals of DAAAM and Proc. International DAAAM Symposium (2011) 1641-1642.

[25] P.A. Laski, Proportional valve with a piezoelectric actuator, EPJ Web of Conf. 143 (2017) art. 02064. https://doi.org/10.1051/epjconf/201714302064

[26] J. Pietraszek et al., Modeling of errors counting system for PCB soldered in the wave soldering technology, Advanced Materials Research 874 (2014) 139-143. https://doi.org/10.4028/www.scientific.net/AMR.874.139

[27] P. Wos et al., Bricklaying robot lifting and levelling system, Communications - Scientific Letters of the University of Žilina 23 (2021) B257-B264. https://doi.org/10.26552/COM.C.2021.4.B257-B264

[28] I. Drach et al., Design Principles of Horizontal Drum Machines with Low Vibration, Adv. Sci. Technol. Res. J. 15 (2021) 258-268. https://doi.org/10.12913/22998624/136441

[29] S. Blasiak et al., Rapid prototyping of pneumatic directional control valves, Polymers 13 (2021) art. 1458. https://doi.org/10.3390/polym13091458

[30] M. Krynke, R. Ulewicz, Analysis of the influence of slewing bearing mounting on their static load capacity, Transp. Res. Proc. 40 (2019) 745-750. https://doi.org/10.1016/j.trpro.2019.07.105

[31] A. Wronska et al., Effect of tool pin length on microstructure and mechanical strength of the FSW joints of Al 7075 metal sheets, Communications - Scientific Letters of the University of Žilina 21 (2019) 40-47.

[32] L. Adamczyk et al., Electrochemical preparation of composite coatings of 3,4-etylenodioxythiophene (EDOT) and 4-(pyrrole-1-yl) benzoic acid (PyBA) with heteropolyanions, Mater. Chem. Phys. 144 (2014) 418-424. https://doi.org/10.1016/j.matchemphys.2014.01.012

[33] A. Dudek, C. Kolan, Assessments of shrinkage degree in bioceramic sinters HA+ZrO2, Solid State Phenomena 165 (2010) 25-30. https://doi.org/10.4028/www.scientific.net/SSP.165.25

[34] J. Pietraszek et al., The parametric RSM model with higher order terms for the meat tumbler machine process, Solid State Phenomena 235 (2015) 37-44. https://doi.org/10.4028/www.scientific.net/SSP.235.37

[35] J. Pietraszek, The modified sequential-binary approach for fuzzy operations on correlated assessments, Lecture Notes in Computer Science 7894 LNAI (2013) 353-364. https://doi.org/10.1007/978-3-642-38658-9_32

[36] J. Zwierzchowski et al., Model tests of wind turbine with a vertical axis of rotation type Lenz 2, EPJ Web of Conf. 143 (2017) art. 02150. https://doi.org/10.1051/epjconf/201714302150

[37]J.M. Djoković et al., Selection of the Optimal Window Type and Orientation for the Two Cities in Serbia and One in Slovakia, Energies 15 (2022) art. 323. https://doi.org/10.3390/en15010323

[38] M. Zenkiewicz et al., Electrostatic separation of binary mixtures of some biodegradable polymers and poly(vinyl chloride) or poly(ethylene terephthalate), Polimery/Polymers 61 (2016) 835-843. https://doi.org/10.14314/polimery.2016.835

[39] M. Ulewicz et al., Ion flotation of zinc(II) and cadmium(II) in the presence of side-armed diphosphaza-16-crown-6 ethers, Separation Science and Technology 38 (2003) 633-645. https://doi.org/10.1081/SS-120016655

[40] V. Ptashnyk et al., Impedance investigation of activated carbon material modified by ultrasound treatment, Przeglad Elektrotechniczny 94 (2018) 186-189. https://doi.org/10.15199/48.2018.05.33

[41] J. Lorkowski et al., Effects of breast and prostate cancer metastases on lumbar spine biomechanics: Rapid in silico evaluation, Advances in Experimental Medicine and Biology 1096 (2018) 31-39. https://doi.org/10.1007/5584_2018_192

[42] L. Wojnar et al., On the role of histomorphometric (stereological) microstructure parameters inthe prediction of vertebrae compression strength, Image Analysis and Stereology 38 (2019) 63-73. https://doi.org/10.5566/ias.2028

[43] E. Skrzypczak-Pietraszek, J. Pietraszek, Chemical profile and seasonal variation of phenolic acid content in bastard balm (Melittis melissophyllum L., Lamiaceae), Journal of Pharmaceutical and Biomedical Analysis 66 (2012) 154-161. https://doi.org/10.1016/j.jpba.2012.03.037

Materials Research Proceedings 62 (2026) 43-51 | https://doi.org/10.21741/9781644904015-6

Mechanical Properties of Camouflage Coating Systems Used in Military Technology

Marek MICHALSKI[1,a*], Norbert RADEK[1,2,b] and Marcin SZCZEPANIAK[3,c]

[1]Barwa Kielce, ul. Warkocz 3-5, 25-253 Kielce, Poland

[2]Kielce University of Technology, Faculty of Mechatronics and Mechanical Engineering, Al. 1000-lecia P. P. 7, 25-314 Kielce, Poland

[3]Military Institute of Engineer Technology, ul. Obornicka 136, 50-961 Wrocław, Poland

[a]mmichalski@barwa.kielce.pl, [b]norrad@tu.kielce.pl, [c]szczepaniak@witi.wroc.pl

Keywords: Military, Camouflage, Coating System, Military Equipment And Vehicles, Erosion Wear, Abrasion, Scratch

Abstract. This article presents the results of experimental studies on the performance properties of camouflage coating systems intended for military applications. Three types of paint systems were tested: a reference system (SP1), a system modified with carbon nanotubes (SP2), and a system modified with glass microspheres (SP3). The coatings were applied to low-carbon steel substrates and tested for abrasion, erosion, and scratch resistance according to Polish and ISO standards. The results showed that the unmodified coating (SP1) demonstrated the highest abrasion and erosion resistance, while the system containing glass microspheres (SP3) achieved the best scratch resistance (10.78 N). Modification with carbon nanotubes (SP2) did not significantly improve the mechanical durability of the coatings. The studies confirm that the use of selected modifiers can improve some functional parameters of paint systems, but not necessarily all performance properties. The results provide a basis for further optimization of camouflage coatings used in military applications. Their purpose is to increase their service life and mechanical strength under demanding environmental and operational conditions.

Introduction

The modern battlefield is characterized by the rapid development of target detection and identification technologies, which presents armed forces with the challenge of effectively concealing their forces. Camouflage, as a key element of military tactics, has evolved from simple visual camouflage methods such as leaves, branches, and camouflage netting to advanced coating systems capable of countering a variety of detection systems across various electromagnetic radiation ranges. Camouflage coating systems play a crucial role, enabling adaptation to changing conditions and also playing a fundamental role in protecting military equipment from harmful external factors. They extend component life, minimize maintenance costs, and ensure operational safety. They protect substrates from the harmful effects of environmental factors [1] such as protection against corrosion, changing weather conditions, and solar radiation. In the case of camouflage coating systems used in military technology, they fulfil additional functions, including camouflage properties, resistance to warfare agents and decontamination agents, protection against specific operating conditions, such as seawater, contamination, abrasion, scratching, etc. [2]. Camouflage coating systems typically consist of several layers, each of which performs a specific function. The primer provides good adhesion to the substrate and corrosion resistance. The topcoat provides resistance to external factors. Intermediate layers can also be used, as necessary to ensure effective camouflage in various electromagnetic radiation ranges. Figure 2 shows a typical coating system.

High resistance to mechanical damage allows the coating to maintain its protective and masking properties, preventing the exposure of the metal substrate and the initiation of corrosion processes. In military practice, equipment is exposed to factors such as friction between moving parts, contact with sand, mud, small stones, and consumables, which requires the use of coatings with increased tribological resistance.

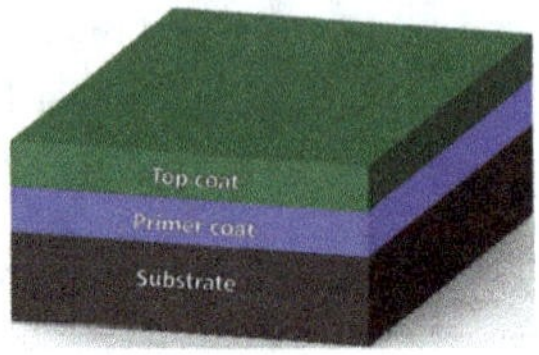

Fig. 1. Diagram of a typical coating system [3].

Scratch and abrasion resistance testing is an essential element in assessing the performance of paint systems. These tests determine the durability of coatings under dynamic surface loading. Mechanical resistance is assessed using methods such as the Taber method, which involves abrading the surface using rotating grinding discs of a specified hardness, and the erosion test, in which the coating surface is exposed to an abrasive jet [4]. Furthermore, scratch resistance testing according to the PN-EN ISO 1518-1 standard allows for the assessment of the damage limit of the coating under concentrated pressure.

Materials and Parameters

The specimens with dimensions of 150 mm x 100 mm x 1 mm were made of low-carbon steel DC01. The steel samples were first washed with XPA10006 remover to degrease the surfaces. Then, a grinding operation was performed using a rotary machine and P80 grit sandpaper. The final stage of surface preparation was washing the surface with XPA10006 solvent. Masking coating systems were applied in a Blowtherm spray booth, using SATA guns. Masking coating systems were applied by air spray in three options:

- paint system (SP1): primer coating, masking coating (green),
- paint system (SP2): primer coating, masking coating (green) + carbon nanotube modification,
- paint system (SP3): primer coating, masking coating (green) + modification with glass microspheres.

Results and Discussion

Abrasion test. The abrasion resistance test was performed using a DT-523 Rotary Abrasion Tester. The test employed Calibrase CS-17 wheels – flexible rubber discs embedded with aluminium oxide abrasive particles. To maintain consistent performance, the abrasive wheels were regenerated every 500 cycles using an S-11 refacing disc. A 1000 g load was applied to each arm during testing.

Prior to testing, each sample was weighed, and the total coating thickness was measured. The thickness measurements were repeated after every 500 abrasion cycles. When the topcoat layer was fully removed, the sample was weighed again. The final weight was recorded once the entire coating system had been abraded from the surface.

Fig.2 illustrates the appearance of samples with SP1, SP2, and SP3 coating systems after the abrasion test. The abrasion resistance test, performed using the Taber Abraser, made it possible to evaluate the mechanical wear resistance of the three camouflage coating systems.

Fig.3 shows the mass loss as a function of the number of abrasion cycles in the Taber tester. The curves show an inflection point, which reflects the difference in abrasion resistance between the primer and topcoat within each coating system.

Based on the results, the unmodified coating system (SP1) demonstrated the highest abrasion resistance, as evidenced by the lowest mass loss per abrasion cycle. Table 1 summarizes the average mass loss per cycle for each coating system.

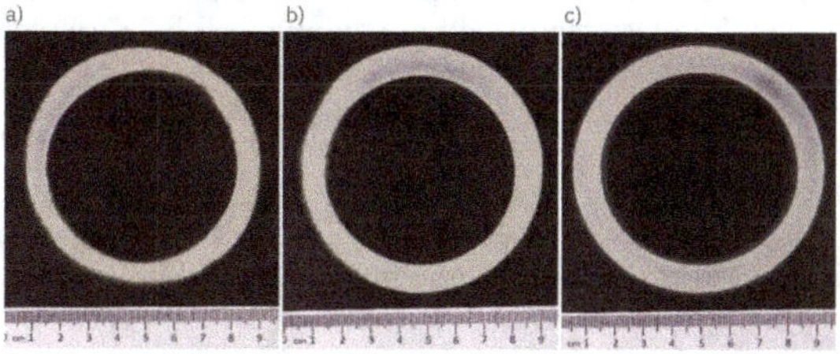

Fig. 2. Coating system after Taber test. a) SP1, b) SP2, c) SP3.

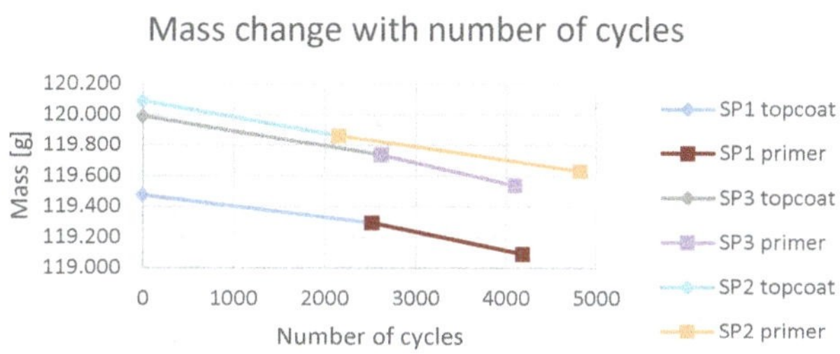

Fig. 3. Weight change with numbers of cycles.

Table 1. The average mass loss per cycle for the entire coating system

Systems / Parameters	SP1	SP2	SP3
Weight loss, mg	384	463	455
Weight loss per 1 cycle, mg	0.092	0.096	0.111

Tests of Resistance to Erosion Wear

The erosive wear process is a highly undesirable phenomenon that leads to progressive degradation of machine components, often in ways that are difficult to predict at the design stage. The intensity of erosive wear depends on several factors, including the size, hardness, and shape of the abrasive particles, as well as the pressure, speed, temperature of the particle-medium mixture, and the angle of impact on the surface. Erosion causes gradual abrasion and scratching of the coating until it is ultimately removed. Repeated impacts of hard particles lead to fatigue spalling and detachment of coating fragments. The resulting surface defects, such as scratches and microcracks, promote corrosive wear and can cause blistering and loss of coating adhesion. The erosive wear resistance of masking coating systems was tested using a device according to the PN-C-81516:1976 standard. The method involved removing the paint coating using an abrasive material jet. The tests were conducted on SP1, SP2, and SP3 coating systems. Three measurements were taken for each system, and the average erosion wear value was determined.

The erosive wear of each tested sample was defined as the ratio of the mass of abrasive material required to remove the coating to the average coating thickness, expressed in kg/μm. During the test, the abrasive material was poured into a hopper, from which it was directed through a vertically positioned pipe and fell by gravity onto the sample surface at a 45° angle. The abrasive was introduced in 3.5 kg portions. The abrasive material used was noble fused alumina 99A with a grain size of 0.5–0.6 mm. The test continued until an elliptical hole was formed in the coating, with a major axis of approximately 3.6–3.7 mm. After every ten portions of abrasive, its mass was replenished to maintain 3.5 kg, and after every hundred portions, the entire batch was replaced with fresh abrasive material.

Figure 4 presents a comparison of the erosion resistance of the tested coating systems. The erosion resistance result was expressed as the mass of fused alumina (kg) required to reduce the coating thickness by one micrometre, as described by the following equation:

$$E = \frac{K}{T} \tag{1}$$

where: E – erosion resistance, kg/μm, K – mass of the abrasive material used for measurement, kg, T – average thickness of the tested coating system, μm.

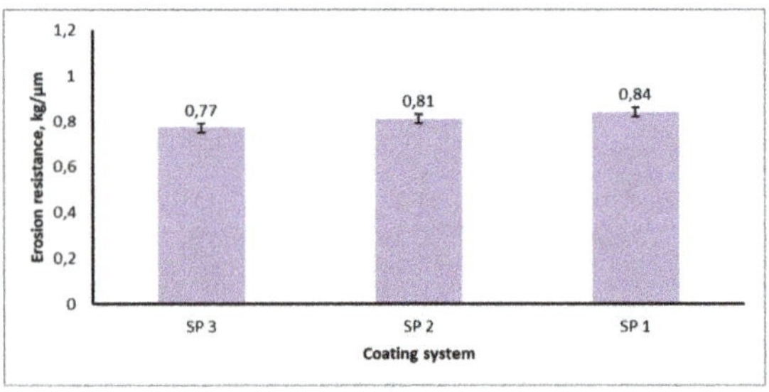

Fig. 4. Comparison of erosion resistance of camouflage coating systems.

The paint coating systems (SP1, SP2, SP3) showed no significant differences in erosion resistance, with results ranging from 0.77 to 0.84 kg/μm. The highest erosion resistance was recorded for the SP1 system (0.84 kg/μm), while the lowest value was observed for the SP3 system (0.77 kg/μm). The erosion resistance of the tested coatings was primarily influenced by the type of masking topcoat used. In practical applications, the goal is to enhance the erosion resistance of the outer (masking) layer, while maintaining its good flexibility, since it is the layer most exposed to contact with abrasive materials such as sand or small stones during service. Damage to the topcoat, followed by the primer layer, leads to a loss of protection against aggressive environmental factors, which can accelerate corrosion processes. It should be noted that the wear intensity of paint coatings increases under exposure to aggressive environments, such as salt mist or acid rain. Scratch Resistance. Scratch resistance testing was conducted in accordance with the PN-EN ISO 1518-1:2019 standard, which specifies the constant load method for determining the resistance of single- or multi-layer coatings to stylus penetration. The principle of the method involves moving a stylus with a specified force across the surface of the coated test panel at a constant speed until visible damage or penetration occurs. The test load was applied gradually until the coating penetrated the substrate. After each test, the scratch mark was examined under 4x magnification to determine the minimum load required to penetrate the coating.

Fig.5 shows the appearance of the tested coating systems after the scratch test. Table 2 summarizes the measured minimum penetration loads for each coating. The results indicate that the SP3 coating, modified with glass microspheres, demonstrated the highest scratch resistance of 10.78 N, while the SP1 and SP2 systems showed slightly lower scratch resistance values of 9.48 N.

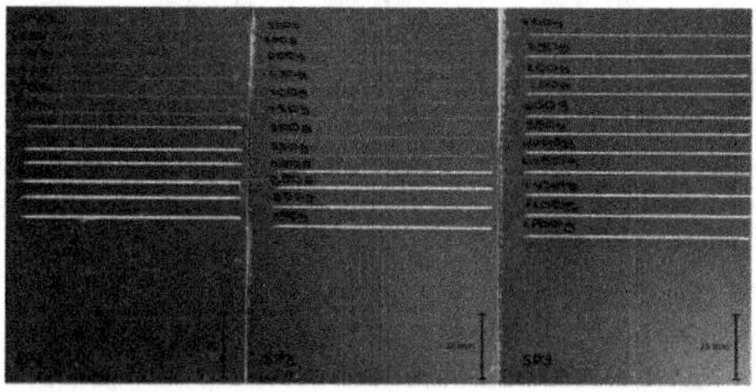

Fig. 5. Coating system after test: SP1, SP2, SP3.

Table 2. Results of scratch resistance

Coating system	Critical force causing cracking, N			Average value and standard deviation, N
SP1	9.32	9.81	9.32	9.48 ±0.28
SP2	9.81	9.32	9.32	9.48 ±0.28
SP3	10.79	11.28	10.30	10.78 ± 0.49

Technological, Economic, Environmental, and Social Impact

The development of new camouflage coatings with enhanced mechanical resistance is crucial for the modernization of military technology, generating significant implications in four key areas: technological, economic, environmental, and social.

Research on the mechanical properties of coatings, particularly resistance to abrasion, erosion, and scratching, represents a milestone towards increasing the durability and reliability of military equipment under extreme operating conditions. The validation of various modifiers, such as carbon nanotubes and glass microspheres, is consistent with a broader trend in materials science, focused on the design of advanced mechanical [5, 6] and material systems [7, 8]. A key element enabling precise evaluation of modifications is the use of advanced statistical methods, such as RSM (Response Surface Methodology) [9, 10] and bootstrap [11, 12], which allows for optimal experiment and model selection [9] and effective quantification of modeling uncertainty [10]. Precise analysis of results, including the assessment of the effect of modifications on the microstructure and strength of metals [6] and minimizing errors in quality control processes [13, 14], is essential for implementing innovations. Investigating the effect of coating modifications on low-carbon steel substrates requires consideration of basic structural mechanics, such as analysis of torsional vibrations of thin-walled beams [15], as well as the influence of external factors, such as steel corrosion [16] or oxygen content on its fatigue strength [17]. In the context of advanced systems where coatings are used, multiphysics modeling and control of machines [18] and manipulator position control systems [19, 20] also become important. To achieve high reliability of systems, including military systems, statistical methods are used to ensure the stability of solutions to inverse problems [21, 22], as well as to analyze complex multi-stage models [23], and to model and predict parameters of technological processes, such as forging [24] and bonding [6, 25]. Increased coating durability thanks to improved abrasion and scratch resistance (e.g., using glass microspheres in the SP3 system) translates directly into reduced equipment operating and maintenance costs. The introduction of advanced statistical models, including fuzzy regression

 Materials Research Forum LLC
https://doi.org/10.21741/9781644904015-6

[26, 27] and RSM [28], enables optimization of production processes and shortens the development time of new formulations, which in turn increases computational efficiency [13]. This efficiency is enhanced by the ability to precisely determine the causes of problems and defects [29] and precise prediction of technological parameters [24]. These actions minimize losses associated with material incompatibility, which is crucial in the face of global economic challenges, including tax regulations such as the harmonization of excise duty in the EU [30]. Economic analysis also addresses the broader legal context, implying the importance of the regulatory framework for technology development [31]. Extending service life thanks to improved coating durability reduces the need for frequent replacement of equipment and consumables, leading to reduced waste and resource consumption throughout the product life cycle. This has a positive environmental impact, which is important in the context of broader safety analyses of human-technical-environmental systems [32]. Although the research is strictly military-specific, the methods used (e.g., statistical modeling of processes and materials) have civilian applications, for example, in biomedical engineering [33] and wind energy [34]. Analysis of mechanics and vibrations (including the effect of ground vibrations on particle accelerators [35]) and the development of robotics and control systems [19, 20] provide a common foundation for defense and civilian technologies, increasing the overall level of engineering and safety knowledge. Even the use of advanced analyses, such as those used in phytochemistry [36], contributes to the development of methods that can ultimately be used for quality control and parameter stability in every industrial sector [37].

Summary

The research focused on assessing the performance of camouflage coating systems intended for military applications. Three paint systems were analyzed: the reference system (SP1), a system modified with carbon nanotubes (SP2), and a system modified with glass microspheres (SP3). Experimental testing included abrasion, erosion, and scratch resistance tests, in accordance with relevant standards.

The results showed that the unmodified SP1 system demonstrated the highest abrasion and erosion resistance, while the SP3 system modified with glass microspheres demonstrated the best scratch resistance (10.78 N). The addition of carbon nanotubes did not significantly improve the mechanical properties of the coatings. Overall, the results indicate that modifying the camouflage coatings affects specific mechanical parameters but does not uniformly improve all performance properties.

References

[1] M. Michalski, Paint coatings in military applications general characteristics. System Safety: Human - Technical Facility - Environment 5 (2023) 285-293. https:// DOI:10.2478/czoto-2023-0031

[2] D. Kotnarowska, Destruction of polymer coatings under the influence of operating factors, UTH Publishing House in Radom, 2013. (in Polish)

[3] N. Radek, M. Michalski, J. Kamiński, M. Szczepaniak, S. Kowalkowski, Analysis of the performance properties of coating systems for military applications, Advances in Science and Technology Research Journal19, 5 (2025) 200–212. https://doi.org/10.12913/22998624/201155

[4] S. Rossi, F. Deflorian, E Scrinzi, Comparison of different abrasion mechanisms on aesthetic properties of organic coatings, Wear 267 (2009) 1574-1580. https://doi.org/10.1016/j.wear.2009.06.021

[5] W. Jasiński et al., Static Mechanical Force Amplifier on the Example of a Fastener with an Electromagnetic Bolt, Materials Research Proceedings 34 (2023) 246-251. https://doi.org/10.21741/9781644902691-29

[6] A. Wronska et al., Effect of tool pin length on microstructure and mechanical strength of the FSW joints of Al 7075 metal sheets, Communications – Scientific Letters of the University of Žilina 21 (2019) 40-47.

[7] J. Takosoglu et al., A Design of a 2 DoF Planar Parallel Manipulator with an Electro-Pneumatic Servo-Drive – Part 2, Energies 16 (2023) art. 2970. https://doi.org/10.3390/en16072970

[8] D.S. Pietrala et al., Design and Control of a Pneumatic Muscle Servo Drive Applied to a 6-DoF Parallel Manipulator, Applied Sciences 14 (2024) art. 5329. https://doi.org/10.3390/app14125329

[9] J. Pietraszek, A. Goroshko, The heuristic approach to the selection of experimental design, model and valid pre-processing transformation of DoE outcome, Advanced Materials Research 874 (2014) 145-149. https://doi.org/10.4028/www.scientific.net/AMR.874.145

[10] J. Pietraszek et al., The non-parametric approach to the quantification of the uncertainty in the design of experiments modelling, UNCECOMP 2017 – Proc. 2nd Int. Conf. Uncertainty Quant. Comp. Sci. Eng. (2017) 598-604. https://doi.org/10.7712/120217.5395.17225

[11] J. Pietraszek, L. Wojnar, The bootstrap approach to the statistical significance of parameters in RSM model, ECCOMAS Congress 2016 – Proc. 7th Europ. Congr. Comput. Methods in App. Sci. Eng. 1 (2016) 2003-2009. https://doi.org/10.7712/100016.1937.9138

[12] J. Pietraszek et al., The bootstrap approach to the statistical significance of parameters in the fixed effects model, ECCOMAS Congress 2016 – Proc. 7th Europ. Congr. Comput. Methods in App. Sci. Eng. 3 (2016) 6061-6068. https://doi.org/10.7712/100016.2240.9206

[13] J. Pietraszek et al., Factorial approach to assessment of GPU computational efficiency in surrogate models, Advanced Materials Research 874 (2014) 157-162. https://doi.org/10.4028/www.scientific.net/AMR.874.157

[14] J. Pietraszek et al., Modeling of errors counting system for PCB soldered in the wave soldering technology, Advanced Materials Research 874 (2014) 139-143. https://doi.org/10.4028/www.scientific.net/AMR.874.139

[15] E. Augustyn, M.S. Kozien, Analytical solution of excited torsional vibrations of prismatic thin-walled beams, Journal of Theoretical and Applied Mechanics (Poland) 53 (2015) 991-1004. https://doi.org/10.15632/jtam-pl.53.4.991

[16] T. Lipiński, J. Pietraszek, Corrosion of the S235JR Carbon Steel after Normalizing and Overheating Annealing in 2.5% Sulphuric Acid at Room Temperature, Materials Research Proceedings 24 (2022) 102-108. https://doi.org/10.21741/9781644902059-16

[17] T. Lipinski et al., Influence of oxygen content in medium carbon steel on bending fatigue strength, Engineering for Rural Develop. 21 (2022) 351-356. https://doi.org/10.22616/ERDev.2022.21.TF116

[18] A. Goroshko et al., Multiphysics Modeling of Direct-Start Induction Motor Considering Rotor Unbalance, Materials Research Proceedings 45 (2024) 267-276. https://doi.org/10.21741/9781644903315-31

[19] J. Takosoglu, Angular position control system of pneumatic artificial muscles, Open Engineering 10 (2020) 681-687. https://doi.org/10.1515/eng-2020-0077

[20] P. Wos et al., The Electro-Hydraulic Lifting and Leveling System for the Bricklaying Robot, Lecture Notes in Mechanical Engineering 24 (2021) 216-227. https://doi.org/10.1007/978-3-030-59509-8_19

[21] V. Royzman, A. Goroshko, Multiple inverse problem, J. Vibroeng. 14 (2012) 1417-1424.

[22] A.V. Goroshko, V.P. Royzman, Statistical methods for providing the stability of the solutions of inverse problems and their application to decrease rotor vibroactivity, J. Machinery Manuf. Rel. 44 (2015) 232-238. https://doi.org/10.3103/S1052618815030073

[23] A. Goroshko et al., Construction and practical application of hybrid statistically-determined models of multistage mechanical systems, Mechanika 20 (2014) 489-493. https://doi.org/10.5755/j01.mech.20.5.8221

[24] R. Dindorf et al., Prediction of the parameters and the hot open die elongation forging process on an 80 MN hydraulic press, Open Eng. 11 (2021) 528-534. https://doi.org/10.1515/eng-2021-0056

[25] V. Ptashnyk et al., Impedance investigation of activated carbon material modified by ultrasound treatment, Przeglad Elektrotechniczny 94 (2018) 186-189. https://doi.org/10.15199/48.2018.05.33

[26] J. Pietraszek, Fuzzy regression compared to classical experimental design in the case of flywheel assembly, Lecture Notes in Computer Science 7267 LNAI (2012) 310-317. https://doi.org/10.1007/978-3-642-29347-4_36

[27] J. Pietraszek, The modified sequential-binary approach for fuzzy operations on correlated assessments, Lecture Notes in Computer Science 7894 LNAI (2013) 353-364. https://doi.org/10.1007/978-3-642-38658-9_32

[28] J. Pietraszek et al., The parametric RSM model with higher order terms for the meat tumbler machine process, Solid State Phenomena 235 (2015) 37-44. https://doi.org/10.4028/www.scientific.net/SSP.235.37

[29] A. Pacana et al., Method of precise determination of the main causes of the problems on example of the pin connecting discs in an engine gear, Transportation Research Procedia 55 (2021) 683-690. https://doi.org/10.1016/j.trpro.2021.07.036

[30] A. Kozień, N. Kozłowska, Harmonization and Deharmonization of Excise Duty in the European Union as Contemporary Challenges of the EU Tax Law, WSEAS Transactions on Business and Economics 19 (2022) 816-824. https://doi.org/10.37394/23207.2022.19.71

[31] E. Kozień, A. Kozień, Is ordoliberalism democratic? – Legal approach, System Safety: Human - Technical Facility - Environment 2 (2020) 33-38. https://doi.org/10.2478/czoto-2020-0005

[32] R. Dwornicka et al., The smoothed bootstrap fine-tuning, System Safety: Human - Technical Facility - Environment 1 (2019) 716-723. https://doi.org/10.2478/czoto-2019-0091

[33] E. Skrzypczak-Pietraszek, Phytochemistry and biotechnology approaches of the genus exacum, The Gentianaceae – Volume 2: Biotechnology and Applications (2015) 383-401. https://doi.org/10.1007/978-3-642-54102-5_16

[34] J. Zwierzchowski et al., Model tests of wind turbine with a vertical axis of rotation type Lenz 2, EPJ Web of Conf. 143 (2017) art. 02150. https://doi.org/10.1051/epjconf/201714302150

[35] Ł. Łacny et al., Selected overview of the impact of ground motion on the vibrations of particle accelerators, AIP Conf. Proc. 2239 (2020) art. 020025. https://doi.org/10.1063/5.0008950

[36] E. Skrzypczak-Pietraszek, J. Pietraszek, Seasonal changes of flavonoid content in Melittis melissophyllum L. (Lamiaceae), Chemistry and Biodiversity 11 (2014) 562-570. https://doi.org/10.1002/cbdv.201300148

[37] J. Pietraszek, A. Gadek-Moszczak, The smooth bootstrap approach to the distribution of a shape in the ferritic stainless steel AISI 434L powders, Solid State Phenomena 197 (2013) 162-167. https://doi.org/10.4028/www.scientific.net/SSP.197.162

Drive Synchronization System for a Laser-Mechanical Tube Bending Device

Grzegorz WITKOWSKI[1,a]

[1]Kielce University of Technology, Faculty of Mechatronics and Mechanical Engineering, Al. 1000-lecia Państwa Polskiego 7, 25-314 Kielce, Poland

[a]gwitkowski@tu.kielce.pl

Keywords: Laser Forming, Laser Technologies, Control Systems, Electric Drives, Stepper Motor, Axes Synchronizing

Abstract. The drive control system of bending equipment for thin-walled pipes has been presented. The design of the mechanical system was presented, and the control system was briefly characterized. The advantages and disadvantages of the presented solution are discussed. Possible modifications of the system have been proposed in order to improve its functionality.

Introduction

Laser forming is classified as one of the methods of thermal forming of structural elements. During this process, deformations arise due to the phenomenon of thermal expansion of the material [1-3]. Adequate and controlled heating of selected material zones by the laser beam causes it to assume the assumed shape [4]. Despite the many advantages of laser forming, this process is energy-intensive and runs relatively slowly compared to classical methods (e.g. plastic shaping). Due to these limitations, research was undertaken to develop a hybrid shaping technology, consisting in heating the bent element with a laser beam with simultaneous mechanical support. For the purpose of this task, a special device was developed and made which was integrated with the TRUMPF TLF 6000 laser machining center.

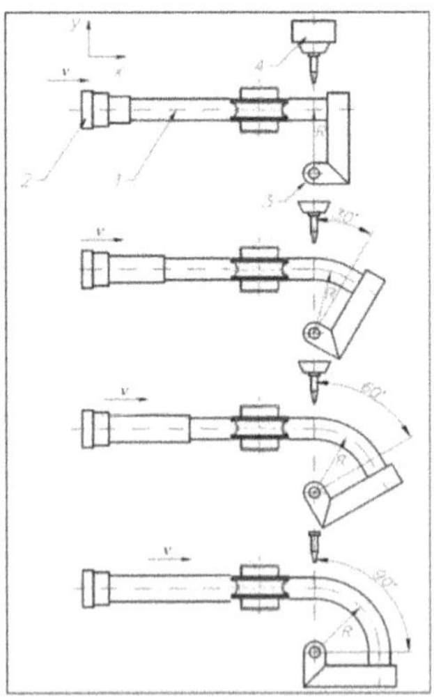

Fig. 1. Kinematics of the bending process: 1 - tubular element subject to bending, 2 - pushing actuator, 3 - free bending arm, 4 - laser head.

Principles of Device Operation

Designed device is intended for mechanically assisted, single-plane laser shaping of thin-walled tubes. The kinematics of the bending process is shown in Fig.1. The device is driven by a linear electric drive (consisting of a motor and gear), hereinafter referred to as an actuator.

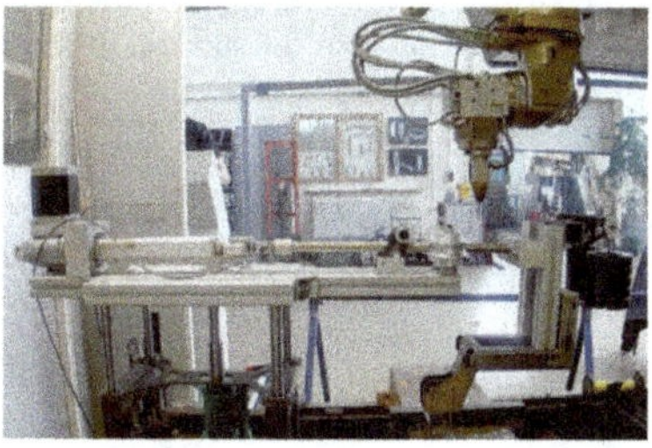

Fig. 2. Laser station.

According to the developed concept, the bent element (1) is placed between the actuator (2) and the free bending arm (3). The laser head (4) peripherally heats the element in the bending plane [5]. At the same time, the cylinder rod is extended along the X axis and the bent element is put into linear motion. The end of this element is attached to the holder located on the bending arm. The arm has one degree of freedom (the ability to rotate around the Z axis of the coordinate system). The bending radius R is determined by the dimension of the bending arm and the distance of its pivot from the axis of the bent tube. The piston rod of the actuator is equipped with a force transducer to determine the forces achieved during the bending process. A digital inclinometer was installed on the bending arm to measure the obtained bending angle.

In order to ensure the precision of the bending process, a device control system has been developed that allows controlling the actuator's extension to a given distance with a specified speed of movement. The linear actuator drive is an induction motor controlled indirectly by the frequency inverter. The control system is based on the MyRio-1900 controller, which uses a real time RT system with a frequency of 1000 Hz. The system uses a classic speed controller with the PI algorithm. The feedback signal for the control system comes from the incremental rotary-pulse transducer mounted on the motor shaft with a resolution of 1024 pulses/revolution. The developed control system makes it possible to force the movement of the actuator's piston rod with a given feed in the range of 0÷400 mm/min, and also to carry out the piston rod homing procedure and reset its current position in any position. The system has been protected by contact limit switches, which prevent excessive extension or retraction of the piston rod. The system allows you to observe the force and bending angle in real time in the LabView environment. The stand is shown in Fig.2.

Problems Encountered

During the operation of the device, it was found that it works correctly, which allows the implementation of the developed hybrid technology. It was observed that during the process of tube bending to an angle not exceeding 45°, the device does not exhibit any dysfunction or disturbing behaviour, and the bending process proceeds correctly. However, when bending pipes to an angle in the range of 45÷90°, a gradual increase in force recorded on the actuator's piston in the direction of extension, and at the same time a gradual elastic deflection of the fastening along the Y axis. These phenomena intensified after exceeding the angle of bending 45° and while reducing the bending radius. This was due to a change in the distribution of the components of the force exerted by the cylinder piston. In order to eliminate these effects, a new device concept has been developed.

Materials Research Forum LLC
https://doi.org/10.21741/9781644904015-7

System with Drive Synchronization

As a method of eliminating the observed undesirable phenomena, it was proposed to use a second drive forcing the rotating arm to rotate. As a drive, a linear actuator with electric drive was used. The concept of the modified device is shown in Figure 3. According to the drawing, the movement of the bending arm will be forced by the simultaneous movement of both actuators. However, there arises the problem of mutual synchronization of the motion of these drives.

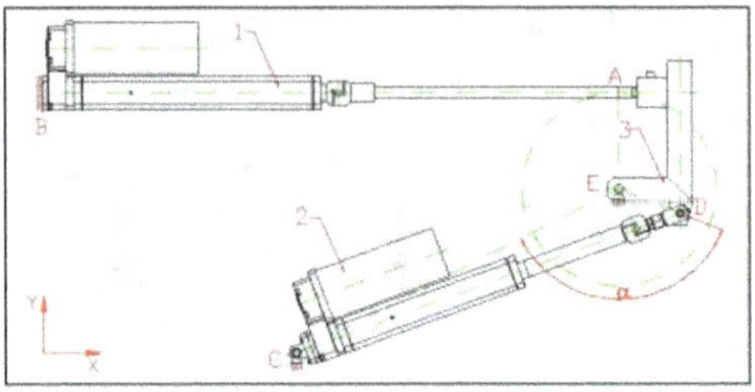

Fig. 3. Modified version of the device kinematics: 1 - first actuator, 2 - second actuator, 3 - bending arm.

Lack of synchronization could cause incorrect operation of the device and, consequently, even damage of the elements of the structure. Control system for synchronized device drives was developed with LabVIEW environment using a real-time MyRIO controller and high performance industrial PC Host . According to the developed concept, the movement of the actuators is to be realized with the use of stepper motors. In order to control them, two signals should be generated for each of the drives. The DIR signal is responsible for the selection of the direction of rotation, and the STEP signal - for determining the rotational speed. The number of generated STEP signal impulses determines the angular displacement of the drive shaft. It should be noted that according to the adopted concept, the speed of extension of the piston rod of the first actuator is fixed and defined by the user. The speed of the second actuator will be variable during the process. It is therefore necessary to designate STEP signals for both actuators and for each moment of pulse of the main loop of the program before starting the system or in real time.

The procedure for determining the control signals for the actuators of the cylinders was started from the determination of the course of changes in the position and speed of the piston rods of the cylinders. Based on the desired process parameters - such as feed and bending angle - the total displacement of the piston rod of the first actuator x1 and the total movement time t were calculated:

$$x_1 = \phi \cdot R \tag{1}$$

$$t = \frac{x_1}{V} \tag{2}$$

where:

ϕ - bending angle, R - bending radius, V - velocity (feed rate).

Using the iterative methods, displacement values of the first actuator were determined for the total time of movement with a step dt of 0.001 s – a displacement signal was obtained at time x(t). The speed of movement of the first actuator is consistent with the feed rate set by the user. The displacement of the second actuator is related to the displacement of the first actuator by

Materials Research Proceedings 62 (2026) 52-57 https://doi.org/10.21741/9781644904015-7

trigonometric relations resulting from the rotational movement of the bending arm. According to Figure. 3, a triangle with CDE vertices can be determined, with an angle α between the arms CE and ED. In the initial position of the process, the measure of this angle is 131°, and then this angle is reduced by the current value of the bending angle. The extending of the second actuator x2 can therefore be determined using the cosine theorem:

$$x_2 = \sqrt{|CE|^2 + |ED|^2 - 2 \cdot |CE| \cdot |ED| \cdot \cos\alpha} \tag{3}$$

where $\alpha = (131° - \phi)$.

After substituting $\phi(t) = \frac{x_1(t)}{R}$, the formula (3) has the form:

$$x_2(t) = \sqrt{|CE|^2 + |ED|^2 - 2 \cdot |CE| \cdot |ED| \cdot \cos(131 - \frac{x_1(t)}{R})} \tag{4}$$

Using the relationship (4), the course of changes in the position of the second actuator for the total time of movement was determined. The course obtained was differentiated sequentially with a time step of 0.001 s and in this way the course of the speed changes of the second actuator was obtained. The system of calculation of control signals, data acquisition and visualization of force waveforms and bending angle was built using the LabView environment. The received courses of displacements and actuator speeds were determined for linear motion. The stepper motors used in the device perform one revolution per 200 control pulses. The stroke of the actuator piston rod is 5 mm. The controllers use a micro step of 4. According to these parameters, a factor z has been determined that allows generating STEP signals for stepper motors in accordance with the formula:

$$z = \frac{k \cdot i}{sp} \tag{5}$$

$$STEP = V \cdot z \tag{6}$$

where:

k - motor impulse/revolution, i - motor micro step, sp - screw pitch mm/revolution, V - desired actuator velocity (feed rate) mm/s.

Experiments

The tests were carried out with various process parameters such as feed rate and bending angle. Control STEP signals for stepper motor controllers were generated in accordance with the way presented in article. The aim was to assess the performance of the developed system for synchronized device drives for laser-mechanical tube bending. In order to verify the correctness of the device's operation, an inclinometer indicating changes in the angle was installed on the free arm. The linear change of the angle of the free arm may indicate the proper control of the drive. Fixed setting control was analyzed. The experiment was carried out with a set feed rate of 10 mm and a bending angle of 90 degrees. Waveforms calculated velocity and displacement of the axis are shown in Figures 4 a, b, c, d respectively. During the experiment also changes in the angle of the free bending arm were observed. The angle signal read from the inclinometer during second experiment is shown in Figure 5.

The analysis of the waveforms (Fig. 4 and Fig. 5) shows that the proposed control system ensures desirable performance of the axis movements. Free arm angle changes are linear with constant slope, indicating proper synchronization X2 axis drive. During the operation of the device, it was found that it works correctly. The device does not exhibit any dysfunction or disturbing behavior, and the bending process proceeds properly.

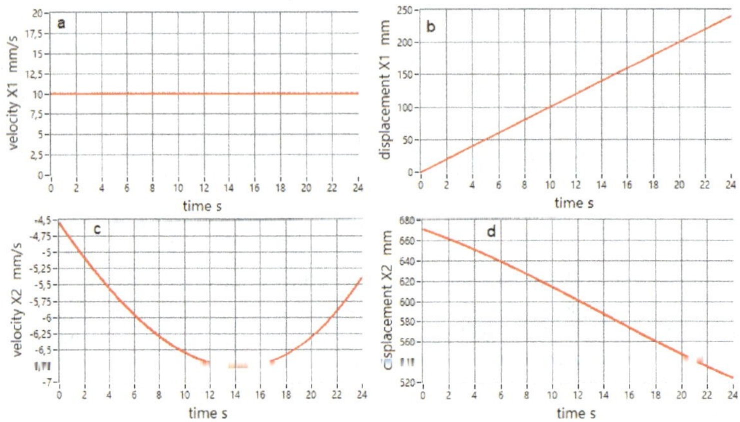

Fig. 4. Kinematic parameters for feed rate 10 mm, bending angle 90 deg.: a - velocity of axis X1, b- displacement of axis X1, c - velocity of axis X2, d - displacement of axis X2.

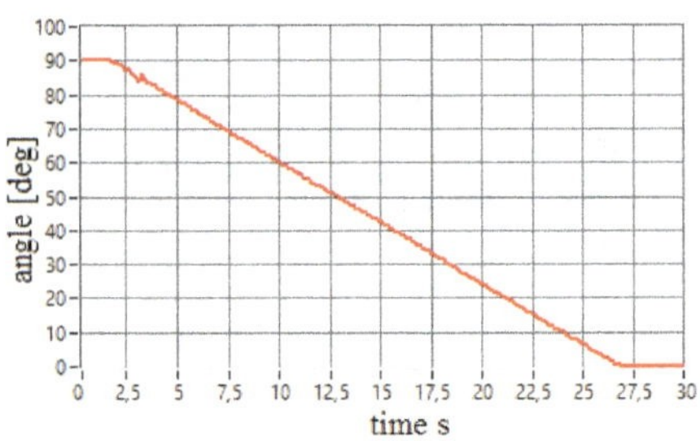

Fig. 5. Angle changes of free bending arm for feed rate 10 mm, bending angle 90 deg.

Conclusions

The proposed method of synchronization consists in controlling two drives using control signals, determined based on the kinematic analysis of the station. Testing of other devices has shown that due to the lack of regulation errors, this method is more effective than the Master-Slave control. An advantage of the proposed method are simple calculations of control signals which may be performed in real time according to the frequency of the control loop. The experimental results show the correct synchronization of linear motion of the piston rod to the rotation of the free bending arm of presented equipment.

References

[1] N. B. Dahotre, Laser Fabrication and Machining of Materials, Springer USA, Boston, 2010.

[2] S. Joshi, U. Dixit, Laser Based Menufacturing, Springer India, New Delhi, 2015.

[3] W. Steen, J. Mazumder, Laser Material Processing, Springer Verlag, London, 2010.

[4] Z. Mucha, J. Widłaszewski, P. Kurp, Mechanically assisted laser forming of thin beams, in: Proceedings of SPIE 10159 (2016), XIth Symposium on Laser Technology, 2016, Jastarnia, Poland. https://doi.org/10.1117/12.2262114

[5] L. Płonecki, G. Witkowski, K. Mulczyk, P. Kurp, The method and implementation of automatic, fuzzy surface temperature control for laser processing, in: Proceedings of SPIE 10974 (2018), Thirteenth Symposium on Laser Technology, 2018, Jastarnia, Poland. https://doi.org/10.1117/12.2516664

Materials Research Proceedings 62 (2026) 58-63
https://doi.org/10.21741/9781644904015-8

Acoustic Drone Detection for Technical Infrastructure Security and Defense Purposes

Jarosław ZIÓŁKOWSKI[1,a], Patrycja GUZANEK[2,b], Mateusz MAZUR[3,c,*], Izabella KĘSY[4,d] and Robert SZCZUR[2,e]

[1]Faculty of Civil Engineering and Geodesy, Military University of Technology, Warsaw, Poland

[2]Doctoral School, Military University of Technology, Warsaw, Poland

[3]3rd School Battalion, Military University of Technology, Warsaw, Poland

[4]Pomeranian Higher School in Starogard Gdanski, Institute of Management, Economics and Logistics, Poland

[a]jaroslaw.ziolkowski@wat.edu.pl, [b]patrycja.guzanek@student.wat.edu.pl, [c]mateusz.mazur02@student.wat.edu.pl, [d]izabella.kesy@twojestudia.pl, [e]robert.szczur@wat.edu.pl

Keywords: Drone Detection, Acoustic Signature, Technical Safety, Background Noises

Abstract. With the widespread use of drones, new challenges have arisen in the field of technical security of facilitates and defense. This article discusses passive detection systems for unmanned aerial vehicles (UAVs) based on acoustic signatures. It presents an overview of methods for detecting drones using sound - from classical signal analysis and machine learning methods to modern deep learning algorithms - in the context of applications in critical infrastructure protection and military systems. Available solutions, their effectiveness and detection range were compared, and the difficulties of background noise and distinguishing between drone types were discussed. Research results were presented indicating high acoustic detection efficiency under favorable conditions (even above 90% using algorithms based on neural networks), as well as limitations caused by a decrease in signal-to-noise ratio with distance from the drone. The conclusion emphasizes the role of acoustic systems as a complement to other UAV detection techniques and their importance for maintaining safe operation of technical systems.

Introduction

The rapid development of Unmanned Aerial Vehicles (UAVs, or drones) has expanded their civil and military applications, including monitoring, inspections, logistics, rescue, detection of harmful substances and photography, but also introduced new security risks [1], [2], [3], [4]. Incidents such as the 2018 Gatwick Airport disruption and a drone attack in Venezuela have shown the dangers of hostile drone use [5]. Drones can be employed for terrorism, sabotage, smuggling, or illegal surveillance, making rapid detection and identification crucial for security agencies. Current detection methods – optical, radar, radiofrequency (RF), and acoustic – each have limitations. Small or slow drones may evade optical and radar systems, while RF methods fail if no active transmission is present or in areas with strong interference [6]. In contrast, acoustic detection is passive, hard to jam, and capable of identifying drones by their unique sound signatures. Drone acoustics are dominated by tonal noise from rotor blade pass frequencies (~100-300 Hz) and harmonics, along with broadband turbulence noise [7]. These signatures vary by drone type, rotor configuration, and flight mode, enabling both detection and potential model classification. However, effectiveness decreases with distance and ambient noise. Despite this, acoustic sensors are valuable components of multi-sensor anti-drone systems protecting critical infrastructure and military sites. This paper reviews current advances, algorithms, and operational challenges in acoustic drone detection.

Literature Review

Acoustic sensors and measurement systems. The simplest solution for listening to drones is a single listening microphone operating in the audible range (20 Hz - 20 kHz). Typical measurement microphones used in acoustic detection research are characterized by a flat frequency response and high sensitivity (in the order of several dozen mV/Pa) so that even weak signals from long distances can be recorded. A single microphone can detect the presence of a characteristic drone sound in the signal but cannot determine the direction from which the sound is coming. For this purpose, multi-element microphone arrays (so-called acoustic cameras) consisting of a series of sensors arranged in a known geometry are used. By using beamforming techniques (directional amplification of the signal from a selected azimuth and elevation), it is possible to indicate the location of the sound source to improve the signal-to-noise ratio. An example is the acoustic camera presented by Busset et al. (2015), consisting of 120 microphones and a video camera, which creates a real-time image of the sound level from all directions [8]. Such a system can simultaneously track the movement of multiple sound sources and selectively isolate the acoustic signal assigned to each of them. In experiments, Busset et al. (2015) demonstrated the ability to track small drones at distances of 160-250 m, depending on their type and emission volume [8]. Benyamin and Holdman from the Army Research Lab, on the other hand, configured a compact antenna with four microphones (arranged in a tetrahedron shape with an edge length of ~27 cm), their system was able to detect small Class I drones with a probability of 99,5% at a distance of up to 600 m in favorable acoustic conditions [9]. The use of acoustic beamforming in this configuration increased the effective detection range and enabled the determination of the bearing (azimuth and elevation) of drone. However, it should be emphasized that such long ranges are only possible with low background noise levels - the presence of strong interference/noise (e.g. traffic, machinery) significantly reduces the range and reliability of detection and can also cause false alarms [10].

Signal Processing and Detection Algorithms

The first stage of acoustic UAV detection is audio signal processing to extract characteristic features. The signal from drone rotors is periodic (tonal) with harmonic components, which is why frequency domain analysis (spectrogram, Fourier transform) is commonly used to reveal the frequency pattern generated by drone. Both one-dimensional methods - time domain or spectrum analysis - and two-dimensional time-frequency approaches (e.g. short-time Fourier transform, STFT) are used. The obtained acoustic signatures are compared with patterns or classified using machine learning algorithms. Earlier works were dominated by classical methods known from acoustic event detection and speech recognition. A frequently used feature set was Mel-frequency Cepstral Coefficients (MFCC), describing the spectral envelope of sound. Based on these, classifiers such as Gaussian Mixture Models (GMM) or Support Vector Machines (SVM) were trained to decide whether a detected audio segment contained the sound of a drone or was it just a background noise. For example, Jeon et al. (2017) compared the GMM model with two popular deep architectures - a Convolutional Neural Network (CNN) and a Recurrent Neural Network (RNN) - in real world conditions [11], [12]. Their goal was to detect drones in an urban street environment as a binary classification problem (drone vs. other sounds). In order to increase the diversity of the training data, they used data augmentation - recordings of various typical background noises (cars, voices, city noises) at different levels were added to the recorded drone noises to simulate real-world conditions. The results of Jeon et al. (2017) experiments showed that in such difficult conditions, the RNN model achieved the highest detection efficiency (F-score ~0.7 with high recall reaching 0.96, at the cost of precision ~0.55). The RNN outperformed CNN and GMM in noisy environments [11]. In another study, Al-Emadi and Al-Ali (2021) showed that convolutional CNNs trained on raw spectrograms can achieve comparably high performance if they have a sufficiently large and diverse training set [13]. In order to increase the data set, Al-

Materials Research Proceedings 62 (2026) 58-63 https://doi.org/10.21741/9781644904015-8

Emadi et al. innovatively used Generative Adversarial Networks (GAN) to artificially generate realistic samples of drone sounds, which were then added to the training data. This improved the detection and identification of drone models that were not present in the original training set, increasing the overall reliability of the system [13].

Nowadays, deep learning methods are increasingly used for automatic drone signal recognition. CNNs are used to analyze spectrograms treated as 2D images, while RNNs (e.g. LSTM [14]) are used to analyze time series or to complement CNNs in a hybrid CRNN (Convolutional Recurrent Neural Network) architecture. Casabianca and Zhang (2021) compared these approaches, pointing out that for acoustic signals presented in the form of spectrograms, pure CNN models perform best, surpassing CRNN and RNN in detection accuracy [15]. However, conflicting conclusions can be found in the literature - for example, the aforementioned work by Jeon et al. (2017) indicated the superiority of RNN over CNN, which shows that the final effectiveness depends on many factors, including the structure and parameters of the network, the type of training data, and the evaluation metrics [11]. In order to further improve the results, Casabianca and Zhang (2021) proposed the use of late fusion of multiple deep models. An ensemble consisting of several CNNs with different architectures was used, combining their results at the decision stage (weighted majority voting). This approach yielded a combined effectiveness of ~94,7% (average accuracy) in drone detection, outperforming each model individually. It was thus noted that ensembles can increase detection reliability by reducing the impact of specific errors of individual classifiers [15].

Impact of Distance and Interference on Detection Effectiveness

The range at which acoustic detection of a drone is possible depends primarily on the signal-to-noise ratio (SNR) – the volume of the drone's sound compared to the acoustic background level – and the frequency characteristics of that background. The attenuation of sound waves in the atmosphere causes the drone signal to weaken with the square of the distance, and for high frequencies the attenuation is even greater. The simplified equation without considering air absorption is $I(r) = \frac{I_0}{r^2}$., where: I_0 - intensity at reference distance r_0, r - distance from source.

Typically, the dominant low frequencies (hundreds of Hz) travel further than the high-frequency components. In the silence of the night or in rural areas, a drone can be heard from much further away than in the middle of the day in a city, where the background noise drowns out the signal. In addition, wind can significantly hinder listening - wind noise on the microphone and the deflection of acoustic waves by turbulent fluctuations reduce the effective detection range.

Research by Tejera-Berengue et al. (2024) systematically analyzed the impact of distance on the effectiveness of acoustic detection algorithms [16]. A database of drone sound recordings at distances of 50-100 m was used, and various classifiers (from simple LDA and SVM to the deep YAMNet) were tested by training them on data from different distances. It was found that if the algorithm is trained only at close distances, its effectiveness drops dramatically when detecting drones further than 200 m - the ROC fields for 300 m and 500 m were only slightly better than random (AUC approx. 0.5-0.7). However, including samples from greater distances (with lower SNR) in the training data significantly extends the maximum detection range. According to Tejera-Berengue et al. (2024), with the right choice of learning strategy, classical methods could effectively detect drones up to ~300 m, and state-of-the-art neural models (YAMNet transfer learning) even up to ~500 m. At the same time, however, for distances >500 m, all tested detectors rapidly lost their effectiveness due to too low SNR [16]. To visualize this the Fig. 1 is presented.

Materials Research Proceedings 62 (2026) 58-63 https://doi.org/10.21741/9781644904015-8

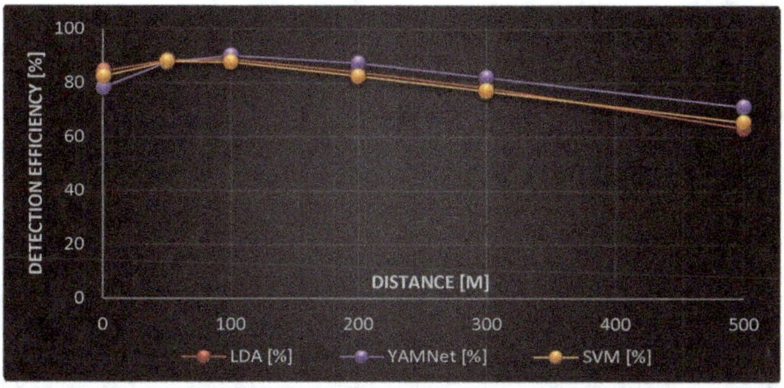

Fig. 1. Effectiveness of acoustic drone detection as a function of distance – based on data from Tejera-Berengue et al. (2024)[12].

In practice, false alarms are a significant problem in acoustic detection - many sound sources (e.g., lawnmowers, motorcycles, airplanes, and even bees) can mimic certain acoustic characteristics of drones. To reduce misclassifications, signal filtering is used (e.g., focusing on narrow frequency bands characteristic of drones) and more complex signal features are taken into account, such as amplitude modulations or non-stationary sound patterns emitted during maneuvers. The use of directional antennas is also helpful, as it reduces the impact of noise sources outside the observation area by appropriately directing the sensitivity of reception.

Another important issue is the identification of drone types based on acoustic signals. This task is more difficult than detection (binary classification) itself, as it requires distinguishing between many classes (models) of drones, often with quite similar sound characteristics. Nevertheless, there are solutions that can recognize specific models or at least categories of UAVs (e.g. small quadcopter vs large multi-rotor) with a certain level or certainty. CNN/CRNN networks can learn specific spectral patterns for individual drones, for example, slightly different harmonic frequencies resulting from different motor speeds. Al-Emadi et al. (2021) report high effectiveness in identifying specific drone models using deep networks, provided that the training set contains sufficient data representations for these models [13]. Generalization to drones not present in the training remains problematic - hence the proposal to use generative networks to create artificial representations of new types of drones. Despite these advances, the reliability of recognising a drone model solely by sound is still limited and may require support from other techniques (e.g. visual verification) in field conditions.

Practical Applications and Operational Challenges

Acoustic drone detection systems are used wherever continuous monitoring of airspace for unauthorized objects is required and the use of traditional sensors is difficult or insufficient. The passive nature of acoustic detectors means that they do not emit any radiation (unlike other radars or lidars), which is beneficial from the point of view of electromagnetic safety (important, for example, near airports, where strong emissions are undesirable) and operational discretion. In military applications acoustic listening can be carried out by networks of sensors deployed in the field of carried by soldiers for early warning of approaching reconnaissance or armed drones. In battlefield conditions, especially when advanced radars are not available, acoustic detectors are a valuable complement to observation, particularly against small, low-flying drones. Furthermore, acoustic systems can also detect other acoustic threats (gunshots, vehicles, etc.), integrating into broader situational monitoring systems. However, the limitations and operational requirements of such systems must be taken into account. Microphones and listening stations must be properly positioned to cover the protected area – the range of a single sensor is usually several hundred meters, so a network of sensors is needed in a larger facility. The equipment must be resistant to environmental conditions (wind, precipitation, temperature changes), as these factors can affect both the electronics and the acoustic distribution itself. As part of the maintenance of acoustic system, it is important to calibrate the microphones, perform periodic tests and update the detection software. The system must be reliable 24/7, which means it must be capable of automatic operation, self-diagnostics and minimizing false alarms that could undermine operator's confidence. Integrating acoustic detection with other sensors as part of data fusion can significantly increase the effectiveness and reliability of drone detection. For example, after initial detection of a drone's acoustic signal, a PTZ (pan-tilt-zoom) camera can be directed at the suspicious area for visual confirmation, or the acoustic bearing can be correlated with a short-range radar reading. The aim is to create multi-sensor anti-drone systems, which, due to the redundancy and complementarity of the sensors, achieve high detection reliability and low probability of missing a threat.

Conclusions

Acoustic drone detection systems are an important element of the anti-drone technologies currently under development, fitting into the broader context of technical and operational security of protected facilities. Sound-based detection is particularly effective against small drones that are hard to spot with radar or optical methods. Under quiet conditions, such systems can detect drones from several hundred meters away, giving security services time to react, while advanced algorithms help maintain effectiveness even in noisy urban environments. Multi-element microphone arrays further improve performance by determining the drone's direction and tracking its movement, though they increase system complexity and cost, which are acceptable trade-offs for stationary applications requiring high reliability.

Modern detection algorithms increasingly rely on machine learning, including deep neural networks such as CNNs and RNNs, achieving over 90% detection accuracy in tests. However, these models require extensive training data, making the expansion of drone sound databases and use of data augmentation essential [17]. The main challenges remain background noise and false alarms, highlighting the need to adapt systems to specific environments and to continuously update acoustic models. While full identification of drone type or model by sound is not yet fully reliable, distinguishing drones from birds or other objects already provides a major advantage and reduces false alerts. In summary, acoustic detection is a promising complement to existing security systems, enhancing protection against the growing threat of drones. Ongoing research focuses on improving algorithm robustness, integrating data from distributed sensors, and developing standarised testing and certification methods to ensure reliable performance in critical applications.

References

[1] P. Gładysz et al., Reliability of Unmanned Aerial Vehicles in the Context of Selected Factors, Eksploatacja i Niezawodność 28 (2026). https://doi.org/10.17531/ein/210312

[2] A. Jaroń et al., UAV-based Statistical Analysis of Seasonal Traffic Pollutant Diffusion: a Case Study of Poland's A4 Highway, Transport Problems 20 (2025) 223–235. https://doi.org/10.20858/tp.2025.20.2.18

[3] R. Altawy, A. M. Youssef, Security, Privacy, and Safety Aspects of Civilian Drones: A Survey, ACM Transactions on Cyber-Physical Systems 1 (2017) art.7. https://doi.org/10.1145/3001836

[4] A. Jaroń et al., An Assessment of the Possibility of Using Unmanned Aerial Vehicles to Identify and Map Air Pollution from Infrastructure Emissions, Energies 17 (2024) 577. https://doi.org/10.3390/en17030577

[5] A. Vozella et al., Solution Set-up for Airport Protection from Intruder Drones, Proceedings of the 30th European Safety and Reliability Conference and 15th Probabilistic Safety Assessment and Management Conference, Research Publishing Services (2020) 4469–4476. https://doi.org/10.3850/978-981-14-8593-0_4189-cd

[6] S. Al-Emadi, F. Al-Senaid, Drone Detection Approach Based on Radio-Frequency Using Convolutional Neural Network, in: 2020 IEEE International Conference on Informatics, IoT, and Enabling Technologies (ICIoT), Doha, IEEE, 2020, 29–34. https://doi.org/10.1109/ICIoT48696.2020.9089489

[7] O. Jokisch, D. Fischer, Drone Sounds and Environmental Signals – A First Review, in: ESSV Konferenz Elektronische Sprachsignalverarbeitung (2019) 212-220.

[8] J. Busset et al., Detection and tracking of drones using advanced acoustic cameras, Proc. Of SPIE 9647 (2015) 96470F. https://doi.org/10.1117/12.2194309

[9] M. Benyamin, G. H. Goldman, Acoustic Detection and Tracking of a Class I UAS with a Small Tetrahedral Microphone Array, Adelphi, MD : Army Research Laboratory, September 2014.

[10] Y. Jiang, Z. Lu, A Feature Extraction Method for Parameter Fault Diagnosis of DC-DC Converters Under Strong Noise and Small Sample Conditions, Eksploatacja i Niezawodność 28 (2026). https://doi.org/10.17531/ein/207882

[11] S. Jeon et al., Empirical Study of Drone Sound Detection in Real-Life Environment with Deep Neural Networks, 25th European Signal Processing Conference, EUSIPCO 2017, 1858-1862. https://doi.org/10.23919/EUSIPCO.2017.8081531

[12] F. Jiang et al., Experimental Research on Vibration Feature Recognition for Bearing Reliability Based on Improved Convolutional Neural Networks, Eksploatacja i Niezawodność 28 (2026). https://doi.org/10.17531/ein/211279

[13] S. Al-Emadi et al., Audio-Based Drone Detection and Identification Using Deep Learning Techniques with Dataset Enhancement through Generative Adversarial Networks, Sensors 21 (2021) 4953. https://doi.org/10.3390/s21154953

[14] M. Zastępa, The statistical-based diagnosis with usage of acoustic sound decomposition and projected LSTM network of induction motors, Eksploatacja i Niezawodność 27 (2025). https://doi.org/10.17531/ein/205651

[15] P. Casabianca, Y. Zhang, Acoustic-Based UAV Detection Using Late Fusion of Deep Neural Networks, Drones 5 (2021) 54. https://doi.org/10.3390/drones5030054

[16] D. Tejera-Berengue et al., Analysis of Distance and Environmental Impact on UAV Acoustic Detection, Electronics 13 (2024) 643. https://doi.org/10.3390/electronics13030643

[17] D. Ma et al., Electromagnetic Relay Contact Resistance Prediction Based on TimeGAN with CNN-LSTM-Attention, Eksploatacja i Niezawodność 27 (2025). https://doi.org/10.17531/ein/204610

Materials Research Proceedings 62 (2026) 64-71 | https://doi.org/10.21741/9781644904015-9

Applicative Potential of Bioactive Glass Materials in the Context of Military Medicine

Olga KĘDZIA[1,a *], Małgorzata LUBAS[1,b], Agata DUDEK[1,c]

[1]Częstochowa University of Technology, Faculty of Production Engineering and Materials Technology, Dąbrowskiego 69, 42-201 Czestochowa, Poland

[a]olga.kedzia@pcz.pl, [b]malgorzata.lubas@pcz.pl, [c]agata.dudek@pcz.pl

Keywords: Bioglass, Bioactivity, Military Medicine, Bioactive Scaffolds

Abstract. The chapter highlights the significance and potential of bioactive glass in military medicine, discussing its applications in both clinical and field practice. Bioglass is a material that forms strong bonds with biological tissues, exhibiting both bioactivity and biocompatibility. By adjusting its chemical composition, it is possible to obtain materials that support the regeneration of bones, soft tissues, and blood vessels, as well as possess antibacterial and anti-inflammatory properties. Examples of bioglass applications in dressings, surgical tools, implants, and biodegradable scaffolds are discussed. The experimental section presents the outcomes of investigations on porous 45S5 bioglass-based materials derived from bioresources, demonstrating high porosity, low density, and promising potential as temporary scaffolds for bone tissue regeneration.

Introduction

„A war benefits medicine more than it benefits anybody else. It's terrible, of course, but it does." – Mary Merritt Crawford, the only woman doctor at the American Hospital during World War I [1]. Although the Middle Ages brought regression, the Renaissance, thanks to figures such as Ambroise Paré, reintroduced a pragmatic approach to surgery – introducing antiseptic agents and tourniquets [2]. Each subsequent war brought innovations: from sanitary wagons and the concept of triage from the Napoleonic era, through the advances in reconstructive surgery during World War I, to antibiotics, blood transfusions, and trauma stabilization techniques of World War II. Contemporary military medicine continues to adapt to various operational environments. Naval medicine focuses on injuries associated with exposure to cold water, the army improves trauma surgery in challenging land conditions, while space medicine investigates the physiological and psychological effects of microgravity and isolation [3].

In 1967, a young professor, Larry L. Hench from the University of Florida, was studying the resistance of glass to nuclear radiation. In the summer of the same year, he attended a scientific conference sponsored by the United States Army in Sagamore, New York. During the trip, he talked with an officer who had just returned from Vietnam, where he was responsible for supplying fifteen mobile surgical units (MASH). The colonel, listening to the properties of radiation-resistant glass, asked Hench: „If you can make a material that will resist gamma rays, why not make a material the body won't resist?" Two years later, Hench and his research team developed the first entirely synthetic material capable of forming a durable bond with human bone – Bioglass® 45S5 [4].

Two years later, Hench and his research team developed the first entirely synthetic material capable of forming a durable bond with human bone – Bioglass® 45S5. In the context of military and field medicine, its unique properties, i.e., bioactivity, biocompatibility, tissue regeneration ability, and antibacterial effect, open completely new perspectives. In conditions where every minute and every piece of equipment matter, bioglass may become a key element of future tactical medicine.

Materials Research Proceedings 62 (2026) 64-71 | https://doi.org/10.21741/9781644904015-9

Characteristics of Bioactive Glasses

The fundamental chemical composition of silicate-based bioactive glass comprises four major oxides: SiO_2 (silica), Na_2O (sodium oxide), CaO (calcium oxide), and P_2O_5 (phosphorus pentoxide). The silica network constitutes its primary structure, nd the silica source may derive from organic precursors [5] or naturally occurring bio-based materials [6, 7]. The incorporation of Na^+ and Ca^{2+} ions into the network introduces non-bridging oxygens (NBOs), disrupting the silicate matrix and enhancing the glass reactivity and degradation rate in physiological fluids [8]. Phosphorus, in the form of P_2O_5, forms complexes with calcium and sodium ions, which promote the nucleation of phosphate and apatite phases [9].

The presence of bioglass in the physiological environment induces a change in pH in its surroundings, leading to ion exchange. Specifically, Na^+ and Ca^{2+} ions are released from the surface, and hydroxyl groups appear in their place, resulting in the formation of a thin layer of silica gel [10]. Subsequently, an amorphous calcium phosphate layer precipitates, followed by the development of carbonate hydroxyapatite (HCA), which corresponds to the biological apatite present in bones and teeth [11]. This process marks the onset of chemical bonding between bioglass and biological tissue. In the next phase, biological bonding occurs, in which collagen fibres and extracellular matrix proteins connect with the HCA layer through electrostatic, hydrogen, and ionic bonds. In this way, a durable, biocompatible interface is formed between the material and the tissue [9].

The chemical composition of bioactive glass is modifiable, allowing for the design of materials with differentiated reactivity and functionality. Increased calcium oxide content, coupled with a decrease in silica concentration, accelerates bioactivity and degradation. In contrast, the addition of stabilizing oxides (Al_2O_3, B_2O_3, Ta_2O_5, ZrO_2) enhances structural integrity but reduces reactivity [6]. Furthermore, compositional modifications can endow bioactive glasses with targeted biological functionalities. Silver, copper, and zinc ions confer antimicrobial activity, while strontium and magnesium promote osteogenesis and mineralization. Additionally, boron and cobalt stimulate angiogenesis—the formation of new vascular networks [12].

Selected Bioactive Medical Materials for Field Application

Traditional surgical instruments, such as scalpels, forceps, or scissors, are made of metals, primarily stainless steel or titanium. A potential direction of development includes specialized tools, such as tweezers, hooks, or rasps, as well as single-use surgical instruments with bioactive coatings. Several studies confirm the feasibility of coating the surface of instruments with thin layers of bioactive glass. Such coatings may exhibit antibacterial activity by reducing microorganism adhesion and biofilm formation, while also improving biocompatibility and reducing the risk of infection during surgical procedures with a high risk of contamination [13].

Bioglass has found its broadest application in wound dressing materials. In advanced wound dressings and hemostatic compresses, fibers or powders derived from bioactive glass are utilized to promote tissue regeneration, provide antibacterial protection, and accelerate wound and burn healing. A representative example is mesoporous bioactive glass (MBG) doped with Ga^{3+} ions, which accelerates coagulation and clot formation and exhibits antimicrobial properties [14]. Composites based on Ga-MBG combined with chitosan [15] and MBG doped with Ta_2O_5 [16] showed similar hemostatic properties. Nanoporous bioglass containing silver (n-BGS) achieved 99% antibacterial efficacy within 12 hours and significantly shortened coagulation time, indicating its potential as an effective antibacterial and hemorrhage-controlling dressing [17]. Another example includes commercial gauze impregnated with chitosan suspension and 45S5 bioglass containing Sr and Mg or Zn, which exhibited improved cell adhesion, fibroblast proliferation, and accelerated wound closure in vitro compared to chitosan-only controls [18]. A borate-based bioactive glass fiber matrix (BBGF) has also shown high efficacy in treating chronic wounds [19].

Hydrogels constitute another group of products utilizing bioglass, which, thanks to their regenerative and antibacterial properties, are employed in the treatment of burns and hard-to-heal wounds [20, 21]. Bioglass may be incorporated into specialized dressings or hemostatic powders [22], which are commonly included in standard tactical first aid kits. Furthermore, occlusive dressings incorporating bioactive glass layers are being investigated for their potential to enhance wound healing and reduce infection risks in delayed surgical intervention scenarios [23].

Application of Bioglass-Based Scaffolds in Field Implantology

In field or battlefield conditions, where access to specialized reconstructive surgery is limited, bioglass- and bioactive composite–based implants provide a practical and effective alternative. Their primary advantage lies in the temporary replacement or mechanical support of damaged bone or soft tissues while promoting tissue regeneration. In contrast to conventional metallic implants, bioactive scaffolds can be biodegradable, which eliminates the need for subsequent removal once the patient reaches a hospital setting. This feature is particularly significant in military environments, where surgical procedures must be concise and opportunities for secondary interventions are minimal.

Bioglass scaffolds stimulate the growth of bone cells (osteoblasts) and support the formation of the extracellular matrix. In the physiological fluid environment, bioactive glass gradually dissolves, releasing calcium, phosphorus, and silicon ions, as well as intentionally introduced ions of silver, zinc, or copper, which not only promote mineralization but also act as anti-inflammatory and antibacterial agents [24].

In military medicine, lightweight, easily handled, and transport-resistant scaffold forms—such as granules, sponges, foams, and flexible bioglass–polymer composites—are especially valuable. Experimental composite scaffolds incorporating polycaprolactone (PCL) and bioglass yield structures with improved flexibility and mechanical stability. These systems are applicable in the regeneration of both hard and soft tissues. Laboratory evidence indicates that the inclusion of bioglass within the polymeric matrix enhances biocompatibility and reduces the risk of infection [25], making them promising candidates for use in military medicine, where wound infections are a primary cause of mortality. In recent years, concepts of scaffolds printed using 3D technology have also been developed, which can be produced on-site (e.g., in a field hospital) from bioglass powder or bioactive composites [26].

Experimental Research

One of the promising methods for obtaining porous bioactive materials is the foaming of bioactive glass through the sintering of glass powder in the presence of a foaming agent. The proper selection of the foaming agent is crucial for achieving uniformity and pore size, and thus for the final mechanical and biological properties of the scaffold. In this study, the foaming of 45S5 bioglass was carried out using plant-derived biosources as eco-friendly foaming agents. Dried, crushed, and milled avocado and horse chestnut seeds acted as gas-generating components during sintering. The use of natural precursors not only reduces environmental impact but may also enhance the biocompatibility and bioactivity of the material due to the incorporation of biogenic elements from the combustion ash. During the foaming process, biomass combustion within the furnace atmosphere induces thermal decomposition, releasing gases that form a cellular structure throughout the sample.

The first stage of the developed process was the synthesis of bioglass and preparation of 45S5 bioglass powder after melting. The glass powder was then sieved to obtain a fraction with a uniform grain size, facilitating the uniform distribution of the foaming agent. The foaming medium, prepared from plant-based biosources, was combined with water glass and water to form pellets. Special attention was paid to the proper selection of sintering time–temperature parameters. Temperature and time were selected to ensure that the foaming agent decomposed

Materials Research Proceedings 62 (2026) 64-71 https://doi.org/10.21741/9781644904015-9

only after the glass powder grains had sintered together, thereby allowing for the formation of a stable skeleton and a uniform pore distribution throughout the material. The result of the foaming process was the production of light, glass-crystalline foams with an open-cell structure (Fig. 1), characterized by a varied pore morphology depending on the type of bioresource used. Foams produced with avocado seeds (Fig. 1a) show a structure with relatively small, regular pores, while those with horse chestnut seeds (Fig. 1b) are characterized by more variable cell size.

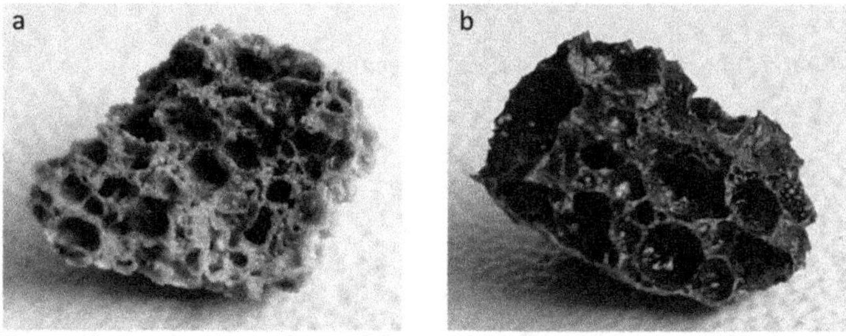

Fig. 1. Porous biomaterials synthesized from 45S5 bioglass using (a) dried avocado seeds and (b) dried horse chestnut seeds as natural pore-forming agents.

Both types of foams showed high porosity and low bulk density, making them potentially useful in field medicine applications as temporary scaffolds filling bone or tissue defects. Additionally, potassium present in the chemical composition of the obtained materials, originating from bioresources, positively influences bioactivity. Potassium reduces the degree of silicate network cross-linking, increasing its reactivity in physiological fluids and thus accelerating the biodegradation process and HCA formation.

Social, Environmental, and Economic Impact

Bioglass has potentially significant social significance, as it aligns with the trend of using biodegradable and naturally derived materials [27,28], allowing, through its production mechanism, the reduction of costly methods for deactivating industrial waste [29,30]. It represents the next stage in the development of materials with increased corrosion resistance [31] and biocompatibility [32,33].

The wider use of bioglass will cause significant changes in the industry, not only the pharmaceutical one, as it is a strong development impulse for both materials engineering [34-36] and the technological area [37-39], as it opens up potential possibilities of shaping surface features [40-42] using laser texturing [43,44], as such changes in the technological features of the surface layer have been observed in the modification of industrial materials [45,46] and structural objects [47,48]. This will undoubtedly lead to significant changes in the areas of process management [49, 50] and quality [51, 52].

Such material and technological development is always supported by appropriate experimental methodology [53,54] and data analysis methods, both non-parametric [55,56] and fuzzy [57,58], due to frequent cases of data incompleteness.

Summary

Bioactive glasses exhibit tunable properties that can be tailored through chemical composition modification, enabling control over solubility, reactivity, and biological performance. These materials can be fabricated in various forms, including powders, scaffolds, coatings, and nanoparticles, allowing for versatile medical applications. The incorporation of bioglass into wound dressing materials is particularly valuable in military medicine, where rapid hemostasis and control of infection are crucial. Due to their antibacterial, hemostatic, and regenerative properties, bioactive glass can be a key component of future medical solutions intended for rapid response in difficult combat conditions. Additionally, bioglass scaffolds represent one of the most promising directions in the field of implantology, as they combine bioactivity, biodegradability, and antibacterial action with the ability to adapt to different types of tissues. Their implementation could markedly improve the management of combat-related injuries where surgical infrastructure is limited.

References

[1] Information on https://www.theatlantic.com/health/archive/2017/02.

[2] C. Van Way 3rd, War and Trauma: A History of Military Medicine, Mo. Med. 113(4) (2016) 260–263

[3] R. Ho, D. Varner, and J. P. Beissinger, Military Medicine Outside Four Walls, Physician Assist. Clin., 9(2) (2024) 165–173. https://doi.org/10.1016/j.cpha.2023.10.002

[4] B.D. Ratner, G. Zhang, A History of Biomaterials, in: E. Yaszemski (Ed.), Biomaterials Science (Fourth Edition), Academic Press, 2020, 21-34. https://doi.org/10.1016/B978-0-12-816137-1.00002-7

[5] D.E. Abulyazied et al., Synthesis, Structural and Biomedical Characterization of Hydroxyapatite/Borosilicate Bioactive Glass Nanocomposites, J. Inorg. Organomet. Polym. Mater. 31 (2021) 4077-4092. https://doi.org/10.1007/s10904-021-02070-6

[6] L.A. Adams et al., Bioactive glass 45S5 from diatom biosilica, J. Sci. Adv. Mater. Devices, 2(4) (20117) 476–482. https://doi.org/10.1016/j.jsamd.2017.09.002

[7] M. Pereira et al., Influence of the use of rice husk as source of silica on the sol-gel synthesis of bioglass, Cerâmica 67 (2021) 333–337. https://doi.org/10.1590/0366-69132021673833134

[8] A. Haider et al., Strontium- and Zinc-Containing Bioactive Glass and Alginates Scaffolds, Bioengineering, 7(1) (2020). https://doi.org/10.3390/bioengineering7010010

[9] M. Cannio et al., Bioactive Glass Applications: A Literature Review of Human Clinical Trials., Materials 14(18) (2021) art. 5440. https://doi.org/10.3390/ma14185440

[10] J.R. Jones, Reprint of: Review of bioactive glass: From Hench to hybrids, Acta Biomater. 23 (2015) S53-S82. https://doi.org/10.1016/j.actbio.2015.07.019

[11] T. Kono et al., Confusion between Carbonate Apatite and Biological Apatite (Carbonated Hydroxyapatite) in Bone and Teeth, Minerals 12(2) (2022) art. 170. https://doi.org/10.3390/min12020170

[12] U. Pantulap et al., Bioactive glasses incorporating less-common ions to improve biological and physical properties, J. Mater. Sci. Mater. Med., 33(1) (2021) 3. https://doi.org/10.1007/s10856-021-06626-3

[13] O. Kędzia et al., Bioactive Coatings and the Safety of Using Metal Implants, Syst. Saf. Hum.-Tech. Facil.-Environ. 6(1) (2024) 227–238. https://doi.org/10.2478/czoto-2024-0025

[14] S. Pourshahrestani et al., Gallium-containing mesoporous bioactive glass with potent hemostatic activity and antibacterial efficacy., J. Mater. Chem. B, 4(1) (2016) 71–86. https://doi.org/10.1039/c5tb02062j

[15] S. Pourshahrestani et al., Potency and Cytotoxicity of a Novel Gallium-Containing Mesoporous Bioactive Glass/Chitosan Composite Scaffold as Hemostatic Agents., ACS Appl. Mater. Interfaces, 9(37) (2017) 31381-31392. https://doi.org/10.1021/acsami.7b07769

[16] M. Nagrath et al., Tantalum-containing mesoporous bioactive glass powder for hemostasis., J. Biomater. Appl. 35(8) (2021) 924-932. https://doi.org/10.1177/0885328220965150

[17] G. Hu et al., Antibacterial hemostatic dressings with nanoporous bioglass containing silver., Int. J. Nanomedicine, 7 (2012) 2613-2620. https://doi.org/10.2147/IJN.S31081

[18] R. Sergi et al., Chitosan-Based Bioactive Glass Gauze: Microstructural Properties, In Vitro Bioactivity, and Biological Tests., Materials 13(12) (2020) art.2819. https://doi.org/10.3390/ma13122819

[19] D.W. Buck 2nd, Innovative Bioactive Glass Fiber Technology Accelerates Wound Healing and Minimizes Costs: A Case Series., Adv. Skin Wound Care 33(8) (2020) 1-6. https://doi.org/10.1097/01.ASW.0000672504.15532.21

[20] S. Tomar et al., In vivo evaluation of bioactive glass-reinforced multifunctional composite hydrogel for thermal burn and chronic diabetic wound repair., Burns 51(8) (2025) art.107618. https://doi.org/10.1016/j.burns.2025.107618

[21] K. Ma et al., Bioactive glass suspension hydrogel promotes wound healing by modulating fibroblasts., Nanotechnology 36(7) (2024) art.075101. https://doi.org/10.1088/1361-6528/ad9610

[22] Y. Wang et al., Multi-layer-structured bioactive glass nanopowder for multistage-stimulated hemostasis and wound repair, Bioact. Mater. 25 (2023) 319-332, https://doi.org/10.1016/j.bioactmat.2023.01.019

[23] Information on https://patents.justia.com/patent/20250049981?utm_.

[24] O. Kędzia et al., Glass and Glass-Ceramic Porous Materials for Biomedical Applications, Syst. Saf. Hum.-Tech. Facil.-Environ. 5(1) (2023) 302-310. https://doi.org/10.2478/czoto-2023-0033

[25] A.S. Pádua et al., Polycaprolactone/Doped Bioactive Glass Composite Scaffolds for Bone Regeneration, J. Funct. Biomater. 16(6) (2025) art.200. https://doi.org/10.3390/jfb16060200

[26] E. Daskalakis et al., Novel 3D Bioglass Scaffolds for Bone Tissue Regeneration., Polymers 14(3) (2022) art.445. https://doi.org/10.3390/polym14030445

[27] E. Skrzypczak-Pietraszek, J. Pietraszek, Chemical profile and seasonal variation of phenolic acid content in bastard balm (Melittis melissophyllum L., Lamiaceae), Journal of Pharmaceutical and Biomedical Analysis 66 (2012) 154-161. https://doi.org/10.1016/j.jpba.2012.03.037

[28] E. Skrzypczak-Pietraszek, A.G. Pietraszek, Phytoremediation and Allelopathy as an Element of Sustainable Circular Economy to Prevent Environmental Pollution, Materials Research Proceedings 45 (2024) 140-145. https://doi.org/10.21741/9781644903315-17

[29] E. Radzyminska-Lenarcik, R. Ulewicz and M. Ulewicz, Zinc recovery from model and waste solutions using polymer inclusion membranes (PIMs) with 1-octyl-4-methylimidazole, Desalination and Water Treatment 102 (2018) 211-219. https://doi.org/10.5004/dwt.2018.21826

[30] A. Pietrzak et al., Application of a Mixture of Fly Ash and Solid Waste from Gas Treatment from Municipal Solid Waste Incineration in Cement Mortar, Materials 18 (2025) art. 481. https://doi.org/10.3390/ma18030481

[31] D. Klimecka-Tatar et al., Corrosion characteristics in acid, alkaline and the ringer solution of Fe68-xCoxZr10Mo5W2B 15 metallic glasses, J. Balkan Tribol. Assoc. 20 (2014) 124-130.

[32] A. Dudek, Microstructure and properties of the composites: Hydroxyapatite with addition of zirconia phase, Journal of Engineering Materials and Technology 133 (2011) art. 021006. https://doi.org/10.1115/1.4003104

[33] A. Dudek, L. Adamczyk, Properties of hydroxyapatite layers used for implant coatings, Optica Applicata 43 (2013) 143-151. https://doi.org/10.5277/oa130118

[34] A. Szczotok et al., Metallographic Study and Repeatability Analysis of γ' Phase Precipitates in Cored, Thin-Walled Castings Made from IN713C Superalloy, Archives of Metallurgy and Materials 62 (2017) 595-601. https://doi.org/10.1515/amm-2017-0088

[35] A. Szczotok et al., The Impact of the Thickness of the Ceramic Shell Mould on the (γ + γ′) Eutectic in the IN713C Superalloy Airfoil Blade Casting, Archives of Metallurgy and Materials 62 (2017) 587-593. https://doi.org/10.1515/amm-2017-0087

[36] A. Szczotok et al., Effect of the induction hardening on microstructures of the selected steels, METAL 2018 – 27th Int. Conf. Metall. Mater. (2018) 1264-1269.

[37] N. Radek et al., The effect of laser treatment on operational properties of ESD coatings, METAL 2021 – 30th Int. Conf. Metall. Mater. (2021) 876-882. https://doi.org/10.37904/metal.2021.4212

[38] N. Radek et al., Operational properties of DLC coatings and their potential application, METAL 2022 – 31st Int. Conf. Metall. Mater., (2022) 531-536. https://doi.org/10.37904/metal.2022.4491

[39] N. Radek et al., Technology and applications of ESD coatings before and after laser processing, METAL 2023 – 32nd Int. Conf. Metall. Mater. (2024) 500-505. https://doi.org/10.37904/metal.2023.4727

[40] D. Klimecka-Tatar, Electrochemical characteristics of titanium for dental implants in case of the electroless surface modification, Archives of Metallurgy and Materials 61 (2016) 923-926. https://doi.org/10.1515/amm-2016-0156

[41] D. Przestacki, T. Chwalczuk, The analysis of surface topography during turning of Waspaloy with the application of response surface method, MATEC Web of Conferences 136 (2017) art. 02006. https://doi.org/10.1051/matecconf/201713602006

[42] D. Nowakowski et al., Application of machine learning in the analysis of surface quality - the detection the surface layer damage of the vehicle body, METAL 2021 – 30th Int. Conf. Metall. Mater. (2021) 864-869. https://doi.org/10.37904/metal.2021.4210

[43] A. Kalinowski et al., Laser surface texturing: characteristics and applications, System Safety: Human - Technical Facility - Environment 5 (2023) 240-248. https://doi.org/10.2478/czoto-2023-0026

[44] A. Kalinowski et al., The effect of laser infrared radiation on the surface modification of DLC coatings, AIP Conf. Proc. 3130 (2024) art. 020014. https://doi.org/10.1063/5.0204985

[45] V. Zhelykh et al., Analysis of the processes of heat exchange on infrared heater surface, Diagnostyka 17 (2016) 81-85.

[46] N. Radek et al., Properties of Steel Welded with CO_2 Laser, Lecture Notes in Mechanical Engineering (2020) 571-580. https://doi.org/10.1007/978-3-030-33146-7_65

[47] Ł.J. Orman et al., Laser Treatment of Surfaces for Pool Boiling Heat Transfer Enhancement, Materials 16 (2023) art. 1365. https://doi.org/10.3390/ma16041365

[48] Ł.J. Orman et al., Application of laser treatment technology for boiling heat transfer augmentation, Production Engineering Archives 30 (2024) 259-265. https://doi.org/10.30657/pea.2024.30.25

[49] D. Klimecka-Tatar, Value Stream Mapping as Lean Production tool to improve the production process organization - Case study in packaging manufacturing, Production Engineering Archives 17 (2017) 40-44. https://doi.org/10.30657/pea.2017.17.09

[50] R. Ulewicz, M. Ulewicz, Problems in the Implementation of the Lean Concept in the Construction Industries, Lecture Notes in Civil Engineering 47 (2020) 495-500. https://doi.org/10.1007/978-3-030-27011-7_63

[51] M. Ingaldi, Overview of the main methods of service quality analysis, Production Engineering Archives 18 (2018) 54-59. https://doi.org/10.30657/pea.2018.18.10

[52] R. Ulewicz, B. Krstić and M. Ingaldi, Mining Industry 4.0 – Opportunities and Barriers, Acta Montanistica Slovaca 27 (2022) 291-305. https://doi.org/10.46544/AMS.v2i2.02

[53] R. Dwornicka, J. Pietraszek, The outline of the expert system for the design of experiment, Production Engineering Archives 20 (2018) 43-48. https://doi.org/10.30657/pea.2018.20.09

[54] J. Pietraszek et al., Challenges for the DOE methodology related to the introduction of Industry 4.0, Production Engineering Archives 26 (2020) 190-194. https://doi.org/10.30657/pea.2020.26.33

[55] J. Pietraszek, A. Gadek-Moszczak, The smooth bootstrap approach to the distribution of a shape in the ferritic stainless steel aisi 434l powders, Solid State Phenomena 197 (2013) 162-167. https://doi.org/10.4028/www.scientific.net/SSP.197.162

[56] R. Dwornicka et al., The smoothed bootstrap fine-tuning, System Safety: Human - Technical Facility - Environment 1 (2019) 716-723. https://doi.org/10.2478/czoto-2019-0091

[57] J. Pietraszek, Fuzzy regression compared to classical experimental design in the case of flywheel assembly, Lecture Notes in Computer Science 7267 LNAI (2012) 310-317. https://doi.org/10.1007/978-3-642-29347-4_36

[58] J. Pietraszek, The modified sequential-binary approach for fuzzy operations on correlated assessments, Lecture Notes in Computer Science 7894 LNAI (2013) 353-364. https://doi.org/10.1007/978-3-642-38658-9_32

Terotechnology XIV | Materials Research Forum LLC
Materials Research Proceedings 62 (2026) 72-79 | https://doi.org/10.21741/9781644904015-10

Effect of Powder Morphology on Surface Topography and Compositional Homogeneity of Oxide Coatings Deposited by Atmospheric Plasma Spraying (APS)

BAŁAGA Zbigniew[1,a], OPYDO Michał[1,b], BEZGIN Sezin[2,c], DUDEK Agata[1,d *]

[1]Czestochowa University of Technology, Czestochowa, Poland

[2]Eskisehir Technical University, Eskişehir, Türkiye

[a]zbigniew.balaga@pcz.pl, [b]michal.opydo@pcz.pl, [c]sezindenkli@gmail.com, [d]agata.dudek@pcz.pl

Keywords: Steel 41Cr4, Steel 316L, APS

Abstract. This study aimed to assess the 3D surface roughness and chemical composition of APS coatings deposited on steel substrates. Two coatings were examined: TiO_2 on 41Cr4 steel and a TiO_2+ZrO_2+Al_2O_3 mixture on 316L steel. Microscopic observations using a KEYENCE VHX-6000 with 3D topography analysis showed lower Sa and Sq values for the TiO_2 coating compared with the composite coating. EDS analysis confirmed the dominance of Ti and O in the TiO_2 coating, as well as a low content of Al and Zr at the surface of the composite coating. This effect was linked to the morphology of the powders and differences in the degree of particle melting. Cross-sectional EDS maps revealed local enrichments of Al and Zr as well as porosity and microcracks typical of the lamellar microstructure produced by APS.

Introduction

Thermal spraying is a surface-engineering technology used to impart resistance to wear, corrosion, and thermal loading. These methods are classified, among others, by the type of energy supplied to the feedstock material (Fig. 1): electrical, chemical, and kinetic [1–6]. Within this group, atmospheric plasma spraying (APS) stands out for its ability to deposit oxide and oxide-composite coatings on structural and corrosion-resistant steels with high throughput and good controllability of process parameters such as arc power density, plasma-gas flow rates, and spray distance.

APS coatings exhibit a lamellar splat-based microstructure formed by the melting of powder particles, their acceleration in the plasma jet, impact on the substrate, and rapid solidification. Interlamellar boundaries and pores are typical features of such coatings, and microcracks may also occur, depending on the conditions. Particle morphology and thermal properties, including specific surface area, thermal conductivity, and heat of fusion, together with particle temperature and velocity, determine the degree of flattening in splats, the crystallographic texture, and the compositional uniformity of the coating.

Rapid quality assessment commonly employs 3D surface-topography metrology (Sa, Sz, Sq, Ssk) in conjunction with energy-dispersive X-ray spectroscopy. The 3D measurements enable a quantitative comparison of surface development and symmetry in the height distribution, while EDS verifies the composition and its spatial uniformity.

In this study, two material–substrate systems were compared: TiO_2 on 41Cr4 steel and a TiO_2+ZrO_2+Al_2O_3 mixture on 316L steel. The scope covered 2D and 3D surface observations, 3D roughness measurements, and EDS analyses of the surface and cross-section. The aim was to demonstrate differences in topography and compositional uniformity arising from the distinct powder particle morphologies and their degree of melting. Interpretation of composition accounted for the limited quantitative accuracy of EDS for light elements and for the influence of surface topography and specimen geometry.

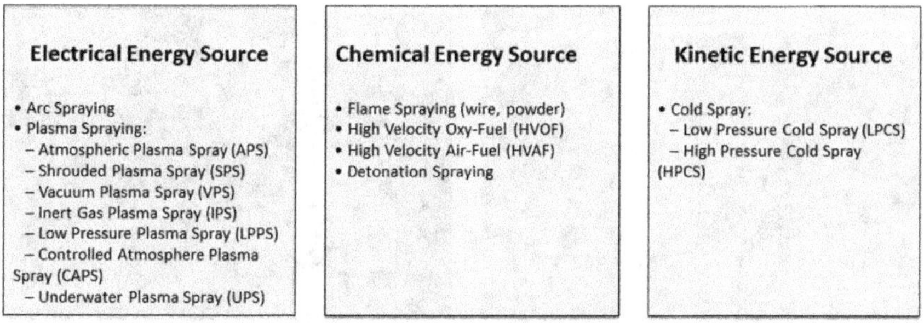

Figure 1. *Classification of thermal spraying processes by energy source*

Materials used in the Study

Two systems were analyzed that differed in substrate grade and in the composition of the APS-deposited coating. The first variant was a TiO_2 coating on 41Cr4 steel; the second was a composite TiO_2+ZrO_2+Al_2O_3 coating on 316L steel. Powders with varied morphology were used: Al_2O_3 formed angular, crushed grains; TiO_2 occurred as flaky and elongated particles; ZrO_2 exhibited spheroidal particles. Substrates were prepared by abrasive blasting and degreasing to ensure adequate anchor roughness and surface cleanliness before deposition.

Spraying was carried out in ambient air using a standard plasma torch. The exact process parameter set was applied to both variants: a plasma-gas mixture of Ar (60 SLPM) and H_2 (12 SLPM), a current of 600 A, a voltage of 60 V, a spray distance of 110 mm, and a torch angle of 90° to the surface. The number of passes was selected to achieve a thickness of approximately 80–100 μm (see specimen designations in Table 1).

Table 1. *Designations of the specimens used in the study.*

Sample ID	Substrate (steel grade)	As-sprayed coating composition
1	41Cr4	TiO_2
2	316L	TiO_2+ZrO_2+Al_2O_3

The coating surfaces were examined using a KEYENCE VHX-6000 digital microscope. Two-dimensional images and three-dimensional topography models were recorded, and the surface roughness parameters Sa, Sz, Sq, and Ssk were also recorded. The results are presented in Figs. 2–3 and Table 2.

Table 2. *Surface roughness parameters*

Sample ID	Surface roughness parameters			
	Sa [μm]	Sz [μm]	Sq [μm]	Ssk
1	4.68	40.44	5.98	0.41
2	9.62	59.49	11.27	-0.14

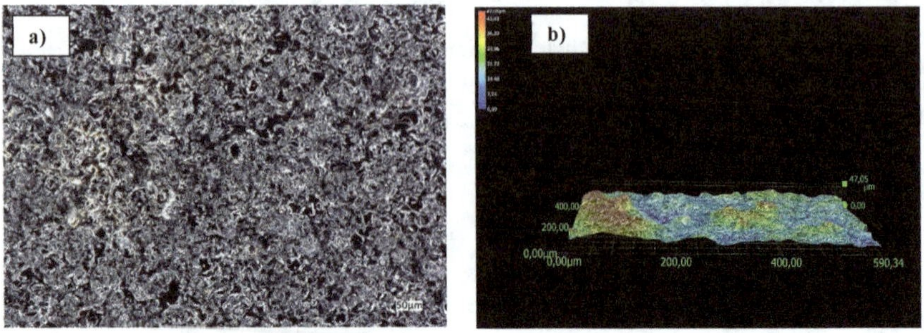

Figure 2. *Coating surface (sample 1): (a) 2D view; (b) 3D view*

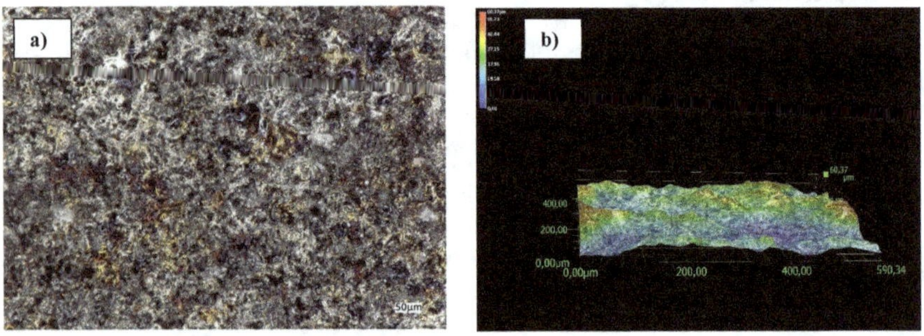

Figure 3. *Coating surface (sample 2): (a) 2D view; (b) 3D view.*

The examinations were conducted using a JEOL 6610LV scanning electron microscope equipped with an energy-dispersive X-ray spectroscopy (EDS) system (Oxford Instruments), which was utilized to verify the chemical composition of the coatings. The results are presented in Fig. 4. As the surface EDS analysis of sample 2 indicated low Al and Zr contents, cross-sectional EDS mapping was performed to assess the homogeneity of their distribution. The results are shown in Fig. 5.

Technological, Economic, and Environmental Impact

The conducted research provides an interesting starting point for potentially more advanced work using designed experiment methodologies [7,8], resampling methods for small samples [9,10], a fuzzy approach for incomplete data [11-13], and shaping the proper dimensionality of analyzed data [14]. The obtained results are of great importance for coating processes, including alternative methods such as ESD [15-17], DLC [18,19], laser sintering [20,21], and saturation with nanoparticles [22,23], as they indicate potential problems that may arise at the microstructural level. The properties of the surface layer [24,25] have a significant impact on both corrosion resistance [26,27] and microhardness [28,29], as well as on the radar signature [30] and visual perception [31,32]. At the same time, it is the geometric parameters of the surface layer that determine possible further laser microprocessing [33,34] of the produced layer and its qualitative assessment [35-37]. Ultimately, the obtained materials with high corrosion and wear resistance have potentially great practical importance in fields exposed to severe biocorrosion, including biotechnology [38] and cleaning of contaminated areas using natural methods [39].

Materials Research Proceedings 62 (2026) 72-79 https://doi.org/10.21741/9781644904015-10

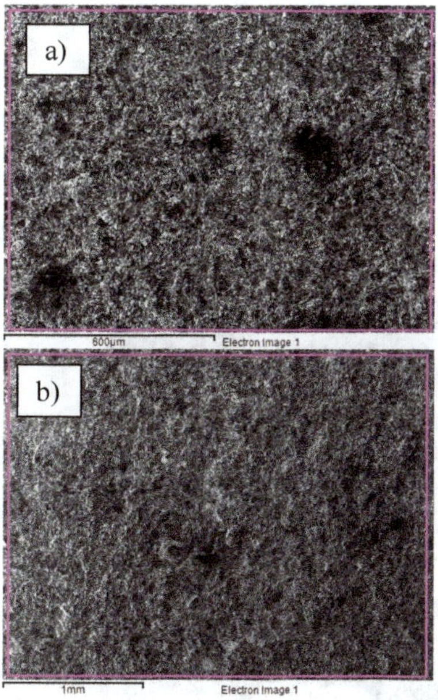

Element	Weight [%]	Atomic [%]
O K	48.45	73.78
Ti K	51.55	26.22
Totals	100.00	

Element	Weight [%]	Atomic [%]
O K	28.50	54.86
Al K	1.44	1.65
Ti K	64.95	41.76
Zr L	5.12	1.73
Totals	100.00	

Figure 4. *Chemical composition analysis of the coatings: (a) sample 1; (b) sample 2.*

Conclusion

Two atmospheric plasma-sprayed coatings were produced and characterized: TiO_2 on 41Cr4 steel and TiO_2+ZrO_2+Al_2O_3 on 316L steel. The powders differed in particle shape. Al_2O_3 formed angular, crushed grains. TiO_2 occurred as flaky and elongated particles. ZrO_2 exhibited spheroidal particles. Three-dimensional topography confirmed distinct surface development. The TiO_2 coating on 41Cr4 exhibited lower Sa and Sq values, at 4.68 μm and 5.98 μm, respectively, and Ssk = +0.41, indicating a predominance of peaks in the height distribution. The composite coating on 316L exhibited higher roughness, with Sa of 9.62 μm, Sq of 11.27 μm, and Ssk = −0.14, indicating a predominance of valleys. This response is consistent with the influence of powder morphology and thermal properties on particle heating, melting, and splat flattening, and thus on the lamellar microstructure of coatings produced by APS.

Surface EDS of the TiO_2/41Cr4 coating showed dominant contributions of titanium and oxygen. The oxygen result has limited quantitative accuracy. For the TiO_2+ZrO_2+Al_2O_3 coating, the surface contributions of Al and Zr were low. EDS maps on cross sections revealed local enrichments of Al and Zr within the Ti–O matrix, confirming compositional nonuniformity through the thickness and suggesting incomplete melting of some particles. Cross sections also revealed pores and microcracks typical of APS coatings, which arise from rapid splat solidification and thermal stresses during cooling in the coating/substrate system. The results indicate that particle morphology and process settings govern the surface topography and compositional uniformity of coatings deposited by APS. Verification of performance implications requires separate functional tests carried out under application-relevant conditions.

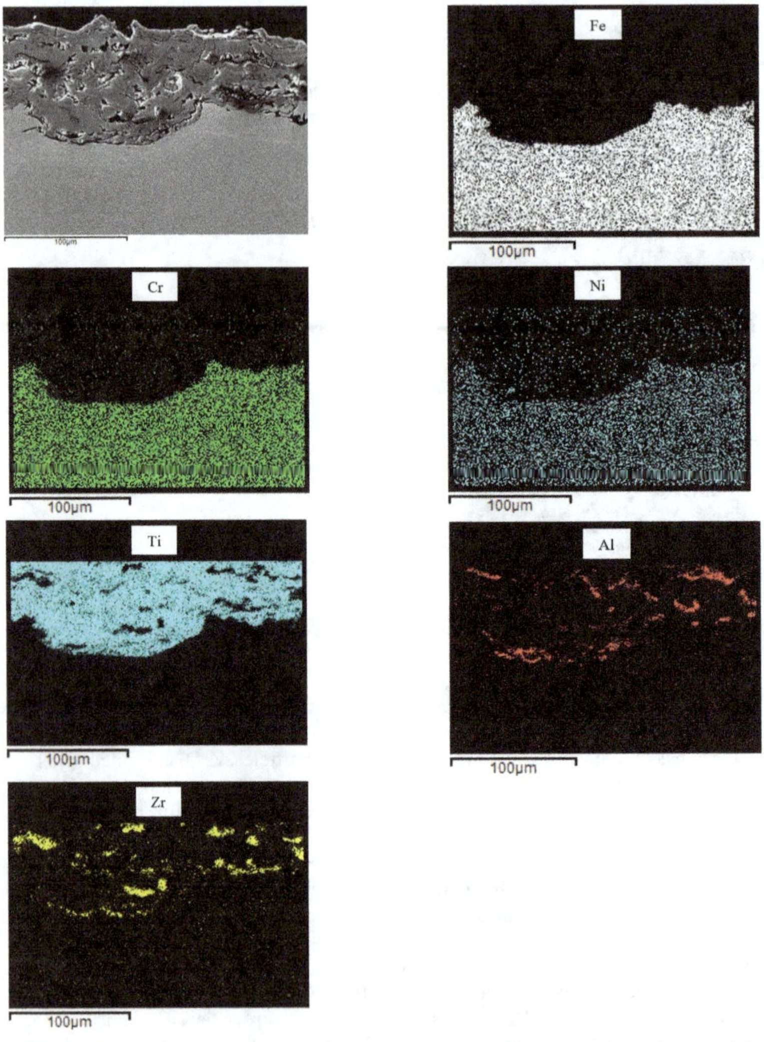

***Figure 5.** Cross-section of sample 2: coating micrograph and EDS elemental distribution maps for Fe, Cr, Ni (substrate), and Ti, Al, Zr (coating).*

References

[1] J.G. Odhiambo et al., Porosity and its significance in plasma-sprayed coatings. Coatings 9 (2019) art. 460. https://doi.org/10.3390/coatings9070460

[2] S. Mahade et al., Erosion performance of atmospheric plasma sprayed thermal barrier coatings with diverse porosity levels. Coatings 11 (2021) art. 86. https://doi.org/10.3390/coatings11010086

[3] K. Mehmood et al., Microstructure and corrosion behavior of atmospheric plasma sprayed NiCoCrAlFe high entropy alloy coating. Materials 15 (2022) art. 1486. https://doi.org/10.3390/ma15041486

[4] K. Płatek et al., The effects of atmospheric plasma spray parameters on structure and mechanical properties of titanium dioxide coatings. Advances in Science and Technology Research Journal 19 (2025) 381–389. https://doi.org/10.12913/22998624/195755

[5] H. Memon et al., Active learning-driven uncertainty reduction for in-flight particle characteristics of atmospheric plasma spraying of silicon. Engineering Applications of Artificial Intelligence 128 (2024) art. 107465. https://doi.org/10.1016/j.engappai.2023.107465

[6] Y. Li et al., Control of the pore structure of plasma-sprayed thermal barrier coatings through the addition of unmelted porous YSZ particles. Coatings 11 (2021) art. 360. https://doi.org/10.3390/coatings11030360

[7] J. Pietraszek, A. Goroshko, The heuristic approach to the selection of experimental design, model and valid pre-processing transformation of DoE outcome, Advanced Materials Research 874 (2014) 145-149. https://doi.org/10.4028/www.scientific.net/AMR.874.145

[8] J. Pietraszek, N. Radek and A.V. Goroshko, Challenges for the DOE methodology related to the introduction of Industry 4.0, Production Engineering Archives 26 (2020) 190-194. https://doi.org/10.30657/pea.2020.26.33

[9] J. Pietraszek, A. Gadek-Moszczak, The smooth bootstrap approach to the distribution of a shape in the ferritic stainless steel AISI 434L powders, Solid State Phenomena 197 (2013) 162-167. https://doi.org/10.4028/www.scientific.net/SSP.197.162

[10] J. Pietraszek, E. Skrzypczak-Pietraszek, The optimization of the technological process with the fuzzy regression, Advanced Materials Research 874 (2014) 151-155. https://doi.org/10.4028/www.scientific.net/AMR.874.151

[11] J. Pietraszek, Fuzzy regression compared to classical experimental design in the case of flywheel assembly, Lecture Notes in Computer Science 7267 LNAI (2012) 310-317. https://doi.org/10.1007/978-3-642-29347-4_36

[12] J. Pietraszek, The modified sequential-binary approach for fuzzy operations on correlated assessments, Lecture Notes in Computer Science 7894 LNAI (2013) 353-364. https://doi.org/10.1007/978-3-642-38658-9_32

[13] G. Filo, A Review of Fuzzy Logic Method Development in Hydraulic and Pneumatic Systems, Energies 16 (2023) art. 7584. https://doi.org/10.3390/en16227584

[14] J. Pietraszek, E. Skrzypczak-Pietraszek, The uncertainty and robustness of the principal component analysis as a tool for the dimensionality reduction, Solid State Phenomena 235 (2015) 1-8. https://doi.org/10.4028/www.scientific.net/SSP.235.1

[15] N. Radek et al., The effect of laser treatment on operational properties of ESD coatings, METAL 2021 – 30th Int. Conf. Metall. Mater. (2021) 876-882. https://doi.org/10.37904/metal.2021.4212

[16] N. Radek, R. Dwornicka and D. Gontarski, The impact of laser processing on the performance properties of electro-spark coatings, World Congress in Computational Mechanics and ECCOMAS Congress 1000 (2021) 1-10. https://doi.org/10.23967/wccm-eccomas.2020.336

[17] N. Radek et al., Technology and applications of ESD coatings before and after laser processing, METAL 2023 – Int. Conf. Metall. Mater. (2024) 500-505. https://doi.org/10.37904/metal.2023.4727

[18] N. Radek et al., Operational properties of DLC coatings and their potential application, METAL 2022 – 31st Int. Conf. Metall. Mater., (2022) 531-536. https://doi.org/10.37904/metal.2022.4491

[19] A. Kalinowski et al., The effect of laser infrared radiation on the surface modification of DLC coatings, AIP Conf. Proc. 3130 (2024) art. 020014. https://doi.org/10.1063/5.0204985

[20] A. Bartkowska et al., Microstructure, microhardness, corrosion resistance and chemical composition of Mo, B and Mo-B coatings produced using laser processing, Materials 13 (2020) art. 3249. https://doi.org/10.3390/MA13153249

[21] N. Radek et al., Formation of coatings with technologies using concentrated energy stream, Production Engineering Archives 28 (2022) 117-122. https://doi.org/10.30657/pea.2022.28.13

[22] J. Korzekwa, A. Gądek-Moszczak and M. Bara, The influence of sample preparation on SEM measurements of anodic oxide layers, Practical Metallography 53 (2016) 36-49. https://doi.org/10.3139/147.110367

[23] J. Korzekwa, A. Gądek-Moszczak and M. Zubko, Influence of the Size of Nanoparticles on the Microstructure of Oxide Coatings, Materials Science 53 (2018) 709-716. https://doi.org/10.1007/s11003-018-0127-x

[24] A. Wrońska, A. Dudek, Characteristics of surface layer of sintered stainless steels after remelting using GTAW method, Archives of Civil and Mechanical Engineering 14 (2014) 425-432. https://doi.org/10.1016/j.acme.2013.10.011

[25] M. Szymański, D. Przestacki and P. Szymański, Tool Wear and Surface Roughness in Turning of Metal Matrix Composite Built of Al2O3 Sinter Saturated by Aluminum Alloy in Vacuum Condition, Materials 15 (2022) art. 8375. https://doi.org/10.3390/ma15238375

[26] O. Kierat, A. Dudek and L. Adamczyk, The effect of the corrosion medium on silane coatings deposited on titanium grade 2 and titanium alloy Ti13Nb13Zr, Materials 14 (2021) art. 6350. https://doi.org/10.3390/ma14216350

[27] T. Lipiński, J. Pietraszek, Corrosion of the S235JR Carbon Steel after Normalizing and Overheating Annealing in 2.5% Sulphuric Acid at Room Temperature, Materials Research Proceedings 24 (2022) 102-108. https://doi.org/10.21741/9781644902059-16

[28] M. Kukliński et al., Influence of microstructure and chemical composition on microhardness and wear properties of laser borided monel 400, Materials 13 (2020) art. 5757. https://doi.org/10.3390/ma13245757

[29] A. Bartkowska et al., Laser processing of diffusion boronized layer produced on monel® alloy 400—microstructure, microhardness, corrosion and wear resistance tests, Materials 14 (2021) art. 7529. https://doi.org/10.3390/ma14247529

[30] W. Przybył et al., Microwave absorption properties of carbonyl iron-based paint coatings for military applications, Defence Technology 22 (2023) 1-9. https://doi.org/10.1016/j.dt.2022.06.013

[31] N. Radek, R. Dwornicka, Fire properties of intumescent coating systems for the rolling stock, Communications – Scientific Letters of the University of Žilina 22 (2020) 90-96. https://doi.org/10.26552/com.C.2020.4.90-96

[32] D. Nowakowski, A. Gądek-Moszczak and P. Lempa, Application of machine learning in the analysis of surface quality - the detection the surface layer damage of the vehicle body, METAL 2021 – 30th Int. Conf. Metall. Mater. (2021) 864-869. https://doi.org/10.37904/metal.2021.4210

[33] M. Kawalec et al., Laser assisted machining of aluminium composite reinforced by SiC particle, ICALEO 2008 – 27th International Congress on Applications of Lasers and Electro-Optics (2008) 895-900. https://doi.org/10.2351/1.5061278

[34] A. Kalinowski et al., Laser surface texturing: characteristics and applications, System Safety: Human-Technical Facility-Environment 5 (2023) 240-248. https://doi.org/10.2478/czoto-2023-0026

[35] N. Radek et al., Properties of Steel Welded with CO2 Laser, Lecture Notes in Mechanical Engineering (2020) 571-580. https://doi.org/10.1007/978-3-030-33146-7_65

[36] A. Gądek-Moszczak et al., Nano X-ray Tomography Application for Quantitative Surface Layer Geometry Analysis after Laser Beam Modification, Materials 15 (2022) art. 5935. https://doi.org/10.3390/ma15175935

[37] A. Pacana, D. Siwiec, Method of Fuzzy Analysis of Qualitative-Environmental Threat in Improving Products and Processes (Fuzzy QE-FMEA), Materials 16 (2023) art. 1651. https://doi.org/10.3390/ma16041651

[38] E. Skrzypczak-Pietraszek, J. Pietraszek, Chemical profile and seasonal variation of phenolic acid content in bastard balm (Melittis melissophyllum L., Lamiaceae), Journal of Pharmaceutical and Biomedical Analysis 66 (2012) 154-161. https://doi.org/10.1016/j.jpba.2012.03.037

[39] E. Skrzypczak-Pietraszek, A.G. Pietraszek, Phytoremediation and Allelopathy as an Element of Sustainable Circular Economy to Prevent Environmental Pollution, Materials Research Proceedings 45 (2024) 140-145. https://doi.org/10.21741/9781644903315-17

Effect of Deformation on the Microstructure and Hardness of X5CrNi18-10 Wire

BAŁAGA Zbigniew[1,a], OPYDO Michał[1,b], BEZGIN Sezin[2,c], DUDEK Agata[1,d *]

[1]Czestochowa University of Technology, Czestochowa, Poland

[2]Eskisehir Technical University, Eskişehir, Türkiye

[a]zbigniew.balaga@pcz.pl, [b]michal.opydo@pcz.pl, [c]sezindenkli@gmail.com, [d]agata.dudek@pcz.pl

Keywords: X5CrNi18-10, Spring Wire, Cold Deformation, Strain Hardening, Austenite

Abstract. This study presents the characterization of cold-drawn spring wire made of X5CrNi18-10 (AISI 304) steel. The investigations comprised phase analysis by X-ray diffraction, microstructural observations in longitudinal and transverse sections, and Vickers hardness measurements (HV2). XRD confirmed austenite as the main phase. The measured hardness, HV2 = 549 ± 8, is high compared with values typical of the annealed condition and results from strain hardening induced by cold drawing. The microstructure shows a banded grain structure characteristic of intense plastic deformation, as well as sulfide inclusions (MnS). Revealing the microstructure required oxidizing/tint etchants suitable for chromium–nickel steels in the heavily deformed state. The results are consistent with the requirements of EN 10270-3 and can serve as a reference for testing and quality control of austenitic wires used for spring elements.

Introduction

Austenitic steels of the X5CrNi18-10 (AISI 304) type are among the most widely used materials for spring applications and components operating in corrosive environments. Their high corrosion resistance and good plasticity enable cold forming and substantial strain hardening, resulting in increased strength without the need for additional heat treatment. For this reason, 18/10 steels are broadly used in spring and fastening systems as well as in medical technology.

In recent years, advanced microstructural characterization techniques have gained importance in the analysis of plastically deformed steels. Coupling energy-dispersive X-ray spectroscopy (EDS) with scanning electron microscopy (SEM) enables the local identification of non-metallic inclusions and the mapping of micrometer-scale elemental distributions [1,2]. X-ray diffraction (XRD) allows the determination of crystalline phases and the detection of any strain-induced martensitic transformations. Modern digital (3D) microscopy enables the three-dimensional observation of deformation bands, while Vickers hardness testing provides a straightforward quantitative measure of the degree of work hardening [4, 5].

This study aimed to apply these techniques to evaluate austenitic X5CrNi18-10 wire after cold drawing. The chemical and phase composition, microstructure, and hardness were analyzed, with particular emphasis on the occurrence and characteristics of sulfide inclusions. The results provide a basis for assessing the effect of deformation on the structure and service properties of austenitic spring wires.

Material Used for the Study

The investigations were conducted on a 2 mm diameter wire made of X5CrNi18-10 steel. Its chemical composition, as specified in EN 10270-3, is given in Table 1.

Materials Research Proceedings 62 (2026) 80-85 https://doi.org/10.21741/9781644904015-11

Table 1. Chemical composition of the tested steel wire according to the standard.

C [%]	Si [%] max	Mn [%] max	P [%] max	S [%] max	Cr [%]	Ni [%]
≤0.07	1.0	2.0	0.045	0.015	17.5-19.0	8.0-10.5

Chemical composition analysis was performed using a JEOL 6610LV scanning electron microscope equipped with an energy-dispersive X-ray spectroscopy (EDS) detector (Oxford Instruments). The contents of the principal alloying elements are within the ranges specified by EN 10270-3. EDS spectra revealed localized sulfur enrichment associated with manganese sulfide (MnS) inclusions (Fig. 1). It should be emphasized that EDS is not a reference method for determining the total sulfur content of the material, because point measurements reflect the composition of inclusions rather than the bulk (matrix-averaged) sulfur level.

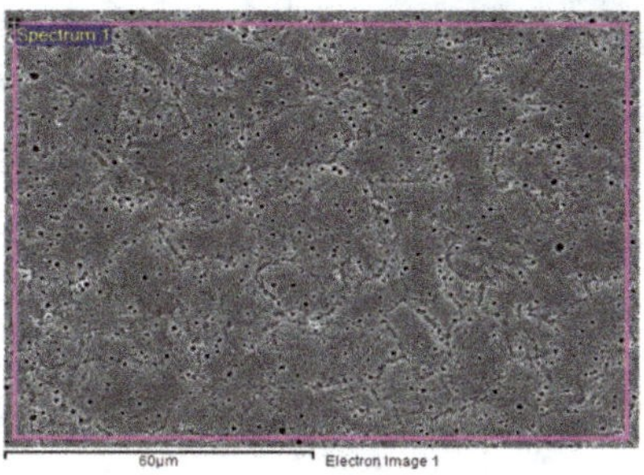

Element	Weight [%]
Si K	0.57
S K	0.19
Cr K	17.92
Mn K	1.22
Fe K	71.90
Ni K	8.20
Totals	100.00

Fig. 1. SEM/EDS chemical analysis of X5CrNi18-10 steel

a)

b)

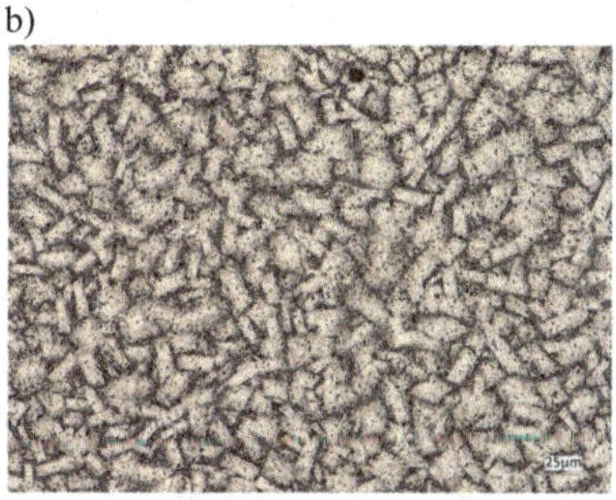

Fig. 2. Microstructure of the X5CrNi18-10 wire after cold drawing: (a) longitudinal section, (b) transverse section.

The specimens were prepared following a metallographic procedure (sequential grinding followed by polishing to obtain a smooth, homogeneous surface). Microstructural analysis was conducted using oxidizing/tint etchants specifically designed for Cr–Ni steels following severe cold deformation. Owing to the high tendency of the surface to passivate and the presence of deformation bands, general-purpose etchants did not provide sufficient contrast.

Microstructural analysis was also performed using a Keyence VHX-6000 optical digital microscope on sections oriented both longitudinally and transversely to the wire axis. In the longitudinal section, banding typical of the drawing process is visible. In the transverse section, contrast variations arising from post-deformation anisotropy were observed. Representative images are shown in Fig. 2.

The longitudinal section exhibits pronounced deformation banding typical of the drawing process. Microstructural revelation on polished sections was hindered by the high degree of strain hardening, which reduced the effectiveness of the etching process. SEM observations further confirmed the presence of sulfide inclusions (see Fig. 3).

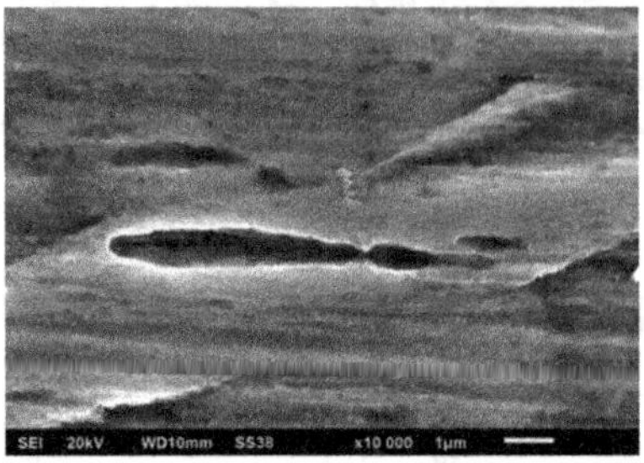

Fig. 3. Sulfide inclusion in the microstructure of the investigated wire.

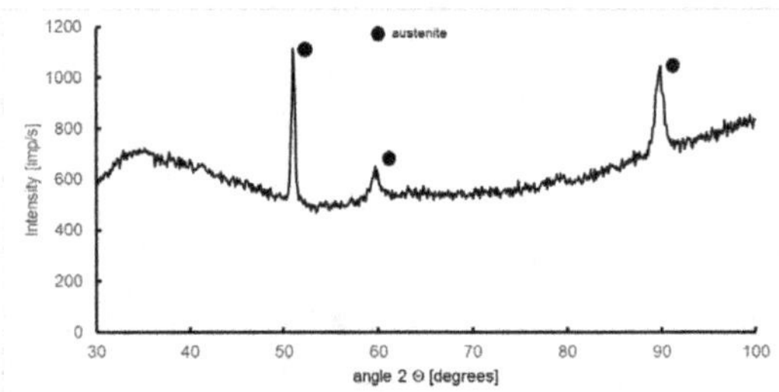

Fig. 4. X-ray diffraction pattern acquired from the investigated wire sample.

The steel under investigation exhibits an austenitic structure, as confirmed by X-ray diffraction and microscopic observations. The XRD pattern (Fig. 4) shows diffraction peaks characteristic of the austenite (fcc) phase.

Vickers hardness measurements at a 2 kgf load (HV2) were performed on polished metallographic sections. The results are summarized in Table 2. The measured hardness, HV2 = 549 ± 8, is high compared to values typical of austenitic steels in the annealed condition but remains characteristic of cold-drawn wire in accordance with EN 10270-3 (spring applications). The elevated hardness does not indicate a phase transformation (XRD confirmed the presence of the austenite phase). The result reflects strain hardening and the likely contribution of residual stresses after severe deformation.

Table 2. Vickers hardness HV2 of the cold-drawn X5CrNi18-10 wire

Test No.	HV2
1	555
2	543
3	559
4	547
5	541
HV mean	**549±8**

Technological and Environmental Impact

The conducted research is of great importance for other areas of materials engineering, including wear issues [5,6], hardening of selected alloys [7-9] and coatings [10], corrosion resistance of steel [11,12], and special coatings [13,14]. This also influences technological issues, including laser micromachining [15,16], heavy metal waste reduction [17,18], and maintaining an appropriate level of quality in technological processes [19,20], at the same time becoming an impulse for the development of process optimization methods [21-23] and design methods [24]. Due to such activities, the final environmental burden is also reduced [25,26],

Conclusion

After cold drawing, the examined X5CrNi18-10 wire exhibited an austenitic structure, as confirmed by diffraction studies. The Vickers hardness HV2 = 549 ± 8 is high relative to levels typical of the annealed condition, yet corresponds to a material strongly strengthened by cold work, in line with the characteristics of spring wires specified in EN 10270-3. In the longitudinal section, the microstructure shows a pronounced banded arrangement, confirming anisotropy introduced by plastic working. SEM images revealed the presence of MnS sulfide inclusions. The sulfur signals recorded by EDS are local in nature and reflect the composition of inclusions rather than the total sulfur content in the metallic matrix. Revealing structural details required oxidizing/tint etchants appropriate for Cr–Ni steels after heavy cold deformation. The resulting picture of phase, hardness, and morphology is consistent with the expected condition of material intended for elastic elements and confirms the proper character of strengthening.

The investigations have practical relevance. They clarify the relationships between microstructure, hardness, and delivery condition in austenitic spring wires, guiding the interpretation of local EDS indications in the presence of MnS. The applied observation and etching methodology facilitates assessment of the uniformity of strengthening and anisotropy, which can be used in quality control and comparative analyses. The results also provide a reference point for further work on fatigue life and the stability of properties in service for wires subjected to intensive cold drawing.

References

[1] K. Kim et al., Improvement of Strength and Impact Toughness for Cold-Worked Austenitic Stainless Steels Using a Surface-Cracking Technique. Metals 8 (2018) art. 932. https://doi.org/10.3390/met8110932

[2] H.A. Tinoco et al., Experimental and numerical study of micromechanical damage induced by MnS-based inclusions, Materials Science and Engineering: A 856 (2022) art.144009. https://doi.org/10.1016/j.msea.2022.144009

[3] Y. Song et al., A review of research on MnS inclusions in high-quality steel. Engineering Reports 6 (2024) art. e12892. https://doi.org/10.1002/eng2.12892

[4] G. Monrrabal et al., Influence of the cold working induced martensite on the electrochemical behavior of AISI 304 stainless steel surfaces, Journal of Materials Research and Technology 8 (2019) 1335-1346. https://doi.org/10.1016/j.jmrt.2018.10.004

[5] D. Bartkowski et al., Microstructure, microhardness, corrosion and wear resistance of B, Si and B-Si coatings produced on C45 steel using laser processing, Metals 10 (2020) art. 792. https://doi.org/10.3390/met10060792

[6] A. Pacana et al., Influence of natural aggregate crushing process on crushing strength index, Sustainability 13 (2021) art. 8353. https://doi.org/10.3390/su13158353

[7] T. Lipiński, The structure and mechanical properties of Al-7%SiMg alloy treated with a homogeneous modifier, Solid State Phenomena 163 (2010) 183-186. https://doi.org/10.4028/www.scientific.net/SSP.163.183

[8] T. Lipiński, A. Wach, Dimensional structure of non-metallic inclusions in high-grade medium carbon steel melted in an electric furnace and subjected to desulfurization, Solid State Phenomena 223 (2015) 46-53. https://doi.org/10.4028/www.scientific.net/SSP.223.46

[9] A. Szczotok et al., Effect of the induction hardening on microstructures of the selected steels, METAL 2018 – 27th Int. Conf. Metall. Mater. (2018) 1264-1269.

[10] A. Bartkowska et al., Microstructure, microhardness, corrosion resistance and chemical composition of Mo, B and Mo-B coatings produced using laser processing, Materials 13 (2020) art. 3249. https://doi.org/10.3390/MA13153249

[11] M. Scendo et al., Purine as an effective corrosion inhibitor for stainless steel in chloride acid solutions, Corrosion Reviews 30 (2012) 33-45. https://doi.org/10.1515/CORRREV-2011-0039

[12] T. Lipiński, J. Pietraszek, Corrosion of the S235JR Carbon Steel after Normalizing and Overheating Annealing in 2.5% Sulphuric Acid at Room Temperature, Materials Research Proceedings 24 (2022) 102-108. https://doi.org/10.21741/9781644902059-16

[13] M. Scendo, J. Trela and N. Radek, Influence of laser power on the corrosive resistance of WC-Cu coating, Surface and Coatings Technology 259 (2014) 401-407. https://doi.org/10.1016/j.surfcoat.2014.10.062

[14] O. Kierat et al., The effect of the corrosion medium on silane coatings deposited on titanium grade 2 and titanium alloy Ti13Nb13Zr, Materials 14 (2021) art. 6350. https://doi.org/10.3390/ma14216350

[15] A. Bartkowska et al., Laser processing of diffusion boronized layer produced on Monel® alloy 400 – microstructure, microhardness, corrosion and wear resistance tests, Materials 14 (2021) art. 7529. https://doi.org/10.3390/ma14247529

[16] A. Kalinowski et al., Laser surface texturing: characteristics and applications, System Safety: Human - Technical Facility - Environment 5 (2023) 240-248. https://doi.org/10.2478/czoto-2023-0026

[17] M. Ulewicz et al., Transport of lead across polymer inclusion membrane with p-tert-butylcalix[4]arene derivative, Physicochemical Problems of Mineral Processing 44 (2010) 245-256.

[18] E. Radzyminska-Lenarcik, M. Ulewicz, Selective Transport of Cu(II) across a Polymer Inclusion Membrane with 1-Alkylimidazole from Nitrate Solutions, Separation Science and Technology (Philadelphia) 47 (2012) 1113-1118. https://doi.org/10.1080/01496395.2011.646378

[19] D. Klimecka-Tatar, R. Dwornicka, The assembly process stability assessment based on the strength parameters statistical control of complex metal products, METAL 2019 – 28th Int. Conf. Metall. Mater. (2019) 709-714.

[20] T. Lipiński, R. Ulewicz, The effect of the impurities spaces on the quality of structural steel working at variable loads, Open Engineering 11 (2021) 233-238. https://doi.org/10.1515/eng-2021-0024

[21] J. Pietraszek, A. Gadek-Moszczak, The smooth bootstrap approach to the distribution of a shape in the ferritic stainless steel AISI 434L powders, Solid State Phenomena 197 (2013) 162-167. https://doi.org/10.4028/www.scientific.net/SSP.197.162

[22] J. Pietraszek et al., Challenges for the DOE methodology related to the introduction of Industry 4.0, Prod. Eng, Arch. 26 (2020) 190-194. https://doi.org/10.30657/pea.2020.26.33

[23] A. Goroshko et al., Multiphysics Modeling of Direct-Start Induction Motor Considering Rotor Unbalance, Materials Research Proceedings 45 (2024) 267-276. https://doi.org/10.21741/9781644903315-31

[24] J.M. Djoković et al., Selection of the Optimal Window Type and Orientation for the Two Cities in Serbia and One in Slovakia, Energies 15 (2022) art. 323. https://doi.org/10.3390/en15010323

[25] A. Pacana, D. Siwiec, Method of Fuzzy Analysis of Qualitative-Environmental Threat in Improving Products and Processes (Fuzzy QE-FMEA), Materials 16 (2023) art. 1651. https://doi.org/10.3390/ma16041651

[26] E. Skrzypczak-Pietraszek, A.G. Pietraszek, Phytoremediation and Allelopathy as an Element of Sustainable Circular Economy to Prevent Environmental Pollution, Materials Research Proceedings 45 (2024) 140-145. https://doi.org/10.21741/9781644903315-17

Materials Research Proceedings 62 (2026) 86-93 https://doi.org/10.21741/9781644904015-12

Modeling of Material Selection for Product Improvement Considering Quality, Innovation, and Life Cycle Assessment

Dominika SIWIEC[1,a], Andrzej PACANA[2,b *] and Renata DWORNICKA[3,c]

[1]Rzeszow University of Technology, Faculty of Mechanical Engineering and Aeronautics, al. Powstańców Warszawy 12, 35-959 Rzeszów, Poland, ORCID: 0000-0002-6663-6621

[2]Rzeszow University of Technology, Faculty of Mechanical Engineering and Aeronautics, al. Powstańców Warszawy 12, 35-959 Rzeszów, Poland, ORCID: 0000-0003-1121-6352

[3]Cracov University of Technology, Faculty of Mechanical Engineering, ul. Warszawska 24, 31-155 Kraków, Poland, ORCID: 0000-0002-2979-1614

[a]d.siwiec@prz.edu.pl, [b]app@prz.edu.pl, [c]renata.dwornicka@pk.edu.pl

Keywords: Quality, LCA, Material Selection, Decision-Making, Sustainable Product Development, Mechanical Engineering, Production Engineering

Abstract. The product development process can be considered from a "cradle to grave" perspective. Making rational material choices that drive efficiency in production, use, and end-of-life seems crucial. Therefore, the aim of this article was to propose a modeling framework to support the materials selection process in product improvement, taking into account aspects of quality, innovation, and environmental impact throughout the life cycle. The model operates in six main stages, expanded with selected techniques to support their implementation. It is based on the identification of product usability, innovation, and environmental impact criteria within the life cycle. Material prototypes are analyzed to assess the importance of these criteria and their fulfillment based on the adopted assumptions. The entire process is presented in a flowchart illustrating a systematic approach. The materials selection model can be applied to materials selection for any product, for example, in the electrical machinery industry. It can be used by designers, engineers, and managers in the early stages of product development, as well as during product improvement for sustainability.

Introduction

Product development encompasses various aspects. A common one is improving product quality while maintaining the cost-effectiveness of the solutions offered [1,2]. Currently, it is becoming increasingly important to produce eco-friendly products that support sustainable development [3,4]. At the same time, to meet market demands, the solutions offered must be innovative [5]. This makes the entire product design and development process a complex decision-making problem [6].

The product development process can be translated into a product life cycle, for example, from cradle to grave (material acquisition and extraction, production, use, and end of life) [7]. Therefore, selecting the appropriate material is considered crucial. As the first step, it generates subsequent decisions at the remaining stages of the product life cycle. Various studies have been conducted in this area; for example, in the work [8], ecological criteria were primarily analyzed when selecting materials for fatigue products. Similarly, in the article [9], the possibilities of improving products were examined by modifying materials in terms of environmental aspects, as well as by changes in production processes. In turn, using the example of the work [10], it was observed that conventional material selection methods are not well-suited for analyzing large databases, especially when searching for material innovations. Therefore, computer-aided expert systems, for example, are proposed. In the aforementioned work, the authors analyze the possibilities of implementing such systems in the selection of biocomposite materials. In contrast, the study [11] presents a problem based, among other things, on standard portfolio materials. Ecodesign is applied, taking into account disassembly design, dematerialization, material refinement, and part reduction. Another example is the work [12], which analyzes children's search for material innovations and then compares these tool material choices with those made by an expert. The concept was based on supporting learning and exploration.

The process of selecting materials for sustainable product development remains an open area of research, with limited work in this field. Therefore, the aim of this article was to propose a modeling approach that supports the material selection process in product improvement, taking into account aspects of quality, innovation, and environmental impact throughout the life cycle. This paper presents the model's concepts and main assumptions, a proposed modeling framework, and its synthetic description. The results are presented in the subsequent sections.

Concept of the Model and Assumptions

The model's concept aims to support the product improvement process in its early stages, particularly during the selection and evaluation of materials used in them. This applies to the materials acquisition and extraction phase, which, from a product life cycle perspective, is the first phase of a product's life. Although the materials selection process is practiced as part of product design and improvement, it is difficult to conduct it while considering other aspects, such as quality, innovation, and environmental impact.

Quality is understood as customer satisfaction with product use, which, in the case of material selection, influences the properties of raw materials, which are important for functionality, safety, and other product specifications. Environmental impact refers to the negative effects of a product on the environment. In the case of materials, this refers to their acquisition and processing to obtain the final product. In turn, innovation can be understood in the context of introducing product innovations. When considering materials, these innovations are evident, for example, through innovative properties such as improved environmental performance or enhanced functionality. The concept of material modeling in the product improvement process, considering quality, innovation, and the environment, is presented in Fig. 1.

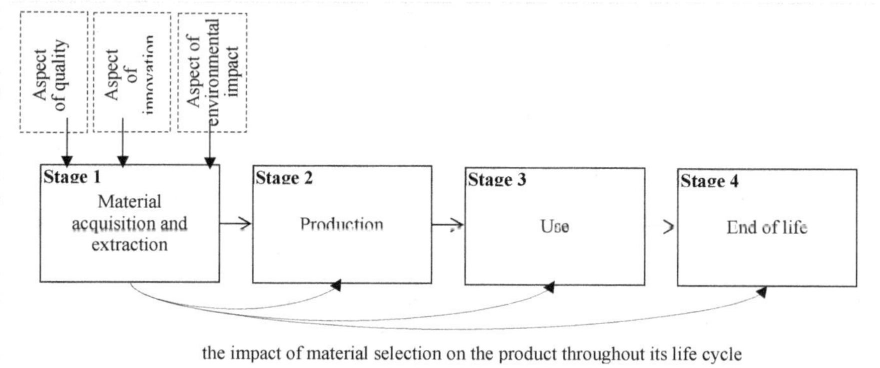

Figure 1. *A general concept for material modeling during product development placed within the cradle-to-grave lifecycle approach.*

It is anticipated that the selection of materials, taking into account aspects of quality, innovation, and environmental impact, will impact the product's life cycle. Materials should be recyclable during the design process. Similarly, the level of quality, innovation, and environmental impact affects customer satisfaction during use. Materials are also related to techniques that ensure the effective recycling or disposal of a product at the end of its life. Therefore, ensuring an efficient materials selection process early in product development is crucial.

To ensure proper modeling, it is essential to adopt key assumptions. These assumptions were based on the authors' previous research: the minimum number of criteria analyzed is three [13]. It is beneficial to consider at least five to nine production alternatives [14]. The proposed material modifications should be optimally more favorable than the current ones [15]. As a result, the model can function in five main stages. These stages are presented in the next section of the article.

Results

Based on the specific research concept and considering the adopted assumptions, a modeling framework was developed. It applies to modeling material selection in product improvement, considering quality (satisfaction with use), innovation, and environmental impact throughout the life cycle. The framework is presented in Fig.2.

A synthetic description of the model is presented. Stage (1) – selecting a product for analysis – involves determining the product to be improved. Any product can be analyzed. The choice depends on the entity that will use the proposed model. This can be based on sales trends or life cycle analysis, where these products are in the maturity phase. Stage (2) – identifying material criteria from the quality, innovation, and environmental impact groups. Criteria closely related to customer satisfaction, environmental impact, and achieving product innovations should be identified. These criteria are determined, for example, during brainstorming sessions (BM) or based on the product catalog. According to the literature, examples of quality criteria [16], environmental criteria [17] and innovation criteria [18] are presented. Typically, approximately 10-15 criteria are selected for a product of medium complexity. Stage (3) – developing alternative material solutions – involves proposing different criteria states, i.e., modifying the current states of these criteria. The states are offered as descriptions or values, or a range of values, expressed in international metric units. These can be parameters above or below the current criterion state. Each selected criterion should be modified, offering up to nine alternative design solutions. Stage (4) validates and evaluates the alternative material solutions in terms of meeting the quality, innovation, and environmental criteria. All criteria are evaluated on a Likert scale or another scale, where the weights of these criteria are determined. Separate weights can be assigned to criteria within the quality, innovation, and environmental aspects. Next, the product prototypes offered (developed in stage 3 of the model) are evaluated, and scores are assigned on a Likert scale or another scale. The quality, innovation, and environmental impact aspects are assessed, considering the states of the analyzed criteria. Stage (5), the analysis of alternative material solutions, involves processing the weighting and aspect feasibility assessments. This can be accomplished using techniques such as MCDM, simplified quality assessment methods, innovation assessment methods, and environmental impact assessment methods, including computationally aided database programs like OpenLCA. The procedure is appropriate to the selected technique, such as [19, 20]. Depending on the needs, it is possible to aggregate the results from the quality, innovation, and environmental assessments, for example, as an arithmetic mean. In such cases, it is necessary to first normalize the results to a single scale, e.g., from 0 to 1. This produces an aggregated decision indicator.

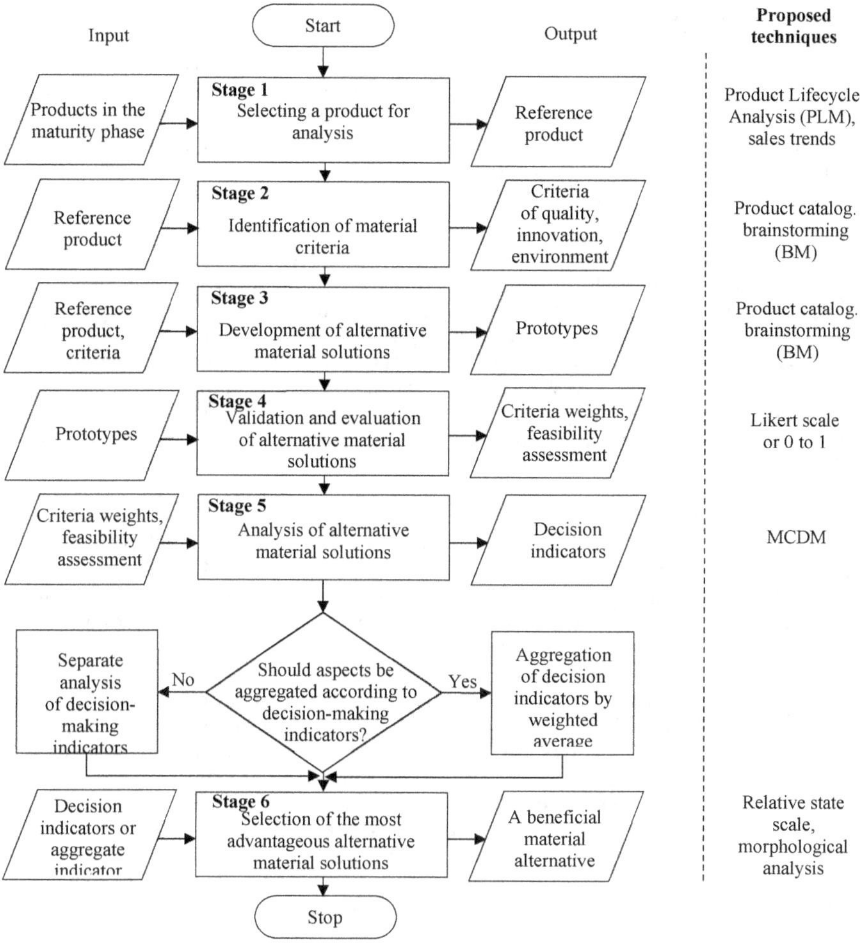

Figure 2. *A modeling scheme for material selection in product improvement considering quality, innovation, and life cycle assessment.*

Stage (6), the selection of the most favorable alternative material solutions, involves analyzing the decision indicator from the previous stage. The higher the indicator, the more favorable the material, considering its qualitative, innovative, and environmental aspects simultaneously. Based on this, it is possible to select the most satisfactory materials that balance these aspects in a single production solution.

Economic and Technological Impact

Product development, approached holistically through all phases, from design concept to final disposal, is analogous to the "robust design" concept proposed by G. Taguchi [21, 22] and subsequently implemented in the engineering education process [23]. By tracing the subsequent stages, the impact of this concept becomes apparent. This approach has a significant impact on management systems [24,25] and quality assurance systems in industry [26-28] and logistics [29]. This has become a development impetus for the development and implementation of special coatings, DLC [30,31] and ESD [32,33], subjected to laser micromachining [34,35], as they significantly increase the durability of the product's surface layer [36-38] and thus reduce the amount of waste [39-41]. Such process chain shaping requires optimization [42], even in cases of incomplete data or knowledge [43, 44]. A similar procedure is evident in the field of biotechnology [45,46], which is also supported by appropriate statistical methods [47,48].

Summary

Addressing the challenges posed by sustainable product development requires clarity of decision-making at every stage of product development (new products) and during their improvement (existing products). Due to the lack of methodologies supporting this process that consider various aspects of sustainability, it was deemed appropriate to propose a methodology for modeling material selection in product improvement, considering quality, innovation, and life cycle assessment. The aim of this article is to propose a modeling approach that supports the material selection process in product improvement, considering aspects of quality, innovation, and environmental impact throughout the product's life cycle. The model operates in six main stages, and during its development, appropriate assumptions were taken into account, along with various supporting techniques, including brainstorming, MCDM, the Likert scale, the relative states scale, and morphological analysis.

The main benefits of the proposed model include supporting the early product development process by providing an in-depth analysis of materials that can be used in the product. Other benefits include facilitating decision-making regarding the selection of materials for final products, considering quality, innovation, and environmental considerations; a straightforward process for evaluating alternative material solutions without the need to create real product prototypes; and paying special attention to key aspects of product sustainability.

However, certain limitations include the potential for subjectivity in expert opinions, which often rely primarily on their knowledge and experience. The model also focuses on selected aspects, where, depending on the industry and product being analyzed, it may be necessary to consider others, such as costs, society, or risk. Another limitation is the omission of the indicator of the impact of selected materials on other stages of the product life cycle. Therefore, future research will focus on expanding material selection modeling to include other aspects of product sustainability. Testing on various product types and generalizing the model are also planned. However, the relatively neutral concept means that the proposed materials selection model can be applied to any product, especially those commonly used. It will find applications, for example, in the electrical machinery industry.

References

[1] M. Ashby, K. Johnson, Materials and Design: The Art and Science of Material Selection in Product Design, Butterworth 2014.

[2] T.A. Saleh, Materials: Types and General Classifications, in: Polymer Hybrid Materials and Nanocomposites, Elsevier (2021) 27–58.

[3] E.A.M. Gómez et al., Exploring a Material-Focused Design Methodology: An Innovative Approach to Pressure Vessel Design, Machines 12 (2024) art.81. https://doi.org/10.3390/machines12010081

[4] T. Li et al., Developing Fibrillated Cellulose as a Sustainable Technological Material, Nature 590 (2021) 47–56. https://doi.org/10.1038/s41586-020-03167-7

[5] S. Sbahieh et al., Comprehensive Analysis of Geopolymer Materials: Properties, Environmental Impacts, and Applications, Materials 16 (2023) art.7363. https://doi.org/10.3390/ma16237363.

[6] M. Korzynski, A. Pacana, Centreless burnishing and influence of its parameters on machining effects. Journal of Materials Processing Technology 210 (2010) 1217-1223. https://doi.org/10.1016/j.jmatprotec.2010.03.008.

[7] J. Gasia et al., Life Cycle Assessment and Life Cycle Costing of an Innovative Component for Refrigeration Units, J. Clean Prod. 295 (2021) art.126442. https://doi.org/10.1016/j.jclepro.2021.126442

[8] V. Chaves, Ecological Criteria for the Selection of Materials in Fatigue, Fatigue Fract. Eng. Mater. Struct. 37 (2014) 1034-1042. https://doi.org/10.1111/ffe.12181

[9] G. Lou et al., Nanometer Montmorillonite Modified Fly Ash Ecological Slope Protection Material and Its Preparation and Application, J. Chem. 2020 (2020) 1-9. https://doi.org/10.1155/2020/6953594

[10] B.A. Ahmed Ali et al., Expert Material Selection for Manufacturing of Green Bio Composites. In: Green Biocomposites (2016) 1-12. https://doi.org/10.1007/978-3-319-46610-1_1

[11] S. Prendeville et al., Material Selection for Eco-Innovation: SPICE Model, J. Clean Prod. 85 (2014) 31-40. https://doi.org/10.1016/j.jclepro.2014.05.023

[12] K.M. Sheridan et al., The Effects of Environment and Ownership on Children's Innovation of Tools and Tool Material Selection. Philosophical Transactions of the Royal Society B: Biological Sciences. 371 (2016) art. 20150191. https://doi.org/10.1098/rstb.2015.0191

[13] D. Siwiec et al., Open Eco-Innovations in Sustainable Product Development: Model Framework of Design Thinking in Quality Life Cycle Assessment (DT-QLCA), J. Open Innov.: Technol. Market Complex. 11 (2025) art. 100480. https://doi.org/10.1016/j.joitmc.2025.100480

[14] D. Siwiec, A. Pacana, A Framework for Total Quality Product in Life Cycle Assessment (TQLCA), Res. Eng. Des. 36 (2025) art.16. https://doi.org/10.1007/s00163-025-00457-x

[15] A. Pacana, D. Siwiec, Method of Determining Sequence Actions of Products Improvement. Materials 15 (2022) art.6321. https://doi.org/10.3390/ma15186321

[16] B. Gajdzik, Frameworks of the Maturity Model for Industry 4.0 with Assessment of Maturity Levels on the Example of the Segment of Steel Enterprises in Poland. J. Open Innov.: Technol. Market Complex. 8 (2022) art. 77. https://doi.org/10.3390/joitmc8020077

[17] A. Pacana et al., Decision support in the production of packaging films by cost-quality analysis. Przemysł Chemiczny 95 (2016) 1042-1044, https://doi.org/10.15199/62.2016.5.31

[18] S. Altuntas et al., Innovative New Product Development: A Case Study. Procedia Comput. Sci. 158 (2019) 214-221. https://doi.org/10.1016/j.procs.2019.09.044

[19] A. Pacana et al., Improving the Process of Product Design in a Phase of Life Cycle Assessment (LCA). Processes 11 (2023) art.2579. https://doi.org/10.3390/pr11092579

[20] D. Siwiec, A. Pacana, A New Model Supporting Stability Quality of Materials and Industrial Products. Materials 15 (2022), art. 4440. https://doi.org/10.3390/ma15134440

[21] J. Pietraszek, A. Goroshko, The heuristic approach to the selection of experimental design, model and valid pre-processing transformation of DoE outcome, Advanced Materials Research 874 (2014) 145-149. https://doi.org/10.4028/www.scientific.net/AMR.874.145

[22] J. Pietraszek et al., Challenges for the DOE methodology related to the introduction of Industry 4.0, Prod. Eng. Arch. 26 (2020) 190-194. https://doi.org/10.30657/pea.2020.26.33

[23] R. Ulewicz, Application of servqual method for evaluation of quality of educational services at the university of higher education, Polish Journal of Management Studies 9 (2014) 254-264.

[24] M. Nowicka-Skowron, R. Ulewicz, Quality management in logistics processes in metal branch, METAL 2015 – 24th Int. Conf. Metall. Mater. (2015) 1707-1712.

[25] R. Ulewicz, F. Nový, Quality management systems in special processes, Transportation Research Procedia 40 (2019) 113-118. https://doi.org/10.1016/j.trpro.2019.07.019

[26] R. Ulewicz, Quality management system operation in the woodworking industry, The Path Forward for Wood Products: A Global Perspective – Proc. Sci. Papers (2016) 51-56.

[27] T. Lipiński, R. Ulewicz, The effect of the impurities spaces on the quality of structural steel working at variable loads, Open Eng. 11 (2021) 233-238. https://doi.org/10.1515/eng-2021-0024

[28] R. Ulewicz et al., Sustainable Vehicle Design Considering Quality Level and Life Cycle Environmental Assessment (LCA), Energies 16 (2023) art. 8122. https://doi.org/10.3390/en16248122

[29] R. Ulewicz et al., Logistic controlling processes and quality issues in a cast iron foundry, Materials Research Proceedings 17 (2020) 65-71. https://doi.org/10.21741/9781644901038-10

[30] N. Radek et al., Formation of coatings with technologies using concentrated energy stream, Production Engineering Archives 28 (2022) 117-122. https://doi.org/10.30657/pea.2022.28.13

[31] N. Radek et al., Operational properties of DLC coatings and their potential application, METAL 2022 – 31st Int. Conf. Metall. Mater., (2022) 531-536. https://doi.org/10.37904/metal.2022.4491

[32] N. Radek, J. Konstanty, Cermet ESD coatings modified by laser treatment, Archives of Metallurgy and Materials 57 (2012) 665-670. https://doi.org/10.2478/v10172-012-0071-y

[33] N. Radek et al., The effect of laser treatment on operational properties of ESD coatings, METAL 2021 – 30th Int. Conf. Metall. Mater. (2021) 876-882. https://doi.org/10.37904/metal.2021.4212

[34] A. Kalinowski et al., Laser surface texturing: characteristics and applications, System Safety: Human-Technical Facility-Environment 5 (2023) 240-248. https://doi.org/10.2478/czoto-2023-0026

[35] Ł.J. Orman et al., Laser Treatment of Surfaces for Pool Boiling Heat Transfer Enhancement, Materials 16 (2023) art. 1365. https://doi.org/10.3390/ma16041365

[36] N. Radek et al., The influence of plasma cutting parameters on the geometric structure of cut surfaces, Mater. Res. Proc. 17 (2020) 132-137. https://doi.org/10.21741/9781644901038-20

[37] A. Dudek et al., Laser Surface Alloying of Sintered Stainless Steel, Materials 15 (2022) art. 6061. https://doi.org/10.3390/ma15176061

[38] A. Kalinowski et al., The effect of laser infrared radiation on the surface modification of DLC coatings, AIP Conference Proceedings 3130 (2024) art. 020014. https://doi.org/10.1063/5.0204985

[39] M. Ulewicz, A. Pietrzak, Properties and structure of concretes doped with production waste of thermoplastic elastomers from the production of car floor mats, Materials 14 (2021) art. 872. https://doi.org/10.3390/ma14040872

[40] E. Skrzypczak-Pietraszek, A.G. Pietraszek, Phytoremediation and Allelopathy as an Element of Sustainable Circular Economy to Prevent Environmental Pollution, Materials Research Proceedings 45 (2024) 140-145. https://doi.org/10.21741/9781644903315-17

[41] A. Pietrzak et al., Application of a Mixture of Fly Ash and Solid Waste from Gas Treatment from Municipal Solid Waste Incineration in Cement Mortar, Materials 18 (2025) art. 481. https://doi.org/10.3390/ma18030481

[42] J. Pietraszek, A. Gadek-Moszczak, The smooth bootstrap approach to the distribution of a shape in the ferritic stainless steel aisi 434l powders, Solid State Phenomena 197 (2013) 162-167. https://doi.org/10.4028/www.scientific.net/SSP.197.162

[43] J. Pietraszek, Fuzzy regression compared to classical experimental design in the case of flywheel assembly, Lecture Notes in Computer Science 7267 LNAI (2012) 310-317. https://doi.org/10.1007/978-3-642-29347-4_36

[44] J. Pietraszek, The modified sequential-binary approach for fuzzy operations on correlated assessments, Lecture Notes in Computer Science 7894 LNAI (2013) 353-364. https://doi.org/10.1007/978-3-642-38658-9_32

[45] E. Skrzypczak-Pietraszek, J. Pietraszek, Chemical profile and seasonal variation of phenolic acid content in bastard balm (Melittis melissophyllum L., Lamiaceae), Journal of Pharmaceutical and Biomedical Analysis 66 (2012) 154-161. https://doi.org/10.1016/j.jpba.2012.03.037

[46] E. Skrzypczak-Pietraszek, J. Pietraszek, Seasonal changes of flavonoid content in Melittis melissophyllum L. (Lamiaceae), Chemistry and Biodiversity 11 (2014) 562-570. https://doi.org/10.1002/cbdv.201300148

[47] J. Pietraszek, E. Skrzypczak-Pietraszek, The optimization of the technological process with the fuzzy regression, Advanced Materials Research 874 (2014) 151-155. https://doi.org/10.4028/www.scientific.net/AMR.874.151

[48] J. Pietraszek, E. Skrzypczak-Pietraszek, The uncertainty and robustness of the principal component analysis as a tool for the dimensionality reduction, Solid State Phenomena 235 (2015) 1-8. https://doi.org/10.4028/www.scientific.net/SSP.235.1

Materials Research Proceedings 62 (2026) 94-101 https://doi.org/10.21741/9781644904015-13

Analysis of Coolant Selection for Machining Processes in the Context of Sustainable Development

Dominika SIWIEC[1,a], Andrzej PACANA[2,b *] and Renata DWORNICKA[3,c]

[1]Rzeszow University of Technology, Faculty of Mechanical Engineering and Aeronautics, al. Powstańców Warszawy 12, 35-959 Rzeszów, Poland, ORCID: 0000-0002-6663-6621

[2]Rzeszow University of Technology, Faculty of Mechanical Engineering and Aeronautics, al. Powstańców Warszawy 12, 35-959 Rzeszów, Poland, ORCID: 0000-0003-1121-6352

[3]Cracov University of Technology, Faculty of Mechanical Engineering, ul. Warszawska 24, 31-155 Kraków, Poland, ORCID: 0000-0002-2979-1614

[a]d.siwiec@prz.edu.pl, [b]app@prz.edu.pl, [c]renata.dwornicka@pk.edu.pl

Keywords: Sustainable Product Development, Decision Support, TOPSIS, Machining, Production Engineering, Mechanical Engineering

Abstract. Sustainable product development is a key activity in the dynamic development of enterprises. For some products, achieving satisfactory quality while limiting negative environmental impact and optimizing production costs remains a significant challenge. An example of such products is cutting fluids. Although widely used, they are relatively expensive and have a negative impact on human health and the environment. Therefore, techniques are sought to support the selection process. Therefore, the aim of this study is to propose a method to support decision-making in product selection, using cutting fluids as an example. The TOPSIS (Technique for Order of Preference by Similarity to Ideal Solution) method was presented and tested. Five cutting fluids were considered and analyzed based on criteria related to quality, environmental impact, and cost. These criteria included: 5% emulsion pH, hand refractometer coefficient, kinematic viscosity at 40°C [mm²/s], density at 15°C [g/ml], capacity (l), purchase price (PLN), and environmental impact. As a result, a ranking of the analyzed products was developed, which allows for selecting a coolant that meets the assumed quality, environmental, and cost requirements. The proposed method can be used to analyze other products as part of their sustainable selection.

Introduction

Manufacturers strive to establish sustainable production methods that address economic and environmental concerns. Various solutions are sought that address not only the final product but also support its production [1]. Among these, lubricants are commonly used in machining operations. However, they are not environmentally beneficial. Achieving sustainable production requires achieving a balance between energy demand, quality, and productivity [2]. Reducing production costs and improving product quality are crucial to increasing product competitiveness [3]. High friction during dry machining causes tool overheating, thus generating wear and chip formation problems. Dry machining poses a challenge for many machining methods [4]. One way to avoid this problem is to use coolants, such as cutting fluids. Although widely used, they are relatively expensive and have a negative impact on human health and the environment. According to [3,5], dry machining is 35% more advantageous in terms of sustainability analyses due to the lack of lubrication or air compressor requirements. Cutting fluids are widely used and developed in metal cutting, including water-based cutting fluids, gaseous cutting fluids, and oil-based cutting fluids. Recent studies, such as [6] have shown that approximately 80% of occupational health problems were caused by skin contact with cutting fluids. Following the concept of sustainable

Materials Research Proceedings 62 (2026) 94-101 https://doi.org/10.21741/9781644904015-13

development, including green manufacturing, researchers have sought methods to help mitigate the risks associated with cutting fluids in various aspects. The main benefits of using cutting fluids during machining include, for example, reducing chemical diffusion, evacuating chips from the rake face, reducing tool wear, reducing energy consumption, and protecting the machined surface from corrosion [7]. Therefore, product quality, performance, and cost depend primarily on the efficiency of machining operations. In addition, various parameter properties influence machining performance, including cutting fluids, cutting tools, machining conditions, and the machine tool itself [8,9]. Therefore, the problem of selecting cutting fluids for machining is complex, and the final decision in this regard is a multi-criteria problem. Therefore, the aim of this study is to propose a method to support decision-making in product selection, using cutting fluids as an example. These decisions will take into account key aspects of product sustainability, such as quality, environmental impact, and cost.

Method

To select a favorable product while taking into account sustainability aspects, a decision-support method is proposed [10]. This is the TOPSIS method (Technique for Order of Preference by Similarity to Ideal Solution) [11], characterized by the analysis of positive and negative distances from the ideal solution. The method analyzes various criteria for different product alternatives. It was assumed that the criteria for selecting products used in production would include aspects of quality, environmental impact, and cost. Quality criteria influence customer satisfaction with the product's usefulness; therefore, they are particularly concerned with product efficiency [12]. Environmental criteria address the impact on the natural environment (including the life and health of employees) [13]. Cost criteria take into account the actual costs of purchasing and/or selling products. A total of up to 10 criteria are selected, covering quality, environment, and cost [14]. A team of experts selects the criteria during a brainstorming session (BM), for example, based on a product catalog or literature review. Various product solutions, so-called alternative product solutions, are analyzed. Similarly to criteria, their number should not exceed 10 for effective comparison [15]. The TOPSIS method is used according to these assumptions. Initially, a decision matrix is created, with alternative product solutions recorded in columns and the criteria for analysis in rows. The offered alternatives are presented according to their actual states (values or verbal descriptions). In the case of criteria presented verbally (e.g., color), the team of experts assigns ratings, for example, on a five-point Likert scale. The states of the criteria represented by the values remain unchanged (there is no need to label them on a numerical scale). Next, vector normalization (1) is performed [16]:

$$\bar{x}_{ij} = \frac{x_{ij}}{\sqrt{\sum_{j=1}^{n} x_{ij}^2}} \tag{1}$$

where: x – state of the i-th criterion according to the j-th alternative, i,j = 1,2,...n.

Then, a normalized decision matrix is obtained. Next, the criteria's weights are considered by dividing them by normalized values. To achieve this, all verified product criteria must be weighted (the importance must be determined) by a team of experts. Weights are assigned on a scale from 0 to 1, where 0 indicates a criterion of no importance (a criterion of little importance), and 1 indicates the absolutely most important criterion. The sum of the weights assigned to all criteria should be 1. This produces a normalized, weighted decision matrix. Next, it is necessary to determine which criteria are benefit criteria and which are cost criteria. Benefit criteria are those whose condition is higher, the more favorable, e.g., efficiency. Cost criteria are those whose condition is lower, the more favorable, e.g., cost. According to these assumptions, minimum and maximum values are determined, depending on the benefit or cost criterion, respectively. Next, the Euclidean distance from the ideal best solution (2) and from the ideal worst solution (3) is estimated [17]:

$$S_i^+ = \left[\sum_{j=1}^{m}(V_{ij} - V_j^+)^2\right]^{0.5} \tag{2}$$

where: V – normalized weighted value of the i-th criterion for the j-th alternative, V^+ – normalized weighted value of benefits according to the i-th criterion for the j-th alternative, i, j = 1,2…, m.

$$S_i^- = \left[\sum_{j=1}^{m}(V_{ij} - V_j^-)^2\right]^{0.5} \tag{3}$$

where: V – normalized weighted value of the i-th criterion for the j-th alternative, V^- – normalized weighted value of benefits according to the i-th criterion for the j-th alternative, i, j = 1,2…, m.

Finally, the performance index is estimated as in formula (4) [18]:

$$P_i - \frac{S_i^-}{S_i^+ + S_i^-} \tag{4}$$

where: S^+ - distance from the ideal positive solution, S^- - distance from the ideal negative solution, i = 1,2,…n.

A product ranking is developed based on the P index. The higher the value, the better the product is in terms of quality, the environment, and cost. Based on this, a product is selected that simultaneously meets usability requirements, is environmentally friendly, and is also cost-effective. This approach promotes sustainable business development. Therefore, the final decision regarding the preferred product depends on the entity using the proposed method.

Results

Five cutting fluids were analyzed. These products are water-soluble metalworking fluids that are free of mineral oil and boron. They form a clear, colorless solution upon dilution with water. They provide effective cooling, wetting, and rinsing during machining processes. These products were characterized based on six quality, cost, and environmental criteria: pH (5% emulsion), handheld refractometer index, kinematic viscosity at 40°C [mm²/s], density at 15°C [g/ml], capacity (l), purchase price (PLN), and environmental impact. The cutting fluids were selected from those available on the market and characterized based on a publicly available product catalog, as shown in Table 1.

Table 1. *Characteristics of selected cutting fluids.*

Products	C1	C2	C3	C4	C5	C6	C7
P1	9.50	1.00	52.00	0.89	4.00	127.00	Medium
P2	9.60	2.10	45.00	1.01	20.00	479.00	Medium
P3	9.70	2.40	15.40	0.85	20.00	507.00	Large
P4	9.40	1.60	20.00	1.01	20.00	737.00	Medium
P5	9.60	2.10	35.00	1.01	20.00	954.00	Large

where: P1-P5 – cutting fluids, C1 – emulsion pH 5%, C2 – hand refractometer coefficient, C3 – kinematic viscosity at 40°C [mm²/s], C4 – density at 15°C [g/ml], C5 – capacity (l), C6 – purchase price (PLN), C7 – environmental impact.

The assessment was conducted using the TOPSIS method. Labels were assigned to the environmental impact criterion, where a medium impact is a score of 3, and a high impact is a score of 5. Based on the adopted values, vector normalization was performed according to formula (1). Next, the criterion weights were considered, with the most important being density at 15°C. Weights were assigned on a scale of 0 to 1, where the sum of these weights was reduced to 1. A normalized weighted decision matrix was developed by estimating the quotient of the criterion weights and the vector normalization values. Next, the benefit and cost criteria were estimated, and based on the adopted assumptions, the minimum and maximum values were adopted from the considered values. The results of these activities are presented in Table 2.

Table 2. *Results from the evaluation of cutting fluids using the TOPSIS method.*

Products/Criteria	**C1**	**C2**	**C3**	**C4**	**C5**	**C6**	**C7**
Weights	0.20	0.10	0.10	0.25	0.10	0.15	0.10
Benefit/Cost	B	B	NB	NB	B	NB	NB
P1	9.50	1.00	52.00	0.89	4.00	127.00	3.00
P2	9.60	2.10	45.00	1.01	20.00	479.00	3.00
P3	9.70	2.40	15.40	0.85	20.00	507.00	5.00
P4	9.40	1.60	20.00	1.01	20.00	737.00	3.00
P5	9.60	2.10	35.00	1.01	20.00	954.00	5.00
V^{+}	0.09	0.06	0.02	0.10	0.05	0.01	0.03
V^{-}	0.09	0.02	0.06	0.12	0.01	0.10	0.06

where: P1-P5 – cutting fluids, C1 – emulsion pH 5%, C2 – hand refractometer coefficient, C3 – kinematic viscosity at 40°C [mm²/s], C4 – density at 15°C [g/ml], C5 – capacity (l), C6 – purchase price (PLN), C7 – environmental impact, B – beneficial, NB – non-beneficial (cost).

Next, the Euclidean distance from the ideal best solution and from the ideal worst solution was estimated. Formulas (2-3) were used to estimate the performance index, as in formula (4). A product ranking was then created based on this estimate. The results are presented in Table 3.

Table 3. *Euclidean Distances and Performance Index with Ranking of Cutting Coolants.*

Products	S^{+}	S^{-}	**P**	**Ranking**
P1	0.07	0.09	0.57	2
P2	0.06	0.07	0.57	2
P3	0.05	0.09	0.65	1
P4	0.07	0.07	0.48	3
P5	0.10	0.05	0.35	4

where: P1-P5 – cutting fluids.

Based on the analysis of cutting fluids using the TOPSIS method, it was observed that the most favorable product is P3. Its index reached the maximum value (0.65). Less favorable are P1 and P2 (0.57), followed by P4 (0.48). The least preferred cutting fluid is P5 (0.35), which achieved a lower decision index. According to the proposed methodology, the choice should be based on the coolant conventionally designated P3. However, the final decision may differ, for example, due to the availability of company resources.

Economic and Technological Impact

Sustainable product development guided by the TOPSIS method is analogous to the suggestion once proposed by G. Taguchi within the "robust design" concept [19,20] and de facto implemented in the design education of engineers [21]. Following the subsequent stages of improvement, one can observe the impact of this concept, which has a very significant impact on management systems [22,23] and quality assurance systems in industry [24-26] and logistics [27]. This kind of product improvement, through the constant pursuit of an otherwise unattainable ideal, became the impetus for the development and implementation of special coatings, DLC [28,29] and ESD [30,31], refined by laser micromachining [32,33], which significantly increases the durability of the product's surface layer [34-36]. By extending its service life, it also reduces the amount of waste [37-39]. Such process chain shaping requires continuous optimization [40], even in the case of incomplete data or knowledge [41,42]. A similar procedure is visible in the field of biotechnology [43,44] also supported by appropriate statistical methods [45,46].

Summary

The aim of this study was to propose a method to support decision-making in product selection, using cutting fluids as an example. This method, the TOPSIS method, takes into account various criteria, each characterized by its own set of parameters. Five cutting fluids were analyzed, randomly selected based on an analysis of their availability on the market. They were characterized based on six main sustainability criteria. Using the proposed method, a ranking of cutting fluids was developed. Product P3 (0.65) proved to be the most favorable in this analysis. Although it has a significant negative environmental impact, its other criteria were found to be overwhelmingly more favorable compared to other products. This is due to the fact that environmental impact was considered a minor aspect. If different assumptions are made, the analysis results may differ. The proposed TOPSIS method was found to be effective in solving the problem of selecting coolants for machining. It may also be applicable to decision-making problems involving other products.

References

[1] U.S. Dixit, D. K. Sarma, J.P. Davim, Environmentally Friendly Machining (2012). Springer US. https://doi.org/10.1007/978-1-4614-2308-9

[2] P. Kumar et al., Sustainable Machining Using Eco-Friendly Cutting Fluids: A Review. Adv. Mater. Sci. Eng. 2022 (2022) 1-16. https://doi.org/10.1155/2022/5284471

[3] K. Hassan, Comparative life cycle analysis of environmental and machining performance under sustainable lubrication techniques, Hybrid Advances 1 (2022) art. 100004. https://doi.org/10.1016/j.hybadv.2022.100004

[4] G.M. Minquiz et al., Sustainable assessment of a milling manufacturing process based on economic tool life and energy modeling, Journal of the Brazilian Society of Mechanical Sciences and Engineering 45(7) (2023) art. 365. https://doi.org/10.1007/s40430-023-04189-8

[5] H.A. Kishawy et al., Sustainability Assessment during Machining Ti-6Al-4V with Nano-Additives-Based Minimum Quantity Lubrication, Journal of Manufacturing and Materials Processing 3 (2019) art. 61. https://doi.org/10.3390/jmmp3030061

[6] X. Luo et al., Sustainable development of cutting fluids: The comprehensive review of vegetable oil. J. Clean. Prod. 473 (2024) 143544. https://doi.org/10.1016/j.jclepro.2024.143544

[7] A. Kumar et al., State-of-the-Art in Sustainable Machining of Different Materials Using Nano Minimum Quality Lubrication (NMQL), Lubricants 11 (2023) art. 64. https://doi.org/10.3390/lubricants11020064

[8] M. Korzynski et al., Fatigue strength of chromium coated elements and possibility of its improvement with ball peening. Surf. Coat. Technol. 204 (2009) 615-620, https://doi.org/10.1016/j.surfcoat.2009.08.049.

[9] F.Z. El Abdelaoui et al., Optimization techniques for energy efficiency in machining processes—a review, Int. J. Adv. Manuf. Technol. 125 (2023) 2967-3001. https://doi.org/10.1007/s00170-023-10927-y

[10] A. Pacana et al., Effect of selected factors of the production process of stretch film for its resistance to puncture. Przemysł Chemiczny 93 (2014) 2263-2264.

[11] M. Bertolini et al., A TOPSIS-based approach for the best match between manufacturing technologies and product specifications, Expert Systems with App. 159 (2020) art. 113610. https://doi.org/10.1016/j.eswa.2020.113610

[12] D. Siwiec, A. Pacana, A New Model Supporting Stability Quality of Materials and Industrial Products. Materials 15 (2022) art. 4440. https://doi.org/10.3390/ma15134440

[13] A. Pacana, D. Siwiec, Method of Determining Sequence Actions of Products Improvement. Materials 15 (2022) art.6321. https://doi.org/10.3390/ma15186321

[14] D. Siwiec, A. Pacana. A framework for total quality product in life cycle assessment (TQLCA), Res. Eng. Design 36 (2025) art.16. https://doi.org/10.1007/s00163-025-00457-x

[15] A. Pacana et al., Analysis of the incompatibility of the product with fluorescent method, Metalurgija 58 (2019) 337-340.

[16] A. Zadeh Sarraf et al., Developing TOPSIS method using statistical normalization for selecting knowledge management strategies, Journal of Industrial Engineering and Management 6 (2013) 860-875. https://doi.org/10.3926/jiem.573

[17] C.-T. Chen, Extensions of the TOPSIS for group decision-making under fuzzy environment. Fuzzy Sets and Systems 114 (2000) 1-9. https://doi.org/10.1016/S0165-0114(97)00377-1

[18] S. Vinodh et al., Sustainable concept selection using modified fuzzy TOPSIS: a case study, Int. J. Sust. Eng. 6 (2013) 109-116. https://doi.org/10.1080/19397038.2012.682100

[19] J. Pietraszek, A. Goroshko, The heuristic approach to the selection of experimental design, model and valid pre-processing transformation of DoE outcome, Advanced Materials Research 874 (2014) 145-149. https://doi.org/10.4028/www.scientific.net/AMR.874.145

[20] J. Pietraszek et al., Challenges for the DOE methodology related to the introduction of Industry 4.0, Prod. Eng. Arch. 26 (2020) 190-194. https://doi.org/10.30657/pea.2020.26.33

[21] R. Ulewicz, Application of servqual method for evaluation of quality of educational services at the university of higher education, Polish Journal of Management Studies 9 (2014) 254-264.

[22] M. Nowicka-Skowron, R. Ulewicz, Quality management in logistics processes in metal branch, METAL 2015 – 24th Int. Conf. Metall. Mater. (2015) 1707-1712.

[23] R. Ulewicz, F. Nový, Quality management systems in special processes, Transportation Research Procedia 40 (2019) 113-118. https://doi.org/10.1016/j.trpro.2019.07.019

[24] R. Ulewicz, Quality management system operation in the woodworking industry, The Path Forward for Wood Products: A Global Perspective – Proc. Sci. Papers (2016) 51-56.

[25] T. Lipiński, R. Ulewicz, The effect of the impurities spaces on the quality of structural steel working at variable loads, Open Eng. 11 (2021) 233-238. https://doi.org/10.1515/eng-2021-0024

[26] R. Ulewicz et al., Sustainable Vehicle Design Considering Quality Level and Life Cycle Environmental Assessment (LCA), Energies 16 (2023) art. 8122. https://doi.org/10.3390/en16248122

[27] R. Ulewicz et al., Logistic controlling processes and quality issues in a cast iron foundry, Materials Research Proceedings 17 (2020) 65-71. https://doi.org/10.21741/9781644901038-10

[28] N. Radek et al., Formation of coatings with technologies using concentrated energy stream, Prod. Eng. Arch. 28 (2022) 117-122. https://doi.org/10.30657/pea.2022.28.13

[29] N. Radek et al., Operational properties of DLC coatings and their potential application, METAL 2022 – 31st Int. Conf. Metall. Mater., (2022) 531-536. https://doi.org/10.37904/metal.2022.4491

[30] N. Radek, J. Konstanty, Cermet ESD coatings modified by laser treatment, Archives of Metallurgy and Materials 57 (2012) 665-670. https://doi.org/10.2478/v10172-012-0071-y

[31] N. Radek et al., The effect of laser treatment on operational properties of ESD coatings, METAL 2021 – 30th Int. Conf. Metall. Mater. (2021) 876-882. https://doi.org/10.37904/metal.2021.4212

[32] A. Kalinowski et al., Laser surface texturing: characteristics and applications, System Safety: Human-Technical Facility-Environment 5 (2023) 240-248. https://doi.org/10.2478/czoto-2023-0026

[33] J. J. Orman et al., Laser Treatment of Surfaces for Pool Boiling Heat Transfer Enhancement, Materials 16 (2023) art. 1365. https://doi.org/10.3390/ma16041365

[34] N. Radek et al., The influence of plasma cutting parameters on the geometric structure of cut surfaces, Mater. Res. Proc. 17 (2020) 132-137. https://doi.org/10.21741/9781644901038-20

[35] A. Dudek et al., Laser Surface Alloying of Sintered Stainless Steel, Materials 15 (2022) art. 6061. https://doi.org/10.3390/ma15176061

[36] A. Kalinowski et al., The effect of laser infrared radiation on the surface modification of DLC coatings, AIP Conf. Proc. 3130 (2024) art. 020014. https://doi.org/10.1063/5.0204985

[37] M. Ulewicz, A. Pietrzak, Properties and structure of concretes doped with production waste of thermoplastic elastomers from the production of car floor mats, Materials 14 (2021) art. 872. https://doi.org/10.3390/ma14040872

[38] E. Skrzypczak-Pietraszek, A.G. Pietraszek, Phytoremediation and Allelopathy as an Element of Sustainable Circular Economy to Prevent Environmental Pollution, Materials Research Proceedings 45 (2024) 140-145. https://doi.org/10.21741/9781644903315-17

[39] A. Pietrzak et al., Application of a Mixture of Fly Ash and Solid Waste from Gas Treatment from Municipal Solid Waste Incineration in Cement Mortar, Materials 18 (2025) art. 481. https://doi.org/10.3390/ma18030481

[40] J. Pietraszek, A. Gadek-Moszczak, The smooth bootstrap approach to the distribution of a shape in the ferritic stainless steel AISI 434L powders, Solid State Phenomena 197 (2013) 162-167. https://doi.org/10.4028/www.scientific.net/SSP.197.162

[41] J. Pietraszek, Fuzzy regression compared to classical experimental design in the case of flywheel assembly, Lecture Notes in Computer Science 7267 LNAI (2012) 310-317. https://doi.org/10.1007/978-3-642-29347-4_36

[42] J. Pietraszek, The modified sequential-binary approach for fuzzy operations on correlated assessments, Lecture Notes in Computer Science 7894 LNAI (2013) 353-364. https://doi.org/10.1007/978-3-642-38658-9_32

[43] E. Skrzypczak-Pietraszek, J. Pietraszek, Chemical profile and seasonal variation of phenolic acid content in bastard balm (Melittis melissophyllum L., Lamiaceae), Journal of Pharmaceutical and Biomedical Analysis 66 (2012) 154-161. https://doi.org/10.1016/j.jpba.2012.03.037

[44] E. Skrzypczak-Pietraszek, J. Pietraszek, Seasonal changes of flavonoid content in Melittis melissophyllum L. (Lamiaceae), Chemistry and Biodiversity 11 (2014) 562-570. https://doi.org/10.1002/cbdv.201300148

[45] J. Pietraszek, E. Skrzypczak-Pietraszek, The optimization of the technological process with the fuzzy regression, Advanced Materials Research 874 (2014) 151-155. https://doi.org/10.4028/www.scientific.net/AMR.874.151

[46] J. Pietraszek, E. Skrzypczak-Pietraszek, The uncertainty and robustness of the principal component analysis as a tool for the dimensionality reduction, Solid State Phenomena 235 (2015) 1-8. https://doi.org/10.4028/www.scientific.net/SSP.235.1

Materials Research Proceedings 62 (2026) 102-110 https://doi.org/10.21741/9781644904015-14

Analysis of the Adaptation of Agile Methods in Management in the Foundry Industry

PACANA Andrzej[1,a *], CZERWIŃSKA Karolina[2,b] and Renata DWORNICKA[3,c]

[1]Rzeszow University of Technology, The Faculty of Mechanical Engineering and Aeronautics al. Powstańców Warszawy 8, 35-959 Rzeszów, Poland, ORCID: 0000-0003-1121-6352

[2]Rzeszow University of Technology, The Faculty of Mechanical Engineering and Aeronautics al. Powstańców Warszawy 8, 35-959 Rzeszów, Poland, ORCID: 000-0003-2150-0963

[3]Cracow University of Technology, Faculty of Mechanical Engineering, ul. Warszawska 24, 31-155 Kraków, Poland, ORCID: 0000-0002-2979-1614

[a]app@prz.edu.pl, [b]k.czerwinska@prz.edu.pl, [c]renata.dwornicka@pk.edu.pl

Keywords: Agile Manufacturing, Production Management, Foundry Industry, Manufacturing Flexibility, Operational Efficiency, Mechanical Engineering, Empirical Study

Abstract. In view of the increased volatility of the environment, the issue of agile management in foundry companies plays an important role, based on flexible procedures and adaptive strategic and organizational solutions. The study aimed to analyze development trends and assess the level of knowledge of agile methods, techniques, and tools in the management field of foundries. The objective also included identifying areas of foundry companies' operations where the implementation of an agile approach could increase their efficiency, with particular emphasis on project and production process management. Knowledge of the Agile approach and tools in the foundries surveyed was relatively low, both in project management and manufacturing process management. The knowledge of the agile approach is most often used in the area of project management and manufacturing process improvement. PDCA and DMAIC were most often used to support improvement in the project area, while PDCA and Kanban were used to improve the efficiency of manufacturing processes. Future research will focus on further analysis of the implementation of the Agile approach to enhancing management in the context of digital transformation in foundries.

Introduction

Currently, industrial companies operate in highly competitive markets, where growth and achieving a leading position depend on whether the company can meet the requirements and operating conditions imposed upon it and whether it can adapt immediately to changes [1, 2]. In today's market economy, project and process management have evolved, and the ability to efficiently optimize and improve processes is a testament to a company's strength. These elements are components of change management in an industrial enterprise [3].

Industrial companies must seek ways to create and maintain a competitive advantage to keep pace with a market characterized by growing potential. Competitive advantage is increasingly influenced not only by mature management, experience, knowledge, and innovation, but mainly by the ability to respond to buyer expectations and market requirements, which is the result of effective and efficient implementation of changes, and especially understanding the need to implement them [4-6]. In the change management process, managers are required to make decisions about the subject of change (i.e., determining what needs to be changed) first, and then take actions to overcome employee resistance during the implementation of changes [7].

Metal casting, one of the earliest forms of manufacturing, is currently transforming the context of advanced technology applications and Industry 4.0 solutions, such as automation and robotization, digital simulation, and computer-aided design (CAD), which promote increased quality, efficiency, and precision in processes and reduce manufacturing costs. Despite dynamic technological development, this sector remains relatively resistant to innovation in management [8-10]. The specific nature of the foundry industry, characterized by the significant complexity of technological processes, high capital intensity, and the need to ensure production continuity despite high order volatility, further hinders the implementation of flexible management methods [11, 12]. In this respect, project management, a crucial area of organizational improvement, becomes particularly significant. The waterfall approach, established in project management practice since the 1970s, is associated with a low level of flexibility and limited ability to adapt to changing market conditions. An alternative to this management practice is the agile approach, which is based on delivering incremental value, ongoing performance monitoring, and increased communication between team members and stakeholders [13-15]. This approach is based on cooperation and self-organization rather than rigid procedures, which contributes to increased work efficiency and product quality. An important developmental stage in the concept of agile management was the emergence of the Scrum method, which provides a framework for addressing complex problems and producing products that meet customer requirements [16,17].

The literature on the subject indicates a growing interest in agility in the management of manufacturing companies, as well as the benefits of its implementation [18-20]. The agile approach to manufacturing, known as agile manufacturing, focuses on flexibility, rapid response to change, and the integration of technologies that enable the adaptation of production processes to individual customer needs [21-23]. The key elements of agile management in manufacturing include: leadership support, human resource management, customer relationship management, and technology management [24-27]. However, there is a lack of studies on the adaptation of the agile approach in industrial enterprises and the benefits it offers.

The study aimed to analyze development trends and assess the level of knowledge of agile methods, techniques, and tools in the management area of foundry companies. The objective also included identifying areas of foundry companies' operations where the implementation of an agile approach could increase their efficiency, with particular emphasis on project and production process management. The study is also intended to contribute to the development of a cognitive framework for the further analysis and practical implementation of Agile in the foundry industry, within the context of specific industry conditions.

Methodology of Research

The research presented in this study was conducted as a preliminary pilot study. This type of research was conducted to gain preliminary knowledge about the implications of agile methods and the specifics of the analyzed community. Due to the pilot nature of the study, the research sample consisted of 50 companies from the foundry industry, which were selected at random. The study covered companies with headquarters located in southeastern Poland. The research was conducted in accordance with the mixed-methods research concept, integrating both qualitative and quantitative approaches. The use of methodological triangulation aimed to increase the reliability of the research, enhance the accuracy of conclusions, and reduce the potential errors resulting from the use of only one method. The research procedure involved content analysis, logical classification, exemplification, and multiple case studies. In accordance with the adopted methodological assumptions, the following research questions were formulated:

- To what extent do the management and employees of foundries have the knowledge and skills to apply agile management methods, techniques, and tools?
- To what extent do foundries use agile methods, techniques, and tools in project portfolio and production process management?

Materials Research Proceedings 62 (2026) 102-110 https://doi.org/10.21741/9781644904015-14

The study was conducted in the first and second quarters of 2025. Data was obtained from: survey forms, in-depth interviews with management personnel, and information from foundries in the form of authorized statements and internal documents.

Results

Manufacturing companies in the foundry industry striving to achieve a leading position and agility should have the ability to respond quickly to changing market requirements, customer expectations, and opportunities and threats by flexibly adapting their technological operations, production processes, and business relationships. In this respect, agility is a key adaptive mechanism that enables effective adaptation to changes in both internal and external environments. For this reason, there is a need to assess the level of knowledge of management and employees in the field of agile methods and the degree of their implementation in foundry companies, which will allow the identification of areas where the implementation of these methods brings the greatest benefits. An analysis of the current level of knowledge of agile methods, tools, and techniques in foundries was conducted. The results are illustrated in Fig.1.

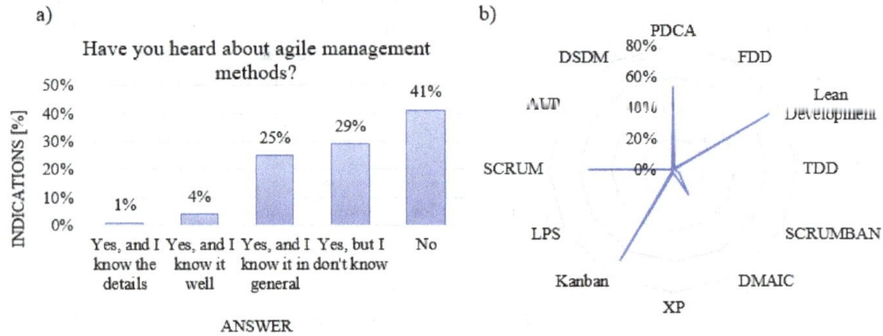

Fig. 1. Knowledge of agile management: a) overall, b) specific methods, techniques, and tools. (source: own study)

As shown in Fig. 1a, the concept of an agile approach proved to be completely unfamiliar to just over 40% of the respondents. At the same time, almost one-third of respondents admitted to having heard of this concept and its methods, techniques, and tools, yet declared that they were unaware of them. One in four respondents reported having a good understanding of this concept. Those who declared a very good knowledge of this approach are in a clear minority (1%). Among the most recognizable methods, tools, and techniques (albeit with a relatively low level of recognition) were Lean Development, Kanban, PDCA, and SCRUM (Fig.1b). These methods are also the best known and most widely used in the world, thanks to their practical effectiveness, versatility, simplicity of use, and wide adaptation in various sectors of industry and services. Their level of familiarity with the surveyed companies is consistent with the generally low level of knowledge of agile methods.

The study also included an assessment of the current level of use of agile methods, tools, and techniques in foundries in the context of project management (Figure 2a) and manufacturing process improvement (Figure 2b).

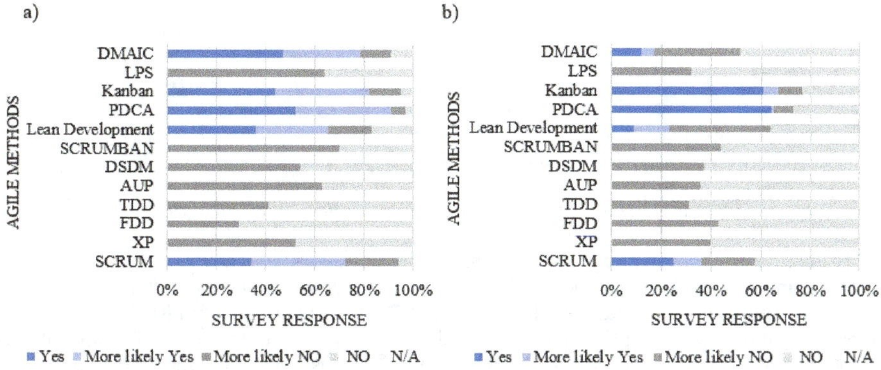

Fig. 2. Degree of implementation of agile methods, tools, and techniques in foundry companies in the management of a) projects, b) production processes. Source: own study.

The interpretation of the results illustrated in Figure 2a) suggests that the use of agile management tools in foundries for project management was at a very low level. In most cases, respondents answered "more likely no" or "no." Of the 12 agile methods, tools, and techniques analyzed, companies indicated the use of five: DMAIC, Kanban, PDCA, Lean Development, and Scrum. Although respondents indicated Lean Development as the most familiar method, it scored lower in terms of applicability than methods such as DMAIC, Kanban, and PDCA.

In the context of adapting agile tools to improve manufacturing processes, the study's results indicate the same set of methods used in project management. This may be due to the relatively low level of knowledge about available agile methods, techniques, and tools, which limits development opportunities at this level. Methods such as PDCA, Kanban, SCRUM, and Lean Development are the foundation of agile management. These methods enable quick response to changes, improved process efficiency, and better communication within teams.

Respondents also assessed the benefits of implementing the Agile approach in foundry companies. The following benefits were assessed in the context of project management: shorter implementation cycles (P1), increased stakeholder interest (P2), increased organizational flexibility (P3), improved communication and cooperation (P4), and better risk management (P5). In the context of manufacturing process improvement, respondents assessed the importance of benefits such as: improved process and product quality (W1), optimal use of resources (W2), stimulation of innovation (W3), better adaptation to technological changes (Industry 4.0) (W4), and increased operational efficiency (W5). The result is shown in Figure 3.

Among the benefits of project management, respondents attached the greatest importance to communication and teamwork, which are essential for the effective application of Agile in foundry companies. In the area of manufacturing processes, however, the key benefit was found to be improved process and product quality. Iterative improvement of foundry processes reduces the number of defects and complaints, bringing additional benefits.

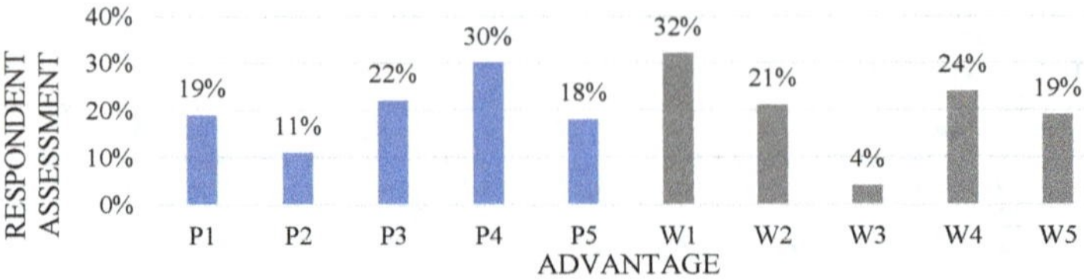

Fig. 3. Assessment of the benefits achieved from implementing the Agile approach

The use of agile management tools in foundries is still a relatively new phenomenon, but one that is gaining in importance due to the need for continuous development and improvement, flexibility, and innovation. Basic methods such as PDCA, Lean, Scrum, and Kanban support the improvement of management processes, particularly those related to the design and streamlining of production processes. Their use brings several benefits, including increased operational flexibility, shorter project cycles, and enhanced communication and teamwork. However, to fully exploit the potential of the Agile approach, it is necessary to raise awareness and further adapt it to the specific conditions of foundry companies.

Economic and Technological Impact

Adapting agile methodology in foundry management [28,29] is a promising development path that can also be effective in related energy-intensive industries such as the power industry [30,31]. Positive effects should also be achievable in the machinery industry, including in the processes of refining surface layers [32] using DLC coatings [33,34] modified by laser micromachining [35,36]. Appropriate optimization of the mentioned processes using the DOE methodology [37,38] supported by resampling techniques [39] and the fuzzy approach [40,41] significantly improves efficiency [42] and increases safety [43,44]. Thus, in subsequent stages, thanks to increased product durability, industrial waste is reduced [45] and environmental burden is reduced [46]. Agile methodologies must be introduced at the initial stage of engineering education [47] and integrated, along with other advanced optimization techniques [48-50], into engineering practice [51].

Summary

The study aimed to analyze development trends and assess the level of knowledge of agile methods, techniques, and tools in the management area of foundry companies. The objective also included identifying areas of foundry companies' operations where the implementation of an agile approach could increase their efficiency, with particular emphasis on project and production process management.

The study provided answers to the research questions adopted in the study. Knowledge of the Agile approach and tools in the surveyed companies was relatively low, both in project management and manufacturing process management. The concept of an agile approach proved to be completely unfamiliar to slightly more than 40% of respondents. Among the recognizable methods, tools, and techniques, the most well-known were Lean Development, Kanban, PDCA, and SCRUM, which are the basic set of agile methods. Knowledge of the agile approach is most commonly applied in the areas of project management and manufacturing process improvement. PDCA and DMAIC were most often used to support improvement in the design area, while PDCA and Kanban were used to improve the efficiency of manufacturing processes.

The research conducted indicates that without a significant increase in knowledge in the field of agile project and manufacturing process management, as well as the application of specific methods in Polish companies, it will be difficult to improve the effectiveness of projects implemented in a dynamically developing environment.

Future research will focus on further analysis of the implementation of the Agile approach to enhancing management in the context of digital transformation within foundry companies. The research will concentrate on the issue of synergy and interdependence between the Agile approach and the implementation of tools characteristic of Industry 4.0.

References

[1] T. Dimitrova et al., Strategic innovative management and sustainable development of industrial enterprises, Economics Ecology Socium 9 (2025) 77-89. https://doi.org/10.61954/2616-7107/2025.9.1-7

[2] L. Filipishyna et al., Strategic management of innovative development of an industrial enterprise, Baltic Journal of Economic Studies 11 (2025) 269-275. https://doi.org/10.30525/2256-0742/2025-11-4-269-275

[3] A. Pacana, K. Czerwińska, Model of diagnosing and searching for incompatibilities in aluminium castings, Materials 14 (2021) art. 6497. https://doi.org/10.3390/ma14216497

[4] M. Dubyna et al., Innovative development and investment advancement of industrial enterprises in deriving conditions of digital economy, Pacific Business Rev. Int. 17 (2024) 40-49.

[5] I. Klosok-Bazan et al., Environmental aspects of innovation and new technology implementation in metallurgy industry, Metalurgija 54 (2015) 433-436.

[6] NV. Shandova, The preventive analysis of risk factors of the development of industrial enterprises, Marketing and Management of Innovations 1 (2018) 317-326. https://doi.org/10.21272/mmi.2018.1-24

[7] K. Czerwińska, A. Pacana, Analysis of the maturity of process monitoring in manufacturing companies, Prod. Eng. Arch. 28 (2022) 246-251. https://doi.org/10.30657/pea.2022.28.30

[8] RR. Ward, C. Monroe, Assessing Process Control in the Foundry, International Journal of Metalcasting 18 (2024) 2048-2058. https://doi.org/10.1007/s40962-024-01375-9

[9] K. Knop, P. Gejdos, Analysis of the impact of modernization of machinery on the quality of castings using quality management tools, Manuf. Technol. 24 (2024) 207-218. https://doi.org/10.21062/mft.2024.036

[10] A. Pacana, K. Czerwińska, Analysis of the causes of control panel inconsistencies in the gravitational casting process by means of quality management instruments, Prod. Eng. Arch. 25 (2019) 12-16. https://doi.org/10.30657/pea.2019.25.03

[11] J.E. Sordan et al., Lean manufacturing for reducing lead time in foundry processes: a design science approach, Int. J. Lean Six Sigma 16 (2025) 328-345. https://doi.org/10.1108/IJLSS-01-2024-0018

[12] RG. Cooper, P. Furst, Agile development in manufacturing companies: best practices and pitfalls, IEEE Eng. Manag. Rev. 51 (2023) 65-76. https://doi.org/10.1109/EMR.2023.3304792

[13] A. Vanichchinchai, The effects of the Toyota Way on agile manufacturing: an empirical analysis, J. Manuf. Technol. Manag. 33 (2022) 1450-1472. https://doi.org/10.1108/JMTM-02-2022-0053

[14] T. Huikkola, M. Kohtamaki, Agile New Solution Development in Manufacturing Companies, Technol. Innov. Manag. Rev. 10 (2020) 16-23. https://doi.org/10.22215/timreview/1333

[15] Sy. Wu et al., Agile strategy adaptation in semiconductor wafer foundries: an example from Taiwan, Technological Forecasting and Social Change 73 (2006) 436-451. https://doi.org/10.1016/j.techfore.2005.01.004

[16] A. Pacana et al., Decision support in the production of packaging films by cost-quality analysis. Przemysł Chemiczny 95 (2016) 1042-1044. https://doi.org/10.15199/62.2016.5.31

[17] E. Michels, MGG. Ferreira, Agile management in development of innovative products: a systematic literature review, Revista de Gestao e Projetos 4 (2013) 52-76. https://doi.org/10.5585/gep.v4i1.119

[18] D. Banas, H.H. Chovanova, Agile maturity assessment for industrial companies in slovakia: enhancing project management efficiency, MM Science Journal 2025 (2025) 8253-8261. https://doi.org/10.17973/MMSJ.2025_06_2025017

[19] M. Korzynski et al., Fatigue strength of chromium coated elements and possibility of its improvement with ball peening. Surface & Coatings Technology 204 (2009) 615-620. https://doi.org/10.1016/j.surfcoat.2009.08.049

[20] X. Sigma et al., Agility-based knowledge management method for small and medium-sized enterprises, Knowledge Management Research & Practice 22 (2024) 234-246. https://doi.org/10.1080/14778238.2023.2192428

[21] M. Leite, V. Braz, Agile manufacturing practices for new product development: industrial case studies, J. Manuf. Technol. Manag. 27 (2016) 560-576. https://doi.org/10.1108/JMTM-09-2015-0073

[22] S.A.M. Elmoselhy, Hybrid lean-agile manufacturing system technical facet, in automotive sector, J. Manuf. Syst. 32 (2013) 598-619. https://doi.org/10.1016/j.jmsy.2013.05.011

[23] M. Yli-Ojanpera et al., Adapting an agile manufacturing concept to the reference architecture model Industry 4.0: A survey and case study, J. Industr. Inform. Integr., 15 (2019) 147-160. https://doi.org/10.1016/j.jii.2018.12.002

[24] A. Pacana et al., Effect of selected factors of the production process of stretch film for its resistance to puncture. Przemysl Chemiczny 93 (2014) 2263-2264.

[25] B.G. Tavares et al., Practices to improve risk management in agile projects, International J. Soft. Eng. Knowl. Eng. 29 (2019) 381-399. https://doi.org/10.1142/S0218194019500165

[26] V.M.B. Garcia et al., Contributions of entrepreneurial orientation in the use of agile methods in project management, Innovation & Management Review 18 (2021) 17-33. https://doi.org/10.1108/INMR-01-2019-0002

[27] M. Korzynski, A. Pacana, Centreless burnishing and influence of its parameters on machining effects. Journal of Materials Processing Technology 210 (2010) 1217-1223, https://doi.org/10.1016/j.jmatprotec.2010.03.008

[28] A. Kozień, Efficient management of cultural heritage by local government bodies, Wiadomosci Konserwatorskie 2020 (2020) 7-16. https://doi.org/10.48234/WK64HERITAGE

[29] N. Baryshnikova, O. Kiriliuk and D. Klimecka-Tatar, Management approach on food export expansion in the conditions of limited internal demand, Polish Journal of Management Studies 21 (2020) 101-114. https://doi.org/10.17512/pjms.2020.21.2.08

[30] Ł.J. Orman et al., Laser Treatment of Surfaces for Pool Boiling Heat Transfer Enhancement, Materials 16 (2023) art. 1365. https://doi.org/10.3390/ma16041365

[31] Ł.J. Orman et al., Application of laser treatment technology for boiling heat transfer augmentation, Prod. Eng. Arch. 30 (2024) 259-265. https://doi.org/10.30657/pea.2024.30.25

[32] M. Szymański et al., Tool Wear and Surface Roughness in Turning of Metal Matrix Composite Built of Al_2O_3 Sinter Saturated by Aluminum Alloy in Vacuum Condition, Materials 15 (2022) art. 8375. https://doi.org/10.3390/ma15238375

[33] M. Kuklinski, A. Bartkowska and D. Przestacki, Laser alloying monel 400 with amorphous boron to obtain hard coatings, Materials 12 (2019) art. 3494. https://doi.org/10.3390/ma12213494

[34] N. Radek et al., Operational properties of DLC coatings and their potential application, METAL 2022 – 31st Int. Conf. Metall. Mater., (2022) 531-536. https://doi.org/10.37904/metal.2022.4491

[35] M. Kukliński et al., Influence of microstructure and chemical composition on microhardness and wear properties of laser borided monel 400, Materials 13 (2020) art. 5757. https://doi.org/10.3390/ma13245757

[36] A. Kalinowski et al., Laser surface texturing: characteristics and applications, System Safety: Human - Technical Facility - Environment 5 (2023) 240-248. https://doi.org/10.2478/czoto-2023-0026

[37] J. Pietraszek, A. Goroshko, The heuristic approach to the selection of experimental design, model and valid pre-processing transformation of DoE outcome, Advanced Materials Research 874 (2014) 145-149. https://doi.org/10.4028/www.scientific.net/AMR.874.145

[38] J. Pietraszek et al., Challenges for the DOE methodology related to the introduction of Industry 4.0, Prod. Eng. Arch. 26 (2020) 190-194. https://doi.org/10.30657/pea.2020.26.33

[39] J. Pietraszek, A. Gadek-Moszczak, The smooth bootstrap approach to the distribution of a shape in the ferritic stainless steel AISI 434L powders, Solid State Phenomena 197 (2013) 162-167. https://doi.org/10.4028/www.scientific.net/SSP.197.162

[40] J. Pietraszek, Fuzzy regression compared to classical experimental design in the case of flywheel assembly, Lecture Notes in Computer Science 7267 LNAI (2012) 310-317. https://doi.org/10.1007/978-3-642-29347-4_36

[41] J. Pietraszek, The modified sequential-binary approach for fuzzy operations on correlated assessments, Lecture Notes in Computer Science 7894 LNAI (2013) 353-364. https://doi.org/10.1007/978-3-642-38658-9_32

[42] N. Radek et al., Properties of Steel Welded with CO_2 Laser, Lecture Notes in Mechanical Engineering (2020) 571-580. https://doi.org/10.1007/978-3-030-33146-7_65

[43] R. Ulewicz et al., Quality and work safety in metal foundry, METAL 2020 – 29th Int. Conf. Metall. Mater. (2020) 1287-1293. https://doi.org/10.37904/metal.2020.3649

[44] A. Kozień, N. Kozłowska, Harmonization and Deharmonization of Excise Duty in the European Union as Contemporary Challenges of the EU Tax Law, WSEAS Transactions on Business and Economics 19 (2022) 816-824. https://doi.org/10.37394/23207.2022.19.71

[45] A. Deja et al., Green technologies in smart city multifloor manufacturing clusters: A framework for additive manufacturing management, Prod. Eng. Arch. 29 (2023) 428-443. https://doi.org/10.30657/pea.2023.29.48

[46] B. Gajdzik et al., Approaching open innovation in customization frameworks for product prototypes with emphasis on quality and life cycle assessment (QLCA), Journal of Open

Innovation: Technology, Market, and Complexity 10 (2024) art. 100268. https://doi.org/10.1016/j.joitmc.2024.100268

[47] M. Radek et al., Matching Computational Tools to User Competence Levels in Education of Engineering Data Processing, Materials Research Proceedings 34 (2023) 453-459. https://doi.org/10.21741/9781644902691-52

[48] J. Pietraszek, E. Skrzypczak-Pietraszek, The optimization of the technological process with the fuzzy regression, Advanced Materials Research 874 (2014) 151-155. https://doi.org/10.4028/www.scientific.net/AMR.874.151

[49] E. Skrzypczak-Pietraszek, J. Pietraszek, Seasonal changes of flavonoid content in Melittis melissophyllum L. (Lamiaceae), Chemistry and Biodiversity 11 (2014) 562-570. https://doi.org/10.1002/cbdv.201300148

[50] J. Pietraszek, E. Skrzypczak-Pietraszek, The uncertainty and robustness of the principal component analysis as a tool for the dimensionality reduction, Solid State Phenomena 235 (2015) 1-8. https://doi.org/10.4028/www.scientific.net/SSP.235.1

[51] Ł.J. Orman et al., Comparative Analysis of Indoor Environmental Quality and Self-Reported Productivity in Intelligent and Traditional Buildings, Energies 16 (2023) art. 6663. https://doi.org/10.3390/en16186663

Terotechnology XIV | Materials Research Forum LLC
Materials Research Proceedings 62 (2026) 111-119 | https://doi.org/10.21741/9781644904015-15

Materials Engineering in the Context of Ballistic Resistance

Agnieszka SZCZOTOK[1,a *]

[1]Silesian University of Technology, Z. Krasinskiego Str. 8, 40-019 Katowice, Poland, ORCID: 0000-0003-0680-0837

[a]agnieszka.szczotok@polsl.pl

Keywords: Materials Engineering, Ballistic Resistance, Modern

Abstract. The escalating complexity of modern threats—from high-velocity armor-piercing projectiles to blast-induced fragments and improvised explosive devices (IEDs)—has driven continuous innovation in ballistic-resistant materials. Effective protection must balance high energy absorption capacity with minimal areal density to preserve operational mobility, particularly in dismounted personnel and lightweight vehicle applications. Central to this challenge is the systematic classification of armor materials based on composition, micro/macro-structure, and dominant energy dissipation mechanisms. This work is a proposal for this classification.

Introduction

Ballistic-resistance materials are materials specifically designed, selected, or engineered to withstand high-velocity impacts from projectiles, such as bullets, fragments, shrapnel, or blast debris. The primary functions of ballistic-resistance materials are to: 1) stop or decelerate the projectile, 2) prevent penetration, 3) minimize rear-face deformation (backface signature), 4) protect against secondary fragments and blast-driven debris, and 5) maintain structural integrity under impact.

Ballistic performance refers to a material or system's measurable ability to resist high-velocity projectile impacts. It is quantified through standardized testing under rigorous, controlled conditions using key metrics such as: ballistic limit velocity (V_{50}), specific energy absorption (SEA), areal density (AD), backface signature (BFS), and multi-hit capability (Table 1). These standards ensure reproducible, threat-specific benchmarks for interoperability, procurement reliability, and user safety across military, law enforcement, and civilian applications. Key testing standards include: NIJ 0101.06 (USA, body armor) [1], ASTM E3004-22 (clay blocks used in ballistic-resistance testing of torso body armor) [2], STANAG 2920 (NATO, ballistic test method for personal armor) [3], STANAG 4569 (NATO, protection levels for armored vehicles) [4], VPAM BSW 2006 (Germany, ballistic protective vests) [5], VPAM APR 2006 (Germany, general basis for ballistic materials) [6], ASTM F1233 (security glazing materials and systems) [7], MIL-STD-662F (USA, V_{50} ballistic testing for armor) [8], HOSDB (UK, ballistic protection) [9], etc.

The defense sector faces growing demands for enhanced corrosion and wear resistance, reduced weight, and minimized component footprint [10,11]. The U.S. Department of Defense (DoD) SBIR/STTR programs continue to emphasize lightweight materials and advanced coatings for defense applications. Lightweight fiber-reinforced composites for vehicle and personnel armour represent a leading priority in defense material innovation [12]. Atomic Defense's research reveals that ceramic plates weigh 30% less than steel equivalents while stopping identical threats [13].

Historically, ballistic protection relied on basic materials such as wood, leather, and metals [10,14]. However, rapid advancement in weapon lethality now requires parallel innovation in protective systems. The escalating need for effective personal and structural protection against evolving threats has positioned materials engineering at the core of defense technology [15]. Ballistic resistance, explained earlier, remains a critical performance metric for military and

Materials Research Proceedings 62 (2026) 111-119 https://doi.org/10.21741/9781644904015-15

civilian applications [16,17]. Modern materials science has transformed this domain by shifting from bulk density to optimized energy absorption mechanisms at the micro- and nanoscale [18,19].

The modern challenge extends beyond merely stopping a projectile to balancing critical trade-offs: 1) weight and mobility (armor must be lightweight to preserve personnel agility and vehicle fuel efficiency), 2) cost-effectiveness (practical solutions must enable scalable mass production), and 3) multi-threat protection (materials must resist a wide range of threats, from small-arms ammunition to armor-piercing projectiles).

Table 1. *Ballistic performance metrics*

Metric	*Definition*	*Formula*	*Importance*	*Source*
V_{50} (Ballistic Limit)	The velocity at which a projectile has a 50% chance of penetrating body armor and a 50% chance of being stopped	—	Primary resistance indicator	[20]
Specific Energy Absorption (SEA)	Energy absorbed per unit areal density of the armor panel	SEA = E / AD ($J*m^2/kg$ or J/kg for a 1 m^2 panel) E – absorbed kinetic energy, AD – areal density	Weight efficiency (higher SEA enables lighter armor at equivalent protection levels)	[21]
Areal Density (AD)	Mass per unit area of a material	AD = m/A (kg/m^2) m – total mass of the armor panel A - facial (projected) area of the panel	Mobility vs. protection, key design parameter in evaluating the mass–protection relationship	[8]
Backface Signature (BFS) or ***Backface Deformation (BFD)***	The deepest depth of the depression made in the backing material, created by a nonpenetrating projectile impact, measured from the reference plane	Measured in mm	Blunt trauma caused by the rapid inward bulge of the armor material.	[1]
Multi-Hit Capability	Minimum distance between hits without penetration (performance under repeated impacts)	—	Real-world reliability (Multi-hit capability is what keeps you alive for the second, third, and sixth).	[22]

The preparation and dissemination of a short literature review on modern ballistic resistance materials presented in this work is scientifically, technologically, and strategically justified for the following interconnected reasons: 1) rapid evolution of threats and technologies, 2) knowledge transfer and education, and 3) identification of research gaps.

Classes of Ballistic Resistance Materials and Their Resistance Mechanisms

Ballistic-resistant materials are classified based on composition, structure, and primary energy dissipation mechanisms. The four major classes are metals, ceramics, fiber-reinforced polymers, matrix composites, and hybrid systems.

Metallic Armor Materials

Metals serve as foundational components in ballistic protection systems, functioning as load-bearing structures, energy-absorbing layers, and structural frameworks. Table 2 contains the main metallic material classes for ballistic resistance.

Table. 2. Main metallic material classes for ballistic resistance [source: 23-32]

Class	***Key alloy***	***Properties***	***Applications***	***Resistance mechanism***
Ultra-High-Hardness (UHH) Steels	Armox® 600T, Miilux Protection 600	Max hardness, limited ductility, hardness/toughness balance	Thin plates, anti-RPG screens (RPG: Rocket-Propelled Grenade)	Shear plugging resistance (thin targets)
High-Hardness Armour (HHA) Steels	MIL A 46100, AR500, Armox® 500T	Optimized vs. AP rounds, balanced ductility	Add-on vehicle armor, APCs	Projectile blunting + delayed plugging
High-Strength Low-Alloy (HSLA) Armour Steels	MIL A 12560, AR400, AR450	Moderate hardness, high toughness, weldable	Vehicle hulls, structural shields	Ductile hole enlargement + plastic work
Bainitic & Nanostructured Steels	Super Bainite	enhanced strength and ductility	Next-generation lightweight vehicle armor	Grain boundary pinning; delayed adiabatic shear banding (ABS)
Maraging Steels	18Ni(300), 1.2709 18Ni(350), Vascomax® C350	Ultra-high strength (18Ni(300): 1900 MPa), high toughness, ductility, and impact strength	Aerospace armor, tank armor, and firearms (e.g., gun barrels)	Martensitic transformation toughening
Aluminum Alloys	5083-H131, 7085	Low density (2.7 g/cm³), corrosion-resistant, protection against impact	Light vehicles (JLTV), helicopter skins, military armored cars, tanks, and bulletproof vehicles	Strain rate hardening + spall resistance
Titanium Alloys	Ti-6Al-4V	High specific strength, ballistic efficiency	Aircraft armor, personnel plates	Twinning + dynamic recrystallization
Dual-Phase (DP) & TRIP Steels	DP1000, TWIP-assisted steels	High work hardening rate, balance of strength, ductility, and formability	Hybrid armor backing	Phase transformation energy sink

Materials Research Proceedings 62 (2026) 111-119 https://doi.org/10.21741/9781644904015-15

Ballistic performance is critically governed by metallurgical purity and grain refinement, which suppress shear band initiation and propagation, thereby reducing susceptibility to plugging failure.

Ballistic Ceramics

Ceramics are brittle, high-hardness materials used as strike faces in armor to erode, fracture, and blunt projectiles before energy is absorbed by a ductile backing.

Ceramics currently constitute the primary material used in hard plates and vehicle armor systems designed to defeat projectiles. The most commonly employed ballistic ceramics are: alumina Al_2O_3, silicon carbide SiC, and boron carbide B_4C [33]. In the group of top 6 bulletproof ceramic materials in 2025 are additionally: silicon nitride (Si_3N_4), zirconia (ZrO_2), and aluminum titanate (Al_2TiO_5) [34]. Usually, ceramic armor systems consist of a monolithic ceramic or composite ceramic–metal body covered by ballistic nylon and bonded with a high tensile strength fiber lining or laminated polyethylene, such as Kevlar™, Twaron™, Spectra™, Dyneema™, placed on the back of the ceramic or ceramic–metal composite [35]. Ballistic effectiveness of ceramics derives from mechanisms such as brittle fracture, erosion of the projectile core, energy dissipation through fragmentation, and the formation of a characteristic conical cracking zone.

The choice of ceramic material reflects a balance among cost, mass, and performance: Al_2O_3 – the least expensive but also the densest, SiC – offering a more favorable hardness-to-mass ratio, B_4C – the lightest and hardest, yet costly and susceptible to catastrophic damage under very high-energy impact.

The most important limitations of ballistic ceramics are the following: 1) brittle failure under multi-hit, 2) high cost (premium-grade boron carbide for armor costs \$50–\$100/kg [36]), 3) requires backing — never standalone, and 4) thermal shock sensitivity.

In armor systems, transparent ceramics such as aluminum oxynitride (ALON) or polycrystalline magnesium aluminate spinel $MgAl_2O_4$ (PMAS), and glass–ceramics are also used, enabling significant reductions in mass and improved multi-hit capability compared with conventional glass–polymer laminates [37].

Fiber-Reinforced Polymer (FRP) Matrix Composites

Fiber-reinforced polymer (FRP) matrix composites are prominent defensive armor materials owing to their excellent strength-to-weight ratio and design flexibility. High-strength fibers such as aramid, carbon, or glass embedded in a polymer matrix exhibit excellent impact resistance and energy absorption, which significantly improves ballistic protection [38]. Ultra-high-molecular-weight polyethylene (UHMWPE) fibers (e.g., Dyneema, Spectra) exhibit exceptionally high specific strength and very low density, while aramid fibers (Kevlar, Twaron [39]) are distinguished by their high impact resistance and good thermal stability.

The primary protective mechanisms in FRP matrix composites include: tension-dominated yarning—fiber stretching and load transfer within the yarn bundle, inter-fiber friction, and mechanical interlocking of the weave, rapid in-plane propagation and dissipation of stress waves, and delamination, which controls the size of the deformation zone.

Advanced Materials and Hybrid Systems

Growing interest in military applications is directed toward systems incorporating shear-thickening fluids (STFs) and shape-memory polymers (SMPs) as impregnating media that enhance dynamic stiffness, improve penetration resistance, and increase wearer comfort. The STF, which is a soft material with the characteristics of intelligent material, bears a higher external force when compared with liquid, and exhibits the characteristics of variable shape, unlike solids. Therefore, STF exhibits wide application prospects in shock absorption and human body protection [40].

Functionally Graded Materials (FGMs) are composite systems with continuously or stepwise varying composition, microstructure, or properties (e.g., hardness, density, modulus) across thickness to optimize stress wave propagation, delay failure, and maximize energy dissipation. The

Materials Research Proceedings 62 (2026) 111-119 https://doi.org/10.21741/9781644904015-15

results of the depth of penetration-based ballistic evaluation profoundly establish the superiority of FGM armors over homogeneous composites [40]. FGMs provide a smooth transition of mechanical properties (hardness, modulus, density) through the thickness of the panel. Functionally graded material (FGM) is replacing conventional and composite materials very rapidly due to its advanced characteristics. Lightweight FGMs can significantly improve the efficacy of armor against ballistic impact by progressively changing the material compositions and porosities over thickness without sacrificing flexibility or durability [41].

Sandwich structures incorporating foam or lattice cores show distinct behaviors under quasi-static and ballistic loading. Sandwich structures using composite materials are foreseen as armor solutions, due to their stiffness and lightweight properties. Either associations of dry fabrics or interlock structures and ceramic layers, or metallic structures (mostly steel thick plates) are often used in ballistic studies [42]. Sandwich armor structures with aluminum foam can be utilized to protect a military vehicle from harmful blast loads, such as a landmine explosion [43].

Advanced materials – including nanomaterials, novel alloys, metamaterials, adaptive self-healing polymers, and 4D-printed smart structures – are increasingly recognized as a key Emerging Disruptive Technology (EDT) in military affairs [44].

Economic and Technological Impact

Anti-ballistic protection materials are of not only military but also potentially industrial importance due to the protection they provide [45,46] against potential injuries. These include processes of applying special coatings [47], which involve the use of concentrated energy beams (DLC [48,49], ESD [50,51], laser micromachining [52,53]) or the application of high pressures [54,55]. Research on such materials is a significant impetus for the development of textile strength issues [56] and provides an ideological basis for the development of new, much stronger implants [57]. Improving production processes related to these materials requires appropriate optimization tools [58,59], also using methodologies that take into account incomplete knowledge [60,61] and appropriate support in the field of microscopy [62,63] and surface layer analytics [64,65].

Summary

Modern battlefields and counter-terrorism operations are characterized by a wide spectrum of threats, ranging from high-kinetic-energy projectiles to improvised explosive devices (IEDs). Concurrently, maintaining user mobility and ergonomics imposes a fundamental requirement for reducing shield mass, quantified through areal density. In practice, this necessitates the development of novel materials and armor architectures capable of delivering equivalent or superior protection at substantially lower total mass.

Ballistic-resistant materials described in this work are classified into four major classes based on composition, structure, and primary energy dissipation mechanisms:

- Metals (e.g., high-hardness armour steels, aluminum, titanium): absorb energy via plastic deformation, projectile blunting, and adiabatic shear banding delay.
- Ceramics (e.g., B_4C, SiC, Al_2O_3): defeat projectiles through dwell, erosion, and comminution of the penetrator, requiring ductile backing.
- Fiber-Reinforced Polymer Composites (e.g., Kevlar®, Dyneema®): dissipate energy via tensile fiber stretching, delamination, and back-face deformation control.
- Advanced/Hybrid Systems (e.g., ceramic-faced FRC, metal-composite laminates): combine sequential mechanisms — ceramic erosion followed by composite/metal absorption — for optimal lightweight, multi-hit performance.

The summary of the principal categories of ballistic materials, their properties, and roles in protective systems is presented in Table 3.

Table 3. Main material classes for ballistic resistance

Material class	***Key properties***	***Mechanisms***	***Typical use***
Metals	Strength, ductility	Plastic deformation, adiabatic shear banding delay	Vehicle armour, structural shields
Ceramics	Extreme hardness	Projectile dwell and erosion, fracture, requires ductile backing	Hard plates, transparent armour
FRP Composites	High tensile resistance	Tensile fiber stretching, delamination, back-face deformation control	Soft armour, ceramic backing
Hybrids	Combined advantages	Layered synergistic effects, sequential mechanisms	Multi-layer armour
Foams/Cellular	Low density, crushability	Energy dissipation	Blast mitigation
Metamaterials	Engineered topology	Wave control, optimized collapse	Advanced lightweight armour
Smart Materials	Adaptive behavior	Rate/power-dependent stiffening	Next-gen adaptive armour

References

[1] M.B. Mukasey, J.L. Sedgwick, D.W. Hagy, NIJ Standard-0101.06: Ballistic Resistance of Body Armor, U.S. Department of Justice, Washington, DC, 2008.

[2] ASTM E3004-22: Standard Specification for Preparation and Verification of Clay Blocks Used in Ballistic-Resistance Testing of Torso Body Armor.

[3] NATO STANAG 2920: Ballistic test method for personal armour materials and combat clothing, 2015.

[4] NATO STANAG 4569: Protection Levels for Occupants of Armoured Vehicles, 2014.

[5] VPAM Vereinigung der Prüfstellen für Angriffshemmende Materialien und Konstruktionen (Association of Testing Bodies for Attack Resistant Materials and Constructions); VPAM BSW 2006: Personal Protective Equipment Ballistic Protective Vests – Requirements, Classifications and Testing Methods, 2009.

[6] VPAM APR 2006: General basis for ballistic material, construction and product testing, 2009.

[7] ASTM F1233: Standard Test Method for Security Glazing Materials and Systems, 2021

[8] MIL-STD-662F Military standard: V50 Ballistic Test for Armor, U.S. Department of Justice, Washington, DC, 1997

[9] UK Home Office Scientific Development Branch (HOSDB): Body Armour Standards for UK Forces: Ballistic Protection, 2017.

[10] R. Fangueiro, S. Rana (Eds.) Advanced Materials for Defense, Development, Analysis and Applications, 1st ed., Springer Cham, 2020. https://doi.org/10.1007/978-3-030-34123-7

[11] Information on http://www.carpentertechnology.com/ [access date: 28-10-2025]

[12] Information on https://armysbir.army.mil/topics/ [access date: 28-10-2025]

[13] Information on https://editverse.com/kevlar-vs-ceramic-body-armor-the-science-that-could-save-your-life/ [access date: 28-10-2025]

[14] W.Zou, Recent advancements in ballistic protection - a review, J. Student Res. 13 (2024) 1-9. https://doi.org/10.47611/jsrhs.v13i1.5936

[15] H. Barhoumi et al., Review of ballistic protection materials: Properties and performances, Journal of Reinforced Plastics and Composites 42/13-14 (2022) 685-699. https://doi.org/10.1177/07316844221137920

[16] M.B. Mukasey et al., NIJ Standard-0101.06: Ballistic Resistance of Body Armor, U.S. Department of Justice, Washington, DC, 2008.

[17] A. Bhatnagar, Lightweight Ballistic Composites. Military and Law-Enforcement Applications, Woodhead Publishing, 2016. https://doi.org/10.1016/C2014-0-03657-X

[18] N. Aanandhan et al., Computational analysis and comparative study on ballistic impact properties of composite materials. in: Talpa Sai, P.H.V.S., Potnuru, S., Avcar, M., Ranjan Kar, V. (eds.) Intelligent Manufacturing and Energy Sustainability. ICIMES 2023. Smart Innovation, Systems and Technologies, 372, Springer, Singapore, 2024, 93-104. https://doi.org/10.1007/978-981-99-6774-2_9

[19] A. Biradar et al., Nanocoatings for ballistic applications: A Review, Nanotechnology Reviews 12 (2023) 1-28. https://doi.org/10.1515/ntrev-2023-0574

[20] Information on https://www.bodyarmornews.com/v-50-test/

[21] Ch.-H. Shih et al., Design and Ballistic Performance of Hybrid Plates Manufactured from Aramid Composites for Developing Multilayered Armor Systems. Polymers 14 (2022) art.5026. https://doi.org/10.3390/polym14225026

[22] B. Russell, Multi-hit ballistic damage characterisation of 304 stainless steel plates with finite elements. Materials & Design 58 (2014) 252-264. https://doi.org/10.1016/j.matdes.2014.01.074

[23] B. Garbarz et al., Ultrahigh-Strength Nanostructured Steels for Armours. Problems of Mechatronics Armament Aviation Safety Engineering 2 (2011) 25-36.

[24] S.H. Atapek, S. Karagoz, Ballistic impact behaviour of a tempered bainitic steel against 7.62 mm armour piercing projectile, Defence Science Journal 61 (2011) 81-87. https://doi.org/10.14429/dsj.61.411

[25] Information on https://defence-industry.eu/maraging-steel-for-defence-industry-unconquerable-defence/

[26] C. Cheng et al., Ballistic response of a high-strength steel. Mechanics of Advanced Materials and Structures 31 (2023) 5377-5388. https://doi.org/10.1080/15376494.2023.2215768

[27] Information on https://www.ssab.com/

[28] Information on https://www.cliftonsteel.com/

[29] Information on https://www.fushunspecialsteel.com/

[30] Information on https://www.chalcoaluminum.com/

[31] A. Janda et al., Ballistic tests on hot-rolled Ti-6Al-4V plates: Experiments and numerical approaches, Defence Technology 26 (2023) 39-53. https://doi.org/10.1016/j.dt.2022.11.012

[32] I.G. Crouch et al., 2 - Armour steels, (Ed.: I G. Crouch), in: Woodhead Publishing in Materials, The Science of Armour Materials, Woodhead Publishing, 2017, 55-115, https://doi.org/10.1016/B978-0-08-100704-4.00002-5

[33] A.B. Dresch et al., Ballistic ceramics and analysis of their mechanical properties for armour applications: A Review, Ceramics International 47 (2021) 8743-8761. https://doi.org/10.1016/j.ceramint.2020.12.095

[34] Information on https://ggsceramic.com/news-item/the-6-best-bulletproof-ceramic-materials-performance-comparison-and-selection-analysis-in-2025

[35] E. Medvedovski, Ballistic performance of armour ceramics: Influence of design and structure. Part 2, Ceramics International 36 (2010) 2117-2127. https://doi.org/10.1016/j.ceramint.2010.05.022

[36] Information on https://www.dragon-abrasives.com/what-is-the-current-price-of-boron-carbide.html

[37] T. Benitez et al., Transparent ceramic and glass-ceramic materials for armor applications, Ceramics International 43 (2017) 13031-13046. https://doi.org/10.1016/j.ceramint.2017.07.205

[38] N.S. Gaikwad et al., Fiber-reinforced polymer matrix composites for improved defence armor – A comprehensive review, Smart Materials in Manufacturing 3 (2025) art. 100084, https://doi.org/10.1016/j.smmf.2025.100084

[39] Information on https://www.difference.wiki/kevlar-vs-twaron/

[40] M. Wei, K. Lin, L. Sun, Shear thickening fluids and their applications, Materials & Design, 216 (2022) art.110570. https://doi.org/10.1016/j.matdes.2022.110570

[41] M. Patel, A comprehensive review of functionally graded materials and their ballistic impact performance: Current status and future challenges, Next Materials 8 (2025) art. 100704. https://doi.org/10.1016/j.nxmate.2025.100704

[42] A. Kolopp et al., Experimental study of sandwich structures as armour against medium-velocity impacts, International Journal of Impact Engineering, 61 (2013) 24-35, https://doi.org/10.1016/j.ijimpeng.2013.05.007

[43] S. Yang et al., Blast-Resistant Improvement of Sandwich Armor Structure with Aluminum Foam Composite, Advances in Materials Science and Engineering (2013) art. 947571. https://doi.org/10.1155/2013/947571

[44] Information on https://defencefinancemonitor.substack.com/p/advanced-materials-as-an-emerging

[45] V. Ptashnyk et al., Impedance investigation of activated carbon material modified by ultrasound treatment, Przeglad Elektrotechniczny 94 (2018) 186-189. https://doi.org/10.15199/48.2018.05.33

[46] R. Caban et al., Work safety during the injection molding process of polymer materials, System Safety: Human - Technical Facility - Environment 6 (2024) 320-329. https://doi.org/10.2478/czoto-2024-0034

[47] A. Dudek, Surface properties in titanium with hydroxyapatite coating, Optica Applicata 39 (2009) 825-831.

[48] N. Radek et al., Influence of laser texturing on tribological properties of DLC coatings, Production Engineering Archives 27 (2021) 119-123. https://doi.org/10.30657/pea.2021.27.15

[49] A. Kalinowski et al., The effect of micro-machining with a single-mode and multi-mode fiber laser on the surface of DLC coatings, METAL 2024 - 33rd Int. Conf. Metall. Mater. (2024) 328-333. https://doi.org/10.37904/metal.2024.4952

[50] N. Radek et al., The impact of laser processing on the performance properties of electro-spark coatings, World Congress in Computational Mechanics and ECCOMAS Congress 1000 (2021) 1-10. https://doi.org/10.23967/wccm-eccomas.2020.336

[51] N. Radek et al., Technology and applications of ESD coatings before and after laser processing, METAL 2023 - Int. Conf. Metall. Mater., 500-505. https://doi.org/10.37904/metal.2023.4727

[52] A. Kalinowski et al., Laser surface texturing: characteristics and applications, System Safety: Human - Technical Facility - Environment 5 (2023) 240-248. https://doi.org/10.2478/czoto-2023-0026

[53] E. Szajna et al., The influence of laser remelting on microstructural changes and hardness level of flame-sprayed NiCrBSi coatings with tungsten carbide addition, Surface and Coatings Technology 478 (2024) art. 130403. https://doi.org/10.1016/j.surfcoat.2024.130403

[54] N. Radek et al., Technology and application of anti-graffiti coating systems for rolling stock, METAL 2019 – 28th Int. Conf. Metall. Mater. (2019) 1127-1132.

[55] N. Radek, R. Dwornicka, Fire properties of intumescent coating systems for the rolling stock, Communications - Scientific Letters of the University of Žilina 22 (2020) 90-96. https://doi.org/10.26552/com.C.2020.4.90-96

[56] P. Krysiak, Microstructural Analysis of Wound Composites with Considerations on the Fiber Winding Force, Materials Research Proceedings 34 (2023) 43-52. https://doi.org/10.21741/9781644902691-6

[57] O. Kędzia et al., Glass and glass-ceramic porous materials for biomedical applications, System Safety: Human - Technical Facility - Environment 5 (2023) 302-310. https://doi.org/10.2478/czoto-2023-0033

[58] J. Pietraszek, A. Goroshko, The heuristic approach to the selection of experimental design, model and valid pre-processing transformation of DoE outcome, Advanced Materials Research 874 (2014) 145-149. https://doi.org/10.4028/www.scientific.net/AMR.874.145

[59] J. Pietraszek et al., Challenges for the DOE methodology related to the introduction of Industry 4.0, Prod. Eng. Arch. 26 (2020) 190-194. https://doi.org/10.30657/pea.2020.26.33

[60] J. Pietraszek, Fuzzy regression compared to classical experimental design in the case of flywheel assembly, Lecture Notes in Computer Science 7267 LNAI (2012) 310-317. https://doi.org/10.1007/978-3-642-29347-4_36

[61] J. Pietraszek, The modified sequential-binary approach for fuzzy operations on correlated assessments, Lecture Notes in Computer Science 7894 LNAI (2013) 353-364. https://doi.org/10.1007/978-3-642-38658-9_32

[62] A. Szczotok, S. Roskosz, New possibilities of light microscopy research resulting from digital recording of images, Materials Science- Poland 23 (2005) 559-565.

[63] A. Gadek-Moszczak, Z. Sabina, Description of 3d microstructure of the composites with polypropylene (pp) matrix and tuf particles fillers, Solid State Phenomena 197 (2013) 186-191. https://doi.org/10.4028/www.scientific.net/SSP.197.186

[64] A. Wrońska, A. Dudek, Characteristics of surface layer of sintered stainless steels after remelting using GTAW method, Archives of Civil and Mechanical Engineering 14 (2014) 425-432. https://doi.org/10.1016/j.acme.2013.10.011

[65] M. Szymański et al., Tool Wear and Surface Roughness in Turning of Metal Matrix Composite Built of Al2O3 Sinter Saturated by Aluminum Alloy in Vacuum Condition, Materials 15 (2022) art. 8375. https://doi.org/10.3390/ma15238375

Materials Research Proceedings 62 (2026) 120-133 | https://doi.org/10.21741/9781644904015-16

Closed Kinematic Devices as the Basis for Military Vehicle Simulators

Paweł A. ŁASKI[a] *, Krzysztof BORKOWSKI[b], Gabriel BRACHA[c], Dawid S. PIETRALA[d], Hubert WIŚNIEWSKI[e], Anna ANNUSEWICZ-MIŚTAL[f], Szczepan KOSTECKI[g], Jakub PAJĄK[h], Adam SIEROŃ[i] and Michał WZOREK[j]

Kielce University of Technology, Faculty of Mechatronics and Mechanical Engineering, Department of Automation and Robotics, Al. 1000-lecia Państwa Polskiego 7, 25-314 Kielce, Poland

[a]pawell@tu.kielce.pl, [b]kborkowski@tu.kielce.pl, [c]gbracha@tu.kielce.pl, [d]dpietrala@tu.kielce.pl, [e]hubert.w@tu.kielce.pl, [f]aannusewicz@tu.kielce.pl, [g]skostecki@tu.kielce.pl, [h]jpajak@tu.kielce.pl, [i]asieron@tu.kielce.pl, [j]mwzorek@tu.kielce.pl

Keywords: Parallel Manipulators, Closed-Loop Kinematic Chains, Military Vehicle Simulators

Abstract. This paper presents the characteristics of parallel structures, their application areas, research results, and practical applications. The most common parallel structures are presented, along with examples of workspace shapes. Manipulators characterized by closed loop kinematic chains are commonly referred to as parallel. However, a more accurate statement, reflecting the concept of their operation, is that such manipulators and robots are structures with simultaneous drives to achieve the desired motion. Parallel manipulators and simulators offer many advantages over serial manipulators, including high accuracy, rigidity, and the ability to manipulate heavy workpieces. The fundamental difference between the design of parallel kinematic simulators and that of a conventional serial kinematic manipulator lies in the coordinate axes of the drives and arms. In contrast, conventional serial structures have a Cartesian x, y, and z axes. Therefore, they have found applications in flight simulators and combat vehicles. Simulators with this kinematic structure are significantly more difficult to control than serial manipulators.

Introduction

Currently, flight simulator designs are mostly based on E.V. Gough or D. Stewart platforms. Typical drive elements include long hydraulic actuators connected to the base and work platform via joints. Movement of the work platform is the result of the controlled extension of the actuators' piston rods. Hydraulic and electric servo drives enable the movement of large masses while maintaining the required dynamics. In flight simulators, the work platform is equipped with platforms that replicate the interior of flying machines.

Simulators

Fig.1 shows two solutions for Boeing 787 flight simulators and a universal platform designed for helicopter pilot training, including military applications. Designed by Thales Group Training & Simulation (France), the simulators are based on E.V. Gough's kinematic structure and feature six hydraulic actuators connected to the base on one side and to the work platform on the other, which houses the simulation platform. The actuators are connected to the base and work platform via universal joints. This simulator's kinematic structure provides six degrees of freedom for the platform relative to the stationary base.

Another example of a hexapod-based flight simulator is the design of the US-based L-3 Link & Simulation Training Division (Figure 2). For over 80 years, L-3 has been progressively introducing new flight simulator designs to the market, most of them based on the hexapod closed-chain structure. L-3 remains a leading systems integrator, providing advanced simulation systems

Materials Research Proceedings 62 (2026) 120-133 https://doi.org/10.21741/9781644904015-16

and training services to civilian and military agencies in the US and to international pilot training companies.

Platform systems enable the simulation of aircraft behavior, not only in terms of forces generated, but also in terms of structural loads. Using hexapod platforms, prototypes of flying machines can be tested without lifting them off the ground. At the University of Bingen in Germany, under the leadership of Professor Michael Schönherr, a parallel simulator with an electric drive was developed (Fig.3) for testing prototypes of smaller aircraft structures.

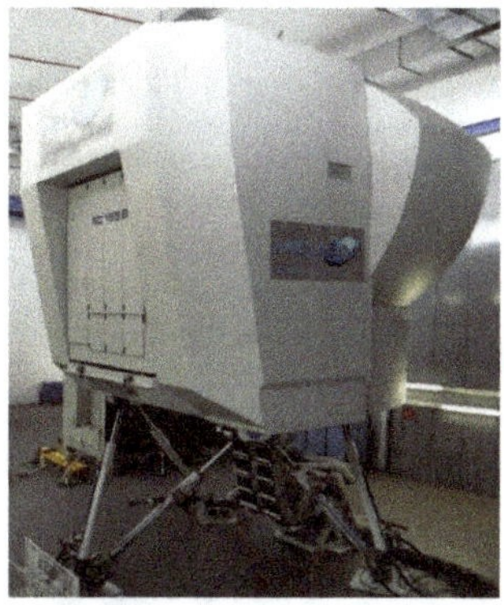

Fig. 1. Flight simulators from Thales Group Training & Simulation, a) Boeing 787 aircraft, b) HELISIM helicopters

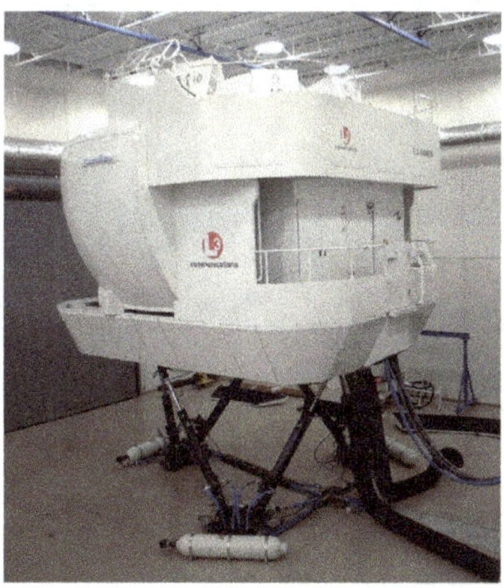

Fig. 2. Flight Simulator by L-3 Communications Corporation

Materials Research Proceedings 62 (2026) 120-133 https://doi.org/10.21741/9781644904015-16

The Military Mechanical Works Consortium in Siemianowice Śląskie (Poland) has developed training simulators for the military. The kinematic design of the ROSOMAK JASKIER simulator (Fig.4a) is based on a hexapod design. The simulator is powered by electric motors, which produce a total power of 800 W. The simulator's approximate dimensions are 3 x 2.8 x 2.1 m (length x width x height), and its unladen weight is 1200 kg.

The use of a hexapod platform with six degrees of freedom enables the realistic modeling of the vehicle interior and turret for real-world operating conditions. The comprehensive TASZNIK crew training simulator (Figure 4b) faithfully replicates the interior and enables live-fire training exercises. It consists of a functional turret mounted on a work platform that reflects the dynamics of a combat vehicle's movement. Among the companies producing simulation platforms is the Automex Technical Consulting and Services Center (ODIUT Automex Polska), founded in 1986 by researchers from the Gdańsk University of Technology (Poland). The developed ATMX6000 platform (Fig.5) is based on a Stewart-Gough 6-DoF hexapod structure and, together with advanced simulation software, creates a system for simulating automotive vehicles and aircraft. The simulator's cabin, located on the work platform, moves to mimic the actual movement of the vehicle operator as closely as possible. The relatively large 500 mm drive stroke contributes to the quality of the simulation. The linear range of motion in the x, y, and z axes for the work platform is ±270 mm, while the angular range around the x, y, and z axes is ±22°. The maximum linear velocity in the x, y, and z axes is 420 mm/s, while the maximum angular velocity around the x, y, and z axes is 35°/s. The maximum linear acceleration in the x, y, and z axes is 6 m/s², and the maximum angular acceleration around the x, y, and z axes is 430°/s². The permissible payload is 1200 kg.

Fig. 3. A simulator of prototypes of flying machines, including drones (UAV)

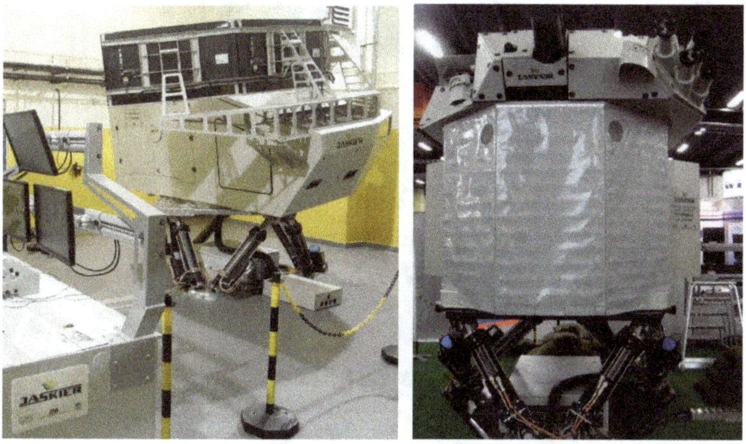

Fig. 4. KTO ROSOMAK training simulator a) JASKIER, b) TASZNIK

Fig. 5. Automex ATMX6000 hexapod simulation platform (Poland)

In 1998, Environmental Tectonics Corporation (USA) ETC acquired a majority stake in MP-PZL Aerospace Industries (Poland), which now operates as ETC-PZL Aerospace Industries Sp. z o.o. and manufactures flight and vehicle simulators for civilian and military use. ETC-PZL Aerospace Industries designed and built, among other things, a comprehensive combat mission simulator for the W-3WA "Sokół" helicopter (Figure 6) for training Army pilots. The hydraulic motion system has six degrees of freedom and a load capacity of 8,500 kg.

Materials Research Proceedings 62 (2026) 120-133 https://doi.org/10.21741/9781644904015-16

Fig. 6. Comprehensive flight simulator of the W-3WA "Sokół" helicopter (Poland)

Hydraulic actuators power the platform with a 1400 mm stroke and has a load capacity of 3000 kg. The device was used, among other things, to build the I-22 aircraft simulator. The platform's angular movements, velocity, and acceleration for the x, y, and z axes are: ± 32° ± 25°/s ± 250°/s², ± 27° ± 25°/s ± 150°/s², ± 36° ± 25°/s ± 225°/s². The ranges of translational movements, speeds and accelerations for the x, y, z axes are, respectively, ± 1.30 m ± 0.86 m/s ± 7m/s², ± 1.37 m ± 0.86 m/s ± 7 m/s², ± 1.80 m ± 0.7 m/s ± 8m/s². ETC-PZL (Poland) has developed a flight and combat simulator for a Russian-made aircraft, featuring a faithful replica of the Su-22M4 attack cockpit (Fig.7). However, the simulator's cockpit design has lost its significance due to the deployment of Su-22M4 aircraft to combat operations in Ukraine. All onboard instruments, including switches and controls, are designed to reflect their actual shape, function, and operating characteristics. All onboard systems have been faithfully modeled and simulated to operate in both normal and emergency modes, including all modes of the navigation system, flight control system, and onboard weapon systems. A six-degree-of-freedom hydraulic motion system is used to replicate the sensations of movement in the simulator. The ambient noise simulation system precisely reproduces all sound effects generated by the simulated aircraft and its surroundings.

Fig. 7. Flight and combat mission simulator of the Su-22 attack aircraft (Poland)

ETC-PZL (Poland), in collaboration with the Polish Air Force Academy in Dęblin and the Military Institute of Aviation Medicine in Warsaw, designed and manufactured a device enabling spatial disorientation training. The device features a seven-degree-of-freedom motion system, including a hexapod platform mounted on a rotating table (Figure 8). Spatial disorientation is defined as an erroneous perception of position and movement relative to the Earth's surface. This perception is influenced by several factors. Pilots who unconsciously experience spatial disorientation, with a distorted perception of reality, often make erroneous decisions, frequently leading to tragic incidents and accidents. Therefore, proper preparation of pilots is crucial, involving familiarization with the phenomenon of spatial disorientation, the ability to recognize it, and developing appropriate habits, as well as prioritizing instrument readings over their own feelings.

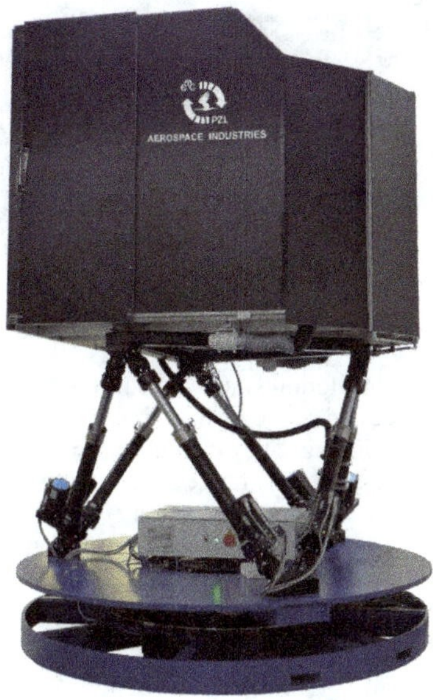

Fig. 8. Spatial disorientation training device.

Motion Systems, located near Wrocław, Poland, specializes in designing two-, three-, and six-degree-of-freedom motion simulators for professional and amateur applications. The simulators are used in various applications, including the entertainment industry, as well as in simulators specifically designed for driving schools and the military. MotionSystems' 2-DoF, 3-DoF, and 6-DoF simulation platforms have been utilized as motion systems in the development of vehicles, such as the Rosomak SK-1 APC (Fig. 9).

Fig. 9. MotionSystems 6-DoF simulator in military use.

The kinematic structures used in MotionSystems solutions are based on hexapod designs and modified delta structures with three and six degrees of freedom. 6-DoF stimulators can manipulate objects weighing from 150 kg to 2500 kg. They can move the work platform along the vertical z-axis and rotate it about the x- and y-axes. The range of motion for the vertical axis depends on the platform model, and the angular ranges for the x-, y-, and z- axes can be customized to suit specific needs and applications. Figure 10 shows one of the 6-DoF platforms.

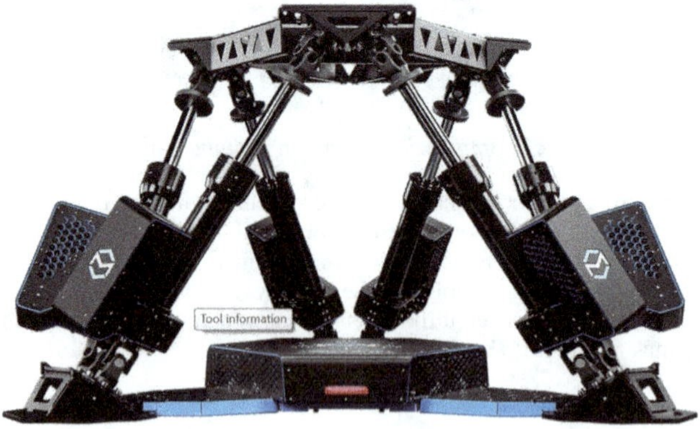

Fig. 10. MotionSystems - 6-DoF simulation platform model PS-6TL-1500

Materials Research Proceedings 62 (2026) 120-133 https://doi.org/10.21741/9781644904015-16

Results and Discussion

The development of kinematic models and analytical solutions for parallel robots and simulators, including those developed by Gough and D. Stewart, is a topic explored by many scientists and engineers. As early as 1947, E.V. Gough built a test stand and tire simulator for Dunlop (Figure 11). The mechanism under analysis has six degrees of freedom and is a structure in which linear drives are grouped in pairs in a triangular configuration. The drives are arranged between the base and the work platform to form triangles. This structure ensures the work platform has high stiffness and a high load capacity. A single kinematic chain has a cylindrical linear drive terminated with universal joints. The work platform is a ring, on the surface of which the ends of the universal joints are mounted. To visualize the shape of the workspace, assumptions were made regarding the geometrical constraints and physical limitations for the joints' ranges of motion. The linear drive operating range was assumed to be 0-1000 mm, and the operating range of the universal joints in the first axis was assumed to be from $\pi/4 \div 3\pi/4$, and the operating range of the universal joints in the second axis was assumed to be from $-\pi/4 \div \pi/4$. The shape of the workspace was determined based on these assumptions (Fig.2). However, it should be remembered that the volume and shape are highly dependent on the geometric dimensions of the remaining structural members.

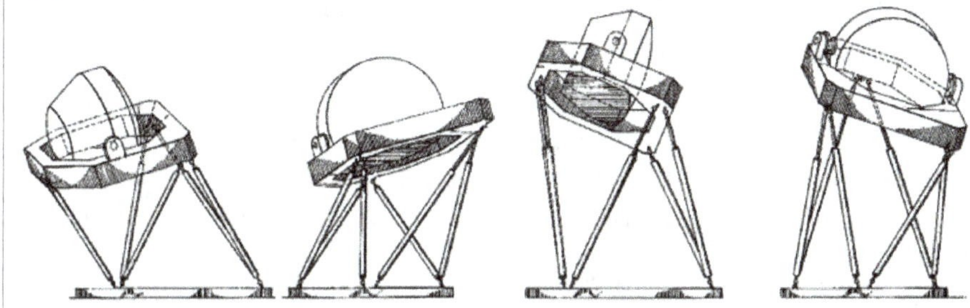

Fig. 11. Hexapod patent from 1967 r. (US 3,295,224) [1]

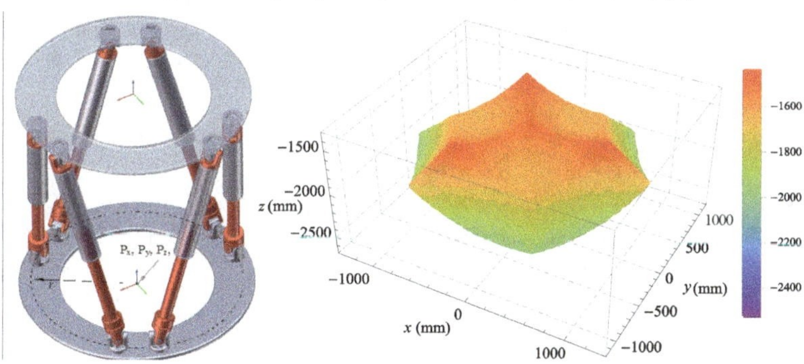

Fig. 12. Kinematic structure model and workspace of the Gough platform – axonometric view for zero orientation of the work platform [3-5].

Materials Research Proceedings 62 (2026) 120-133 https://doi.org/10.21741/9781644904015-16

The second most recognizable structure with closed kinematic chains is the D. Stewart platform. In 1965, D. Stewart published a paper in IMechE describing a parallel platform manipulator with six degrees of freedom, designed as a simulator for the aviation industry. The parallel mechanism consisted of six drives connected by joints to a platform on which the pilot's pulpit was mounted (Fig. 13). This mechanism differs from E.V. Gough's hexapod, which is paradoxically often referred to as the "Stewart Platform." Without a doubt, D. Stewart's innovative design had a significant impact on the further development of parallel kinematics used in flight simulators.

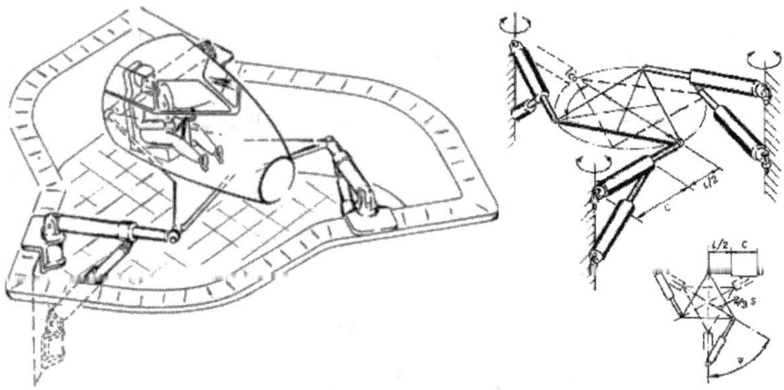

Fig. 13. Symulator lotniczy wg D. Stewarta [2]

The analyzed mechanism has six degrees of freedom and is a structure in which linear drives are grouped in pairs, one above the other. The drives are arranged in an equilateral triangle. A characteristic feature is that some drives support others, creating triangular structures. This structure provides significant stiffness and load capacity for the work platform. In the analyzed structure, a single kinematic chain has two cylindrical linear drives connected to the base by two rotary joints, forming a universal joint. Three universal joints are mounted in the equilateral triangle-shaped work platform. The joints connect the work platform and the three device drives. Fig.14 shows a view of an example solid model and the designated example workspace of the simulator.

Materials Research Proceedings 62 (2026) 120-133 https://doi.org/10.21741/9781644904015-16

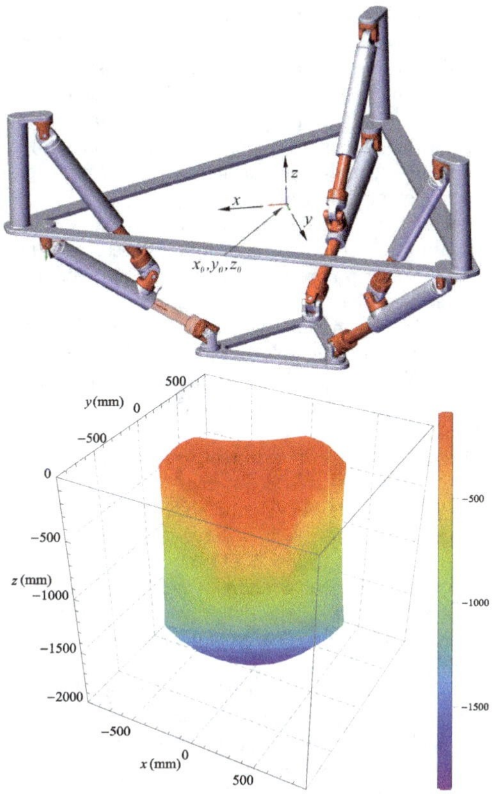

Fig. 14. Solid model of the Stewart platform type, example shape of the workspace [3-5]

The workspace of the analyzed mechanism was determined, taking into account geometric constraints and physical limitations on the joints' ranges of motion. The linear drives were assumed to have a working range of 0-1000 mm, and the working range of the actuator's rotational axis was assumed to be from $-\pi/6 \div \pi/6$

Technological, Economic, Environmental, and Social Impact

The implementation of simulators based on closed kinematic mechanisms (parallel platforms) has a significant impact in all key areas: technological, economic, environmental, and social.

Technological progress in military and aviation simulators is closely linked to the development of advanced drive systems and control methods. The high dynamics, stiffness, and accuracy of parallel mechanisms require precise and intelligent positioning systems, such as those using electro-pneumatic servo drives [6]. Research into pneumatic servo drives, including those using artificial pneumatic muscles, enables the design of new motion systems [7, 8], which directly impacts the capabilities of the 6-DoF platform [9]. The development of advanced drive components, including proportional valves with piezoelectric actuators, enables the construction of smaller and more responsive motion bases [10].

Materials Research Proceedings 62 (2026) 120-133 https://doi.org/10.21741/9781644904015-16

In terms of design itself, prototyping low-cost yet functional parallel robots, such as the Delta 4-DoF type [11], provides a basis for optimizing the production costs of training bases. Research also includes the design and control of smaller, flat, parallel manipulators with electro-pneumatic servo drives, with an emphasis on energy efficiency [12]. The use of advanced mechanical and drive systems, such as electro-hydraulic lifting and leveling systems (e.g., in construction robotics [13]) and general-purpose lifting systems [14], expands the technological base for hydraulic simulators. Rapid prototyping technologies for key components, e.g., control valves [15], accelerate the research and development cycle.

Advanced analytical and computational methods are crucial for reliability and accuracy. Efficient use of GPU computing power for surrogate models is essential for real-time simulations. Statistical methods, such as bootstrap approaches for assessing parameter significance in RSM models [16] and fixed-effects models [17], or nonparametric methods [18] for quantifying uncertainty, are crucial for validating motion dynamics. The use of fuzzy logic for operations on correlated estimates [19] or for evaluating ANOVA results [20] allows for better management of uncertainty and complexity in measured data. Reducing data dimensionality using principal component analysis (PCA), taking into account its robustness and uncertainty [21], is crucial in designing control algorithms.

Parallel simulators, being a costly investment, must demonstrate high training cost-effectiveness. This is achieved through advanced process modelling and optimization. Compared to the classical design of experiments (DoE), methods such as fuzzy regression enable the optimization of manufacturing processes, such as assembly [22] and production (e.g., modeling processes in technological machines using RSM models [23]). Improving manufacturing processes, for example, by modeling error-counting systems in PCB soldering [24] or heuristic optimization [25], directly reduces costs and improves the quality of components used in simulators. The use of computational tools tailored to the competency level of the user (engineers and operators) also impacts work efficiency and personnel training costs [26]. In a broader economic context, legal and regulatory issues (ordoliberalism and EU tax law [27, 28]) influence international trade and the production of advanced training systems.

The environmental and social impact of simulators primarily concerns safety, reliability, and alternative technologies. The use of statistical methods, such as inverse problem analysis [29, 30] and statistical methods for ensuring stability and reducing vibration [31-33], is essential for operator safety and equipment longevity. The context of the system – the Human-Technical Object-Environment relationship – is crucial in risk assessment. Replacing traditional hydraulic drives with pneumatic muscle servo drives [6] and electro-pneumatic drives [7, 10] can reduce noise and oil consumption, which is beneficial for the environment. Furthermore, research on new materials, such as impedance studies on modified activated carbon [34], could potentially lead to improved power supply or filtration systems in simulators [35]. Simulator development also has significant potential for advancing phytochemical and biotechnological research [36-38] due to the research, design, and analytical capabilities provided by coupling VR/AR visualization with motion simulators.

Socially, simulators provide a safe and repeatable platform for training military and aviation personnel, which has a profoundly valuable impact on public safety and combat readiness. Moreover, advanced simulation and numerical techniques used to design kinetic systems can be applied to other fields, such as the rapid assessment of spine biomechanics (in silico simulations) [34], demonstrating the broad reach and societal value of computational engineering. The ability to cope with vibrations or machine reliability issues (e.g., analysis of turbulence, dynamic instability) has a direct impact on the reliability of the simulators themselves, which is crucial for training in critical situations [35].

Summary

The article discusses the application of manipulators with closed kinematic chains, specifically parallel platforms (mainly of the Gough and Stewart type), as motion bases in simulators for military and aviation vehicles. The first part presents an overview of existing solutions – mainly 6-DOF hexapods – used in flight and military simulators, including Thales Group platforms for the Boeing 787 and HELISIM helicopter simulators, L-3 company simulators, a research test bench at the University of Bingen for testing aircraft prototypes, as well as Polish designs: KTO ROSOMAK simulators (JASKIER, TASZNIK), the ATMX6000 platform by Automex, W-3WA "Sokół" and Su-22 simulators developed by ETC-PZL, and a device for spatial disorientation training. Commercial platforms by Motion Systems (3 and 6 DOF) used, among others, in military vehicle simulators are also discussed. In the analytical part, the focus is on two classical parallel structures: the Gough platform and the Stewart platform. In summary, the article presents example workspaces determined for both platforms, assuming specific geometric constraints and ranges of motion of the actuators and joints. The illustrations depict the shape of these workspaces and highlight that their volume and form are strongly dependent on the design dimensions. The article concludes that parallel kinematic mechanisms – despite their higher control complexity – constitute a very effective basis for simulators of military and aviation vehicles, especially where high dynamics, accuracy, and realism of motion reproduction are required.

References

[1] V.E. Gough, S.G. Whitehall, (1962) Universal tyre test machine, Proceedings of 9th International Congress FISITA, May 1962, 117-137.

[2] D. Stewart, A platform with six degrees of freedom, Proceedings of the Institution of Mechanical Engineers 180 (1965) 371-386. https://doi.org/10.1243/PIME_PROC_1965_180_029

[3] D.S. Pietrala et al., Design, Implementation, and Experimental Evaluation of a 6-DoF Parallel Manipulator Driven by Pneumatic Muscles, Applied Sciences 15 (2025) art. 10126. https://doi.org/10.3390/app151810126

[4] D.S. Pietrala et al., Design and Control of a Pneumatic Muscle Servo Drive Applied to a 6-DoF Parallel Manipulator, Applied Sciences 14 (2024) art.5329. https://doi.org/10.3390/app14125329

[5] P.A. Laski, M. Smykowski, Using a development platform with an stm32 processor to prototype an inexpensive 4-DOF delta parallel robot, Sensors 21 (2021) art. 7962. https://doi.org/10.3390/s21237962

[6] J.E. Takosoglu, P.A. Laski, Intelligent positioning system of electro-pneumatic servo-drive, Annals of DAAAM and Proc. International DAAAM Symposium (2011) 1641-1642.

[7] D.S. Pietrala, P.A. Laski, Design and Control of a Pneumatic Muscle Servo Drive Containing Its Own Pneumatic Muscles, Applied Sciences (Switzerland) 12 (2022) art. 11024. [21]

[8] J. Takosoglu, Angular position control system of pneumatic artificial muscles, Open Engineering 10 (2020) 681-687. [30]

[9] D.S. Pietrala, P.A. Laski, Design and Control of a Pneumatic Muscle Servo Drive Containing Its Own Pneumatic Muscles, Applied Sciences 12 (2022) art. 11024. https://doi.org/10.3390/app122111024

[10] P.A. Laski, Proportional valve with a piezoelectric actuator, EPJ Web of Conferences 143 (2017) art. 02064. https://doi.org/10.1051/epjconf/201714302064

[12] J. Takosoglu et al., A Design of a 2 DoF Planar Parallel Manipulator with an Electro-Pneumatic Servo-Drive—Part 2, Energies 16 (2023) art. 2970. https://doi.org/10.3390/en16072970

[13] P. Wos et al., The Electro-Hydraulic Lifting and Leveling System for the Bricklaying Robot, Lecture Notes in Mechanical Engineering 24 (2021) 216-227. [19]

[14] P. Wos et al., Bricklaying robot lifting and levelling system, Communications - Scientific Letters of the University of Žilina 23 (2021) B257-B264. https://doi.org/10.26552/COM.C.2021.4.B257-B264

[15] S. Blasiak et al., Rapid prototyping of pneumatic directional control valves, Polymers 13 (2021) art. 1458. https://doi.org/10.3390/polym13091458

[16] J. Pietraszek, L. Wojnar, The bootstrap approach to the statistical significance of parameters in RSM model, ECCOMAS Congress 2016 – Proc. 7th Europ. Congr. Comput. Methods in App. Sci. Eng. 1 (2016) 2003-2009. https://doi.org/10.7712/100016.1937.9138

[17] J. Pietraszek et al., The bootstrap approach to the statistical significance of parameters in the fixed effects model, ECCOMAS Congress 2016 - Proc. 7th Europ. Congr. Comput. Methods in App. Sci. Eng. 3 (2016) 6061-6068. https://doi.org/10.7712/100016.2240.9206

[18] J. Pietraszek et al., The non-parametric approach to the quantification of the uncertainty in the design of experiments modelling, UNCECOMP 2017 – Proc. 2nd Int. Conf. Uncertainty Quantification in Computational Sciences and Engineering (2017) 598-604. https://doi.org/10.7712/120217.5395.17225

[19] J. Pietraszek, The modified sequential-binary approach for fuzzy operations on correlated assessments, Lecture Notes in Computer Science 7894 LNAI (2013) 353-364. https://doi.org/10.1007/978-3-642-38658-9_32

[20] J. Pietraszek et al., The fuzzy approach to assessment of ANOVA results, Lecture Notes in Computer Science 9875 LNCS (2016) 260-268. https://doi.org/10.1007/978-3-319-45243-2_24

[21] J. Pietraszek, E. Skrzypczak-Pietraszek, The uncertainty and robustness of the principal component analysis as a tool for the dimensionality reduction, Solid State Phenomena 235 (2015) 1-8. https://doi.org/10.4028/www.scientific.net/SSP.235.1

[22] J. Pietraszek, Fuzzy regression compared to classical experimental design in the case of flywheel assembly, Lecture Notes in Computer Science 7267 LNAI (2012) 310-317. https://doi.org/10.1007/978-3-642-29347-4_36

[23] J. Pietraszek et al., The parametric RSM model with higher order terms for the meat tumbler machine process, Solid State Phenomena 235 (2015) 37-44. https://doi.org/10.4028/www.scientific.net/SSP.235.37

[24] J. Pietraszek et al., Modeling of errors counting system for PCB soldered in the wave soldering technology, Advanced Materials Research 874 (2014) 139-143. https://doi.org/10.4028/www.scientific.net/AMR.874.139

[25] J. Pietraszek, A. Goroshko, The heuristic approach to the selection of experimental design, model and valid pre-processing transformation of DoE outcome, Advanced Materials Research 874 (2014) 145-149. https://doi.org/10.4028/www.scientific.net/AMR.874.145

[26] M. Radek et al., Matching Computational Tools to User Competence Levels in Education of Engineering Data Processing, Materials Research Proceedings 34 (2023) 453-459. https://doi.org/10.21741/9781644902691-52

[27] E. Kozień, A. Kozień, Is ordoliberalism democratic? – Legal approach, System Safety: Human - Technical Facility - Environment 2 (2020) 33-38. https://doi.org/10.2478/czoto-2020-0005

[28] A. Kozień, N. Kozłowska, Harmonization and Deharmonization of Excise Duty in the European Union as Contemporary Challenges of the EU Tax Law, WSEAS Transactions on Business and Economics 19 (2022) 816-824. https://doi.org/10.37394/23207.2022.19.71

[29] V. Royzman, A. Goroshko, Multiple inverse problem, Journal of Vibroengineering 14 (2012) 1417-1424.

[30] A.V. Goroshko, V.P. Royzman, Statistical methods for providing the stability of the solutions of inverse problems and their application to decrease rotor vibroactivity, Journal of Machinery Manufacture and Reliability 44 (2015) 232-238. https://doi.org/10.3103/S1052618815030073

[31] A. Goroshko et al., Multiphysics Modeling of Direct-Start Induction Motor Considering Rotor Unbalance, Materials Research Proceedings 45 (2024) 267-276. https://doi.org/10.21741/9781644903315-31

[32] R. Dwornicka et al., The smoothed bootstrap fine-tuning, System Safety: Human - Technical Facility - Environment 1 (2019) 716-723. https://doi.org/10.2478/czoto-2019-0091

[33] V. Ptashnyk et al., Impedance investigation of activated carbon material modified by ultrasound treatment; [Badanie impedancji materiału z węgla aktywowanego zmodyfikowanego metodą ultradźwięków] , Przeglad Elektrotechniczny 94 (2018) 186-189. [14]

[34] J. Lorkowski et al., Effects of breast and prostate cancer metastases on lumbar spine biomechanics: Rapid in silico evaluation, Advances in Experimental Medicine and Biology 1096 (2018) 31-39. https://doi.org/10.1007/5584_2018_192

[35] Ł. Łacny et al., Selected overview of the impact of ground motion on the vibrations of particle accelerators, AIP Conf. Proc. 2239 (2020) art. 020025. https://doi.org/10.1063/5.0008950

[36] E. Skrzypczak-Pietraszek et al., Gentiana tibetica King from tissue culture, Scientia Pharmaceutica 61 (1993) 287-296.

[37] E. Skrzypczak-Pietraszek et al., Biotransformation of hydroquinone to arbutin in plant in vitro cultures – Preliminary results, Acta Physiologiae Plantarum 27 (2005) 79-87. https://doi.org/10.1007/s11738-005-0039-x

[38] E. Skrzypczak-Pietraszek, J. Pietraszek, Seasonal changes of flavonoid content in Melittis melissophyllum L. (Lamiaceae), Chemistry and Biodiversity 11 (2014) 562-570. https://doi.org/10.1002/cbdv.201300148

Terotechnology XIV | Materials Research Forum LLC
Materials Research Proceedings 62 (2026) 134-141 | https://doi.org/10.21741/9781644904015-17

Laser-Made Heat Exchangers for Boiling Heat Transfer Augmentation

Łukasz J. ORMAN[1,a *], Norbert RADEK[1,b] and Stanislav HONUS[2,c]

[1]Kielce University of Technology, al. Tysiaclecia P.P.7, 25-314 Kielce, Poland

[2]VSB - Technical University of Ostrava, 17. listopadu 2172/15, 708 00 Ostrava-Poruba, Czech Republic

[a]orman@tu.kielce.pl, [b]norrad@tu.kielce.pl, [c]stanislav.honus@vsb.cz

Keywords: Boiling, Laser Treatment, Surface Treatment

Abstract. The use of lasers for surface microgeometry modification can be an effective method for improving the performance of heat exchangers, particularly in cases involving boiling. The paper describes the performance of laser–treated surfaces as well as their production. It discusses the morphology and visual characteristics during boiling under ambient pressure. A modification of the model selected from the literature has enabled verification of its usefulness in predicting heat flux from laser–made surfaces.

Introduction

The application of lasers has become a common method for metal processing, used for a wide range of purposes. The broad and significant potential of this technique has created considerable interest in various areas of engineering and technology. Particularly promising is the use of the laser beam for heat transfer augmentation purposes. Anderson et al. [1] noted that increasing the surface area can enhance heat transfer characteristics, and roughening may be achieved with a pulsed laser. When applied to copper, picosecond laser-generated mound-like microstructures. However, the authors stated that when treatment is carried out in ambient air, surface oxides are created, which influence the heat transfer characteristics of the surface. In their paper, an efficient and effective method for removing oxides (from both surface and subsurface levels) was proposed, utilizing citric acid.

In the case of boiling, any modifications of the heater surface can lead to enhanced performance. Aizawa et al. [2] conducted subcooled water boiling tests on micro- and nano-textured copper and aluminum samples. Two types of samples were produced, with the plasma printing and wet plating methods. The first sample type consisted of regular concave micro-textures with square micro-cavities of a certain unit size and pitch created on an aluminum sheet. The second one consisted of acicular iron-nickel (Fe–Ni) alloy micro-textures in the form of pyramids applied to copper. The authors concluded that heat flux can be enhanced by micro- or nano-texturing, which can also control heat transfer between the flowing coolant and the heated surface. Additionally, based on high-speed camera observations, the authors claimed that the mass of the two-phase local flow with fine bubbles departs from the modified interface and moves away into the downstream region of the forced turbulent coolant flow.

The laser beam can be particularly efficient in altering the surface of metals and producing samples with properties favorable for boiling applications. Costa-Greger et al. [3] produced modifications on the aluminum surfaces and claimed an improvement of the maximal values of the heat transfer coefficient by 390% to 680% compared to the untreated surface. The initiation of boiling was also more favorable (superheats reduced by 35% - 53%). The authors also observed that the critical heat flux of the laser–made samples was higher (by approximately fifty percent). Sugumaran et al. [4] tested the boiling of distilled water on step-like microstructured copper surfaces created using a nanosecond laser. Two surfaces were prepared, differing in the width of the cavity at a specific depth. It was concluded that the modified surfaces showed superior performance compared to the untreated surface at heat fluxes below 100 kW/m². The authors attributed this fact to the larger nucleation sites and surface area that enhanced bubble formation, as well as the liquid feeding rate. Nevertheless, the heat transfer performance of the surface with a smaller groove width at the bottom showed a declining trend as the heat flux rose. It was explained by the fact that at lower heat flux values, smaller isolated bubbles were generated at a lower rate. As the heat flux increased, the bubbles merged with adjacent bubbles, creating larger bubbles that encountered resistance when attempting to escape from the heater. This led to a deterioration in the overall boiling heat transfer performance. On the other hand, pool boiling inversion and hysteresis on laser-processed stainless steel surfaces containing hierarchical micro/nanostructures were analyzed in [5]. The authors reported that at heat flux values below the pool boiling inversion point, the superheat value remained constant as boiling times extended, indicating an adequate steady-state condition. The opposite situation occurred at heat flux values above the pool boiling inversion point. As for hysteresis, it was stated to occur during each run.

Laser–made surfaces are also investigated in the flow boiling mode, which might even have broader practical application potential. Pettit et al. [6] carried out experiments on single-channel water flow boiling on steel surfaces treated with a laser. For the mass flow rates of 230 kg m^{-2} s^{-1} and 463 kg m^{-2} s^{-1}, the authors reported enhancements in the heat transfer coefficient of approximately 140% and 164%, respectively, compared to the untreated surface. Similarly, the critical heat flux of the modified surface rose by 22.1% and 11.6% for the same mass flow rates. Strąk and Piasecka [7] conducted experiments in FC-72 flow boiling in a minichannel. Several testing surfaces were modified with a laser (vibration-assisted processes were also used). It was stated that generally, augmentation of heat transfer (in relation to the smooth surface test results) occurred due to the application of the laser. Recently, Mertens et al. [8] experimentally analyzed flow boiling on 3 stainless steel foils modified with a femtosecond laser. The three samples included inclined microgrooves, inclined conical microholes, and laser-induced periodic surface structures. The authors reported that the specimen with inclined conical microholes provided the best results. It turned out to increase the heat transfer coefficients by up to seventy percent and reduce the incipient temperature by up to 13.5 °C compared to the unmodified reference surface. It is also worth noting that the most significant augmentation occurred at low heat fluxes. It was also concluded that further investigations are necessary to optimize surface patterns and gain a better understanding of how the geometrical parameters of the laser-made cavities influence the heat transfer performance.

Despite considerable interest in the subject and numerous scientific projects (both completed and ongoing), knowledge of boiling performance on laser–made surfaces remains incomplete. This paper aims to provide additional information on this issue, with a focus on samples with longitudinal grooves and the problem of heat transfer modeling.

Samples and Methodology

The laser beam was applied onto the substrates of 3 cm in diameter and 3 mm in height. It produced longitudinal grooves of pre-designed dimensions. The treatment lasted about two hours for each sample. The specimens were made of commercial copper. The pulsed fiber laser with a focal spot size of 35 μm and a power of 20 W was used for the procedures. After the treatment, the surface characteristics were analyzed with the optical microscope. The thermal action of the laser beam not only evaporates the metal material, forming grooves, but also alters the morphology of the surface at its bottom parts. It becomes rougher/uneven, which is actually a positive feature for boiling, because it promotes nucleation. Fig. 1 presents a laser-made sample and the details of the groove's surface, whose roughness proved to be at least several times the value of the smooth (untreated) surface.

Materials Research Proceedings 62 (2026) 134-141 https://doi.org/10.21741/9781644904015-17

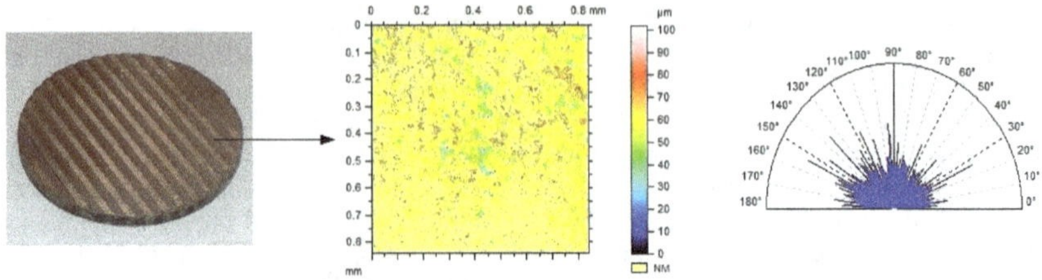

Fig. 1. The laser-made sample and characteristics of the groove's surface.

The experiments of boiling heat transfer cover two boiling agents: water and ethanol. The samples were soldered onto the copper heater and located horizontally. The measurements consisted of determining temperature values with thermocouples and calculating heat flux values based on these data. As a result, boiling curves can be created, which convey information about the performance of each sample (typically, five heat flux values were obtained for this purpose). The details of the experimental methodology were provided by Orman et al. [9].

Experimental

Nucleate boiling is characterized by the development of vapor bubbles and their movement within the pool of liquid. The intensity of bubble generation depends on heat flux. Fig. 2 presents the results of the investigation into the bubble movement at high thermal load with distilled water as the boiling agent. High-speed photos (taken every 0.002 seconds) reveal that at fully developed nucleate boiling, bubbles depart from sites on the entire surface and often merge with each other even before actual departure occurs.

The experimental data on heat flux dissipated from laser–made samples, presented by Orman et al. [9], indicate that a significant improvement can be obtained over a smooth surface. The heat flux values could be much larger than those recorded for the untreated reference surface at the same temperature difference between the surface and the liquid pool. Similarly, boiling could begin at much smaller temperature differences. Moreover, the authors reported that the augmentation was bigger for the sample with a larger fin height (groove depth) as well as a higher roughness of the surface at the groove's bottom. For both cases, the most visible enhancement was observed in the region of small temperature differences – below approximately 5 K. This is explained by the fact that at elevated temperature differences, more nucleation locations become active, and the morphology of the heater plays a less decisive role. When it comes to the groove depth in the fully developed boiling, the most vital mechanism of heat transfer might be the growth as well as departure of the bubbles and not convection without phase change (which can be enhanced with a larger surface area, namely for higher microfins (deeper grooves) [9].

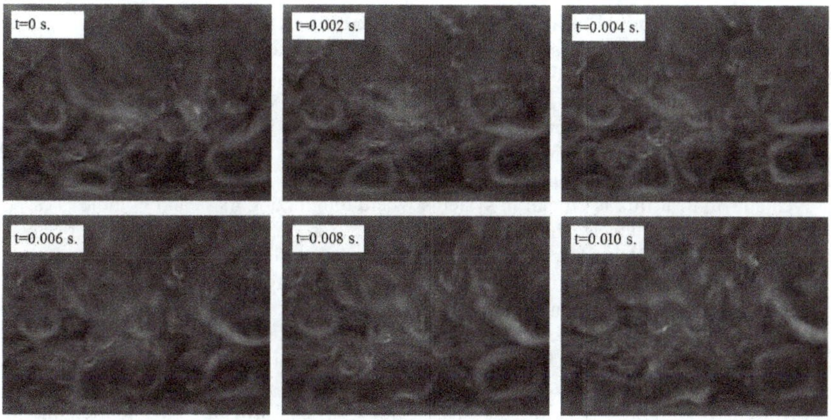

Fig. 2. High-speed images of the boiling process.

A crucial problem in designing efficient phase–change heat exchangers is the inaccuracy of models and correlations used for determining heat flux. Rannenberg and Beer [10] developed a model for enhanced surfaces, but of a different kind (i.e., wire meshes). Recently, a modification of this model was proposed by Orman et al. [11]. Fig. 3 presents a comparison of the experimental data adopted from [9] with the calculation results obtained from the original model [10] – (1) and from this model, modified with two constants [11] – (2).

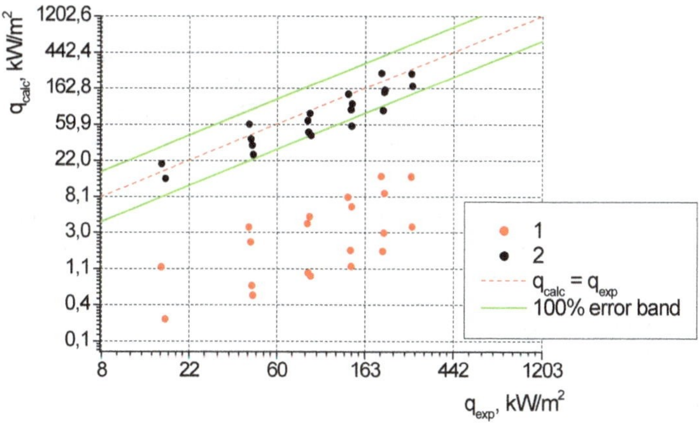

Fig. 3. Comparison of the experimental results from [9] with the calculation result according to the original model [10] – '1' and a modification of this model [11] – '2'.

As can be seen, the modification proved to be much more successful in predicting the heat flux dissipated from the laser–made surfaces. Almost all the experimental points (except two) are within the ±100% error band.

Materials Research Proceedings 62 (2026) 134-141 https://doi.org/10.21741/9781644904015-17

Technological, Economic, Environmental, and Social Impact

The technology of laser-assisted surface modification of heat exchangers, described in this article, has the potential to have a significant impact in many fields. Improving heat transfer efficiency, particularly in boiling processes, can lead to reduced energy consumption, which has a direct impact on environmental and economic aspects.

Technological improvements in boiling processes through laser surface treatment open the door to the design of more compact and efficient heat exchangers. The ability to achieve significant improvements in heat flux is crucial in systems where minimizing mass and volume is essential. Precision techniques, such as those described in research on control systems [12], advanced manipulator design [13], and servo drives [14], can be similarly applied to the control and optimization of laser surface modification, ensuring repeatability and high quality.

In the context of technological development, improving modeling and validation methods is also crucial. Although the abstract indicates that current heat flux prediction methods are not yet sufficiently reliable, the need for further work on a universal boiling model is clearly emphasized. Experience in the field of statistical modeling [15], solving complex inverse problems [16], constructing hybrid models [17], or stability analysis [18] can provide valuable methodological foundations for developing more precise and reliable correlations. Uncertainty in modeling [19] and the use of appropriate computational tools [20] are areas that have direct application in the validation of heat transfer models.

Economically, implementing this technology can contribute to operational cost savings by increasing energy efficiency. This applies to a wide range of applications, from advanced energy systems such as wind turbines [21] and induction motors [22], to the electronics industry, where error modeling [23] is crucial for production quality. Reducing device dimensions can also reduce material and transportation costs. Even in the context of materials, understanding phenomena such as corrosion [24] or material modification [25] is crucial for the longevity and stability of heat exchangers.

Increasing energy efficiency is a key driver for minimizing environmental impact. Lower energy consumption in industrial and refrigeration processes translates into reduced greenhouse gas emissions and reduced demand for primary resources. Although the article focuses on thermal engineering, research in the field of phytochemistry [26, 27] and biotechnology [28, 29] indicates a wide spectrum of interests. Research, including concerns for the sustainable use of natural resources and biotechnology, indirectly highlights the broader environmental awareness within the scientific community. Improving heat exchanger efficiency has broad societal implications, impacting energy quality and costs, as well as infrastructure development. Although not the primary goal, this technology could find applications in medicine, including more efficient cooling systems for diagnostic or therapeutic devices, as well as in research on the biomechanics of the human body [30] and orthopedic analysis [31]. The impact on society is also related to technical safety [32] and technological development, which require an appropriate legal context [33, 34] and cultural environment [35] for scientific progress and innovation. Investing in the development of technologies such as surface treatment lasers also requires support in the form of advanced mechanical components [36].

Summary

Laser-made surfaces can be effectively used for boiling augmentation. They might provide a significant improvement in the values of heat flux. However, prediction methods of heat flux are still not reliable enough. A modification of the correlation, adopted from the literature, showed quite satisfactory agreement with the experimental data; however, further work is necessary in this area to develop a universal model of nucleate boiling on laser–made surfaces, which would be applicable to a wide range of boiling agents as well as various pressure values.

Acknowledgement
This paper was created as part of the project No. CZ.02.01.01/00/22_008/0004631 "Materials and technologies for sustainable development" within the Jan Amos Komenský Operational Program financed by the European Union and from the state budget of the Czech Republic.

References

[1] M. Anderson et al., Shield, Heat transfer behavior of as-processed and cleaned picosecond pulse laser processed copper, Thermal Sci. Eng. Progr. 27 (2022) 101105. https://doi.org/10.1016/j.tsep.2021.101105

[2] T. Aizawa, N. Ono, H. Nakata, Boiling Heat Transfer on the Micro-Textured Interfaces, in: S.N.Kazi (Ed.) Heat Transfer – Fundamentals, Enhancement and Applications, IntechOpen, 2023. http://dx.doi.org/10.5772/intechopen.105872

[3] J. Costa-Greger et al., Augmented Pool Boiling Heat Transfer of PF-5060 Using Femtosecond Laser Surface Processed Aluminum: Effect of Laser Fluence, Proc. of 22nd IEEE Int. Conf. on Thermal and Thermomechanical Phenomena in Electronic Systems (ITherm), Orlando, USA, 2023, pp. 1-9. http://dx.doi.org/10.1109/ITherm55368.2023.10177664

[4] L. Sugumaran et al., Experimental investigation of nucleate pool boiling heat transfer characteristics on modified copper surface via laser-structured microstructures, IOP Conf. Series: Earth Env. Sci. 1281 (2023) 012062. http://dx.doi.org/10.1088/1755-1315/1281/1/012062

[5] J. Costa-Greger et al., Pool Boiling Inversion and Hysteresis with Femtosecond Laser Processed 304 Stainless Steel Surfaces for Heat Transfer Enhancement, Proc. of 19th IEEE Int. Conf. on Thermal and Thermomechanical Phenomena in Electronic Systems (ITherm), Orlando, USA, 2020, pp. 298-305. http://dx.doi.org/10.1109/ITherm45881.2020.9190948

[6] L. Pettit et al. Enhancing Minichannel Flow Boiling with Femtosecond Laser Surface Processed Stainless Steel Surfaces in Water, Proc of 23rd IEEE Int. Conf. on Thermal and Thermomechanical Phenomena in Electronic Systems (ITherm), Aurora, USA, 2024, pp. 1-8, http://dx.doi.org/10.1109/ITherm55375.2024.10709576

[7] K. Strąk, M. Piasecka, Laser Surface Texturing for the Intensification of Boiling Heat Transfer in a Minichannel, Energies 17 (2024) 6481. https://doi.org/10.3390/en17246481

[8] F. Mertens et al., Flow Boiling Heat Transfer Enhancement via Femtosecond Laser-Textured Inclined Microfeatures, Energies 18 (2025) 2732. https://doi.org/10.3390/en18112732

[9] Ł.J. Orman et al., Laser treatment technology of the surface for pool boiling heat transfer enhancement, Materials 16 (2023) 1365. https://doi.org/10.3390/ma16041365

[10] M. Rannenberg, H. Beer, H., Heat transfer by evaporation in capillary porous wire mesh structures. Letters in Heat and Mass Transfer 7 (1980) 425-436. https://doi.org/10.1016/0094-4548(80)90019-3

[11] Ł.J. Orman et al., Thermal Performance of Phase Change Copper Heat Exchangers Produced with the Laser Technology – Development of Model Modification, Rocznik Ochrona Środowiska 27 (2025) 628-634. https://doi.org/10.54740/ros.2025.050

[12] J. Takosoglu, Angular position control system of pneumatic artificial muscles, Open Engineering 10 (2020) 681-687. https://doi.org/10.1515/eng-2020-0077

[13] J. Takosoglu et al., A Design of a 2 DoF Planar Parallel Manipulator with an Electro-Pneumatic Servo-Drive – Part 2, Energies 16 (2023) art. 2970. https://doi.org/10.3390/en16072970

[14] D.S. Pietrala et al., Design and Control of a Pneumatic Muscle Servo Drive Applied to a 6-DoF Parallel Manipulator, Applied Sciences 14 (2024) art. 5329. https://doi.org/10.3390/app14125329

[15] J. Pietraszek, A. Goroshko, The heuristic approach to the selection of experimental design, model and valid pre-processing transformation of DoE outcome, Advanced Materials Research 874 (2014) 145-149. https://doi.org/10.4028/www.scientific.net/AMR.874.145

[16] V. Royzman, A. Goroshko, Multiple inverse problem, Journal of Vibroengineering 14 (2012) 1417-1424.

[17] A. Goroshko et al., Construction and practical application of hybrid statistically-determined models of multistage mechanical systems, Mechanika 20 (2014) 489-493. https://doi.org/10.5755/j01.mech.20.5.8221

[18] A.V. Goroshko, V.P. Royzman, Statistical methods for providing the stability of the solutions of inverse problems and their application to decrease rotor vibroactivity, J. Machinery Manuf. Rel. 44 (2015) 232-238. https://doi.org/10.3103/S1052618815030073

[19] J. Pietraszek et al., The non-parametric approach to the quantification of the uncertainty in the design of experiments modelling, UNCECOMP 2017 – Proc. 2nd Int. Conf. Uncertainty Quantification in Comp. Sci. Eng. (2017) 598-604. https://doi.org/10.7712/120217.5395.17225

[20] M. Radek et al., Matching Computational Tools to User Competence Levels in Education of Engineering Data Processing, Materials Research Proceedings 34 (2023) 453-459. https://doi.org/10.21741/9781644902691-52

[21] J. Zwierzchowski et al., Model tests of wind turbine with a vertical axis of rotation type Lenz 2, EPJ Web of Conf. 143 (2017) art. 02150. https://doi.org/10.1051/epjconf/201714302150

[22] A. Goroshko et al., Multiphysics Modeling of Direct-Start Induction Motor Considering Rotor Unbalance, Materials Research Proceedings 45 (2024) 267-276. https://doi.org/10.21741/9781644903315-31

[23] J. Pietraszek et al., Modeling of errors counting system for PCB soldered in the wave soldering technology, Advanced Materials Research 874 (2014) 139-143. https://doi.org/10.4028/www.scientific.net/AMR.874.139

[24] T. Lipiński, J. Pietraszek, Corrosion of the S235JR Carbon Steel after Normalizing and Overheating Annealing in 2.5% Sulphuric Acid at Room Temperature, Materials Research Proceedings 24 (2022) 102-108. https://doi.org/10.21741/9781644902059-16

[25] V. Ptashnyk et al., Impedance investigation of activated carbon material modified by ultrasound treatment, Przeglad Elektrotechniczny 94 (2018) 186-189. https://doi.org/10.15199/48.2018.05.33

[26] E. Skrzypczak-Pietraszek, J. Pietraszek, Chemical profile and seasonal variation of phenolic acid content in bastard balm (Melittis melissophyllum L., Lamiaceae), Journal of Pharmaceutical and Biomedical Analysis 66 (2012) 154-161. https://doi.org/10.1016/j.jpba.2012.03.037

[27] E. Skrzypczak-Pietraszek, J. Pietraszek, Seasonal changes of flavonoid content in Melittis melissophyllum L. (Lamiaceae), Chemistry and Biodiversity 11 (2014) 562-570. https://doi.org/10.1002/cbdv.201300148

[28] E. Skrzypczak-Pietraszek, Phytochemistry and biotechnology approaches of the genus exacum, The Gentianaceae – Volume 2: Biotechnology and Applications (2015) 383-401. https://doi.org/10.1007/978-3-642-54102-5_16

[29] E. Skrzypczak-Pietraszek et al., Elicitation with methyl jasmonate combined with cultivation in the Plantform™ temporary immersion bioreactor highly increases the accumulation of selected centellosides and phenolics in Centella asiatica (L.) Urban shoot culture, Engineering in Life Sciences 19 (2019) 931-943. https://doi.org/10.1002/elsc.201900051

[30] J. Lorkowski et al., Effects of breast and prostate cancer metastases on lumbar spine biomechanics: Rapid in silico evaluation, Advances in Experimental Medicine and Biology 1096 (2018) 31-39. https://doi.org/10.1007/5584_2018_192

[31] B. Jasiewicz et al., Inter-observer and intra-observer reliability in the radiographic measurements of paediatric forefoot alignment, Foot and Ankle Surgery 27 (2021) 371-376. https://doi.org/10.1016/j.fas.2020.04.015

[32] Ł. Łacny et al., Selected overview of the impact of ground motion on the vibrations of particle accelerators, AIP Conference Proceedings 2239 (2020) art. 020025. https://doi.org/10.1063/5.0008950

[33] A. Kozień, N. Kozłowska, Harmonization and Deharmonization of Excise Duty in the European Union as Contemporary Challenges of the EU Tax Law, WSEAS Transactions on Business and Economics 19 (2022) 816-824. https://doi.org/10.37394/23207.2022.19.71

[34] E. Kozień, A. Kozień, Is ordoliberalism democratic? – Legal approach, System Safety: Human - Technical Facility - Environment 2 (2020) 33-38. https://doi.org/10.2478/czoto-2020-0005

[35] W. Górny, A. Kozień, VI scientific conference legal forms of protection and care of monuments: „cultural heritage components – between the protection of tangible and intangible cultural heritage", Santander Art and Culture Law Review 2021 (2021) 236-239.

[36] W. Jasiński et al., Static Mechanical Force Amplifier on the Example of a Fastener with an Electromagnetic Bolt, Materials Research Proceedings 34 (2023) 246-251. https://doi.org/10.21741/9781644902691-29

Terotechnology XIV Materials Research Forum LLC
Materials Research Proceedings 62 (2026) 142-150 https://doi.org/10.21741/9781644904015-18

RAM Materials and Structures Manufactured Using Additive Techniques for Military Applications

Wojciech PRZYBYŁ

Military Institute of Engineer Technology, ul. Obornicka 136, 50-961 Wrocław, Poland, ORCID: 0000-0002-0384-3463

przybyl@witi.wroc.pl

Keywords: RAM, RCS, Stealth, Polymer Composites, Graphene

Abstract. This paper reviews contemporary radar-absorbing materials (RAM) exhibiting strong microwave attenuation. It discusses fundamental dissipation mechanisms, impedance matching principles and major RAM classes, including carbon composites, ferrites, conductive polymers and metamaterials. The experimental section examines polylactide (PLA) reinforced with flake graphene for fused-filament fabrication, evaluating its microwave absorption efficiency and shielding effectiveness as a structural material for additively manufactured electronic enclosures.

Introduction

Modern electromagnetic reconnaissance systems have elevated the strategic significance of stealth technologies. A key determinant of radar detectability is the Radar Cross-Section (RCS) [1], which depends on object size, geometry and material properties, including the use of radar-absorbing materials [2]. Shaping redirects scattered radiation, whereas RAM dissipates incident electromagnetic energy, reducing radar echo and increasing operational survivability [3].

RAM is employed across air, land and naval domains, functioning either as structural material or as specialized coatings [4–6]. Its effectiveness depends on bandwidth, incidence angle, polarization and environmental conditions. Airborne applications require low mass and thermal stability [7], while land systems demand resistance to mechanical and chemical stressors [8]. Parallel advances in radar sensors, including multi-band and passive systems, aim to counter stealth solutions [9], driving ongoing research on next-generation RAM.

Recent interest in additively manufactured polymer-based composites motivates investigation into their applicability as lightweight RAM solutions.

Physical Properties of RAM Materials

The interaction of electromagnetic waves with RAM is governed by complex permittivity and permeability, whose imaginary components determine loss mechanisms and absorption capacity [10]. Effective absorbers require impedance matching to free space [11], quantified by the reflection loss (RL) parameter; values below −10 dB indicate practical performance, with military applications often targeting −20 dB.

Absorption results from dielectric losses [12], magnetic losses [13], conduction losses [14] and structural scattering. Hybrid systems combining magnetic fillers with dielectric matrices improve impedance matching [15], while multilayer and gradient structures expand effective bandwidth [16]. Wideband absorption across radar bands (L–K) is particularly desirable [17]. Material thickness remains a constraint, motivating research on thin lightweight absorbers.

Classes of Contemporary Radar-Absorbing Materials

- Conductive Carbon Materials and Their Composites. Carbon black, activated carbon, carbon fibers, nanotubes and graphene provide strong dielectric losses due to high conductivity and surface area. They achieve high attenuation (RL below −20 dB), though typically within narrower bands. Hybridization with magnetic fillers broadens effective absorption [18].

Materials Research Proceedings 62 (2026) 142-150 https://doi.org/10.21741/9781644904015-18

- Ferrite and Other Magnetic Absorbers. Ferrites, iron oxides and magnetic alloys rely on magnetic-loss mechanisms and allow high attenuation with reduced thickness. Their performance depends on composition and microstructure [19], but higher density limits use in mass-constrained applications.
- Conductive Polymers (CPs). Polyaniline, polypyrrole and related polymers exhibit tunable conductivity and low density, serving mainly as dielectric absorbers. Their properties, however, are sensitive to environmental conditions and thermal degradation [20], so they are typically used as components of composites.
- Conductive Polymer Composites. Composites combining conductive polymers with carbon or magnetic fillers provide synergistic dielectric and magnetic losses, enabling high RL and wide effective bandwidths. Their limitations include thermal stability and manufacturing complexity [21]
- Metamaterials. Metamaterials employ engineered sub-wavelength structures to achieve resonant absorption phenomena, offering very high attenuation at minimal thickness. Their drawbacks include narrow operational bandwidths and angle sensitivity, though multilayer and tunable architectures alleviate these constraints [22]

Studies on PLA Filament with Flake Graphene Additive

The author conducts research on the use of consumer-grade filaments for additive FFF fabrication of electronic device enclosures with electromagnetic-absorbing properties. A biodegradable polylactide (PLA) filament containing a small fraction of expanded flake graphene (EFG) was selected to assess its potential for electromagnetic shielding and radar-absorption in the 4–18 GHz band.

The experimental work employed a commercial Prografen PLA Graphene Strong filament (1.75 mm) with a declared 5% graphene content. The material, intended for advanced prototyping, exhibits improved mechanical characteristics, slightly enhanced thermal conductivity, reduced shrinkage and higher printing precision compared with standard PLA-G formulations.

Given that effective absorption requires continuous conductive networks or magnetic-loss mechanisms, the chosen PLA–graphene filament provides a relevant case study for assessing the feasibility limits of consumer-grade composites.

A. Research Methodology

The research procedure included the following stages:

- Manufacturing a test sample using 3D printing (FDM – Fused Deposition Modeling) technology with dimensions of 20×20×1 mm;
- Microscopic observation of surface morphology using a Keyence VHX-S750E digital optical microscope and surface roughness measurement;
- Measurement of electrical conductivity to assess the continuity of the conductive network formed by graphene particles in the polymer matrix;
- Testing of absorption properties in the microwave range using a PR-17CXKu reflectometer in the frequency range of 4-18 GHz with a measurement step of 100 MHz.

Microwave absorption measurements were performed three times for each of the two sample orientation directions (transverse and longitudinal) relative to the arrangement of the measuring antennas, which allowed for the assessment of anisotropy of the electromagnetic properties of the tested material. An aluminum plate was adopted as a reference, and the results are presented as the difference between reflection from the tested sample and reflection from the reference, expressed in decibels [dB].

Materials Research Proceedings 62 (2026) 142-150 https://doi.org/10.21741/9781644904015-18

B. Research Results

1) Mechanical Properties and Surface Morphology

The printed sample with a thickness of 1 mm was characterized by significant stiffness, which is an important advantage in the context of the potential application of this material for manufacturing self-supporting enclosures of electronic devices with relatively thin walls (approx. 1 mm). Microscopic observations (Fig. 1) showed that the sample surface has a structure of regular bands with a width of approx. 0.8 mm, resulting from the applied trajectory of the printing head. Profilometric measurements revealed the presence of periodic depressions with a depth of approx. 0.2 mm, being an effect of the technological process and the applied printing parameters (speed, temperature, filling).

(A)

(B)

Sample surface observed under an optical microscope (A), visualization of sample surface changes (B)

2) Electrical Conductivity

Measurements of the electrical resistance of the sample showed that the tested material is characterized by very high resistance (practically infinite), which means no electrical conductivity. This is explained by the low content of flake graphene in the filament (5%) and lack of electrical continuity between its particles, which remain isolated from each other by the surrounding PLA polymer matrix, causing the percolation threshold, which would enable electrical conductivity in the entire volume of the sample, not to be exceeded.

3) Absorption Properties in the Microwave Range

Measurements of electromagnetic wave attenuation in the 4-18 GHz range showed that the tested material is characterized by very small absorption properties, additionally dependent on frequency and to a small extent on the orientation of the sample relative to the incident electromagnetic wave (Fig. 2).

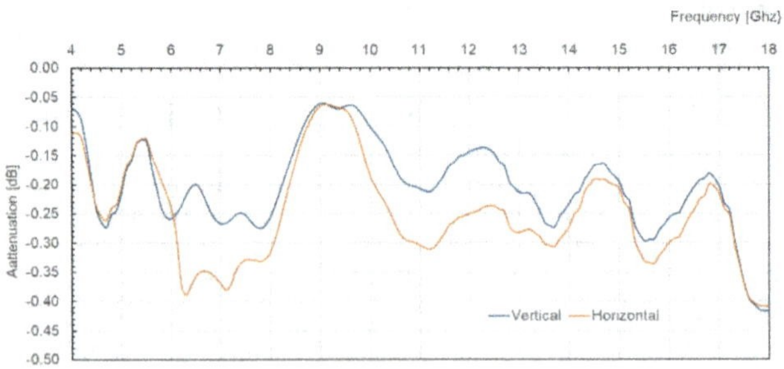

Visualization of average attenuation measurements for samples in horizontal (H) and vertical (V) orientation.

For horizontal orientation, the attenuation level changed as follows:

- In the 6-8 GHz range, an increase in attenuation was observed up to approx. 0.35 dB
- In the 9-17 GHz range, there was variable attenuation – from approx. 0.05 dB to approx. 0.35 dB, reaching a value of approx. 0.41 in the vicinity of 18 dB

For vertical orientation, the attenuation pattern had a similar character, but with significant quantitative differences:

- In the 6-9 GHz range, there was a smaller decrease in attenuation, to a value of approx. 0.25 dB
- The maximum attenuation, also located at a frequency of 18 GHz, reached approx. 0.42 dB

The obtained results clearly indicate a slight anisotropy of the absorption properties of the tested material and may result from the band structure of the sample being an effect of the applied manufacturing technology (filament 3D printing). The maximum attenuation value of 0.42 dB corresponds to a reduction of reflected power by approx. 9.3%, which is a value too low for effective RAM materials used in military applications, where typically attenuation of 10-20 dB (90-99% reduction of reflected power) is required.

Economic and Technological Impact

The development of radar-absorbing materials (RAM) and additive manufacturing (AM) structures for military applications, although currently in the research phase, offers significant long-term technological and economic benefits. Innovation is needed in the materials engineering of RAM composites. Integrating electromagnetic functionality with structural functionality is crucial. To meet military requirements, RAM materials must effectively utilize dielectric, conductive, interfacial polarization, and magnetic loss mechanisms. The study suggests increasing the graphene content and incorporating ferromagnetic additives. Such development is based on:

- multiphase composite engineering; the production of advanced composites requires in-depth knowledge of microstructure, which is the subject of research in materials science [23]. This development includes composites based on titanium alloys [24], ceramics [25], and magnetic materials [26]. It is necessary to create hybrid dielectric-magnetic mechanisms that will provide broadband attenuation [27].
- surface and functional layer modifications; surface modification techniques can be used to achieve better impedance matching and increased absorption. Research on coatings and their properties [28], laser processing [29,30] and electrospark processing [31,32] is important because it allows for the precise shaping of functional layers that influence loss mechanisms.

AM, as a technology enabling the creation of complex geometries, is becoming crucial for the production of integrated military systems. AM enables the creation of structures whose functionality depends on geometry, not just composition (metamaterials). This requires high-precision process control [33]. The development of drive systems, including pneumatic ones [34], has a direct impact on the accuracy of AM machines. AM enables the production of specialized electronic enclosures that must be both durable and electromagnetically functional, often in the context of complex systems [35,36]. Research on the strength of composites [37,38] is essential for RAM structures to meet load-bearing requirements, which is crucial for military applications [39].

The introduction of AM and new materials into military production has the potential to radically change the economics of manufacturing, especially in prototyping, short-run production, and quality management. To achieve the required level of damping and reliability, manufacturing processes must be closely controlled and optimized, especially when using new techniques such as AM.

Technology assessments [40,41], process stability analysis [42], and risk assessment [43] are crucial in military production. The use of advanced statistical and modeling techniques, including regression and reliability analysis [44], enables precise refinement of AM process parameters to achieve repeatable AM properties. This is essential for process and risk management. Integrating AM with the Industry 4.0 concept requires the digitalization of processes [45,46] and the use of advanced computational tools for modeling and simulation [47,48].

The introduction of innovative materials into production, even in the military sector, must take into account the broader environmental and ethical context. Although the military prioritizes functionality, the trend in engineering is toward product life cycle assessment (LCA) [49] and sustainable resource management. Even RAM materials based on polylactide (PLA) must align with broader concepts of responsibility [50] and environmental care [51]. Ultimate economic success depends on the durability and reliability of components. Research on degradation [52], fatigue [53], and corrosion of materials [54] ensures that expensive RAM components manufactured using AM will achieve their expected operational lifespan.

Conclusions

The investigation of PLA filament containing 5% flake graphene indicates that, in its current formulation, the material does not demonstrate functional characteristics required for effective RAM or electromagnetic shielding applications. The graphene content is insufficient to form a continuous conductive network, resulting in negligible dielectric, conductive and interfacial-polarization losses. The lack of magnetic constituents excludes magnetic loss mechanisms, while the 1-mm sample thickness is far below the minimum required for absorption in the 4–18 GHz band.

The measured attenuation (max. 0.42 dB) falls significantly short of levels expected for military applications, and the observed reduction is likely dominated by surface-scattering effects rather than true absorption. Likewise, the electrical conductivity is far too low to ensure meaningful electromagnetic shielding, as neither reflection-based nor absorption-based mechanisms are active at such filler concentration.

The material nonetheless shows developmental potential. Increasing graphene loading, increasing layer thickness and incorporating ferromagnetic additives could enable the formation of hybrid dielectric–magnetic absorption mechanisms more suitable for wideband attenuation. The study contributes to the understanding of consumer-grade composite filaments in additive manufacturing and highlights key directions for advancing 3D-printed RAM materials.

References

[1] E.F. Knott et al., Radar cross section, SciTech Publishing, 2004.

[2] P. Thomas et al., Wideband radar absorbing structure using polyaniline-graphene nanocomposite, C 6 (2020) art. 72. https://doi.org/10.3390/c6040072

[3] I. Plebankiewicz et al., Ochrona krzemowych paneli fotowoltaicznych i ładowarek słonecznych przed rozpoznaniem radarowym, Przegląd Elektrotechniczny 100 (2024) 154-157. https://doi.org/10.15199/48.2024.11.28

[4] I. Huynen et al., Multifunctional hybrids for electromagnetic absorption. Acta Materialia, 59 (2011) 3255-3266. https://doi.org/10.1016/j.actamat.2011.01.065

[5] D. Micheli et al., Broadband electromagnetic absorbers using carbon nanostructure-based composites. IEEE Trans. Microwave Theory and Techniques 59 (2011) 2633-2646. https://doi.org/10.1109/TMTT.2011.2160198

[6] F. Ruiz-Perez et al., Carbon-based radar absorbing materials: A critical review. Journal of Science: Advanced Materials and Devices 7 (2022) art.100454. https://doi.org/10.1016/j.jsamd.2022.100454

[7] A.K. Saha, W.G. Pendleton, Role of Dielectric Loss in Microwave Absorber Design. SoutheastCon, Mobile, AL, USA, (2022) 542-543. https://doi.org/10.1109/SoutheastCon48659.2022.9764067

[8] W. Przybył et al., Microwave absorption properties of carbonyl iron-based paint coatings for military applications. Defence Technology 22 (2023) 1-9. https://doi.org/10.1016/j.dt.2022.06.013

[9] M.I. Skolnik, Introduction to radar systems, Vol. 3, New York: McGraw-Hill, New York, 1980, pp. 81-92.

[10] W. Przybył et al., Badania radarowe paneli fotowoltaicznych w kontekście ich ochrony radiolokacyjnej. Przegląd Elektrotechniczny 65 (2024) 6-10. https://doi.org/10.15199/13.2024.9.1

[11] S. Jovanović et al., A review on graphene and graphene composites for application in electromagnetic shielding, Graphene and 2D Materials 8 (2023) 59–80. https://doi.org/10.1007/s41127-023-00065-3

[12] A. Munir, Microwave radar absorbing properties of multiwalled carbon nanotubes polymer composites: A review, Advanced Electromagnetics 6 (2007) 10-19. https://doi.org/10.1002/adv.21617

[13] H. Bayrakdar, Complex permittivity, complex permeability and microwave absorption properties of ferrite–paraffin polymer composites, Journal of Magnetism and Magnetic Materials 323 (2011) 1882-1885. https://doi.org/10.1016/j.jmmm.2011.02.030

[14] S. Bashir et al., Interactive nanomaterials for energy storage and conversion, in: Nanostructured materials for sustainable energy: design, evaluation, and applications, American Chemical Society, 2022, pp. 27-81. https://doi.org/10.1021/bk-2022-1421.ch002

[15] Y. Du et al., Shell thickness-dependent microwave absorption of core-shell Fe3O4@C composites. ACS Applied Materials & Interfaces 6 (2014) 12997-13006. https://doi.org/10.1021/am502910d

[16] Y.B. Feng et al., Absorbing properties and structural design of microwave absorbers based on carbonyl iron and barium ferrite, Journal of Magnetism and Magnetic Materials 318 (2007) 8-13. https://doi.org/10.1016/j.jmmm.2007.04.012

[17] J. Lyu et al., Nanofibrous kevlar aerogel films and their phase-change composites for highly efficient infrared stealth, ACS Nano 13 (2019) 2236-2245. https://doi.org/10.1021/acsnano.8b08913

[18] D.X. Yan et al., Structured reduced graphene oxide/polymer composites for ultra-efficient electromagnetic interference shielding, Advanced Functional Materials 25 (2015) 559-566. https://doi.org/10.1002/adfm.201403809

[19] Z. Chen et al., Lightweight and flexible graphene foam composites for high-performance electromagnetic interference shielding, Advanced Materials 25 (2013) 1296-1300. https://doi.org/10.1002/adma.201204196

[20] Z. Wang et al., Dielectric and microwave attenuation properties of graphene nanoplatelet-epoxy composites, AIP Advances 4 (2014) art.017139. https://doi.org/10.1063/1.4863687

[21] Y. Sani et al., Enhanced electromagnetic microwave absorbing performance of carbon nanostructures for RAMs: A review, Applied Surface Science Advances 18 (2023) art.100455. https://doi.org/10.1016/j.apsadv.2023.100455

[22] J. Liu et al., Hydrophobic, flexible, and lightweight MXene foams for high-performance electromagnetic-interference shielding. Advanced Materials 29 (2017) art.1702367. https://doi.org/10.1002/adma.201702367

[23] A. Gadek-Moszczak, Z. Sabina, Description of 3d microstructure of the composites with polypropylene (pp) matrix and tuf particles fillers, Solid State Phenomena 197 (2013) 186-191. https://doi.org/10.4028/www.scientific.net/SSP.197.186

[24] A. Dudek, M. Klimas, Composites based on titanium alloy Ti-6Al-4V with an addition of inert ceramics and bioactive ceramics for medical applications fabricated by spark plasma sintering (SPS method), Materialwissenschaft und Werkstofftechnik 46 (2015) 237-247. https://doi.org/10.1002/mawe.201500334

[25] O. Kędzia et al., Glass and glass-ceramic porous materials for biomedical applications, System Safety: Human - Technical Facility - Environment 5 (2023) 302-310. https://doi.org/10.2478/czoto-2023-0033

[26] D. Klimecka-Tatar, The powdered magnets technology improvement by biencapsulation method and its effect on mechanical properties, Manufacturing Technology 14 (2014) 30-36.

[27] W. Przybył et al., Microwave absorption properties of carbonyl iron-based paint coatings for military applications, Defence Technology 22 (2023) 1-9. https://doi.org/10.1016/j.dt.2022.06.013

[28] W. Zórawski et al., Plasma-sprayed composite coatings with reduced friction coefficient, Surface and Coatings Technology 202 (2008) 4578-4582. https://doi.org/10.1016/j.surfcoat.2008.04.026

[29] N. Radek, B. Antoszewski, The influence of laser treatment on the properties of electro-spark deposited coatings, Kovove Materialy 47 (2009) 31-38.

[30] N. Radek et al., The effect of laser beam processing on the properties of WC-Co coatings deposited on steel, Materials 14 (2021) art. 538. https://doi.org/10.3390/ma14030538

[31] N. Radek et al., Electrospark alloying of carbon steel with WC-Co-Al2O3: Deposition technique and coating properties, Advanced Materials Research 874 (2014) 101-106. https://doi.org/10.4028/www.scientific.net/AMR.874.101

[32] J. Pietraszek et al., Advanced statistical refinement of surface layer's discretization in the case of electro-spark deposited carbide-ceramic coatings modified by a laser beam, Solid State Phenomena 197 (2013) 198-202. https://doi.org/10.4028/www.scientific.net/SSP.197.198

[33] P.A. Laski, Proportional valve with a piezoelectric actuator, EPJ Web of Conferences 143 (2017) art. 02064. https://doi.org/10.1051/epjconf/201714302064

[34] D.S. Pietrala, P.A. Laski, Design and Control of a Pneumatic Muscle Servo Drive Containing Its Own Pneumatic Muscles, Applied Sciences 12 (2022) art. 11024. https://doi.org/10.3390/app122111024

[35] I. Krzysztofik et al., Selected methods of control of the scanning and tracking gyroscope system mounted on a combat vehicle, Annual Reviews in Control 44 (2017) 173-182. https://doi.org/10.1016/j.arcontrol.2016.10.003

[36] J. Smyrski et al., Design of the forward straw tube tracker for the PANDA experiment, Journal of Instrumentation 12 (2017) art. C06032. https://doi.org/10.1088/1748-0221/12/06/C06032

[37] P. Krysiak, Microstructural Analysis of Wound Composites with Considerations on the Fiber Winding Force, Mater. Res. Proc. 34 (2023) 43-52. https://doi.org/10.21741/9781644902691-6

[38] P. Krysiak et al., Comparative Analysis of the Strain in Composite Rings Fabricated with Fibre Prestressing, Mater. Res. Proc. 45 (2024) 91-100. https://doi.org/10.21741/9781644903315-12

[39] P. Krysiak et al., Strength testing of a composite mounting frame for a multi-sensor detection system, Mater. Res. Proc. 17 (2020) 165-170. https://doi.org/10.21741/9781644901038-25

[40] D. Klimecka-Tatar, M. Ingaldi, Assessment of the technological position of a selected enterprise in the metallurgical industry, Mater. Res. Proc. 17 (2020) 72-78. https://doi.org/10.21741/9781644901038-11

[41] A. Pacana, K. Czerwińska, Indicator analysis of the technological position of a manufacturing company, Prod. Eng. Arch. 29 (2023) 162-167. https://doi.org/10.30657/pea.2023.29.19

[42] D. Klimecka-Tatar, R. Dwornicka, The assembly process stability assessment based on the strength parameters statistical control of complex metal products, METAL 2019 – 28th Int. Conf. Metall. Mater. (2019) 709-714.

[43] E. Kozień, M.S. Kozień, Ex-ante risk estimation in the production project, System Safety: Human - Technical Facility - Environment 1 (2019) 708-715. https://doi.org/10.2478/czoto-2019-0090

[44] J. Pietraszek, E. Skrzypczak-Pietraszek, The optimization of the technological process with the fuzzy regression, Advanced Materials Research 874 (2014) 151-155. https://doi.org/10.4028/www.scientific.net/AMR.874.151

[45] D. Klimecka-Tatar, Value Stream Mapping as Lean Production tool to improve the production process organization - Case study in packaging manufacturing, Prod. Eng. Arch. 17 (2017) 40-44. https://doi.org/10.30657/pea.2017.17.09

[46] M. Ingaldi, D. Klimecka-Tatar, Digitization of the service provision process - Requirements and readiness of the small and medium-sized enterprise sector, Procedia Computer Science 200 (2022) 237-246. https://doi.org/10.1016/j.procs.2022.01.222

[47] M.S. Kozień, J. Wiciak, Passive structural acoustic control of the smart plate - FEM simulation, Acta Physica Polonica A 118 (2010) 1186-1188. https://doi.org/10.12693/APhysPolA.118.1186

[48] R. Dwornicka, J. Pietraszek, The outline of the expert system for the design of experiment, Production Engineering Archives 20 (2018) 43-48. https://doi.org/10.30657/pea.2018.20.09

[49] A. Pacana et al., Life Cycle Assessment (LCA) of Truck Steel Wheels, Mater. Res. Proc. 45 (2024) 213-222. https://doi.org/10.21741/9781644903315-25

[50] E. Kozień, A. Kozień, The Importance of Corporate Social Responsibility from the Perspective of Customers in the Light of Own Research, Mater. Res. Proc. 45 (2024) 249-257. https://doi.org/10.21741/9781644903315-29

[51] E. Skrzypczak-Pietraszek, A.G. Pietraszek, Phytoremediation and Allelopathy as an Element of Sustainable Circular Economy to Prevent Environmental Pollution, Mater. Res. Proc. 45 (2024) 140-145. https://doi.org/10.21741/9781644903315-17

[52] T. Lipinski, Corrosion effect of 20 % NaCl solution on basic carbon structural S235JR steel, Engineering for Rural Develop. 16 (2017) 1069-1074. https://doi.org/10.22616/ERDev2017.16.N225

[53] R. Ulewicz et al., Fatigue strength of ductile iron in ultra-high cycle regime, Advanced Materials Research 874 (2014) 43-48. https://doi.org/10.4028/www.scientific.net/AMR.874.43

[54] T. Lipinski, J. Pietraszek, Influence of animal slurry on carbon C35 steel with different microstructure at room temperature, Engineering for Rural Develop. 21 (2022) 344-350. https://doi.org/10.22616/ERDev.2022.21.TF115

Materials Research Proceedings 62 (2026) 151-156
https://doi.org/10.21741/9781644904015-19

Application of Laser Interferometers in Form Deviation Measurements

Krzysztof STĘPIEŃ[1,a *], Uros ZUPERL[2,b], Lenka CEPOVA[3,c], Michal HATALA[4,d] and Gunther POSZVEK[5,e]

[1]Kielce University of Technology, Al. 1000-lecia P. P. 7, 25-314 Kielce, Poland

[2]University of Maribor, Faculty of Mechanical Engineering, Smetanova ulica 17, 2000 Maribor, Slovenia

[3]VSB-Technical University of Ostrava, Faculty of Mechanical Engineering, 17. listopadu 2172/15 708 00 Ostrava-Poruba Czech Republic

[4]Technical University of Kosice, Faculty of Manufacturing Technologies, Letná 1/9, 042 00 Košice-Sever, Slovak Republic

[5]TU Wien Institut für Fertigungstechnik und Photonische Technologien, Getreidemarkt 9 / E311/ BA / 8. und 9. Stock, A-1060 Wien, Austria

[a]kstepien@tu.kielce.pl, [b]zuperl@um.si, [c]lenka.cepova@vsb.cz, [d]michal.hatala@tuke.sk, [e]poszvek@ift.at

Keywords: Metrology, Measurement, Form Deviation, Laser Interferometer

Abstract. Nowadays, interferometry is very widely used in many areas of science and technology. In metrology of geometrical quantities, interferometers are primarily used in length measurements (e.g. of gauge blocks). Interferometers are also widely used in the diagnostics of machine tools and coordinate measuring machines. However, once the optics are properly configured, interferometers can also measure form deviations. The paper presents the use of a single-beam laser interferometer for straightness measurement.

Introduction

Interference is a phenomenon characteristic of every wave movement. It occurs as a result of the overlapping of two or more waves, and these waves must be coherent, (i.e. they must be correlated in phase and frequency). As a result of the interference of the component waves, interference bands are formed, i.e. areas in which the component waves are alternately amplified or extinguished.

The use of interferometry in practice is associated with the name of Albert Michelson. Michelson conducted research on the propagation of light waves in different directions and built an optical system, the diagram of which is shown in Figure 1, later called the Michelson interferometer [2]. The system consists of following components: the source of light, the beam splitter and two mirrors: the former is fixed and the latter is moveable one. The light beam comes out of the source and falls on a beam splitter, where it is divided into two beams. The first one runs to the fixed mirror, the second to the moveable mirror. After reflection from the mirrors, the beams return to the detector, where they interfere with each other.

A modification of the Michelson interferometer is the so-called Twyman-Green interferometer, built in 1916 and most often used to check the quality of lenses.

Metrological Applications of Interferometers

The invention and popularization of lasers resulted in the dynamic development of interferometry in the second half of the twentieth century. Lasers turned out to be a much more convenient and less limited source of light than thermal sources previously used in interferometers.

Currently, interferometry is very widely used in many areas of science and technology. In metrology of geometric quantities, interferometers are used, among others, for length measurements (e.g. gauge blocks), measurement of vibration levels, measurements of form deviations (straightness, flatness). Interferometers are also widely used in the diagnostics of machine tools and coordinate measuring machines [3, 4]. In the above measurement tasks, interferometers are most often used, which are in fact modified and modernized versions of the instrument constructed by Michelson.

It should also be emphasized that interferometers are also used in the measurement transducers of contact profilometers. Thanks to this, the vertical resolution of the most accurate instruments of this type reaches a value of less than 1 nanometer while maintaining a measuring range of a few or even a dozen or so micrometers.

Another very important area of application of interferometers is the measurement of surface unevenness on the micro and nano scale using the non-contact method. In particular, this applies to phase-shift interference methods and scanning coherent interference. In addition to solutions based on the Twyman-Green interferometer, designs based on Michelson, Mirau and Linnik interferometers are particularly often used in this type of application.

Apart from Michelson, Mirau, Linnik interferometers and in measurements of surface unevenness, the Fizeau two-beam interferometer is often used [5-8].

Other types of interferometers used in various types of metrology tasks include:

- the Jamin interferometer, used very often to measure the refractive factor in gases as a result of pressure changes in [9];
- the Mach-Zehnder interferometer (which is a modification of the Jamin interferometer), which is well suited, for example, for measuring changes in pressure, temperature, stress distribution in gaseous and liquid media [10]. Inter-ferrometers of this type are also often used in optoelectronics [11];
- The Fabry-Perot interferometer is widely used in metrology for displacement measurements, telecommunications, optics, astronomy and experimental physics. Interferometers of this type are, for example, part of systems that enable very accurate measurement of the speed of light, as well as spectrometers [12];
- Kösters interferometer used in the measurement of gauge blocks [13].

The Experiment

As an example of the application of laser interferometers to measure form deviations, the use of Renishaw's XL-80 measuring system to measure the straightness of an edge standard will be presented.

The XL-80 laser head is capable of producing a beam with wavelength stability of ±0.05 ppm in one year and ±0.02 ppm in one hour. Such high long-wave stability of the laser is possible thanks to the use of dynamic thermal control of the length of the laser tube.

The maximum permissible error of linear measurement and the maximum permissible error of frequency are determined in relation to the measured length and vary proportionally to it. Therefore, these errors are expressed in ppm units rather than length units. High measurement accuracy of the system is possible thanks to the use of a compensator. This compensator is equipped with a number of sensors that measure temperature, pressure and relative humidity with high accuracy. Based on the signals from the sensors, the system modifies the rated value of the laser wavelength, entering the actual value, which is used to correct for measurement errors. This correction takes place automatically, every seven seconds. At the same time, each sensor turns off between readings to minimize the heat dissipation that can affect the measurement signal.

The optical elements included in the XL-80 system allow the following geometric measurements to be performed:

- measurements of linear displacements,
- measurements of angular displacements (including rotational axes),

- straightness measurements,
- flatness measurements,
- perpendicularity measurements.

In addition, the system, with the use of appropriate software, has the ability to collect, view and save dynamic data. This makes it possible to measure vibrations. This is especially useful when performing machine tool diagnostics.

This system was used to measure the straightness of the measuring table (see Fig. 1). For this purpose, a measuring system consisting of a laser head (3), an interferometer (2) and a mirror system (1) was first built. The interferometer and mirror system were placed on the test table (5). During the measurement, the mirror system (1) is moved in the direction determined by the direction of the beam running from the laser head (3). In order to ensure the linear movement of the mirror system, a ruler (4) was used as an auxiliary element. It should be noted that in the present case, the purpose of the measurement was to determine the straightness deviation in the direction determined by the laser beam. Therefore, the position of the mirror system (1) here is vertical, not horizontal.

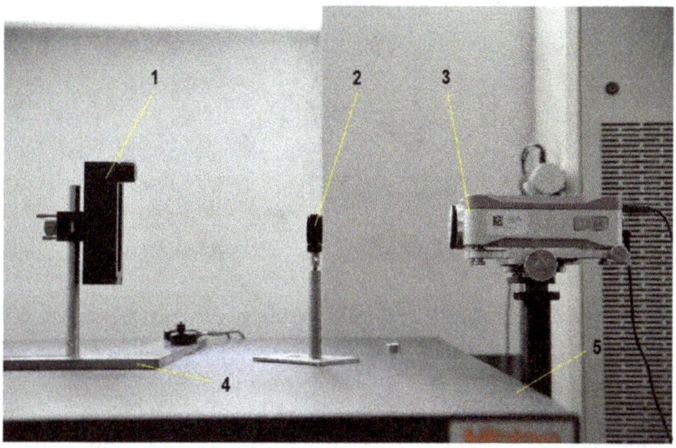

Fig. 1. System configuration for measurements of straightness deviations of the measuring table in the direction determined by the laser beam: 1 – mirror system, 2 – interferometer, 3 – laser head, 4 – scale (auxiliary element to ensure the straightness of the mirror movement), 5 – measuring table.

After setting the system shown in Fig.1, measurements were made to determine the straightness deviation of the table (5). Next, an option was launched in the system control program to measure the straightness deviation.

At the top of the window, there is a drop-down menu with functions for data file operations, loading measurement point coordinates, collecting measurement data, and configuring the system, among other things. The most frequently used functions are also available by clicking on the icons located just below the menu. The program window also indicates the current value of the measurement signal and the most important information about the system status (e.g. whether the laser is switched on, which measurement option is running and in which units the results are given). On the left side of the window there is a bar showing the current signal strength read by the sensor in the head. An additional help is also a large graphic, indicating which program option is currently running.

After starting the program and selecting the Targets option, the data of the measurement points where the values of the measurement signal were to be read were entered. The measurements were made as follows: the mirror system (1) was moved along the scale (4) over a distance of 475 mm. Every 25 millimeters, the data was read and sent to the program. Therefore, the coordinates from 0 to 475, every 25 mm, were entered in the program. The capture function was then used, the mirror system was moved to the desired positions and the values of the measurement signal were read. After taking measurements at all positions, a message appears in the program window that Data Capture Complete has been completed (see Fig.2).

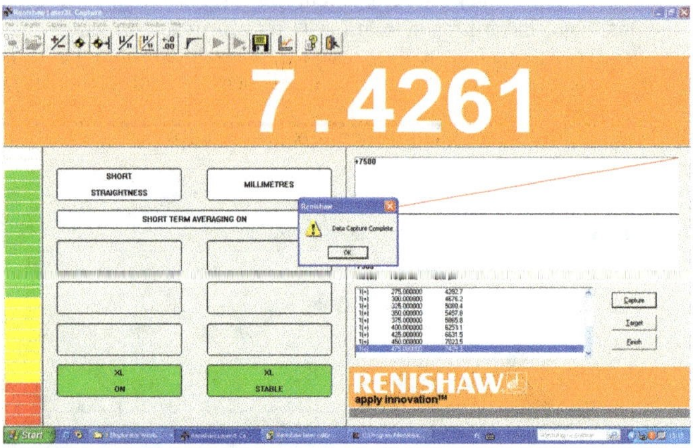

Fig. 2. Program window after the completion of measurement data collection.

As one can see in Fig.2, the program shows all the positions in which the measurement was made along with the values of the signal read in a given position.

The next stage of the study was the analysis of measurement data. To do this, the Analyse option was launched, which appears after clicking on the Date option from the main menu. However, this is not sufficient to obtain a reliable result for measuring the straightness deviation, as the beam running from the laser is usually not parallel to the surface to be measured. This inclination affects the value of the read signal. Therefore, it is necessary to carry out additional processing of the measurement signal by selecting the Least Squares Data option from the drop-down menu that appears after clicking on the Tools option from the main menu. Then a window with the measurement result appears, as shown in Fig.3.

Materials Research Proceedings 62 (2026) 151-156 https://doi.org/10.21741/9781644904015-19

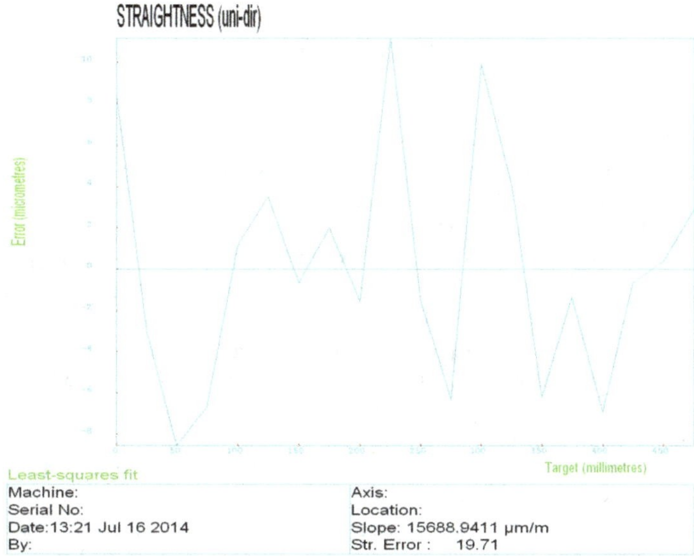

Fig. 3. The results of measuring the straightness of the measured element.

Fig.3 shows that the program displays a graphical representation of the straightness profile of the measured element, as well as gives a numerical value of the straightness deviation (parameter Str. Error located in the last row of the second column containing the measurement data). In addition, the program also provides a number of information identifying the measurement, such as: the name and serial number of the device, the date, the operator's data, as well as the axis along which the measurement was carried out and the inclination of the measured surface in relation to the beam coming out of the laser head. The program also records the numerical values of the signal collected at the individual measurement points, so that they can be used by the user for additional analysis, e.g. by means of mathematical programs

Conclusions

Laser interferometers are currently the most accurate instruments for measuring characteristics such as length, form and position deviations. It is worth noting that they are used as separate devices, but also as components of other instruments, e.g. laser trackers, transducers in contact profilometers, etc. The research has shown that modern interferometers can be easily used in form deviation measurements, allowing reliable measurement results to be obtained quickly.

Summarizing the measurements carried out, it can be concluded that the XL-80 system that was used in the experiment has a clear graphical interface and has a wide range of possibilities for analyzing the results. Referring to the measurement of the table straightness deviation itself, it should be stated that the obtained result could have been influenced by such factors as deviations in the flatness of the interferometer bases and the mirror system, and the deviation of the straightness of the ruler edges. The influence of these factors can be eliminated by ensuring a three-point contact of the interferometer bases and mirrors with the table plane and by identifying the deviation of the straightness of the scale used in the experiment.

References

[1] K. Patorski, Kujawińska M., Sałbut L., Interferometria laserowa z automatyczną analizą obrazu, Wydawnictwo Politechniki Warszawskiej, Warszawa, 2005.

[2] M. Bielenin, Laser system for accurate spatial measurements, PhD Thesis, Wrocław University of Technology, Wrocław 2009.

[3] T. Tadej, B. Acko, Integration of a laser interferometer and a CMM into a measure-ment system for measuring internal dimensions, Measurement 44 (2011) 426-433. https://doi.org/10.1016/j.measurement.2010.11.002

[4] S. Topcu et al., Improving the accuracy of homodyne Michelson interferometers using polarization state measurement techniques, Optics Communications 247 (2005) 133-139. https://doi.org/10.1016/j.optcom.2004.11.040

[5] S. Chatterjee et al., Measurement of surface figure of plane optical surfaces with polarization phase-shifting Fizeau interferometer, Optics & Laser Technology 39 (2007) 268-274. https://doi.org/10.1016/j.optlastec.2005.08.006

[6] S. Chatterjee, Measurement of surface figure of plane optical surfaces using Fizeau interferometer with wedge phase-shifter, Optics & Laser Technology 37 (2005) 43-49. https://doi.org/10.1016/j.optlastec.2004.02.009

[7] T. Nomura et al., On-machine shape measurement of workpiece surface with Fizeau interferometer, Precision Engineering 14 (1992) 155-159. https://doi.org/10.1016/0141-6359(92)90004-G

[8] A. Yamamoto, I. Yamaguchi, Surface profilometry by wavelength scanning Fizeau interferometer, Optics & Laser Technology 32 (2000) 261-266. https://doi.org/10.1016/S0030-3992(00)00055-4

[9] L. Wang et al., Phase-shifting Jamin double-shearing interferometer based on polarization beam splitting film, Optik 124 (2013) 1215-1217. https://doi.org/10.1016/j.ijleo.2012.03.003

[10] Q. Yao et al., Simultaneous measurement of refractive index and temperature based on a core-offset Mach–Zehnder interferometer combined with a fiber Bragg grating, Sensors and Actuators A 209 (2014) 73-77. https://doi.org/10.1016/j.sna.2014.01.017

[11] L. Tan et al., Performance comparison of Fabry–Perót and Mach–Zehnder interferometers for molecular Doppler lidar based on fringe-imaging technique, Optik 125 (2014) 6291-6295. https://doi.org/10.1016/j.ijleo.2014.06.147

[12] Ch.-P. Chang et al., Fabry–Perót displacement interferometer for the measuring range up to 100 mm, Measurement 46 (2013) 4094-4099. https://doi.org/10.1016/j.measurement.2013.06.029

[13] M.G. Nikolić et al., Koester's interferometer modification for gauge blocks calibration, Proc. SPIE 6604, 14th Int. School on Quantum Electronics: Laser Physics and Applications, 66040P (March 05, 2007). https://doi.org/10.1117/12.726899

Treatment of Al-7%SiMg Silumin with an Exothermic Mixture Containing $NaNO_3$ and $NaClO_3$

Tomasz Lipiński[1,a *] and Jacek Pietraszek[2,b]

[1]University of Warmia and Mazury in Olsztyn, Faculty of Technical Sciences, Department of Materials and Machines Technology, ul.Oczapowskiego 11, 10-957 Olsztyn, Poland

[2]Cracow University of Technology, Faculty of Mechanical Engineering, 31-864 Kraków, ul.Warszawska 24, Poland, ORCID: 0000-0003-2851-1606

[a]tomekl@uwm.edu.pl, [b]jacek.pietraszek@pk.edu.pl

Keywords: Silumin, Al-Si Alloy, Modification, Sodium, Exothermic Modifier

Abstract. Hypoeutectic silumins are one of the most popular groups of casting alloys. A problem with the use of hypoeutectic silumins is their microstructure, which negatively impacts their performance. The primary negative impact is exerted by the unmodified eutectic β phase, which constitutes natural microstructural notches. Methods for refining the microstructure of these alloys, thereby enhancing their properties, have been known in industrial practice for many years. One method is the treatment with exothermic mixtures. This method allows not only for modifying the microstructure but also for introducing chemical elements and compounds into the alloy that influence the properties of the silumins. Therefore, the research topic in this area remains partially unexplored. This confirms the need for further research on this modification method. The paper presents the treatment of hypoeutectic silumin containing 7% Si and the addition of magnesium using sodium nitrate ($NaNO_3$). This compound was added to the alloy, which was then combined with aluminum and magnesium. The resulting mixtures produced an exothermic effect. A $NaClO_3$ compound was also added to the alloy, which was expected to strengthen the alloy. The experimental results indicate that sodium nitrate affects the refinement of the eutectic β phase and improves mechanical properties. The addition of the $NaClO_3$ compound increased the alloy's tensile strength and hardness while reducing its ductility.

Introduction

Aluminum-silicon foundry alloys have gained considerable popularity due to several factors. The most important are their good mechanical properties, low density, and resistance to a significant number of environmental factors. These include their low melting point, which allows them to be melted and cast without the need for high-temperature furnaces. Hypoeutectic silumins. Hypoeutectic Al-Si alloys, when unmodified, have a eutectic β phase, which is a solid solution of aluminum in silicon, with the solid solution of silicon in aluminum as the α phase. The hard and non-deformable eutectic β phase occurs in the form of flakes, which constitute microstructural notches. These notches reduce the alloy's ductility and strength. Both the original and the eutectic plastic α phases constitute the matrix of the β phase. The microstructure of such an alloy is not suitable for industrial applications [1]. To improve the functional properties of Al-Si foundry alloys, they are modified [2,3]. Traditionally, silumins were modified with sodium or strontium [4-6]. Sodium modification proved unstable. The precipitate permanently modified the alloy's microstructure. Refining the microstructure and improving the resulting mechanical properties can also be achieved through technological methods. By controlling the crystallization rate, the eutectic β phase can be refined. This process can be achieved using a Bridgman furnace or other methods, such as welding [7], laser processing, plasma treatment, or heat treatment [8,9].

Currently, microstructural changes are achieved in many ways. Industrial applications utilize mixtures of elements and chemical compounds [10,11]. These mixtures are selected not only to refine the eutectic β phase but also to refine the alloy using other mechanisms, such as strengthening the solutions with chemical compounds [12].

A modern solution, although known for many years, is influencing the silumin crystallization process through treatment with exothermic mixtures [13]. Exothermic modification is also successfully applied to other alloys, e.g., iron alloys [14]. The exothermic mixture can be composed so that the reaction product is also a desired component in the resulting alloy. An example is reactions (1) to (4), which can result in the formation of sodium, which modifies silumin.

$$6NaNO_3+10Al \rightarrow 5Al_2O_3+3Na_2O+6N\uparrow \text{ and } \Delta H^0_{293} = -623\frac{kJ}{mol\ Al} \tag{1}$$

$$2NaNO_3+5Mg \rightarrow 5MgO+Na_2O+2N\uparrow \text{ and } \Delta H^0_{293} = -461{,}2\frac{kJ}{molMg} \tag{2}$$

$$NaClO_3+2Al \rightarrow Al_2O_3+NaCl \text{ and } \Delta H^0_{293} = -783\frac{kJ}{molAl} \tag{3}$$

$$3Na_2O+2Al \rightarrow Al_2O_3+5Na \text{ and } \Delta H^0_{293} = -422\frac{kJ}{molAl} \tag{4}$$

The aim of the research was to analyze the effect of sodium nitrate $NaClO_3$ in combination with aluminum and magnesium on the mechanical properties of hypoeutectic silumin.

Materials and Methods

The experiment was conducted on a low-Al 7%SiMg alloy. The real chemical composition of the tested alloy is presented in Table 1.

Table 1. *Chemical composition of the Al-7%SiMg alloy*

Si	Mg	Mn	Fe	Cu	Ti	Ni	Al
7.11	0.34	0.32	0.42	0.10	0.1	0.006	balance

The mixtures were prepared from powders of the chemical compounds presented in equations (1) to (3). These mixtures, prepared in proportions according to the above reactions, were placed after compression in reaction chamber 3 of the casting mold (Fig. 1), which was made of molding sand.

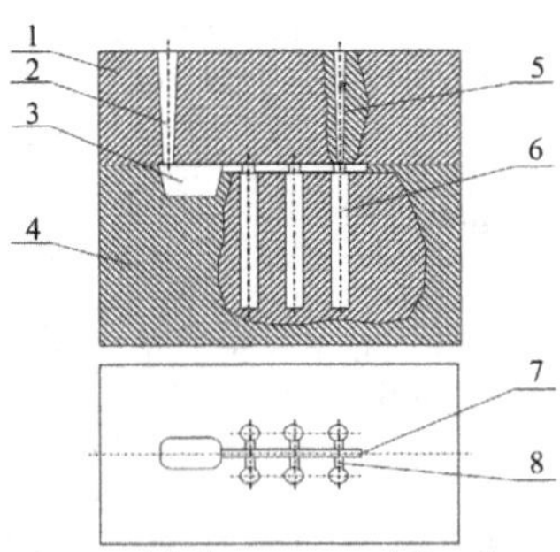

Figure 1. *Section of the casting mold: 1 – top moulding box, 2 – down-gate, 3 – reaction chamber, 4 – down moulding box, 5- strain relief, 6 – sample for strength testing, 7 – cross-gate, 8- running gate.*

The mold was poured with AlSi7Mg silumin at 750°C, melted in a 1 dm³ ceramic crucible placed in an electric resistance furnace. The tests were conducted using a 2^3-factorial design [14] with three variables: X1, NaNO3 + Al; X2, NaNO3 + Mg; and $NaClO_3$ + Al. Based on preliminary tests, the base level was set at 0.2%, and the range of variation was set at 0.1% (resulting in a lower level of 0.1% and a higher level of 0.3%, respectively). The response was obtained as a function of the three variables. The adequacy of the response was verified at t = 0.05. To graphically represent the response, each variable was set to -1 and 1. This resulted in six equations for each response, each of which was graphically illustrated, as shown in Figs. 2-4, respectively.

Results and Discussion

Tensile strength of AlSi7Mg alloy with: $NaNO_3$+Al ϵ <0.1, 0.3> [%], $NaNO_3$+Mg ϵ <0.1, 0.3> [%], $NaClO_3$+Al ϵ <0.1, 0.3>[%] is presented in Fig. 2, elongations in Fig. 3, and Brinell Hardness in Fig. 4.

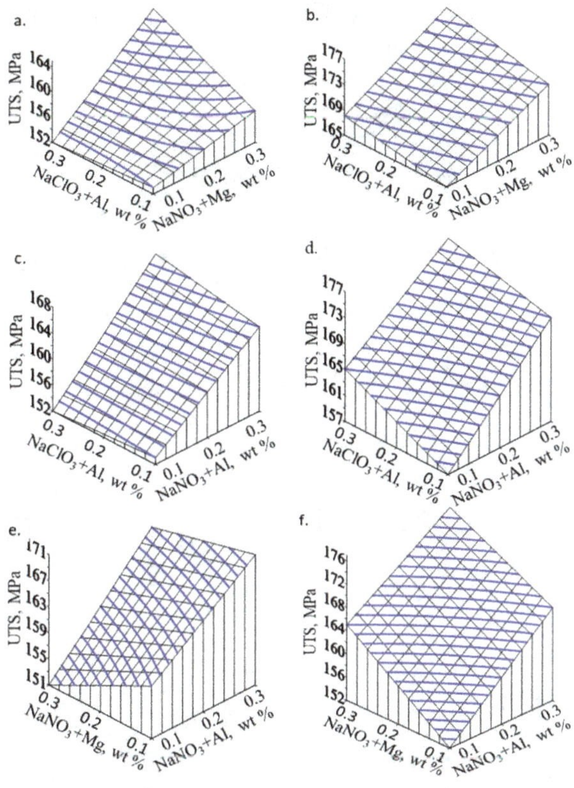

Fig. 2. Tensile strength of AlSi7Mg alloy with: $NaNO_3$+Al ε <0.1, 0.3> [%], $NaNO_3$+Mg ε <0.1, 0.3> [%], $NaClO_3$+Al ε <0.1, 0.3> [%] for: a. $NaNO_3$+Al =0.1%, b. $NaNO_3$+Al =0.3%, c. $NaNO_3$+Mg =0.1%, d. $NaNO_3$+Mg =0.03%, e. $NaClO_3$+Al =0.1%, f. $NaClO_3$+Al =0.3%.

The tensile strength after treatment of AlSi7Mg silumin with an exothermic mixture composed of 0.1%($NaNO_3$+Al)+0.1%($NaNO_3$+Mg)+0.3%($NaClO_3$) increased from 144 MPa for the untreated alloy to 152 MPa (Fig. 2a) and the elongation from 1.4% to 2.3% (Fig. 3a) as well as the hardness from 49 HB to 55 HB. The highest tensile strength Rm=176 MPa was obtained after treatment of silumin 0.3% (NaNO3+Al)+0.3% (NaNO3+Mg)+0.3% (NaClO3+Al), Fig. 2f. For this mixture composition, the highest elongation A=6.7% (Fig. 3f) and the highest hardness 60 HB (Fig. 4f) were also obtained. Comparing the effect of the mixture in which (NaClO3 + Al) was at a higher level of 0.3% and the share of the other two components at a lower level of 0.1% on the tensile strength (152 MPa) with the strength obtained for the composition in which all the mixture components were present at a lower level of 0.1% (152 MPa), it was noted that at the lower level of (NaNO3 + Al) and NaNO3 + Mg), the third component influenced the alloy strength inversely (Fig. 2c), while increasing elongation to 3.2% (Fig. 3c) and hardness to 56 HB (Fig. 4c). For the two mixtures based on NaNO3 introduced at a higher level of 0.3%, it showed a direct proportional effect on strength (Fig. 2d, Fig. 2b) and hardness (Fig. 4b) and inversely proportional effect on elongation (Fig. 4d). Analyzing the interaction of (NaNO3 + Al) and (NaNO3 + Mg) at a constant level of (NaClO3 + Al), a directly proportional effect of both mixture components on tensile strength (Fig. 2e and Fig. 2f) and elongation (Fig. 3e and Fig. 3f) was observed.

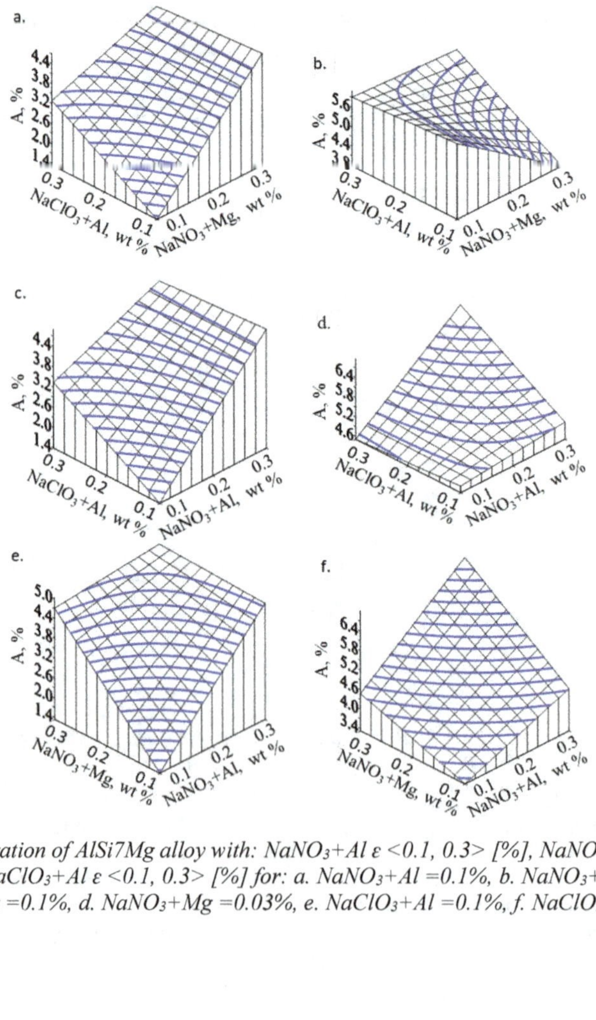

Fig. 3. Elongation of AlSi7Mg alloy with: $NaNO_3$+Al ε <0.1, 0.3> [%], $NaNO_3$+Mg ε <0.1, 0.3> [%], $NaClO_3$+Al ε <0.1, 0.3> [%] for: a. $NaNO_3$+Al =0.1%, b. $NaNO_3$+Al =0.3%, c. $NaNO_3$+Mg =0.1%, d. $NaNO_3$+Mg =0.03%, e. $NaClO_3$+Al =0.1%, f. $NaClO_3$+Al =0.3%.

Materials Research Proceedings 62 (2026) 157-164
https://doi.org/10.21741/9781644904015-20

The interaction between ($NaNO_3$ + Al) was more intense. Hardness after treatment with the mixture containing 0.1% ($NaClO_3$ + Al) did not depend on the change in the share of the other two components on the $NaNO_3$ base (Fig. 4d). For 0.3% ($NaClO_3$ + Al), a direct proportional effect was found only for the ($NaNO_3$ + Al) component on the silumin hardness. The hardness increased from 56 to 60 HB. However, the change in the ($NaNO_3$ + Mg) share did not change the hardness of the alloy (Fig. 4e).

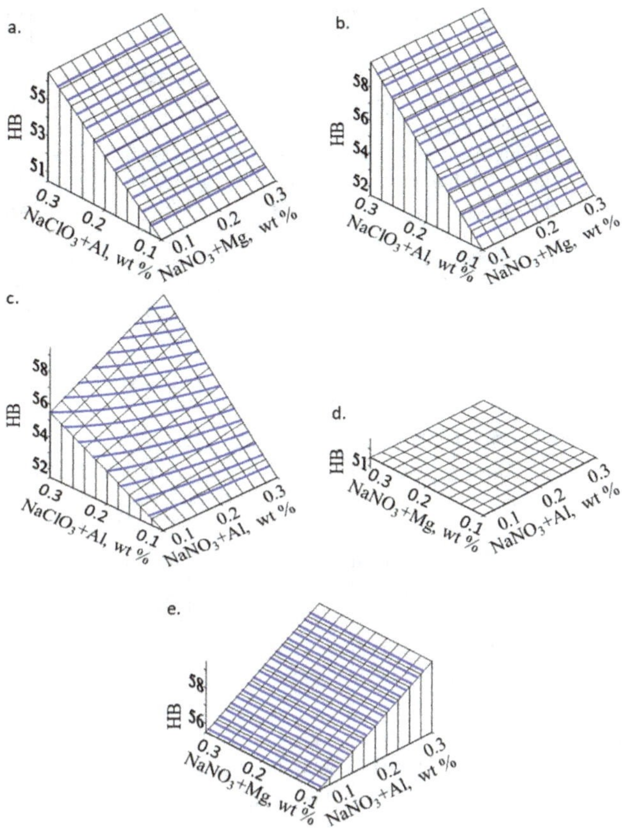

Fig. 4. Brinell hardness of AlSi7Mg alloy with: $NaNO_3$+Al ε <0.1, 0.3> [%], $NaNO_3$+Mg ε <0.1, 0.3> [%], $NaClO_3$+Al ε <0.1, 0.3> [%] for: a. NaNO3+Al =0.1%, b. $NaNO_3$+Al =0.3%, c. $NaNO_3$+Mg =0.1% and 0.3%, d. $NaClO_3$+Al =0.1%, e. $NaClO_3$+Al =0.3%.

Technological Impact

Improving the functional properties of structural materials by eliminating microstructural notches is a key issue for their wider application. The procedure described in the article, which provides quantitative data, opens the possibility of adapting this information in the industrial field [15], in particular with respect to the tested material as a potential substrate for special coatings [16-18], as a structural material for less loaded robotic elements [19,20] with appropriately fine-tuned control [21,22]. This also creates the possibility of substituting other metal materials that are more susceptible to corrosion [23-25], including alloys [26,27], which allows for a reduction in energy demand [28,29]. This approach significantly alters the company's situation by implementing the appropriate material [30], while simultaneously enabling a substantial increase in production quality [31-33] and a notable decrease in wear [34]. Suitable materials, due to their technological features, primarily low density, may be interesting for military purposes with a low radar signature [35]. Regardless of this, the phenomena related to the exothermic refinement of silumins are a strong driving force for the development of microstructure testing methods [36-38].

Conclusions

Based on the results of the studies conducted, it was concluded that:

1. $NaNO_3$ introduced as an exothermic mixture improves the mechanical properties of silumins with both aluminum and magnesium.
2. The $NaNO_3$ and aluminum mixture has a stronger effect than the NaNO3 and magnesium mixture. This may be due to the higher reaction energy of NaNO3 and aluminum.
3. The $NaClO_3$ and aluminum mixture increases tensile strength and hardness, while slightly reducing elongation.
4. The effect of the $NaClO_3$ and aluminum mixture increases when it is introduced into an alloy with a mixture containing $NaNO_3$. This effect increases more significantly with increasing exothermic reaction energy.

References

[1] S. Li et al., Development and applications of aluminum alloys for the aerospace industry. Journal of Materials Research and Technology 27 (2023) 944-983. https://doi.org/10.1016/j.jmrt.2023.09.274

[2] R. Romankiewicz, F. Romankiewicz. Influence of modification on the refinement of primary silicon crystals in hypereutectic silumin AlSi21CuNi. Prod. Eng. Arch. 19 (2018) 30-36. https://doi.org/10.30657/pea.2018.19.07

[3] T. Lipiński. Microstructure and Mechanical Properties AlSi7Mg Alloy with Sr, Al and AlSi7Mg Manufacturing Technology 24 (2024) 227-234. https://doi.org/10.21062/MFT.2024.031

[4] R. Romankiewicz. Changes in the structure of the ak7 alloy in terms of modification. System Safety: Human - Technical Facility – Environment 6 (2024) 384-394. https://doi.org/10.2478/czoto-2024-0040

[5] J. Barrirero et al., Cluster formation at the Si/liquid interface in Sr and Na modified Al–Si alloys. Scripta Materialia 117 (2016) 16-19. https://doi.org/10.1016/j.scriptamat.2016.02.018

[6] N. Náprstková et al., Modification of AlSi7Mg0.3 alloy by strontium. Manufacturing Technology 13 (3) (2013) 373-380.

[7] T. Lipiński. Influence of Surface Refinement on Microstructure of Al-Si Cast Alloys Processed by Welding Method. Manufacturing Technology 15 (2015) 576-581.

[8] N. Naprstkova et al., Modification of AlSi9CuMnNi alloy by antimony and heat treatment and their influence on tool wear after turning. Manufacturing Technology 16 (2016) 209-214.

[9] J. Zou et al., Optimizing the solidification structure and performance of Al–Si–Mg cast alloys subjected to combined Sr–Ca modification and rotating magnetic fields. Journal of Alloys and Compounds 1046 (2025) art. 184889. https://doi.org/10.1016/j.jallcom.2025.184889

[10] R. Romankiewicz. The Study of the Tensile Strength of AlSi21CuNiMg Silumin in the Final Stage of Solidification and the Initial Stage of Cooling, J. Cast. Mater. Eng. 3 (2019) 57-61. https://doi.org/10.7494/jcme.2019.3.3.57

[11] L. Zhang et al., Factors affecting eutectic Si modification in Al-Si hypoeutectic alloy with the addition of Na, Sr, Eu and Yb. Materials Letters 308(B) (2022) art. 131206. https://doi.org/10.1016/j.matlet.2021.131206

[12] T. Lipiński, P. Szabracki, Modification of the hypo-eutectic Al-Si alloys with an exothermic modifier. Arch. Metall. Mater. 58 (2013) 453-458.

[13] T. Lipiński, Use Properties of the AlSi9Mg Alloy with Exothermical Modifier. Manufacturing Technology 11(11) (2011) 44-49.

[14] С.Л. Ахназарова, В.В. Кафаров. Оптимализация эксперимента в химии и химической технологии. Издательство Высшая школа. Москва 1978.

[15] D. Klimecka-Tatar, R. Dwornicka, The assembly process stability assessment based on the strength parameters statistical control of complex metal products, METAL 2019 - 28th Int. Conf. Metall. Mater. (2019) 709-714.

[16] N. Radek et al., The WC-Co electrospark alloying coatings modified by laser treatment, Powder Metallurgy and Metal Ceramics 47 (2008) 197-201. https://doi.org/10.1007/s11106-008-9005-7

[17] L. Adamczyk et al., Electrochemical preparation of composite coatings of 3,4-etylenodioxythiophene (EDOT) and 4-(pyrrole-1-yl) benzoic acid (PyBA) with heteropolyanions, Materials Chemistry and Physics 144 (2014) 418-424. https://doi.org/10.1016/j.matchemphys.2014.01.012

[18] J. Korzekwa et al., Influence of the Size of Nanoparticles on the Microstructure of Oxide Coatings, Materials Science 53 (2018) 709-716. https://doi.org/10.1007/s11003-018-0127-x

[19] P. Wos et al., Bricklaying robot lifting and levelling system, Communications – Sci. Letters Univ. Žilina 23 (2021) B257-B264. https://doi.org/10.26552/COM.C.2021.4.B257-B264

[20] A. Annusewicz-Mistal et al., Autonomous Manipulator of a Mobile Robot Based on a Vision System, Applied Sciences 13 (2023) art. 439. https://doi.org/10.3390/app13010439

[21] J.E. Takosoglu, P.A. Laski, Intelligent positioning system of electro-pneumatic servo-drive, Annals of DAAAM and Proceedings of the International DAAAM Symposium (2011) 1641-1642.

[22] P.A. Laski, M. Smykowski, Using a development platform with an stm32 processor to prototype an inexpensive 4-dof delta parallel robot, Sensors 21 (2021) art. 7962. https://doi.org/10.3390/s21237962

[23] M. Scendo et al., Purine as an effective corrosion inhibitor for stainless steel in chloride acid solutions, Corrosion Reviews 30 (2012) 33-45. https://doi.org/10.1515/CORRREV-2011-0039

[24] D. Klimecka-Tatar et al., Corrosion characteristics in acid, alkaline and the ringer solution of Fe68-xCoxZr10Mo5W2B 15 metallic glasses, Journal of the Balkan Tribological Association 20 (2014) 124-130.

[25] O. Kierat et al., The effect of the corrosion medium on silane coatings deposited on titanium grade 2 and titanium alloy Ti13Nb13Zr, Materials 14 (2021) art. 6350. https://doi.org/10.3390/ma14216350

[26] A. Dudek, M. Klimas, Composites based on titanium alloy Ti-6Al-4V with an addition of inert ceramics and bioactive ceramics for medical applications fabricated by spark plasma sintering (SPS method), Materialwissenschaft und Werkstofftechnik 46 (2015) 237-247. https://doi.org/10.1002/mawe.201500334

[27] A. Dudek et al., The effect of alloying method on the structure and properties of sintered stainless steel, Arch. Metall. Mater. 62 (2017) 281-287. https://doi.org/10.1515/amm-2017-0042

[28] V. Zhelykh et al., Analysis of the processes of heat exchange on infrared heater surface, Diagnostyka 17 (2016) 81-85.

[29] J. Zwierzchowski et al., Model tests of wind turbine with a vertical axis of rotation type Lenz 2, EPJ Web of Conf. 143 (2017) art. 02150. https://doi.org/10.1051/epjconf/201714302150

[30] D. Klimecka-Tatar, M. Ingaldi, Assessment of the technological position of a selected enterprise in the metallurgical industry, Materials Research Proceedings 17 (2020) 72-78. https://doi.org/10.21741/9781644901038-11

[31] A. Pacana et al., Analysis of quality control efficiency in the automotive industry, Transportation Research Procedia 55 (2021) 691-698. https://doi.org/10.1016/j.trpro.2021.07.037

[32] D. Siwiec et al., Analysis of problems with maintaining the desired quality of gas sealant, METAL 2022 – 31st Int. Conf. Metall. Mater. (2022) 848-853. https://doi.org/10.37904/metal.2022.4524

[33] D. Siwiec et al., Improving the Quality of Products from Cast Alloy with the use of Grey Relational Analysis (GRA), Materials Research Proceedings 24 (2022) 204-211. https://doi.org/10.21741/9781644902059-30

[34] A. Pacana et al., Influence of natural aggregate crushing process on crushing strength index, Sustainability (Switzerland) 13 (2021) art. 8353. https://doi.org/10.3390/su13158353

[35] W. Przybył et al., Microwave absorption properties of carbonyl iron-based paint coatings for military applications, Defence Technology 22 (2023) 1-9. https://doi.org/10.1016/j.dt.2022.06.013

[36] A. Szczotok, S. Roskosz, New possibilities of light microscopy research resulting from digital recording of images, Materials Science-Poland 23 (2005) 559-565.

[37] J. Korzekwa et al., The influence of sample preparation on SEM measurements of anodic oxide layers; Practical Metallography 53 (2016) 36-49. https://doi.org/10.3139/147.110367

[38] A. Gądek-Moszczak et al., Nano X-ray Tomography Application for Quantitative Surface Layer Geometry Analysis after Laser Beam Modification, Materials 15 (2022) art. 5935. https://doi.org/10.3390/ma15175935

Terotechnology XIV Materials Research Forum LLC
Materials Research Proceedings 62 (2026) 165-171 https://doi.org/10.21741/9781644904015-21

Supporting the Evaluation of Coating Systems through Quantitative Image Analysis of Their Structures

Aneta GĄDEK-MOSZCZAK[1,a *] and Norbert RADEK[2,b]

[1]Cracow University of Technology, Faculty of Mechanical Engineering, al. Jana Pawła II 32, 31-864, Poland

[2]Kielce University of Technology, Faculty of Mechatronics and Mechanical Engineering, al. Tysiąclecia Państwa Polskiego 7, 25-314 Kielce, Poland

[a]aneta.gadek-moszczak@pk.edu.pl, [b]norrad@tu.kielce.pl

Keywords: Surface Geometry, Image Processing, Stereology

Abstract. This article presents exemplary methods for evaluating the quality of coating systems using image analysis techniques. It discusses the types of information that can be obtained about the analyzed coating systems, considering different observation approaches and the use of various imaging methods. A typical image processing workflow, designed to enable digital measurements of the examined objects, is also described. The data obtained in this manner provides objective, quantitative information characterizing the analyzed coatings. Although the discussed methods for quantitatively assessing the geometric structures of materials are commonly applied in microstructural analyses, they are still not widely used in other areas. Therefore, the authors focus on demonstrating their potential for assessing the quality of various coating systems.

Introduction

Coating systems are applied to engineering materials as a method to enhance their properties for specific applications [1, 2]. An appropriately chosen and applied coating system enhances the coefficient of friction, increases resistance to atmospheric and environmental conditions, and significantly extends the lifespan of machine components. There are numerous systems of coatings, designed according to the base material [3-5] and the working conditions of the components [6-10]. The deposited surface layers may consist of a single coating applied directly to the substrate or form a multilayer coating system, wherein several layers collectively provide the desired functional properties of the base material. Nevertheless, all of them, at the stage of developing the technology and process adjustments, undergo a set of various mechanical, chemical tests, and structural analyses. A crucial aspect of structural analysis is examining a structure that has been imaged using light or scanning microscopy. Expert visual assessment enables the observation of severe failure in layer adhesion or internal discontinuities. The application of automated algorithms for geometric analysis enables objective and quantitative evaluation of the coatings [2, 11].

The examination of images representing the geometric structure of the analyzed coating systems, performed using image processing algorithms and stereological methods, enables the acquisition of substantially more information than can be discerned by the human eye [22]. Furthermore, it allows for a precise quantitative characterization of the identified layers and their defects. Such quantitative analysis facilitates the monitoring of structural changes within the coating systems as a function of the deposition process parameters, thereby significantly enhancing the efficiency of process optimization [2].

Quantitative image analysis of material structures has been well established in materials science for many years, particularly in determining grain size [12], assessing the content of inclusions in steel [13], or evaluating banding and structural orientation [14]. However, numerous areas remain

Materials Research Proceedings 62 (2026) 165-171 https://doi.org/10.21741/9781644904015-21

in which digital image processing and analysis methods are applied as part of metallographic investigations, yet are not regulated by international standards. In the authors' opinion, it is also worthwhile to further develop and apply these methods to structural studies in fields that have traditionally relied on visual, expert-based assessment—such as in the case of coating systems. In addition to providing higher accuracy and quantitative characterization, image analysis enables the rapid acquisition of data describing an entire series of images, while ensuring that measurements remain repeatable and objective.

The rapidly developing methods of artificial intelligence and their applications across various fields of science, business, and everyday life may also find use in designing new materials and optimizing their manufacturing processes. However, for this to be possible, it is necessary to provide AI models with numerical data that describes the material structure, which can be utilized during both training and prediction [15-18]. The potential application of AI in materials science constitutes an additional argument emphasizing the importance of quantitatively characterizing the geometric structure of materials at all levels of observation, and consequently, the need to further develop algorithms for their precise description [2].

Materials and Methods

Quantitative structural image analysis of coating systems is performed independently of the base material or coating materials used. However, it is essential to select an appropriate method and observation scale that fully reveal the relevant structural features. Such analysis can be carried out on microscopic images of samples, which represent a perpendicular projection of the surface, a cross-sectional view, or, when feasible, a reconstructed 3D model obtained from tomographic investigation.

An example of surface analysis, presenting a perpendicular projection of the surface, is shown in Fig.1. Such images provide information on the relative porosity, expressed as the ratio of detected pores to the total analyzed surface area, as well as a description of the shape and surface area of each pore. This information enables the assessment of pore homogeneity in terms of its geometric characteristics [20-22].

The image processing procedure employed to obtain the required dataset involved several sequential steps. Initially, the images underwent pre-processing, followed by object detection. Potential artifacts were subsequently removed using mathematical morphology methods. Quantitative measurements were then performed, and the results of the detection were visualized by overlaying the contours of the identified objects on the original image (Fig. 1b).

Cross-sectional analysis enables the measurement of coating local thickness and its heterogeneity, the assessment of coating adhesion to the base material, and the detection of internal defects within the coating, such as microcracks or pores. The procedure for image processing of this type is analogous to that described above for surface analysis: preprocessing, including any necessary filtering and image enhancement, followed by the detection of objects for analysis. For the automatic analysis of geometrical properties, the image must be binary (with pixel values only 0 or 1), as shown in Fig. 2. Pixels representing objects have a value of 1, while pixels representing background have a value of 0. In contrast, those representing the background have a value of 0. On binary images, information on the geometrical shape of the layer is extracted from the background.

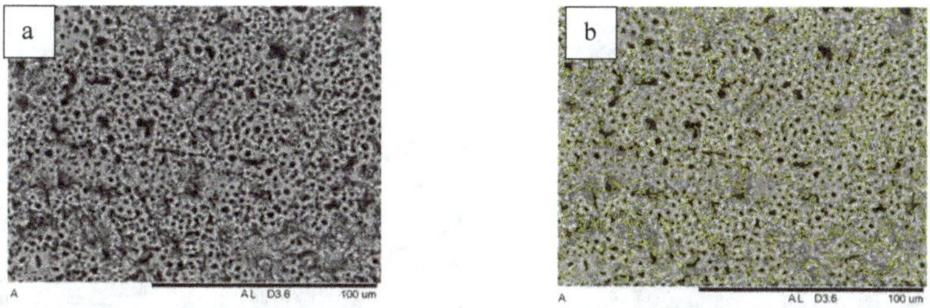

Fig. 1. Analysis of surface porosity a) initial image, b) visualization of detection effects [20].

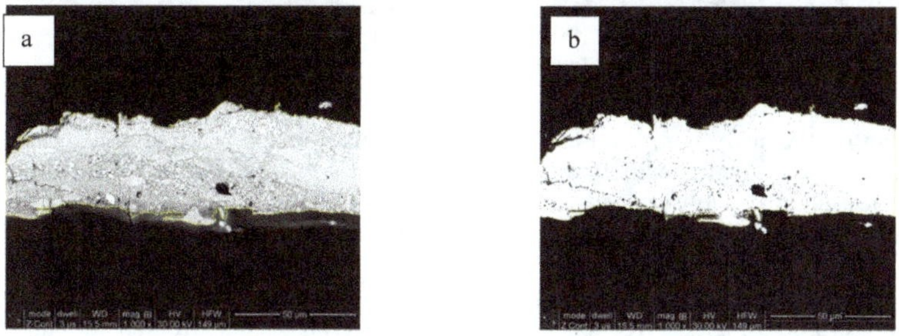

Fig. 2. Analysis of defects in the structure of the WC-Co-Al_2O_3 layer after laser treatment: a) initial image, b) binary image presenting the detected layer.

In a multi-layer coating system (Fig. 3), especially when the image of each layer is difficult to distinguish, visual control analysis of the cross-section profile is helpful, as it presents a profile (Fig. 3). A profile is a chart that displays the change in pixel values along the test line. Analysis of the cure helps to indicate the borders between the layers. Profile analysis may also be used to establish the threshold values for the binarization transform.

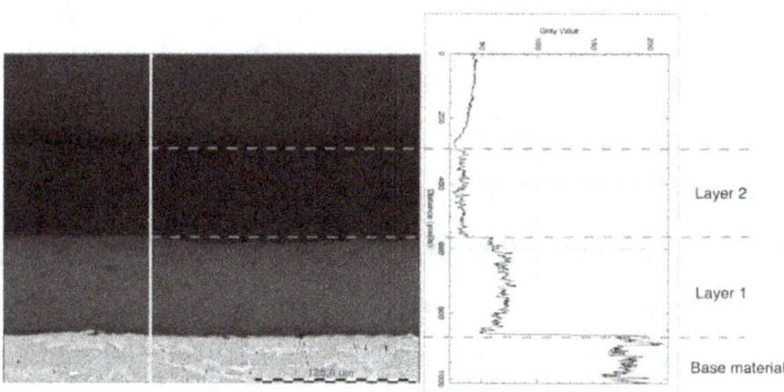

Fig. 3. Multilayer coating system and its profile.

The last method, which is presented in this paper, is a three-dimensional analysis of both the base material and the coatings. Three-dimensional visualization and analysis enable assessment of layer continuity, thickness, and internal defects (Fig. 4) [24].

Materials Research Proceedings 62 (2026) 165-171 https://doi.org/10.21741/9781644904015-21

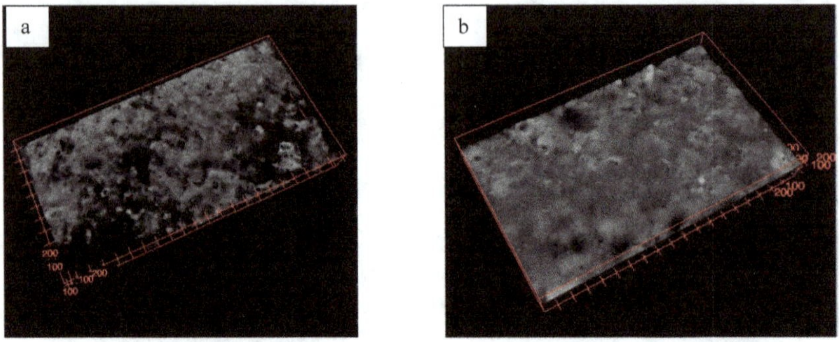

Fig. 4. WC-Co-Al_2O_3 coating before (a) and (b) after laser treatment.

Image Analysis

Depending on the analysis method and research objective, image processing algorithms enable automated detection and quantitative characterization of structural features [21-23]. A standard image processing workflow typically comprises the following stages Fig.5:

1. Image acquisition - The selection of an appropriate observation method should take into account the observation scale and material properties, as well as the way the image is captured as a digital file, with suitably chosen bit depth, color space, spatial resolution, and file format.
2. Image pre-processing – correction of image artifacts using geometric transformations, in the case of optical system-induced distortions, and filtering techniques to reduce noise or enhance contrast between objects and background.
3. Object detection – a set of image transformations aimed at segmenting objects from the background, typically by converting the image to a binary format, in which objects are represented by pixels with a value of 1 and the background by pixels with a value of 0.
4. Digital measurements – selection of measurement algorithms appropriate to the nature of the objects and the specific research goals. Measurements are conducted using stereological methods, which enable the assessment of three-dimensional features based on two-dimensional projections or cross-sections. Standard stereological parameters that deliver information on the analyzed microstructure are: VV - volume fraction of the analyzed phase, NA – number of objects per unit of analyzed area. For three-dimensional image data, direct quantitative information is obtained through volumetric reconstruction of the analyzed objects.

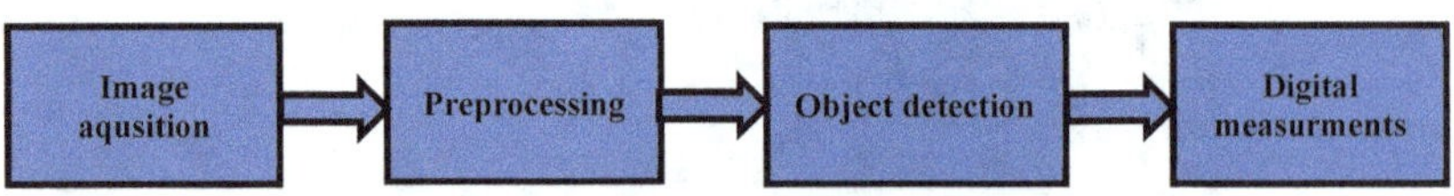

Fig. 5. Scheme of the typical image analysis workflow.

Although the presented general scheme of the image processing algorithm may appear simple, analyzing a specific type of coating system in a given modality often requires the development of a dedicated solution for fully automated processing. While some software packages allow for the interactive adjustment of transformation parameters, which can also yield a dataset that quantitatively describes the objects, human interaction compromises both the objectivity and, most importantly, the repeatability of the analysis. Objectivity, repeatability, and reliability are particularly critical when the analysis aims to generate a dataset from a large number of images for subsequent use in AI-based methods [16-19].

Materials Research Proceedings 62 (2026) 165-171 https://doi.org/10.21741/9781644904015-21

Economic and Technological Impact

The evaluation of coating systems is a key element in quality assessment, both during technological processes [25] and for final products [26]. Human visual assessment always involves an undefined level of subjectivity, which can be reduced by automated image analysis [27,28]. This is particularly important for medium- or high-throughput processes, where subjective visual assessment would be too slow. Examples include processes for applying special coatings [29-31], laser texturing [32,33], or laser remelting of the surface layer [34,35], which result in significant changes to the microgeometry's characteristics [36,37]. Quantitative characteristics obtained from image analysis can serve as observed quantities [38,39] in process optimization methods. Due to the specificity of the data provided, it is sometimes necessary to either support resampling methods [40, 41] or account for the incompleteness of existing knowledge [42, 43]. The end result, however, is a significant improvement in the quality of the final products [44].

Summary

Automated measurements enable the objective evaluation and ensure the repeatability of the measurement process, which is of critical importance during the design phase of coating development and the evolution of deposition technology. Furthermore, they constitute an essential component of quality control in the manufacturing process. The acquired data serve as a valuable source of information regarding the influence of process parameters on the properties of the surface layer, supporting effective modeling of its behavior and performance.

Ongoing research on the application of artificial intelligence for predicting geometric structure aims to optimize the processing parameters of the surface layer while significantly reducing the number of experimental tests required. This approach offers promising prospects for increasing process efficiency and accelerating the development of advanced coating technologies.

References

[1] B. Antoszewski, Inżynieria powierzchni: wybrane zagadnienia. Wydawnictwo Politechniki Świętokrzyskiej, 2011.

[2] D. Shichang, X. Lifeng, High Definition Metrology Based Surface Quality, Control and Applications, Springer, 2019. https://doi.org/10.1007/978-981-15-0279-8

[3] A.R. Yan et al., High Power Diode Laser Clad Cu-WC-Ni Composite Coatings on Brass for Wear Resistance Enhancement. Advanced Materials Research 816-817 (2013) 47-53. https://doi.org/10.4028/www.scientific.net/AMR.816-817.47

[4] K.R.C. Soma Raju et al., Electro-spark coatings for enhanced performance of twist drills. Surf. Coat. Technol. 202 (2008) 1636-1644. https://doi.org/10.1016/j.surfcoat.2007.07.084

[5] A.V. Ribalko, O. Sahin, The use of bipolar current pulses in electrospark alloying of metal surfaces. Surf. Coat. Technol. 168 (2003) 129-135. https://doi.org/10.1016/S0257-8972(02)00877-0

[6] A. Gądek-Moszczak et al., Nano X-ray Tomography Application for Quantitative Surface Layer Geometry Analysis after Laser Beam Modification. Materials 15 (2022) art.5935. https://doi.org/10.3390/ma15175935

[7] S.P. Murzin, V.B. Balyakin, Microstructuring the surface of silicon carbide ceramic by laser action for reducing friction losses in rolling bearings. Optics & Laser Technology 88 (2017) 96-98. https://doi.org/10.1016/j.optlastec.2016.09.007

[8] L. Zhu, W. Gao, Y. Wang, A self-lubricating composite coating on 6061 aluminum alloy surface with an intermediate anodised layer, Surf. Coat. Technol. 513 (2025) art.132483. https://doi.org/10.1016/j.surfcoat.2025.132483

[9] K.A. Chandran et al., Characterization of Surface Engineering and Coatings, in: Advanced Materials and Manufacturing Processes, CRC Press, Boca Raton, 2021, 173-190. https://doi.org/10.1201/9781003093213-11

[10] T. Balusamy et al., Surface nanostructuring of metallic materials for implant applications, in: Nanomaterials and Their Biomedical Applications, Springer, 16, 2021, 465-511. https://doi.org/10.1007/978-981-33-6252-9_16

[11] N. Radek et al., The Morphology and Mechanical Properties of ESD Coatings before and after Laser Beam Machining. Materials 13 (2020) art.2331. https://doi.org/10.3390/ma13102331

12] ASTM ASTM E1382 – 97 (2015) Standard Test Methods for Determining Average Grain Size Using Semiautomatic and Automatic Image Analysis.

[13] ASTM E45-2013: Standard Test Methods for Steel Inclusion Content

[15] A. Akinpelu et al., Discovery of novel materials through machine learning. Journal of Physics Condensed Matter 36 (2024) art.453001. https://doi.org/10.1088/1361-648X/ad6bdb

[16] V. Diwakar et al., Machine learning-based prediction of single clad characteristics and non-destructive characterization of multi-layer deposited FeCoNiCrMo HEA on EN24 via laser cladding. Materials Today Communications 41 (2024) art.110839. https://doi.org/10.1016/j.mtcomm.2024.110839

[17] M. Ghilom, S. Latifi, The Role of Machine Learning in Advanced Biometric Systems. Electronics 13 (2024) art.2667. https://doi.org/10.3390/electronics13132667

[18] K.J. DeMille et al., Materials design using genetic algorithms informed by convolutional neural networks: Application to carbon nanotube bundles. Composites Part B: Engineering 286 (2024) art.111751. https://doi.org/10.1016/j.compositesb.2024.111751

[19] D. Morgan, R. Jacobs, Opportunities and Challenges for Machine Learning in Materials Science. Ann. Rev. Mater. Res. 50 (2020) 71-103. https://doi.org/10.1146/annurev-matsci-070218-010015

[20] A. Gądek-Moszczak, M. Niedźwiedź, Pore Morphology Assessment of Oxide Coatings on AZ31B Magnesium Alloy, ECSIA 2025 – 14th Europ. Congr. Stereology and Image Analysis, Prague, Czech Republic, September 15-18, 2025

[21] L.M. Cruz-Orive (2024). Stereology. Theory and Application, Springer, 2024.

[22] J.C. Russ, F.B. Neal, The Image Processing Handbook, CRC Press, Boca Raton, 2016.

[23] B.L. DeCost, E.A. Holm, Computer vision approach for automated analysis and classification of microstructural image data. Comp. Mater. Sci. 110 (2015) 126-133. https://doi.org/10.1016/j.commatsci.2015.08.011

[24] A. Gądek-Moszczak et al., (2014). Application of 3D Image Analysis Techniques for Assessment of the Quality of Material Surface Layer Before and After Laser Treatment. Adv. Mater. Res. 874 (2014) 133–138. https://doi.org/10.4028/www.scientific.net/AMR.874.133

[25] K. Czerwińska et al., Improving quality control of siluminial castings used in the automotive industry, METAL 2020 – 29th Int. Conf. Metall. Mater. (2020) 1382-1387. https://doi.org/10.37904/metal.2020.3661

[26] M. Ingaldi, Overview of the main methods of service quality analysis, Production Engineering Archives 18 (2018) 54-59. https://doi.org/10.30657/pea.2018.18.10

[27] A. Szczotok, S. Roskosz, New possibilities of light microscopy research resulting from digital recording of images, Materials Science- Poland 23 (2005) 559-565.

[28] A. Szczotok, M. Sozańska, A comparison of grain quantitative evaluation performed with standard method of imaging with light microscopy and EBSD analysis, Practical Metallography 46 (2009) 454-468. https://doi.org/10.3139/147.110043

[29] A. Dudek, Investigations of microstructure and properties in bioceramic coatings used in medicine, Arch. Metall. Mater. 56 (2011) 135-140. https://doi.org/10.2478/v10172-011-0015-y

[30] S. Michna et al., Research the causes of surface stains after eloxal coating for the profile from the AlMgSi alloy using substructural analysis, Manufacturing Technology 15 (2015) 620-624.

[31] N. Radek et al., Technology and application of anti-graffiti coating systems for rolling stock, METAL 2019 – 28th Int. Conf. Metall. Mater. (2019) 1127-1132.

[32] M. Kukliński et al., Influence of microstructure and chemical composition on microhardness and wear properties of laser borided monel 400, Materials 13 (2020) art. 5757. https://doi.org/10.3390/ma13245757

[33] Ł.J. Orman et al., Laser Treatment of Surfaces for Pool Boiling Heat Transfer Enhancement, Materials 16 (2023) art. 1365. https://doi.org/10.3390/ma16041365

[34] N. Radek et al., The effect of laser beam processing on the properties of WC-Co coatings deposited on steel, Materials 14 (2021) art. 538. https://doi.org/10.3390/ma14030538

[35] E. Szajna et al., The influence of laser remelting on microstructural changes and hardness level of flame-sprayed NiCrBSi coatings with tungsten carbide addition, Surface and Coatings Technology 478 (2024) art. 130403. https://doi.org/10.1016/j.surfcoat.2024.130403

[36] D. Przestacki, T. Chwalczuk, The analysis of surface topography during turning of Waspaloy with the application of response surface method, MATEC Web of Conferences 136 (2017) art. 02006. https://doi.org/10.1051/matecconf/201713602006

[37] A. Dudek et al., Laser Surface Alloying of Sintered Stainless Steel, Materials 15 (2022) art. 6061. https://doi.org/10.3390/ma15176061

[38] J. Pietraszek, A. Gadek-Moszczak, The smooth bootstrap approach to the distribution of a shape in the ferritic stainless steel AISI 434L powders, Solid State Phenomena 197 (2013) 162-167. https://doi.org/10.4028/www.scientific.net/SSP.197.162

[39] J. Pietraszek et al., Challenges for the DOE methodology related to the introduction of Industry 4.0, Prod. Eng. Arch. 26 (2020) 190-194. https://doi.org/10.30657/pea.2020.26.33

[40] J. Pietraszek, L. Wojnar, The bootstrap approach to the statistical significance of parameters in RSM model, ECCOMAS Congress 2016 - Proc. 7th Europ. Congr. Comput. Methods in App. Sci. Eng. 1 (2016) 2003-2009. https://doi.org/10.7712/100016.1937.9138

[41] R. Dwornicka et al., The smoothed bootstrap fine-tuning, System Safety: Human - Technical Facility - Environment 1 (2019) 716-723. https://doi.org/10.2478/czoto-2019-0091

[42] J. Pietraszek, Fuzzy regression compared to classical experimental design in the case of flywheel assembly, Lecture Notes in Computer Science 7267 LNAI (2012) 310-317. https://doi.org/10.1007/978-3-642-29347-4_36

[43] J. Pietraszek, The modified sequential-binary approach for fuzzy operations on correlated assessments, Lecture Notes in Computer Science 7894 LNAI (2013) 353-364. https://doi.org/10.1007/978-3-642-38658-9_32

[44] B. Gajdzik et al., Approaching open innovation in customization frameworks for product prototypes with emphasis on quality and life cycle assessment (QLCA), Journal of Open Innovation: Technology, Market, and Complexity 10 (2024) art. 100268. https://doi.org/10.1016/j.joitmc.2024.100268

Terotechnology XIV Materials Research Forum LLC
Materials Research Proceedings 62 (2026) 172-178 https://doi.org/10.21741/9781644904015-22

Experimental Evaluation of Laser-Assisted Machining of Si_3N_4 Ceramics

Damian PRZESTACKI[1,a *], Norbert RADEK[2,b], Krzysztof NAPLOCHA[3,c] and Pandiarajan NARAYANASAMY[4,d]

[1]Poznan University of Technology, Faculty of Mechanical Engineering, Poznan University of Technology, ul.Piotrowo 3, 61-138 Poznan, Poland

[2]Kielce University of Technology, Faculty of Mechatronics and Mechanical Engineering,

Al.1000-lecia Państwa Polskiego 7, 25-314 Kielce, Poland

[3]Wroclaw University of Science and Technology, Faculty of Mechanical Engineering, ul.Smoluchowskiego 25, 50-370 Wrocław, Poland

[4]Department of Mechanical Engineering, Kamaraj College of Engineering and Technology, Madurai-625 701, Tamil Nadu, India

[a]damian.przestacki@put.poznan.pl, [b]norrad@tu.kielce.pl, [c]krzysztof.naplocha@pwr.edu.pl, [d]narayananx5@gmail.com

Keywords: Laser Beam Processing, Tool Wear, Si3N4 Ceramic

Abstract. Silicon nitride (Si_3N_4) ceramics represent a significant class of materials with numerous applications in engineering and biomedical fields. However, their exceptionally high hardness (approximately 2125 HV) and inherent brittleness pose significant challenges during machining. Laser-assisted machining (LAM) has recently emerged as a promising method for enhancing the machinability of hard and brittle materials. This hybrid process combines localized laser heating of the workpiece with conventional cutting, where the softened surface layer is removed by a cutting tool with a defined edge geometry. Compared to conventional machining, this technique substantially extends tool life and enhances process efficiency. The present study investigates the machinability of hard Si_3N_4 ceramics by evaluating the flank wear (VBc) of the cutting tool edge. Both conventional turning and laser-assisted turning were performed, with real-time temperature monitoring of the machined surface. A clear distinction in tool wear behavior and cutting performance was observed between the two processes, demonstrating the advantages of laser-assisted machining over traditional methods.

Introduction

An increasing number of modern engineering products are being manufactured from hard-to-machine materials such as titanium alloys, nickel-based superalloys, advanced ceramics, and metal-ceramic composites. A representative example of such a material is silicon nitride (Si_3N_4), a technical ceramic that, due to its low density, excellent wear and corrosion resistance, and high-temperature strength, finds applications in both engineering and biomedical fields. Typical components include bearings, valves, rotors, cutting tools, and artificial hip joints [1].

These components frequently require precision machining operations to achieve high surface integrity and tight dimensional tolerances. However, machining of ceramic materials remains a considerable challenge due to their high hardness, abrasiveness, and intrinsic brittleness. Currently, diamond wheel grinding is the primary method for manufacturing precision ceramic components, although it incurs relatively high production costs.

Materials Research Proceedings 62 (2026) 172-178
https://doi.org/10.21741/9781644904015-22

Continuous improvement of cutting and abrasive machining processes, therefore, focuses on developing and implementing more efficient techniques and process variants that enhance not only productivity, dimensional accuracy, and surface quality, but also overall cost-effectiveness.

One promising approach for increasing machining efficiency is hot machining, which involves preheating the workpiece using a plasma or laser beam, or, in certain cases, high-frequency electric currents. Among these techniques, laser beam heating offers a distinct advantage, as it enables precise control of the heating process through adjustments to the laser power, as well as the shape, size, and position of the laser spot on the workpiece surface.

Consequently, laser-assisted machining (LAM) of hard ceramics, such as Si_3N_4, has emerged as a promising hybrid manufacturing method. In LAM, the workpiece is locally heated by a laser beam, and the softened material is subsequently removed using cutting tools with a defined geometry. This approach has been shown to significantly extend tool life and improve machining performance compared with conventional cutting processes.

Methodology and Research Technique

Machinability tests of Si_3N_4 ceramic were conducted during longitudinal turning. The experimental setup consisted of a numerically controlled lathe equipped with a TRUMPF diode laser with a power output of P = 3000 W. The schematic of laser-assisted material removal is shown in Fig. 1. The laser head moves along the axis of the turned samples. At the same time, the cutting edge corner is positioned in a plane perpendicular to the turning axis and passing through the center of the laser beam.

The transverse surface roughness of the turned samples was measured using a Hommel Tester profilographometer. Cutting forces were recorded with a strain gauge dynamometer of in-house design, and tool wear measurements were performed using an optical microscope.

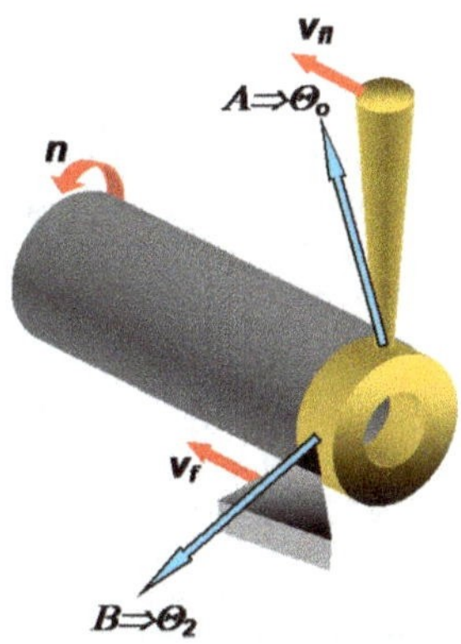

Figure 1. *Schematic diagram of laser-assisted turning of Si3N4 hard ceramics. A - heating area by laser beam, B – zone of machining, θ_0, θ_2 - areas of temperature measurement.*

This research aimed to evaluate the cutting performance of tool wedges made from various cutting tool materials during the laser-assisted turning of hard silicon nitride (Si_3N_4) ceramics, which have a hardness of 2125 HV. The ceramic specimens were supplied by Norton Dias (Czech Republic). During machining, the cutting zone was heated by a laser beam to a temperature of approximately Θ_2 = 1200-1500°C.

The experiments were conducted under constant cutting parameters:
Si_3N_4: vc = 10 m/min, f = 0.04 mm/rev, ap = 0.1 mm.

The following indexable cutting inserts were used in the study:

- SNMG 120408 MS made of fine-grained tungsten carbide grade KC5510, TiAlN-coated using the PVD method (Kennametal, USA);
- SNMG 120408 MS made of fine-grained tungsten carbide grade KC5525 with an increased cobalt content, TiAlN-coated using the PVD method (Kennametal, USA);
- SNGN 120708T 02020 made of oxide ceramics (Al_2O_3 + ZrO_2), grade AC5 (Kennametal, USA);
- SNGN 120408T 02020 made of mixed ceramics (Al_2O_3 + TiCN), grade MC2 (Kennametal Hertel, USA).

Results of Research and Analysis

The results of the study and subsequent analysis were presented in Figures 2 and 3, illustrating the influence of the cutting zone temperature during laser-assisted machining of Si_3N_4 ceramics on the wear behavior of cemented carbide and ceramic cutting inserts.

The highest cutting performance was obtained for the cemented carbide inserts of grade KC5510. Laser-assisted turning of Si_3N_4 ceramics using these inserts at temperatures between Θ_2 = 1300-1500°C resulted in relatively low tool wear. The minimum wear of the cemented carbide tool was observed at a cutting temperature of approximately 1500 °C. For most of the tested tool materials, an increase in temperature (θ_2) resulted in a reduction in cutting-edge wear; however, the temperature corresponding to the lowest wear value varied depending on the specific tool material.

An exception to this trend was observed for the mixed ceramic insert (MC2) (Fig. 4), where an increase in cutting temperature resulted in a progressive rise in tool wear. This behavior makes the MC2 ceramic unsuitable for laser-assisted machining. During turning at approximately 1300 °C, the MC2 tool exhibited a nearly constant wear value ($VBc \approx 0.74$ mm), whereas above this temperature, the wear increased monotonically (Fig. 4).

The wear of the AC5 insert during machining at a temperature of approximately 1300 °C was the lowest, being nearly three times smaller compared with that observed under conventional turning conditions (Fig. 5). The wear of ceramic inserts, when compared with cemented carbide tools, was significantly higher. These ceramic tools were classified as unsuitable for such severe cutting conditions.

Based on the results of the present study, as well as findings reported by other researchers [4], it can be concluded that moderate cutting temperatures during laser-assisted machining of Si_3N_4 ceramics are the most favorable in terms of tool wear. These intermediate temperatures allow sufficient softening of the work material without compromising the strength of the cutting edge. Conversely, lower temperatures do not adequately soften the ceramic material, and this simultaneously leads to intensive tool wear caused by adhesion and diffusion mechanisms.

Materials Research Proceedings 62 (2026) 172-178 https://doi.org/10.21741/9781644904015-22

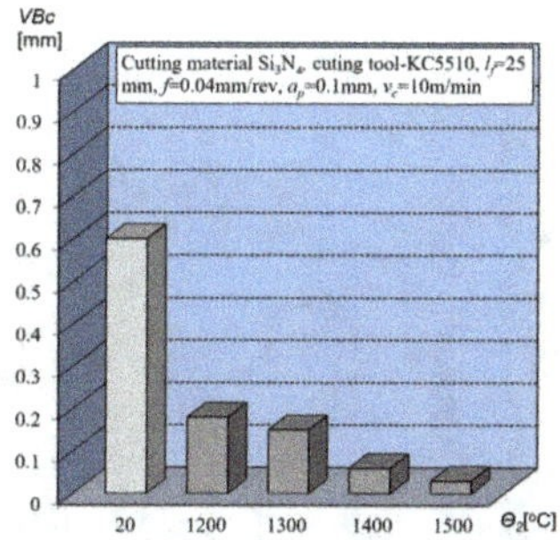

***Figure 2.** Wedge wear for coated cemented carbide KC5510 for various temperatures Θ_2. $t_s = 2.5min$*

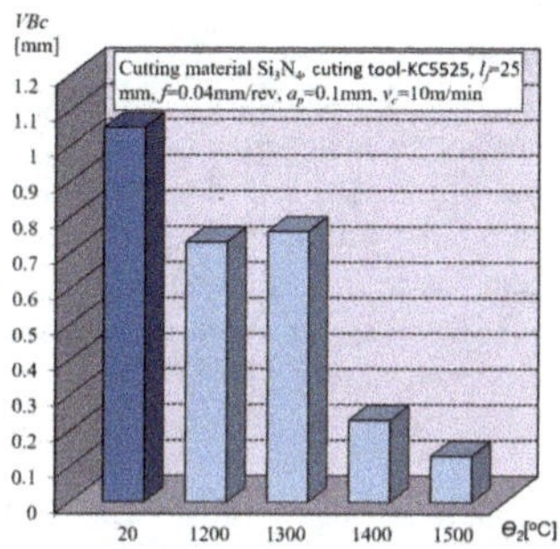

***Figure 3.** Wedge wear for coated cemented carbide KC5525 for various temperatures Θ_2. $t_s = 2.5\ min$*

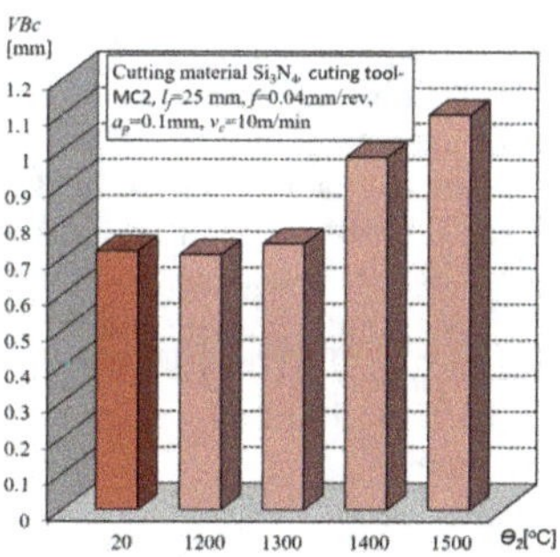

***Figure 4.** Wedge wear for ceramic MC2 with various temperature Θ_2. $t_s = 2.5min$*

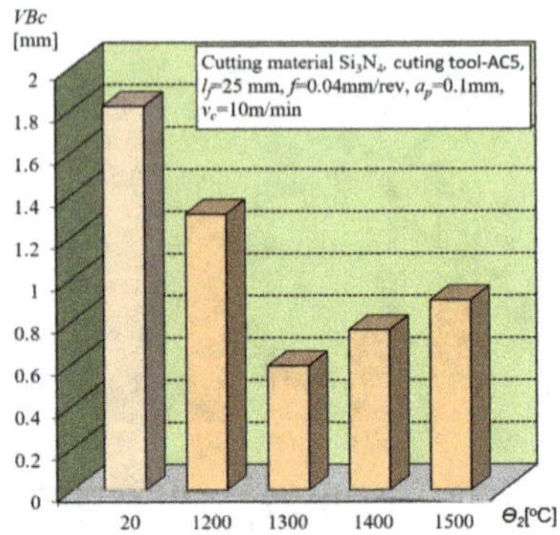

***Figure 5.** Wedge wear for ceramic AC5 with various temperature Θ_2. t_s = 2.5min*

Technological and Economic Impact

This article highlights the technological challenge of machining materials that are both hard and brittle, which are susceptible to cracking during machining [4,5]. The use of laser softening has opened up effective machining possibilities [6]. Similar machining difficulties arise in the case of bioceramics [7-9], further compounded by biocompatibility issues. Appropriate recognition of physical phenomena and subsequent modeling of technological processes [10-12] are crucial for achieving the desired quality [13-15] while maintaining reasonable costs [16,17]. Furthermore, this significantly improves occupational safety [18,19] by minimizing the risk of high-energy chipping. Regardless, the presented methods are inspiring for data analysis methods, both quantitative [20,21] and qualitative [22,23]. They provide very interesting data on the geometry of the surface layer [24, 25] for automated image analysis [26, 27]. This clearly demonstrates the value of implementing the teaching of these data analysis methods in engineering practice [28-30].

Conclusions

Based on the experimental investigations conducted and the analysis of the relevant literature, the following conclusions were drawn:

1. Within the investigated range of parameters, an increase in the temperature of the laser-heated cutting zone during laser-assisted turning of hard technical Si_3N_4 ceramics generally leads to a reduction in tool wear for most of the cutting tool materials tested.
2. The best cutting performance, evaluated using the VBc flank wear criterion, was obtained for coated cemented carbide inserts (KC5510) under laser-assisted machining (LAM) conditions. These tools demonstrated favorable wear resistance and may serve as a viable alternative to polycrystalline diamond (PCD) tools for machining ceramics.
3. Inserts made of oxide ceramics (MC2) and mixed ceramics (AC5) were found to be unsuitable for laser-assisted turning of hard technical Si_3N_4 ceramics due to their limited wear resistance under elevated temperature conditions.

References

[1] X. Cai et al., Effects of laser energy uniformization via static beam shaping on ablation behavior and mechanism of SiCf/SiC composites, Journal of Materials Processing Technology 347 (2026) art.119149. https://doi.org/10.1016/j.jmatprotec.2025.119149

[2] C. Liu et al., Cutting mechanism of reaction-bonded silicon carbide in laser-assisted ultra-precision machining. Int. J. Mach. Tools Manuf. 203 (2024) art. 104219. https://doi.org/10.1016/j.ijmachtools.2024.104219

[3] D. Xu et al., Materials removal mechanism in laser-assisted grinding of SiC fibre-reinforced titanium alloy composite. CIRP Ann. 74 (2025) 417-421. https://doi.org/10.1016/j.cirp.2025.04.054

[4] G. Guidoni et al., Fracture behaviour of thermal barrier coatings after high temperature exposure in air, Materials Science and Engineering: A 397 (2005) 209-214. https://doi.org/10.1016/j.msea.2005.02.022

[5] A. Szczotok et al., The Impact of the Thickness of the Ceramic Shell Mould on the ($\gamma + \gamma'$) Eutectic in the IN713C Superalloy Airfoil Blade Casting, Archives of Metallurgy and Materials 62 (2017) 587-593. https://doi.org/10.1515/amm-2017-0087

[6] N. Radek et al., Properties of Steel Welded with CO2 Laser, Lecture Notes in Mechanical Engineering (2020) 571-580. https://doi.org/10.1007/978-3-030-33146-7_65

[7] A. Dudek, Surface properties in titanium with hydroxyapatite coating, Optica Applicata 39 (2009) 825-831.

[8] A. Dudek, R. Wlodarczyk, Structure and properties of bioceramics layers used for implant coatings, Solid State Phenomena 165 (2010) 31-36. https://doi.org/10.4028/www.scientific.net/SSP.165.31

[9] A. Dudek, Investigations of microstructure and properties in bioceramic coatings used in medicine, Archives of Metallurgy and Materials 56 (2011) 135-140. https://doi.org/10.2478/v10172-011-0015-y

[10] J. Pietraszek, A. Goroshko, The heuristic approach to the selection of experimental design, model and valid pre-processing transformation of DoE outcome, Advanced Materials Research 874 (2014) 145-149. https://doi.org/10.4028/www.scientific.net/AMR.874.145

[11] R. Dwornicka, J. Pietraszek, The outline of the expert system for the design of experiment, Production Engineering Archives 20 (2018) 43-48. https://doi.org/10.30657/pea.2018.20.09

[12] J. Pietraszek et al., Challenges for the DOE methodology related to the introduction of Industry 4.0, Prod. Eng. Arch. 26 (2020) 190-194. https://doi.org/10.30657/pea.2020.26.33

[13] R. Ulewicz, Outsorcing quality control in the automotive industry, MATEC Web of Conf. 183 (2018) 0-0. https://doi.org/10.1051/matecconf/201818303001

[14] A. Pacana, K. Czerwińska, Comparative tests of the quality of the piston combustion chamber for a diesel engine, Tehnicki Vjesnik 27 (2020) 1021-1024. https://doi.org/10.17559/TV-20190112193319

[15] A. Pacana, K. Czerwińska, Improving the quality level in the automotive industry, Production Engineering Archives 26 (2020) 162-166. https://doi.org/10.30657/pea.2020.26.29

[16] D. Siwiec et al., Quality analysis of the mechanical lever and improvement of non-destructive testing process, System Safety: Human - Technical Facility - Environment 2 (2020) 299-306. https://doi.org/10.2478/czoto-2020-0037

[17] D. Siwiec et al., Concept of a model to predict the qualitative-cost level considering customers' expectations, Polish Journal of Management Studies 26 (2022) 330-340. https://doi.org/10.17512/pjms.2022.26.2.20

[18] R. Ulewicz, Practical application of quality tools in the cast iron foundry, Manufacturing Technology 14 (2014) 104-111.

[19] R. Ulewicz et al., Quality and work safety in metal foundry, METAL 2020 - 29th Int. Conf. Metall. Mater. (2020) 1287-1293. https://doi.org/10.37904/metal.2020.3649

[20] J. Pietraszek, A. Gadek-Moszczak, The smooth bootstrap approach to the distribution of a shape in the ferritic stainless steel AISI 434L powders, Solid State Phenomena 197 (2013) 162-167. https://doi.org/10.4028/www.scientific.net/SSP.197.162

[21] A. Gądek-Moszczak et al., The bootstrap approach to the comparison of two methods applied to the evaluation of the growth index in the analysis of the digital X-ray image of a bone regenerate, Studies in Computational Intelligence 572 (2015) 127-136. https://doi.org/10.1007/978-3-319-10774-5_12

[22] J. Pietraszek, Fuzzy regression compared to classical experimental design in the case of flywheel assembly, Lecture Notes in Computer Science 7267 LNAI (2012) 310-317. https://doi.org/10.1007/978-3-642-29347-4_36

[23] J. Pietraszek, The modified sequential-binary approach for fuzzy operations on correlated assessments, Lecture Notes in Computer Science 7894 LNAI (2013) 353-364. https://doi.org/10.1007/978-3-642-38658-9_32

[24] J. Pietraszek et al., Advanced statistical refinement of surface layer's discretization in the case of electro-spark deposited carbide-ceramic coatings modified by a laser beam, Solid State Phenomena 197 (2013) 198-202. https://doi.org/10.4028/www.scientific.net/SSP.197.198

[25] A. Gądek-Moszczak et al., Nano X-ray Tomography Application for Quantitative Surface Layer Geometry Analysis after Laser Beam Modification, Materials 15 (2022) art. 5935. https://doi.org/10.3390/ma15175935

[26] A. Szczotok, S. Roskosz, New possibilities of light microscopy research resulting from digital recording of images, Materials Science- Poland 23 (2005) 559-565.

[27] A. Szczotok, M. Sozańska, A comparison of grain quantitative evaluation performed with standard method of imaging with light microscopy and EBSD analysis, Practical Metallography 46 (2009) 454-468. https://doi.org/10.3139/147.110043

[28] E. Skrzypczak-Pietraszek, J. Pietraszek, Chemical profile and seasonal variation of phenolic acid content in bastard balm (Melittis melissophyllum L., Lamiaceae), Journal of Pharmaceutical and Biomedical Analysis 66 (2012) 154-161. https://doi.org/10.1016/j.jpba.2012.03.037

[29] J. Zwierzchowski et al., Model tests of wind turbine with a vertical axis of rotation type Lenz 2, EPJ Web of Conferences 143 (2017) art. 02150. https://doi.org/10.1051/epjconf/201714302150

[30] M. Radek et al., Matching Computational Tools to User Competence Levels in Education of Engineering Data Processing, Materials Research Proceedings 34 (2023) 453-459. https://doi.org/10.21741/9781644902691-52

Materials Research Proceedings 62 (2026) 179-187 | https://doi.org/10.21741/9781644904015-23

Functional Assessment of Shelter Facilities in Terms of Material and Technical Changes to Basic Structural Elements

Jacek PARTYKA[1,a*] and Piotr KRYSIAK[1,b]

[1]Military Institute of Engineer Technology, Obornicka 136, 50-961 Wrocław, Poland

[a]partyka@witi.wroc.pl, [b]krysiak@witi.wroc.pl

Keywords: Dugout, Cellar, Concrete, Reinforced Concrete, Composite, Plastics, Shelter

Abstract. The aim of this work is to present a comprehensive overview of shelter construction. Particular attention is paid to interesting material and technical solutions and the ecological aspects of the technologies used. The content is presented in terms of historical changes, with the materials used in shelter construction including stone, soil, wood, and steel, and currently, synthetic materials. The study focuses heavily on the so-called PwK technology, which is used in underground construction. It is pointed out that the most popular structural solutions for shelter construction currently utilize prefabricated reinforced concrete materials. Underground structures constructed using this technology provide effective protection for people and property. In addition to the traditional shelter construction methods mentioned, the study presents selected other material solutions for shelter components, such as laminate, high-density polyethylene (HDPE), and composite structures. The study concludes by noting that development opportunities in various technologies should only contribute to ensuring safety in the civil protection and civil defense systems.

Introduction

Currently, there is a significant increase in interest in the development of shelter construction. This stems from the increasing global political tensions and the ongoing conflict in Ukraine, and is reflected in numerous publications. For example, in [1], the author expresses the view that, given the current state of international security, organizational, functional, and technical preparations in collective protection facilities are disproportionate to the threats. Therefore, it is necessary to ensure human safety by providing facilities primarily characterized by resistance to the effects of fire. Such facilities include shelters, which, based on [2], are either buildings recognized as protective structures or constitute part of a closed and airtight structure, equipped with filtering and ventilation systems or regenerative absorbers. Shelter facilities should meet many other requirements and functions, as indicated by the authors in [3] and [4]. These include functional requirements that ensure safe use, as well as parameters such as mechanical strength, impact resistance, fire resistance, resistance to increased humidity, thermal insulation of partitions, and vapor permeability. These are some of the requirements that, when met, impact the comfort and convenience of using modular buildings, ensuring optimal operating conditions throughout their lifespan. It should be noted that, in addition to the requirements listed above, the proper design of basic structural elements, such as walls and ceilings, is crucial.

Based on [5], it can be concluded that shelter partitions can be made of homogeneous materials or composed of components. Their construction, especially in the case of explosion-proof shelters, should be durable, as discussed in [6]. Ceilings, however, are most often made of reinforced, monolithic, or prefabricated concrete. The solutions for the basic structural elements of shelter partitions presented in this study should be assessed for compliance with the requirements before use.

Materials Research Proceedings 62 (2026) 179-187 https://doi.org/10.21741/9781644904015-23

Using Natural Resources in the Design of Shelter Structures

Humans have been using natural rock formations, such as caves, crevices, and rock overhangs, for centuries as habitats and shelters, as well as for defensive and ritual purposes, as confirmed by archaeological finds. Historically, natural structures have served as shelters for humans [7] and animals. They provided protection from various threats thanks to their durability and weather resistance, offering protection from wind, rain, and sun.

Fig. 1. Rock crevices used for shelter – an example [source: https://www.survivalworld.com/outdoors/rock-shelter/]

The rock formations visible in Figure 1 may have provided people with protection from the elements. These naturally occurring shelters were widely used by indigenous peoples and still serve as temporary shelters. Although a rock shelter offers less complete protection than a deep cave, it provides a more practical and comfortable living space with access to daylight and fresh air. Over time, especially in less mountainous regions, shelters have been found in depressions or excavations, creating a suitably shaped and free space. Depressions and excavations represent the artificial process of removing soil to create a trench, recess, or ditch. These types of landforms can be natural or artificial and refer specifically to the artificial creation of these formations for various purposes, such as building foundations, installing underground infrastructure, or archaeological research. Thus, over time, the soil became a material for securing shelters. To ensure effective protection against threats, the structure in the ground was equipped with wooden structural elements, creating a framework. Thanks to this, such a structure will have a load-bearing structure, contributing to the structure's greater strength and durability.

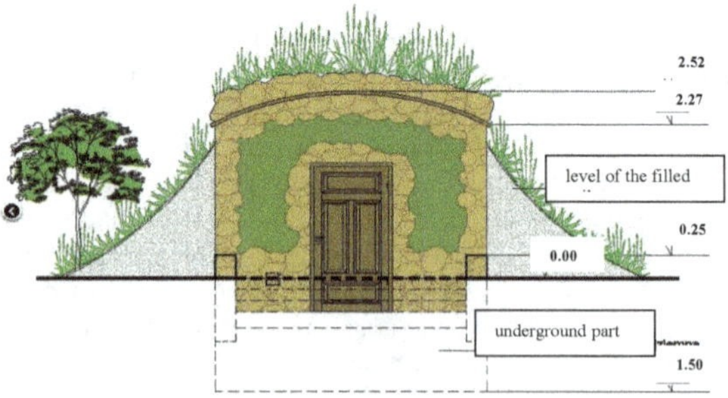

Fig.2. The stone dugout structure sunk below the ground surface [8]

Currently, structures constructed in this manner, called cellars (dugouts), constitute a popular group of investments. The reason for this trend is the ability to appropriately utilize warehouse space for storing various types of products and portable equipment. Dugouts can vary in shape and appearance. They often blend into their surroundings, making aerial identification difficult. An example of such a structure, constructed primarily of fieldstone and cement-lime mortar, is shown in Fig.2. The facility shown in Fig.2 is a storage facility. To effectively protect persons, it should be adapted and equipped with the necessary facilities.

PwK Technology in the Construction of Shelter Structures

Based on the work [8], the innovative PwK technology refers to "Foam in a Grid." This technology is characterized by a geogrid filled with polyurethane foam, creating a lightweight, three-dimensional structure. Thanks to properties such as satisfactory compressive strength and resistance to mechanical stress, the structure is stable and ensures functionality. The work [9] shows that, structurally, the actual structure is a shell composed of prefabricated modules (Fig. 3) on a hardened substrate.

Fig. 3. Cylindrical shell of the dugout creating an arched vault using PwK technology [9]

The supporting shell with the geogrid is filled with a substrate and a plant layer, which limits surface water runoff and the formation of ponds near the structure, ensuring drainage. Figures 4 and 5 show examples of dugouts from the Manufacturer's extensive range.

Fig.4 General view of the dugout situated in a depression [9]

Fig. 5. General view of the cellar with a grass-covered surface [9]

Based on the presented PwK-based facilities, it can be concluded that they do not provide sufficient protection against the threat of explosives and hinder long-term occupancy. Such protection can be provided by a structured facility that meets the requirements for resistance to explosives.

The Use of Concrete and Steel in the Construction of Shelter Structures

Contemporarily, concrete is one of the most popular materials used in the construction of shelter structures. The general characteristics of concrete structural elements of shelters and their resistance to explosions, as well as the threat of nuclear attack and ionizing radiation, are presented in [10] [11] by the authors. The widespread use of concrete as a structural material in the construction of shelter structures is also illustrated in the previously mentioned work [8]. The body of the dugout, as a shelter structure, is placed in a trench and covered with a layer of soil. An example of a concrete dugout body with basic equipment is shown in Fig.4.

Fig. 6. Modular concrete dugout [8]

As can be seen from the shelter shown in Fig.6, it is a modular structure. It is a construction system that utilizes prefabricated, factory-made reinforced concrete (steel-reinforced concrete) modules that are assembled on-site for rapid construction. Another example of a modular reinforced concrete air raid shelter is the one shown in Fig.7.

Fig. 7. Prefabricated air raid shelters [https://protectedunderground.com/products/]

The advantage of prefabricated shelters, in addition to their resistance to explosives, is the rapid assembly of prefabricated modules, which significantly shortens construction time and lowers costs, as well as their ease of assembly. Because the components are manufactured in a factory rather than entirely on site, superior quality and repeatability is ensured. As the primary construction material of the shelters is concrete, these structures are characterized by significant resistance to selected explosives.

Selected Other Material Solutions for Shelter Facilities

There are many other interesting material- and technical solutions related to shelter facilities, which include, among others, facilities made of plastic. A plastic shelter is typically a home bunker, often made of durable high-density polyethylene (HDPE), which serves as a safe place to wait in the event of a threat, such as a bombing. Such a shelter is durable, waterproof, and resistant to external factors, and its interior can be equipped with the necessary components. An example shelter is shown in Fig.8.

Fig. 8. The interior of a laminate dugout-type structure [https://ekobudowa-sklep.pl/schron-350-bezpieczna-piwnica-ukrycie-hdpe]

The shelter in Fig.8 serves as a safe basement, hiding place, and emergency shelter. According to the manufacturer, it is equipped with a steel door, natural ventilation, and four rows of shelves,

Materials Research Proceedings 62 (2026) 179-187 https://doi.org/10.21741/9781644904015-23

providing protection and comfort even in extreme conditions. Shelters made of composite materials are very popular. These are modern solutions that also provide protection against various threats. They utilize composites and polymers for construction, and their modular design allows for flexible configurations for shelters for residents, crisis management centers, or warehouses. These types of shelters are tested for resistance to detonations. Based on [12], it is known that composite materials can be used in air raid shelters as shields against air strikes. An interesting solution is the steel-composite RIDE shelter. It is a modular reinforcement system that allows for the adaptation of rooms in existing buildings to the level of protective structures (e.g., garages, basements). Above-ground modular shelters can be deployed very quickly in areas. where it is not possible to adapt the facility or build a shelter; together they create a coherent set of implementation paths, i.e. adaptation of existing spaces, quick above-ground shelter or a full-fledged underground shelter.

Fig. 9. Shelter made of steel-composite technology RIDEhttps://builderpolska.pl/2025/09/29/budujemy-odpornosc/

Soil steel composite structures, known as SSCS, in modules is proven to be lightweight yet highly durable, facilitating transport and rapid assembly. Compared to reinforced concrete structures, it is more readily available and quicker to implement, resulting in significantly lower overall costs without compromising the performance of the solutions. Interesting views on the properties of steel-composite structures embedded in the ground are presented in [13, 14]. Specifically, the authors state that, unlike rigid structures, the performance of SSCS structures is largely dependent on the quality of the backfill. This applies to the type of soil used, particularly its compaction. There are many other examples of the use of composite materials in the construction of shelters.

Economic and Technological Impact

The issue of shelter construction has unfortunately become increasingly relevant in recent years. This is due to both increasingly violent weather phenomena and the spread of armed conflicts. Given the neglect of new construction and maintenance of existing shelters in recent decades, techniques for the rapid and, equally importantly, cost-effective construction of high-quality shelters deserve special attention. Therefore, it is important to consider the implementation of durable composite materials [15] and welded structures [16-18] using clean laser techniques [19-21]. The potential use of robotic devices [22] allows for high quality [23,24], confirmed by tests [25]. A side effect, but positive, is the minimization of waste, thus saving energy [26,27] while simultaneously reducing the burden on the environment [28,29] and adhering to the principles of a circular economy [30,31]. To effectively achieve the above-mentioned effects, with technologies combining industrial materials and natural resources, it is necessary to use appropriate optimization methods [32-34] supported by resampling techniques [35-37] taking into account the incompleteness of the available data [38-40].

Conclusions

This work provides a general overview of shelter construction. It follows that local materials have historically influenced shelter construction. Nowadays, several alternatives to utilizing local resources are proposed to enhance sustainability, and these include, above all, composite materials, which are extensively discussed in this work. The observed changes in the use of various construction materials for shelters demonstrate their versatility stemming from the availability of these materials and technologies, as well as the time required to complete a given project. The basic material for shelter construction – reinforced concrete – provides strength and resistance to threats. Special composites are also used to reinforce walls, and depending on the project, other materials can also be used, such as bricks, stones, and mineral wool or polyurethane foam for insulation. The development potential of various technologies should only support the overarching goal of ensuring safety within the civil protection and defense system.

References:

[1] D. Krupnik, The Problem of Providing Civilians with Collective Protection Facilities, Roczniki Nauk Społecznych 53 (2025) 71-84. https://doi.org/10.18290/rns2025.0018.

[2] Ustawa z dnia 5 grudnia 2024 r. o ochronie ludności i obronie cywilnej, Dz.U.2024.1907

[3] A. Siniecki, Bezpieczeństwo (w) domu. Schrony w budownictwie mieszkaniowym, Zeszyty Naukowe Politechniki Poznańskiej, Architektura, Urbanistyka, Architektura Wnętrz 17 (2023) 359-369. https://doi.org/10.21008/j.2658-2619.2023.17.22

[4] D. Olszewski et al., Modułowe konstrukcje w budowlach ochronnych – wybrane aspekty technologii, warunków technicznych i wykorzystania, Wiedza Obronna 287 (2024) 83-107. https://doi.org/10.34752/2024-2-4

[5] M. Soltani, E. Akurang, Characteristics of saferoom/shelter wall structures – a state-of-the-art review, Journal of Structural Integrity and Maintenance 7 (2022) 265-270. https://doi.org/10.1080/24705314.2022.2088330

[6] S.M. Anas et al., Reinforced cement concrete (RCC) shelter and prediction of its blast loads capacity, Materials Today Proc. 74 (2023) 547-568. https://doi.org/10.1016/j.matpr.2022.09.125

[7] M. Szruba, Budownictwo wspiera ekologię, Nowoczesne Budownictwo Inżynieryjne 95 (2021) 12-30.

[8] J. Partyka et al., Innowacyjność rozwiązań technologicznych piwniczek ekologicznych, Inżynieria i Budownictwo 79 (2023) 421-427. https://doi.org/10.5604/01.3001.0053.7992

[9] J. Partyka et al., Wybrane aspekty nośności konstrukcji powłokowej podziemnego obiektu magazynowego, Przegląd Budowlany 95 (2023) 39-43. https://doi.org/10.5604/01.3001.0054.4878

[10] F. Yi et al., Considering explosive charge shape and embedded depth in the design of concrete shelter thickness, Defence Technology 20 (2023) 44-57. https://doi.org/10.1016/j.dt.2022.09.015

[11] R. Hristova, S. Simovski, Practical Approach to Nuclear Shelter Wall Thickness Estimation, Journal of Nuclear Engineering and Radiation Science 10 (2024) art.040802. https://doi.org/10.1115/1.4065463

[12] H. May et al., Effect of sustainable/durable composite materials on dynamic performance of aircraft shelters, Journal of Physics: Conference Series 2616 (2023) 012046. https://doi.org/10.1088/1742-6596/2616/1/012046

[13] A.M. Legese et al., Behaviour of soil-steel composite structures during construction and service: a review, Archives of Civil Engineering 69 (2023) 263-292. https://doi.org/10.24425/ace.2023.147659

[14] P. Krysiak et al., (2017). Identification of strains in a multilayer composite pipe. Journal of Science of the Military Academy of Land Forces 49 (2017) 272-282. https://doi.org/10.5604/01.3001.0010.7233

[15] V. Ptashnyk et al., Impedance investigation of activated carbon material modified by ultrasound treatment, Przeglad Elektrotechniczny 94 (2018) 186-189. https://doi.org/10.15199/48.2018.05.33

[16] M. Patek et al., Non-destructive testing of split sleeve welds by the ultrasonic TOFD method, Manufacturing Technology 14 (2014) 403-407.

[17] T. Lipinski, Influence of surface refinement on microstructure of Al-Si cast alloys processed by welding method, Manufacturing Technology 15 (2015) 576-581.

[18] N. Radek et al., Properties of Steel Welded with CO_2 Laser, Lecture Notes in Mechanical Engineering (2020) 571-580. https://doi.org/10.1007/978-3-030-33146-7_65

[19] N. Radek et al., Technology of laser forming, Manufacturing Technology 14 (2014) 428-431.

[20] A. Bartkowska et al., Laser processing of diffusion boronized layer produced on monel® alloy 400—microstructure, microhardness, corrosion and wear resistance tests, Materials 14 (2021) art. 7529. https://doi.org/10.3390/ma14247529

[21] E. Szajna et al., The influence of laser remelting on microstructural changes and hardness level of flame-sprayed NiCrBSi coatings with tungsten carbide addition, Surf. Coat. Technol. 478 (2024) art. 130403. https://doi.org/10.1016/j.surfcoat.2024.130403

[22] R. Ulewicz, M. Mazur, Economic aspects of robotization of production processes by example of a car semi-trailers manufacturer, Manufacturing Technology 19 (2019) 1054-1059. https://doi.org/10.21062/ujep/408.2019/a/1213-2489/MT/19/6/1054

[23] A. Pacana, K. Czerwińska, Comparative tests of the quality of the piston combustion chamber for a diesel engine, Tehnicki Vjesnik 27 (2020) 1021-1024. https://doi.org/10.17559/TV-20190112193319

[24] A. Pacana, K. Czerwińska, Improving the quality level in the automotive industry, Prod. Eng. Arch. 26 (2020) 162-166. https://doi.org/10.30657/pea.2020.26.29

[25] D. Siwiec et al., Improving the Non-Destructive Testing Process of the Outer Bearing Ring, Materials Research Proceedings 24 (2022) 174-180. https://doi.org/10.21741/9781644902059-26

[26] K. Trzewiczek et al., Evaluation of the state for the material of the live steam superheater pipe coils of V degree, Advanced Materials Research 874 (2014) 35-42. https://doi.org/10.4028/www.scientific.net/AMR.874.35

[27] Ł.J. Orman et al., Application of laser treatment technology for boiling heat transfer augmentation, Prod. Eng. Arch. 30 (2024) 259-265. https://doi.org/10.30657/pea.2024.30.25

[28] E. Skrzypczak-Pietraszek, J. Pietraszek, Chemical profile and seasonal variation of phenolic acid content in bastard balm (Melittis melissophyllum L., Lamiaceae), Journal of Pharm. Biomed. Anal. 66 (2012) 154-161. https://doi.org/10.1016/j.jpba.2012.03.037

[29] A. Pacana, D. Siwiec, Method of Fuzzy Analysis of Qualitative-Environmental Threat in Improving Products and Processes (Fuzzy QE-FMEA), Materials 16 (2023) art. 1651. https://doi.org/10.3390/ma16041651

[30] J. Pietraszek, E. Skrzypczak-Pietraszek, The optimization of the technological process with the fuzzy regression, Advanced Materials Research 874 (2014) 151-155. https://doi.org/10.4028/www.scientific.net/AMR.874.151

[31] E. Skrzypczak-Pietraszek, A.G. Pietraszek, Phytoremediation and Allelopathy as an Element of Sustainable Circular Economy to Prevent Environmental Pollution, Materials Research Proceedings 45 (2024) 140-145. https://doi.org/10.21741/9781644903315-17

[32] J. Pietraszek, A. Goroshko, The heuristic approach to the selection of experimental design, model and valid pre-processing transformation of DoE outcome, Advanced Materials Research 874 (2014) 145-149. https://doi.org/10.4028/www.scientific.net/AMR.874.145

[33] R. Dwornicka, J. Pietraszek, The outline of the expert system for the design of experiment, Production Engineering Archives 20 (2018) 43-48. https://doi.org/10.30657/pea.2018.20.09

[34] J. Pietraszek et al., Challenges for the DOE methodology related to the introduction of Industry 4.0, Prod. Eng. Arch. 26 (2020) 190-194. https://doi.org/10.30657/pea.2020.26.33

[35] J. Pietraszek, A. Gadek-Moszczak, The smooth bootstrap approach to the distribution of a shape in the ferritic stainless steel AISI 434L powders, Solid State Phenomena 197 (2013) 162-167. https://doi.org/10.4028/www.scientific.net/SSP.197.162

[36] J. Pietraszek, L. Wojnar, The bootstrap approach to the statistical significance of parameters in RSM model, ECCOMAS Congress 2016 – Proc. 7th Europ. Congr. Comput. Methods in App. Sci. Eng. 1 (2016) 2003-2009. https://doi.org/10.7712/100016.1937.9138

[37] R. Dwornicka et al., The smoothed bootstrap fine-tuning, System Safety: Human - Technical Facility - Environment 1 (2019) 716-723. https://doi.org/10.2478/czoto-2019-0091

[38] J. Pietraszek, Fuzzy regression compared to classical experimental design in the case of flywheel assembly, Lecture Notes in Computer Science 7267 LNAI (2012) 310-317. https://doi.org/10.1007/978-3-642-29347-4_36

[39] J. Pietraszek, The modified sequential-binary approach for fuzzy operations on correlated assessments, Lecture Notes in Computer Science 7894 LNAI (2013) 353-364. https://doi.org/10.1007/978-3-642-38658-9_32

[40] R. Dwornicka et al., Fuzzy Statistics-Aided Inference in Experimental Design, World Congress in Computational Mechanics and ECCOMAS Congress (2024). https://doi.org/10.23967/wccm.2024.131

Development and Testing of an On-Board Mine Detector

Piotr KRYSIAK[1,a*], Wiesław MADEJ[1,b], Cezary ŚLIWIŃSKI[1,c] and Yvan BAUDOIN[2,d]

[1]Military Institute of Engineer Technology, Obornicka 136 Str.,50-961 Wrocław, Poland

[2]International CBRNE Institute (ICI), Rue de la Vacherie 78, B5060 SAMBREVILLE, BELGIUM

[a]krysiak@witi.wroc.pl, [b]madej@witi.wroc.pl, [c]sliwinski.c@witi.wroc.pl, [d]yvan.baudoin@ici-belgium.be

Keywords: Metal Detector, IED Detection, Magnetic Fuse Mine, On-Board Detector

Abstract. This paper presents the results of tests conducted on a plate-type mine detector developed using Pulse Induction (PI) technology. Due to its characteristics, this technology demonstrates high efficiency in detecting objects buried at considerable depths in various environments. Based on experiments conducted using two types of test objects – a solid metallic wheel rim and a surface-type metal can – the proper operation of the detector was confirmed, and the corresponding signal values obtained for both objects were estimated.

Introduction

When considering the problem of modern mine detection methods, both in humanitarian and military demining, it can be concluded that satisfactory detection efficiency is currently achieved primarily by techniques based on two principles:
1) Electromagnetic induction, commonly referred to as induction-based metal detectors [1–5];
2) Radar methods, employing the Ground Penetrating Radar (GPR) technique [6–9].
Presently, a high level of efficiency and user-acceptable detection speed can be obtained only for mines containing relatively large metallic components. Such objects can be detected by numerous types of induction-based detectors that are in service with the armed forces of most developed countries and humanitarian demining organisations.

The detection of mines and other charges containing either small metallic components, elements with minimal metal content, or completely non-metallic ones remains a highly demanding task due to the limited efficiency of existing detection systems. Devices of this kind, available worldwide, are typically research or prototype solutions that represent alternative techniques, which have not yet achieved a sufficient Technology Readiness Level (TRL) or are subject to operational limitations that prevent their practical use.

Metal detectors have found wide application in military technology, as the identification of mines and other hazardous objects is a critical issue. The clearance of areas contaminated with explosive remnants of war poses complex technical, humanitarian, and economic challenges. Electromagnetic metal detectors play a key role in locating landmines.

With the continuous evolution of mine design towards minimising the amount of metal components, the need to modernise detection systems has arisen to enhance their sensitivity to even minimal quantities of metal.

In modern detector designs, the use of microprocessor-based signal processing is becoming indispensable for advanced signal analysis and the identification of detected objects. Efforts are being made to develop systems capable of determining the type of material (e.g. metallic, ferromagnetic, diamagnetic), as well as the shape, size, and burial depth of the object.

Contemporary armed conflicts often cover vast areas, and the safe detection and removal of explosive remnants may take many years. To increase the efficiency of the demining process,

Materials Research Proceedings 62 (2026) 188-195 https://doi.org/10.21741/9781644904015-24

multi-section vehicle-mounted mine detectors are being developed, allowing large areas to be scanned in a short period of time [5, 10].

Magnetic Induction-Based Detection Techniques

Detection techniques of this type are used to identify metallic elements that form part of a mine's construction. This is a relatively simple method based on the generation of a magnetic field by the detector's circuitry (the primary or transmitting coil, through which an electric current flows). The generated magnetic field induces eddy currents in metallic objects located within its range. These eddy currents, in turn, produce a secondary electromagnetic field, which is detected by the receiver circuit of the induction detector (the secondary or receiving coil). Due to their relatively simple design, ease of operation, and lack of significant disadvantages, detectors of this type are widely used. Nevertheless, research into improving this method continues, as it is still burdened by several shortcomings:

1) Detecting the simplest conventional mines with low metal content is very difficult due to the usually insufficient sensitivity of the detectors. Increasing the sensitivity, on the other hand, leads to excessive interference from disturbing factors such as soil inhomogeneity, humidity, and the presence of iron compounds.
2) Detecting mines containing metal components in areas heavily contaminated with metallic debris or deliberately scattered metallic fragments is difficult, time-consuming, and requires significant operator involvement. Mines that contain no metallic components cannot be detected by any currently available induction detector types.

With the increasing use of mines containing only small amounts of metal elements (weighing just a few grams) or none at all, multi-sensor detection systems are becoming more prevalent in military applications. These systems combine several detection techniques, such as magnetic induction, thermal imaging, and ground-penetrating radar (GPR), to increase the probability of detection and classification accuracy.

Experimental Study of the Plate-Type Mine Detector

The experimental study was conducted using the Plate Mine Detector (Polish acronym: PWM), developed at the Military Institute of Engineering Technology (Polish acronym: WITI). The detector operates on the principle of pulse induction (PI). A single PWM panel consists of four coils operating simultaneously. The use of several panels enables the detection of objects over a large surface area within a short time, significantly improving the efficiency of the scanning process.

In the conducted research, a detector configuration consisting of three detection panels was used. The purpose of the tests was to determine signal parameters that could be potentially useful in classifying detected objects.

The size of an object detected by an impulse detector can be estimated based on the ground surface area over which the detection signal is observed. In multi-channel detectors, such as PWM, simultaneous signal responses from adjacent channels indicate the presence of a larger object.

The detector's operation and measurement data acquisition were managed using PWM Tester software, developed specifically for diagnostic purposes and the analysis of detection signals.

For testing purposes, the detector was mounted on a special composite support frame [11, 12] made of polyester-glass composite [13, 14] (Fig.1). The use of non-magnetic materials was essential, as metallic structural components in the immediate vicinity of the detection system could cause interference and affect the accuracy of the results.

Materials Research Proceedings 62 (2026) 188-195
https://doi.org/10.21741/9781644904015-24

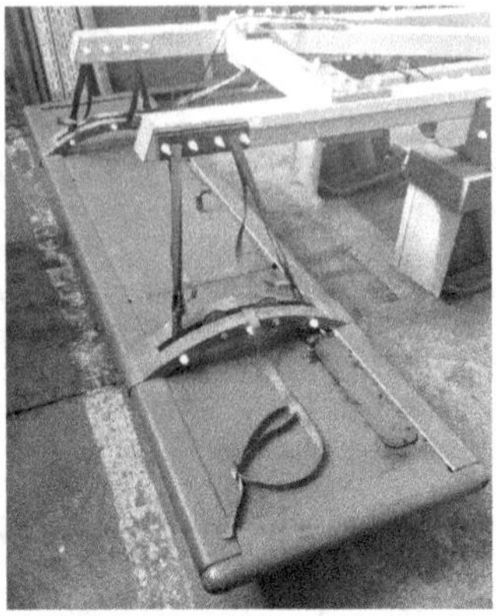

Figure 1. PWM detector during testing

During detector testing, signals were recorded for two sample objects: a steel rim (height: 17 cm, diameter: 39 cm) and a metal can (height: 24 cm, diameter: 17 cm). To simulate the detector's movement over the sought-after objects, these were moved beneath the stationary detector (Fig. 2). Data acquisition was performed for both slow and fast object movement rates. The device tests were conducted for two detector-to-object distances: 23 cm and 50 cm.

Figure 2. Object detection during PWM detector tests

For each tested case, the recorded signals are presented as time-domain waveforms. Figure 3 and Figure 4 present the results for the test objects at a detector-to-object distance of 23 cm.

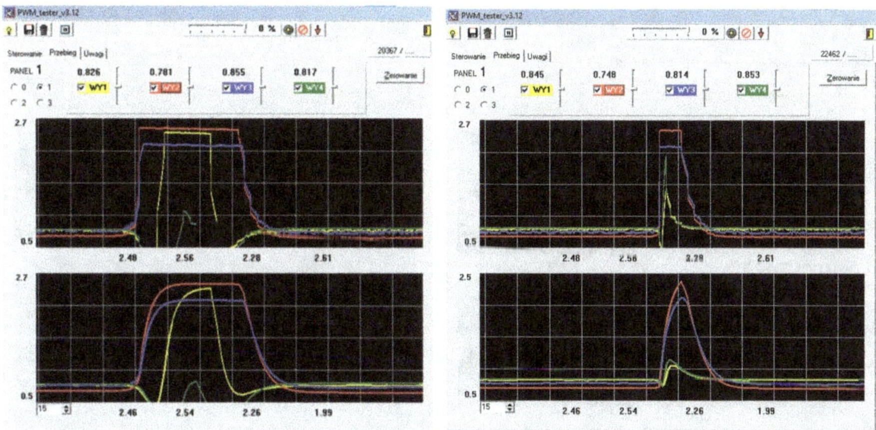

Figure 3. Signal waveforms recorded during slow (left) and fast (right) scanning of the "wheel rim" object at a distance of 23 cm from the detector

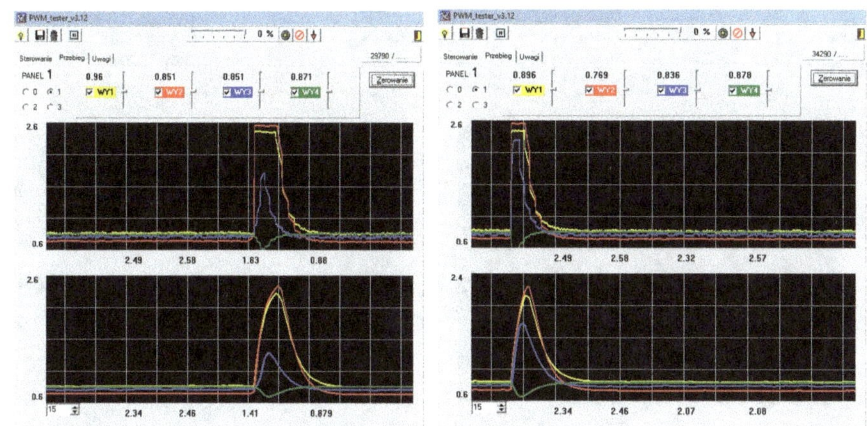

Figure 4. Signal waveforms recorded during slow (left) and fast (right) scanning of the "metal can" object at a distance of 23 cm from the detector

Figures 5 and 6 present the results obtained when the distance between the detector and the tested objects was 50 cm.

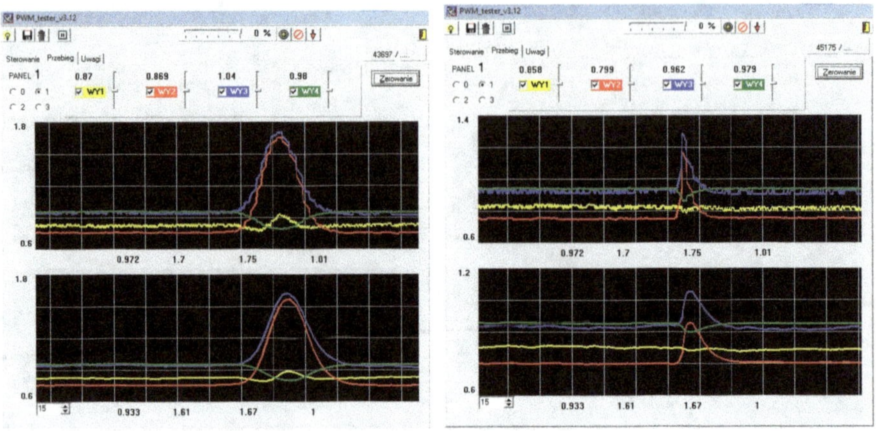

Figure 5. Signal waveforms recorded during slow (left) and fast (right) scanning of the "wheel rim" object at a distance of 50 cm from the detector

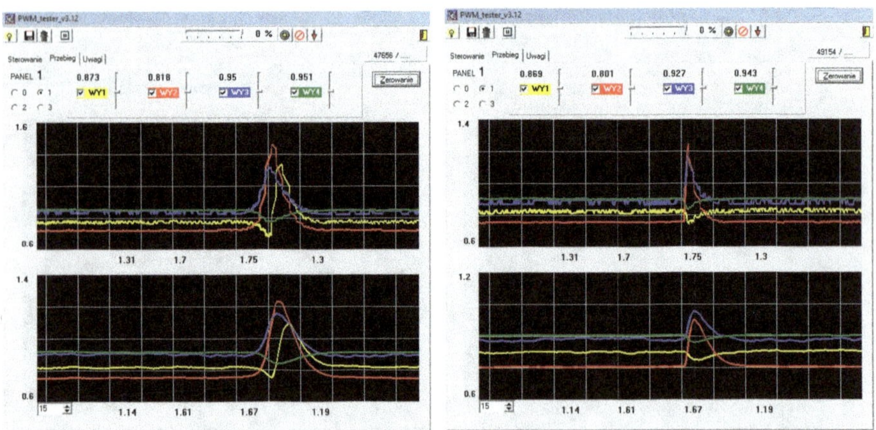

Figure 6. Signal waveforms recorded during slow (left) and fast (right) scanning of the "metal can" object at a distance of 50 cm from the detector

Technological Impact

Regardless of its military applications, the plate mine detector can also be effectively used in civilian environments to detect underground infrastructure. Damage to cable lines and all types of pipelines is a common occurrence on many construction sites, and in the case of gas pipelines, it has repeatedly led to fatal accidents. An interesting issue worthy of further research is the effect of possible special coatings [15-17] on the detection capability of hidden objects, as well as the similar effect of corrosive deposits [18,19] and biological deposits [20,21]. Proper analysis of controlled factors and observed quantities requires the use of appropriate analytical methodology [22-24], which should be supported by resampling [25-27] for low-density datasets and by a fuzzy approach [28-30] for incomplete datasets. Regardless of this, for civilian applications, automation would be valuable [31,32], which, with appropriate mobile assembly [33,34], would allow the implementation of autonomous control algorithms with reduced complexity of the environment description [35,36].

Summary

The conducted tests confirmed that the designed plate-type mine detector, based on magnetic induction technology, meets the assumed requirements for detecting metallic objects. From the presented signal plots, a clear correlation is observed between the distance of the detection coil from the detected object and the speed of movement. It should be noted that for a solid object containing a significant amount of metal (the wheel rim), this dependency is particularly evident (Figures 3 and 5). In contrast, for a surface-type object of low mass relative to its volume (the metal can), this dependency shows only slight differences (Figures 4 and 6). Nevertheless, the signal amplitude remains at a similar level in both cases.

References

[1] M. Szczepaniak et al., Wieloczujnikowy system wykrywania przedmiotów niebezpiecznych – Konferencja Naukowo-Techniczna WITI, Kościerzyna, 2010.

[2] M. Szczepaniak, W. Barnat, Rodzina pojazdów do poszukiwania i zwalczania IED, www.armia24.pl, 2012.

[3] Ł. Czubaty et al., Projekt urządzenia do detekcji przedmiotów niebezpiecznych w oparciu o indukcyjny wykrywacz metali oraz georadar. Systems: Journal of Transdisciplinary Systems Science 16 (2012) 151-159.

[4] I. Arai, Y. Tomizawa, S. Goto, Impulse mine detection radar using an array antenna, Instrument and Control 4/5–6 (2006).

[5] B. Candy, Metal Detector Basics and Theory, Minelab Pty Ltd, 2017.

[6] J. Ishikawa et al., Experimental design for test and evaluation of anti-personnel landmine detection based on vehicle-mounted GPR systems, Detection and Remediation Technologies for Mines and Minelike Targets X, SPIE 5794 (2005). https://doi.org/10.1117/12.603149

[7] R. Pavlik et al., System concept for GPR reflectance measurement method in time and frequency domain, Research report VOP-026 Sternberk, s.p., VTUO Brno Division, 2007.

[8] E.V. Denisova et al., Complex analysis of GPR signals to control contact zone of concrete lining and rock mass, Geohazard Mechanics 3 (2025) 197-205. https://doi.org/10.1016/j.ghm.2025.08.004

[9] T.O. Olowu et al., Multi-objective GPR-based design optimization of high frequency transformers, Power Electronic Devices and Components (2025) art.100122. https://doi.org/10.1016/j.pedc.2025.100122

[10] B. Kim et al., Pulse-induction metal detector with time-domain bucking circuit for landmine detection, Electronics Letters 51 (2015) 159-160. https://doi.org/10.1049/el.2014.3895

[11] P. Krysiak et al., Strength testing of a composite mounting frame for a multi-sensor detection system, Materials Research Proceedings 17 (2020) 165-170. https://doi.org/10.21741/9781644901038-25

[12] M.-X. Tao et al., Experimental study on longitudinal shear resistance in composite frame beams under lateral loads, Engineering Structures 345 (2025) art.121606. https://doi.org/10.1016/j.engstruct.2025.121606.

[13] P. Krysiak et al., Strength Testing and Ring Stiffness Testing of Underground Composite Pressure Pipes. Mater. Res. Proc. 17 (2020) 191-202. https://doi.org/10.21741/9781644901038-29

[14] P. Krysiak et al., Identification of strains in a multilayer composite pipe. Zeszyty Naukowe/Wyższa Szkoła Oficerska Wojsk Lądowych im. gen. T. Kościuszki 49 (2017) 272-282.

[15] M. Scendo et al., Influence of laser treatment on the corrosive resistance of WC-Cu coating produced by electrospark deposition, Int. J. Electrochem. Sci. 8 (2013) 9264-9277.

[16] M. Kuklinski et al., Laser alloying monel 400 with amorphous boron to obtain hard coatings, Materials 12 (2019) art. 3494. https://doi.org/10.3390/ma12213494

[17] N. Radek et al., Formation of coatings with technologics using concentrated energy stream, Prod. Eng. Arch. 28 (2022) 117-122. https://doi.org/10.30657/pea.2022.28.13

[18] A. Dudek et al., Effect of corrosion protection method on properties of RSW and RFSSW lap joints applied in production of thin-walled aerostructures, Materials 13 (2020) art. 1841. https://doi.org/10.3390/MA13081841

[19] T. Lipiński, J. Pietraszek, Corrosion of the S235JR Carbon Steel after Normalizing and Overheating Annealing in 2.5% Sulphuric Acid at Room Temperature, Materials Research Proceedings 24 (2022) 102-108. https://doi.org/10.21741/9781644902059-16

[20] A. Dudek, L. Adamczyk, Properties of hydroxyapatite layers used for implant coatings, Optica Applicata 43 (2013) 143-151. https://doi.org/10.5277/oa130118

[21] T. Lipinski, J. Pietraszek, Influence of animal slurry on carbon C35 steel with different microstructure at room temperature, Engineering for Rural Development 21 (2022) 344-350. https://doi.org/10.22616/ERDev.2022.21.TF115

[22] J. Pietraszek, A. Goroshko, The heuristic approach to the selection of experimental design, model and valid pre-processing transformation of DoE outcome, Advanced Materials Research 874 (2014) 145-149. https://doi.org/10.4028/www.scientific.net/AMR.874.145

[23] R. Dwornicka, J. Pietraszek, The outline of the expert system for the design of experiment, Prod. Eng. Arch. 20 (2018) 43-48. https://doi.org/10.30657/pea.2018.20.09

[24] J. Pietraszek et al., Challenges for the DOE methodology related to the introduction of Industry 4.0, Prod. Eng. Arch. 26 (2020) 190-194. https://doi.org/10.30657/pea.2020.26.33

[25] J. Pietraszek, A. Gadek-Moszczak, The smooth bootstrap approach to the distribution of a shape in the ferritic stainless steel AISI 434L powders, Solid State Phenomena 197 (2013) 162-167. https://doi.org/10.4028/www.scientific.net/SSP.197.162

[26] A. Szczotok et al., The bootstrap analysis of one-way ANOVA stability in the case of the ceramic shell mould of airfoil blade casting, Solid State Phenomena 235 (2015) 24-30. https://doi.org/10.4028/www.scientific.net/SSP.235.24

[27] J. Pietraszek, L. Wojnar, The bootstrap approach to the statistical significance of parameters in RSM model, ECCOMAS Congress 2016 - Proc. 7th Europ. Congr. Comput. Methods in App. Sci. Eng. 1 (2016) 2003-2009. https://doi.org/10.7712/100016.1937.9138

[28] J. Pietraszek, Fuzzy regression compared to classical experimental design in the case of flywheel assembly, Lecture Notes in Computer Science 7267 LNAI (2012) 310-317. https://doi.org/10.1007/978-3-642-29347-4_36

[29] J. Pietraszek, The modified sequential-binary approach for fuzzy operations on correlated assessments, Lecture Notes in Computer Science 7894 LNAI (2013) 353-364. https://doi.org/10.1007/978-3-642-38658-9_32

[30] R. Dwornicka et al., The smoothed bootstrap fine-tuning, System Safety: Human - Technical Facility - Environment 1 (2019) 716-723. https://doi.org/10.2478/czoto-2019-0091

[31] N. Radek et al., Technology of laser forming, Manuf. Technol. 14 (2014) 428-431.

[32] J. Takosoglu, Angular position control system of pneumatic artificial muscles, Open Engineering 10 (2020) 681-687. https://doi.org/10.1515/eng-2020-0077

[33] A. Goroshko et al., Construction and practical application of hybrid statistically-determined models of multistage mechanical systems, Mechanika 20 (2014) 489-493. https://doi.org/10.5755/j01.mech.20.5.8221

[34] A. Goroshko et al., Multiphysics Modeling of Direct-Start Induction Motor Considering Rotor Unbalance, Materials Research Proceedings 45 (2024) 267-276. https://doi.org/10.21741/9781644903315-31

[35] J. Pietraszek et al., Modeling of errors counting system for PCB soldered in the wave soldering technology, Advanced Materials Research 874 (2014) 139-143. https://doi.org/10.4028/www.scientific.net/AMR.874.139

[36] J. Pietraszek, E. Skrzypczak-Pietraszek, The uncertainty and robustness of the principal component analysis as a tool for the dimensionality reduction, Solid State Phenomena 235 (2015) 1-8. https://doi.org/10.4028/www.scientific.net/SSP.235.1

Numerical and Metallographic Analysis of Laser Cladding of 316L, Stainless Steel

Hubert DANIELEWSKI[1,a *], Piotr KURP[1,b], Andrzej SKRZYPCZYK[1,c], Dariusz GONTARSKI[1,d] and Krystian MULCZYK[1,e]

[1]Kielce University of Technology, Faculty of Mechatronics and Mechanical Engineering, Al. 1000-lecia Państwa Polskiego 7, 25-314 Kielce, Poland

[a]hdanielewski@tu.kielce.pl, [b]pkurp@tu.kielce.pl, [c]andrzej.skrzypczyk@gmail.com, [d]gontar@tu.kielce.pl, [e]kmulczyk@tu.kielce.pl

Keywords: Laser Cladding, Numerical Simulation, Metallographic Analysis

Abstract. This study presents research on the laser cladding process of stainless steel grade 316L. This widely used austenitic stainless steel exhibits good corrosion resistance and mechanical properties [1,2]. Maintaining these properties after welding or cladding requires precise control of heat input and the selection of appropriate process parameters. By applying an advanced cladding technique such as laser metal deposition (LMD), these effects can be achieved. However, accurate parameter selection is particularly important; therefore, a finite element method (FEM) analysis was employed to determine the optimal process conditions. Based on the thermal simulation, parameters were estimated to ensure complete melting of the deposited material while minimizing heat input to the base metal [3]. The derived parameters were then used to perform experimental cladding, resulting in a multilayer surface coating. The cladded material was examined using metallographic analysis. A comparison of the numerical results with the metallographic observations showed good agreement, confirming the effectiveness of FEM analysis for predicting suitable laser process parameters.

Introduction

Today, laser technologies, including welding and cladding processes, are widely utilized in various industrial applications. The primary advantage of using a laser beam as a heat source is its ability to precisely melt the material, thereby minimizing thermal impact on the surrounding areas [4]. Accurate estimation of process parameters can be performed using analytical calculations or numerical simulations. Although analytical solutions provide faster results, their accuracy is often limited. As an alternative, the Finite Element Method (FEM) can be used to predict process parameters more accurately and to reduce the need for time-consuming and costly experimental trials. Furthermore, numerical simulations enable the estimation of heat input, thermal gradients, and maximum temperatures in the material, facilitating better process optimization [5,6].

Stainless steels are widely used materials with significant potential to replace carbon steels, which are typically protected with galvanized or painted coatings whose service life is limited by corrosion. The high chromium and nickel content of 316L stainless steel ensures excellent corrosion resistance, while its low carbon content provides good weldability (Table 1) [7].

Table 1. Chemical composition of cladded and base material

Material	Elements (wt %)								
	C	Mn	Si	P	S	Cr	Ni	Al	Fe
316L	0.018	1.57	0.48	0.04	0.002	16.7	11.2	-	balanced

Therefore, other surface modification processes, such as the analyzed cladding process, can be applied. Laser cladding using either filler wire or metal powder offers specific advantages and limitations [8]. The use of filler wire provides better material density; however, it requires higher laser power. As a result, the amount of heat energy absorbed by the material increases, and post-cladding heat treatment may be necessary.

Numerical Analysis

A numerical thermal analysis was performed using the Finite Element Method (FEM) and Simufact Welding software. The numerical model included a geometric model along with defined initial and boundary conditions. Based on the literature review, the size of the finite elements was set to 0.125 mm. Both the base material (BM) and the deposited material were stainless steel grade 316L; the corresponding material properties were taken from the software's material library.

The plate was positioned within the working space by applying appropriate geometric constraints. At the same time, the clad geometry was defined as a filler region with the same mesh size as the base material. The heat source (HS) model was programmed based on a conical laser beam geometry. The dimensions of the HS were defined as follows: an upper diameter of 0.4 mm, a lower diameter of 0.2 mm, and a height of 4 mm. These parameters were selected according to the characteristics of the CO_2 laser used in the experimental cladding process. The results of the single-layer cladding simulation are presented in Fig.1.

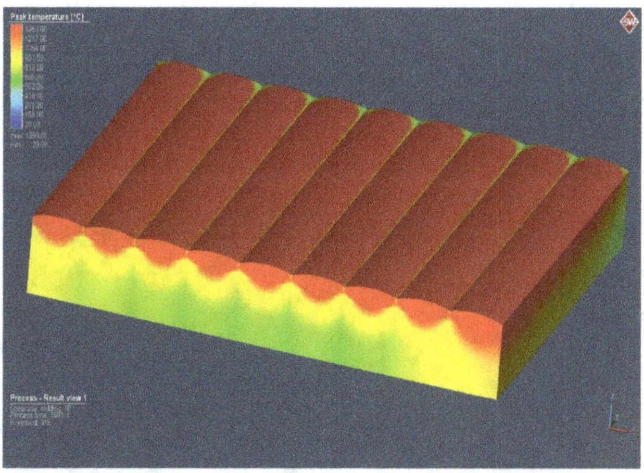

Fig. 1. Isotherm of maximum temperature in laser cladding of 316L steel.

Based on the performed simulations, an estimation of process parameters was conducted, yielding an output power of 2.2 kW, a cladding speed of 1 m/min, and a powder feed rate of 23 g/min.

Metallographic Analysis

Based on the estimated parameters, laser cladding of 316L stainless steel was performed. A specimen for metallographic analysis was extracted, polished, and etched. The metallographic study was carried out using an optical digital microscope (Hirox KH-8700) and a scanning electron microscope (JEOL JSM-7100F). Energy-dispersive spectroscopy (EDS) analysis was performed on the cross-section between the cladded layers and the base material.

Global observations confirmed a proper cladding build, with the fusion line located within the substrate material but at a shallow depth relative to the cladded surface (Figure 2). Both the cladded and base materials exhibited an austenitic structure; therefore, no typical heat-affected zone (HAZ) with noticeable grain size variation was identified. Grain refinement was observed only along the fusion line, where a localized concentration of δ-ferrite occurred [9].

Fig. 2. Macroscopic image of the fusion line of deposited 316L steel.

The obtained cladding exhibited a typical dendritic structure consisting of pillar (I) and cellular (II) dendrites (Figure 3). The dendrites grew from the fusion line toward the cladded surface in accordance with the thermal gradient that developed during the solidification process. Visible structural differences were observed in samples that underwent multiple heat cycles during successive cladding passes. No evidence of overheating was observed [10].

Fig. 3. SEM microstructure of dendrite growth in the obtained clad of 316L steel.

Technological, Economic, Environmental, and Social Impact

Laser cladding (LMD) technology, especially when combined with numerical optimization, represents a significant technological leap in surface engineering. Optimizing technological processes, including cladding, is crucial for achieving the desired microstructure and mechanical properties [11], which directly impact the durability of components.

The introduction of advanced thermal processes, such as normalizing, is crucial for enhancing the mechanical properties of engineering steels, including 38MnVS6 [12, 13]. This expertise is being applied to microstructural control in cladding. The ability to precisely shape the chemical composition and microstructure of coatings is fundamental; for example, studies on the effect of alloying methods on the structure of sintered stainless steel [14] and the effect of ceramic mold thickness on eutectic properties in engine blade castings [15] provide critical knowledge on the control of thermo-material processes that is directly applicable to LMD. Analysis of the effect of CO_2 laser welding on steel properties [16] and the impact of laser welding parameters on the mechanical properties of welds [17] also highlights technological challenges related to laser energy control, similar to those encountered in hardfacing.

As a surface modification tool, lasers enable the production of high-quality WC-Co coatings [18] and those produced by the electro-spark (ESD) method [19], which significantly improve operational properties. The effect of laser processing on ESD coatings, by enhancing their performance [20], highlights the potential of LMD in creating functional layers on wear parts. The economic impact is closely linked to improved efficiency, durability, and quality control in the metallurgical industry. The use of advanced statistical methods, such as fuzzy regression in process optimization [11, 21] and correlation analysis using a modified sequential-binary approach for fuzzy evaluation [22], enables effective management of uncertainty in production conditions. Similarly, the use of a smooth bootstrap approach to particle shape distribution analysis [23] or fixed-effects analysis of the relationship between the primary dendritic distance (SDAS) and carbides in blade castings [24] provides tools for statistical quality control and material characterization, which is essential for critical applications. Hybrid statistically determined models, used to describe multi-stage mechanical systems [25], support the prediction of the behavior of machines and devices whose components undergo repair or surface strengthening.

In the context of production management and maintenance, assessing machine modernity [26] and optimizing the thermal operation of components such as steam gates [27] directly impact plant economic efficiency. The development of hybrid methods for identifying the causes of product nonconformities [28] and the challenges of DOE methodology in the context of Industry 4.0 [29] demonstrate how innovations in process control (e.g., laser) contribute to waste minimization and the implementation of advanced, intelligent manufacturing. The analysis of equipment for producing metal components in construction [30] and the use of fine-structured steel with improved mechanical properties [31] indicate a wider technological market for enhanced materials.

LMD, as a surface regeneration or modification technique, contributes to reducing the environmental footprint by extending the life of expensive components, thus reducing resource consumption and waste. Corrosion resistance is a key aspect of stainless steel's durability in aggressive environments. Studies on the corrosion rate of stainless steel in hot nitric acid [32] and the corrosion resistance of steel in this acid [33] highlight the need for continuous improvement of surface properties to ensure the safety and reliability of industrial installations, such as those in the energy sector. This aspect is particularly important in the context of materials exposed to harmful deposits formed during the co-firing of biomass and coal in fluidized bed boilers, which leads to high-temperature corrosion of steels such as P91 [34, 35]. LMD has the potential to create protective barriers against such phenomena.

The social aspect concerns the safety and durability of infrastructure. The load-bearing capacity assessment of rotating bearings in transport [36] and the fire performance analysis of intumescent coating systems for railway rolling stock [37] are examples of how materials engineering influences the safety and reliability of key infrastructure and transportation components. Furthermore, although in a different context, nonparametric chemical profile analysis methods [38] (based on DOE analysis and advanced statistics) are examples of interdisciplinary applications of analytical tools that are equally important for the verification and validation of metallurgical results in advanced technological processes.

Summary

Based on the numerical analysis performed, the optimal cladding parameters for 316L stainless steel were determined. Using these parameters, a multilayer cladding was successfully produced. Microstructural analysis of the cladded material indicated low heat accumulation in both the deposited and base materials. The structure exhibited a typical dendritic morphology, consisting of regions with pillar and cellular dendrites. Further mechanical testing will be conducted to provide a comprehensive characterization of the cladded material.

References

[1] A. Saboori et al. Microstructure and Mechanical Properties of AISI 316L Produced by Directed Energy Deposition-Based Additive Manufacturing: A Review, Appl. Sci. 10 (2020) art.3310. https://doi.org/10.3390/app10093310

[2] T. Keller et al. Strength, microhardness, and microstructure analysis of 316L stainless steel manufactured via hybrid laser wire and laser powder bed additive, J. Manuf. Process. 150 (2025) 1-9. https://doi.org/10.1016/j.jmapro.2025.06.037

[3] J. Gao et al., Numerical simulation and experimental investigation on three-dimensional modelling of single-track geometry and temperature evolution by laser cladding, Opt. Laser Technol. 129 (2020) art.106287. https://doi.org/10.1016/j.optlastec.2020.106287

[4] C. Li et al., Quantitative analysis and experimental study of the influence of process parameters on the evolution of laser cladding, J. Adhes. Sci. Technol., 36 (2021) 1894-1920. https://doi.org/10.1080/01694243.2021.1991142

[5] X. Chen et al., Numerical simulation and structure properties of laser clad 316 L stainless steel coating, Appl. Phys. A 131 (2025) 111. https://doi.org/10.1007/s00339-025-08243-1

[6] C. Li et al., Numerical simulation method of the multi-field coupling mechanism for laser cladding 316L powder, Weld World 66 (2022) 423-440. https://doi.org/10.1007/s40194-021-01213-0

[7] Z. Mianji et al., Numerical Simulation of Thermal Behavior and Experimental Investigation of Thin Walls during Direct Metal Deposition of 316L Stainless Steel Powder, Lasers Manuf. Mater. Process. 8 (2021) 426-442. https://doi.org/10.1007/s40516-021-00155-1

[8] B. Antoszewski et al., Semi-Hybrid CO_2 Laser Metal Deposition Method with Inter Substrate Buffer Zone, Materials 14 (2021) 720. https://doi.org/10.3390/ma14040720

[9] M.A. Belsvik et al., Microstructural studies of direct-laser-deposited stainless steel 316L-Si on 316L base material, SN Appl. Sci. 2 (2020) 1967. https://doi.org/10.1007/s42452-020-03530-3

[10] H. Yin, W. Sun. Experimental laser cladding of 316L based on thermodynamic coupling and numerical simulation study, J. Phys. Conf. Ser. 3068 (2025) 012011. https://doi.org/10.1088/1742-6596/3068/1/012011

[11] J. Pietraszek, E. Skrzypczak-Pietraszek, The optimization of the technological process with the fuzzy regression, Advanced Materials Research 874 (2014) 151-155. https://doi.org/10.4028/www.scientific.net/AMR.874.151

[12] D. Siwiec et al., Improving the process of achieving required microstructure and mechanical properties of 38MnVS6 steel, METAL 2020 – 29th Int. Conf. Metall. Mater. (2020) 591-596. https://doi.org/10.37904/metal.2020.3525

[13] D. Siwiec et al., An analysis of the influence double normalization process on mechanical properties and microstructure 38MnVS6 steel, METAL 2021 – 30th Int. Conf. Metall. Mater. (2021) 579-584. https://doi.org/10.37904/metal.2021.4162

[14] A. Dudek et al., The effect of alloying method on the structure and properties of sintered stainless steel, Arch. Metall. Mater. 62 (2017) 281-287. https://doi.org/10.1515/amm-2017-0042

[15] A. Szczotok et al., The Impact of the Thickness of the Ceramic Shell Mould on the ($\gamma + \gamma'$) Eutectic in the IN713C Superalloy Airfoil Blade Casting, Arch. Metall. Mater. 62 (2017) 587-593. https://doi.org/10.1515/amm-2017-0087

[16] N. Radek et al., Properties of Steel Welded with CO_2 Laser, Lecture Notes in Mechanical Engineering (2020) 571-580. https://doi.org/10.1007/978-3-030-33146-7_65

[17] N. Radek et al., The impact of laser welding parameters on the mechanical properties of the weld, AIP Conf. Proc. 2017 (2018) art. 020025. https://doi.org/10.1063/1.5056288

[18] N. Radek et al., The WC-Co electrospark alloying coatings modified by laser treatment, Powder Metall. Metal Ceramics 47 (2008) 197-201. https://doi.org/10.1007/s11106-008-9005-7

[19] N. Radek, B. Antoszewski, The influence of laser treatment on the properties of electro-spark deposited coatings, Kovove Materialy 47 (2009) 31-38.

[20] N. Radek et al., The impact of laser processing on the performance properties of electro-spark coatings, World Congress in Computational Mechanics and ECCOMAS Congress 1000 (2021) 1-10. https://doi.org/10.23967/wccm-eccomas.2020.336

[21] J. Pietraszek, Fuzzy regression compared to classical experimental design in the case of flywheel assembly, Lecture Notes in Computer Science 7267 LNAI (2012) 310-317. https://doi.org/10.1007/978-3-642-29347-4_36

[22] J. Pietraszek, The modified sequential-binary approach for fuzzy operations on correlated assessments, Lecture Notes in Computer Science 7894 LNAI (2013) 353-364. https://doi.org/10.1007/978-3-642-38658-9_32

[23] J. Pietraszek, A. Gadek-Moszczak, The smooth bootstrap approach to the distribution of a shape in the ferritic stainless steel AISI 434L powders, Solid State Phenomena 197 (2013) 162-167. https://doi.org/10.4028/www.scientific.net/SSP.197.162

[24] J. Pietraszek et al., The fixed-effects analysis of the relation between SDAS and carbides for the airfoil blade traces, Archives of Metallurgy and Materials 62 (2017) 235-239. https://doi.org/10.1515/amm-2017-0035

[25] A. Goroshko et al., Construction and practical application of hybrid statistically-determined models of multistage mechanical systems, Mechanika 20 (2014) 489-493. https://doi.org/10.5755/j01.mech.20.5.8221

[26] M. Ingaldi, S.T. Dziuba, Modernity evaluation of the machines used during production process of metal products, METAL 2015 – 24th Int. Conf. Metall. Mater. (2015) 1908-1914.

[27] P. Duda, R. Dwornicka, Optimization of heating and cooling operations of steam gate valve, Structural and Multidisciplinary Optimization 40 (2010) 529-535. https://doi.org/10.1007/s00158-009-0370-8

[28] R. Dwornicka et al., Development of a Hybrid Method for Identifying the Causes of Product Incompatibility in Metallurgical Manufacturing, Materials Research Proceedings 34 (2023) 199-206. https://doi.org/10.21741/9781644902691-24

[29] J. Pietraszek et al., Challenges for the DOE methodology related to the introduction of Industry 4.0, Prod. Eng. Arch. 26 (2020) 190-194. https://doi.org/10.30657/pea.2020.26.33

[30] J. Selejdak et al., The evaluation of the use of a device for producing metal elements applied in civil engineering, METAL 2014 – 23rd Int. Conf. Metall. Mater. (2014) 1882-1888.

[31] R. Ulewicz et al., Structure and mechanical properties of fine-grained steels, Periodica Polytechnica Transportation Engineering 41 (2013) 111-115. https://doi.org/10.3311/PPtr.7110

[32] T. Lipiński, D. Karpisz, Corrosion rate of 1.4152 stainless steel in a hot nitrate acid, METAL 2019 – 28th Int. Conf. Metall. Mater. (2019) 1086-1091.

[33] T. Lipiński, Corrosion resistance of 1.4362 steel in boiling 65% nitric acid, Manufacturing Technology 16 (2016) 1004-1009.

[34] M. Opydo et al., The effect of biomass co-combustion in a CFB boiler on solids accumulation on surfaces of P91 steel tube samples, Biomass and Bioenergy 85 (2016) 61-68. https://doi.org/10.1016/j.biombioe.2015.12.011

[35] M. Opydo et al., Characteristics of solids accumulation on steel samples during co-combustion of biomass and coal in a CFB boiler, Biomass and Bioenergy 120 (2019) 291-300. https://doi.org/10.1016/j.biombioe.2018.11.027

[36] M. Krynke, R. Ulewicz, Analysis of the influence of slewing bearing mounting on their static load capacity, Transportation Research Procedia 40 (2019) 745-750. https://doi.org/10.1016/j.trpro.2019.07.105

[37] N. Radek, R. Dwornicka, Fire properties of intumescent coating systems for the rolling stock, Communications – Scientific Letters of the University of Žilina 22 (2020) 90-96. https://doi.org/10.26552/com.C.2020.4.90-96

[38] E. Skrzypczak-Pietraszek, J. Pietraszek, Chemical profile and seasonal variation of phenolic acid content in bastard balm Melittis melissophyllum L., Lamiaceae, Journal of Pharmaceutical and Biomedical Analysis 66 (2012) 154-161. https://doi.org/10.1016/j.jpba.2012.03.037

Terotechnology XIV | Materials Research Forum LLC
Materials Research Proceedings 62 (2026) 203-210 | https://doi.org/10.21741/9781644904015-26

The Role of Laser Technology in Military Defense Systems – An Overview

Artur KALINOWSKI[1,a*]

[1]Kielce University of Technology, Faculty of Mechatronics and Mechanical Engineering, Al. 1000-lecia P.P. 7, 25-314 Kielce, Poland

[a]akalinowski@tu.kielce.pl

Keywords: Laser Technology, High-Energy Systems, Military Defense Systems

Abstract. This article presents the role of laser technology in modern military defense systems, with particular emphasis on its applications as direct energy weapons. The current uses of lasers in targeting, rangefinding, and communication are discussed alongside the prospects for deploying laser systems capable of neutralizing targets such as drones and missiles. The paper analyzes key technological challenges, including power management, thermal regulation, and the impact of atmospheric conditions on laser effectiveness. It also outlines applications of lasers in LiDAR, target designation, dazzling systems, and auxiliary operations such as cutting, welding, and repairs. The further development of advanced materials and optical solutions will be crucial for the successful integration of laser weapons into dynamic combat scenarios of the 21st century.

Introduction

Since their invention, lasers have sparked speculation about their potential as directed-energy weapons, resulting in substantial investments in research aimed at developing them for military applications. Laser weapon systems can fundamentally transform contemporary warfare, as military forces worldwide allocate substantial resources toward their advancement. These sophisticated systems, capable of delivering concentrated energy pulses within milliseconds, have the potential to neutralize targets such as drones and missiles with exceptional speed and precision. Although current military lasers primarily serve in roles of targeting and rangefinding, the forthcoming generation aspires to execute destructive effects by melting the structural components of unmanned aerial vehicles (UAVs) and igniting missiles. In the ongoing international pursuit to refine directed-energy weaponry, lasers distinguish themselves through unparalleled precision, velocity, and cost-efficiency. Unlike conventional munitions, they possess the advantage of effectively limitless "ammunition," contingent solely on a reliable power source. In addition, their ability to sequentially engage multiple hostile targets with high efficiency significantly enhances their strategic utility within contemporary defense frameworks. However, significant technical challenges, including power generation, atmospheric interference, and thermal management, persist as critical factors that must be addressed before the widespread adoption of laser weaponry on active battlefields. From an engineering standpoint, the future of laser weapon systems hinges on advances in beam control, thermal regulation, power generation efficiency, and overcoming actual challenges. High-energy lasers require precise optical alignment and adaptive optics to maintain beam coherence over long distances and through turbulent atmospheric layers. Innovations in compact power storage are equally critical. Thermal management remains one of the most challenging aspects, as the dissipation of waste heat directly impacts operational stability and component lifespan. Progress in materials science, particularly in high-durability optical coatings and heat-resistant composites, therefore plays an essential role in enabling sustained high-power operation. As these technological barriers are progressively overcome, laser weapons are expected to transition from experimental demonstrators to fully integrated platforms capable of real-time engagement in dynamic combat scenarios.

Materials Research Proceedings 62 (2026) 203-210 https://doi.org/10.21741/9781644904015-26

Development of Laser Technology in Military Systems

The applications of laser technology in the military sector can be divided into seven distinct categories, which are presented in Fig. 1.

Fig. 1. Military applications of lasers.

Lasers are widely used for precise distance measurement, terrain profiling, and obstacle detection across multiple industries and scientific fields, and are also used for military purposes. These applications leverage the unique properties of laser light, including directionality, coherence, and the ability to produce narrow, focused beams, which enable highly accurate and rapid data collection compared to traditional methods. LiDAR (Light Detection and Ranging) is a laser-based technology that plays a vital role in military applications due to its ability to emit rapid laser pulses and measure the time it takes for these pulses to reflect from surfaces. LiDAR (Fig. 2) supports navigation for autonomous military vehicles, including aerial, land, and naval platforms, enabling obstacle avoidance and route optimization. It is also used in specialized tasks, such as mine countermeasures and underwater terrain mapping (bathymetric LiDAR) [1, 2]. Laser target designation refers to the process of marking a target with a laser beam, enabling precision-guided munitions, such as bombs, missiles, or artillery shells, to home in on the marked location and strike with exceptional accuracy. Lasers also allow for very reliable and fast communication. This technology enables the secure, high-speed transmission of data, voice, and video across the battlefield. This technology utilizes tightly focused laser beams to establish point-to-point links, typically between mobile units, command centers, satellites, ships, or unmanned vehicles, enabling rapid and reliable information exchange without reliance on traditional radio frequency (RF) communication [3].

There are also many attempts in the military to use lasers as direct-energy weapons. This class of advanced military systems is designed to inflict physical damage or neutralize material targets using concentrated beams of electromagnetic energy generated by lasers. Unlike conventional ammunition, these weapons deliver energy at the speed of light, enabling precise engagement of threats such as drones, missiles, artillery shells, and small vehicles in real-time. Moreover, the laser beam is not affected by gravity [4].

Materials Research Forum LLC
https://doi.org/10.21741/9781644904015-26

Fig. 2. LiDAR system and its example application [2].

Another significant military application of lasers is their use as dazzlers. Laser dazzlers and blinding systems are specialized military devices designed to temporarily impair the function of electro-optical sensors and, in some cases, human vision using intense, directed beams of light. These non-lethal weapons utilize specific wavelengths, often green or infrared lasers, to overwhelm sensors in surveillance equipment, cameras, guided missiles, drones, and other electronic targeting systems [5].

Lasers also play an essential role in auxiliary military applications such as cutting, welding, and rapid repairs on equipment and infrastructure. Their precision and reliability can make them valuable tools for maintenance operations in specialized maintenance facilities.

Military Applications of Laser Technology

Although most laser applications are dedicated to range finding and communication, continuous advances are being made in the development of high-energy laser weapon systems.

The U.S. Navy has initiated the Songbow program (Fig. 3a) to develop shipborne high-energy laser weapon systems designed to counter unmanned aerial threats while reducing dependence on expensive missile interceptors. The program aims to enable rapid, cost-efficient, directed-energy targeting by advancing technologies in beam control, target tracking, and thermal management. The system's output power reportedly approaches 400 kW, allowing it to engage a wide range of emerging threats, including drone swarms, cruise missiles, and hypersonic glide vehicles. Songbow achieves this performance by coherently combining multiple 50kW fiber laser modules into a single, precisely aligned output beam.

Another example of a laser-directed energy weapon is the British DragonFire system (Fig. 3b), which was developed for the Royal Navy. It uses a 50 kW-class fiber-combined solid-state laser with beam-combining technology pioneered in the UK. The laser is integrated into a turret equipped with an advanced electro-optical targeting system and a secondary lower-power laser for imaging and tracking. This setup allows for highly precise tracking and targeting of moving aerial and maritime threats at ranges of several kilometers. The system can engage and defeat a variety of threats, including small unmanned aerial vehicles, mortar and artillery rounds, fast inshore attack craft, and some classes of cruise missile seekers. It neutralizes targets by ablating structural materials, detonating warheads, or dazzling electro-optical sensors. With very high precision, the system can hit objects as small as a coin from a distance of one kilometer, emphasizing the accuracy of the system [7].

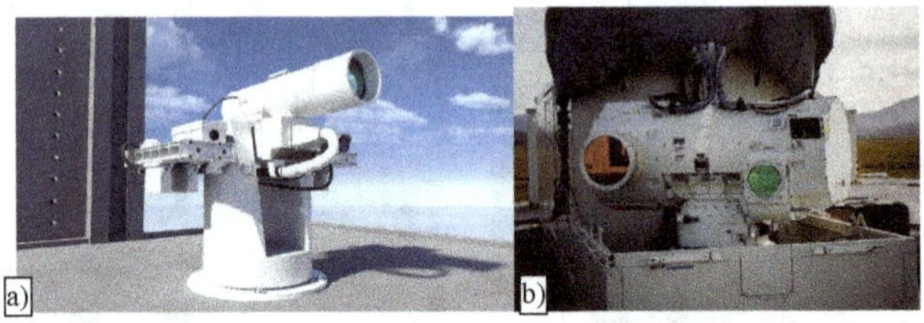

Fig. 3. High-energy laser weapon systems: a) US Navy Songbow laser weapon project, b) British DragonFire [6,7].

Another noteworthy military application of laser technology is conducted by the Turkish Armed Forces (Fig. 4a). The ISIK (IŞIK) project is a high-power laser weapon system developed by TÜBİTAK BİLGEM. The ISIK system features a 20 kW laser capable of dazzling targets up to 5 km away and destroying or melting both stationary and moving targets at distances of up to 1.5 km. This system is designed for integration with various platforms, including land vehicles, naval vessels, and stationary defense towers. The ISIK laser weapon successfully passed critical acceptance tests and is prepared for mass production and field deployment within the Turkish military. With continued development, it is expected to lay the foundation for more powerful and portable laser weapons in the future.

Fig. 4. Turkish weapon laser systems: a) The ISIK project, b) ALKA directed-energy weapon system [8].

The ALKA directed-energy weapon system (Fig. 4b) is a Turkish dual electromagnetic and laser weapon developed by Roketsan. It is designed as a highly integrated hybrid air defense system, primarily targeting asymmetric threats. ALKA can detect, track, and simultaneously engage up to 100 targets such as drones, loitering munitions, mini/micro-UAVs, improvised explosive devices (IEDs), floating mines, unmanned marine vehicles, and drone swarms. It utilizes radar and electro-optical methods for automatic target detection, and is equipped with artificial intelligence for autonomous target recognition, prioritization, and assignment. The system employs both electromagnetic jamming and laser destruction with a two-layered defense architecture. The weapon system operates efficiently at laser power levels of 2.5 kW, with effective ranges of 750 meters, and up to 5 kW, extending up to 1,500 meters for laser engagements. For electromagnetic disturbances, it can affect targets up to 1,000 meters. ALKA operates 24/7 and can be deployed as a mobile, portable, or stationary platform [8].

Technological and Economic Impact

The implementation of laser technology in defense systems is driving significant technological, economic, and strategic changes, while also having social and environmental implications.

Ensuring operational reliability and precision is a key technological challenge, requiring advanced control systems and advanced engineering.

The development of high-energy laser (HEL) weapons is closely linked to advances in materials science and precision engineering. The durability of key components, such as mirrors, optics, and support structures, depends on their resistance to extreme thermal and mechanical conditions. For example, microstructural analysis of nickel-based superalloys used in power-generating components allows for a better understanding of their high-temperature performance [9]. Mobile platforms require lightweight yet robust structures, often based on titanium alloys, whose thermal oxidation kinetics must be carefully studied to ensure the durability of protective coatings [10]. Corrosion protection is equally crucial. Research on the depth profiles of corrosion resistance in nitrided layers on austenitic steel provides a basis for designing naval and land systems [11]. The precision of laser systems depends on advanced positioning systems. Intelligent electropneumatic servo systems have the potential for rapid and accurate target tracking [12]. Implementing these systems in combat vehicles requires integrated control and scanning methods based on gyroscopes and advanced algorithms [13]. Modern engineering also requires control of manufacturing processes. For example, assessing shrinkage in bioceramic sintered materials is crucial in the context of manufacturing composite components [14], while analyzing the composite microstructure is essential for understanding their mechanical properties [15]. The reliability of electronic systems supporting the laser (power supply, control) is critical, requiring precise modeling of error-counting systems, such as those used in PCB wave soldering technology [16]. In auxiliary operations, such as field repairs or component production, lasers can be used to cut various materials, including HSLA steel, which requires the comparison of different machining techniques [17]. The issue of cost-effectiveness is crucial, as laser weapons offer a potentially unlimited number of shots at a low unit cost. The introduction of advanced laser systems is part of the industrial transformation associated with Industry 4.0 [18]. In the context of mass production, effective process management is crucial, for example, by evaluating manufacturing technologies using a 3x3 matrix [19]. Achieving high-quality components requires the use of advanced quality engineering and statistical methods. Process modeling using fuzzy regression offers an alternative to classical design of experiments (DoE) [20]. In another paper, the authors proposed a modified binary approach to fuzzy operations, which applies to the assessment of correlated data and uncertainty [21]. To meet quality requirements, it is necessary to utilize optimization tools, such as expert systems for experimental design [22]. Advanced systems also utilize numerical simulations and computational methods, such as estimating the accuracy of neural network approximations for sintered metals [23] or applying the parametric Response Surface Methodology (RSM) model in the optimization of manufacturing processes [24]. Heuristic approaches to model selection and output data transformation are crucial in the validation of complex systems [25]. Mechanical reliability, for example, in bearing fatigue analysis [26], is crucial for mobile laser platform systems, as is the control and reduction of uncertainty in large datasets using principal component analysis (PCA) [27]. Manufacturing advanced materials for optical components may require the use of powders, the shape distribution of which can be analyzed using the bootstrap method [28]. Even for advanced biometric analyses, such as phenolic acid analysis, the use of advanced statistical tools, including nonparametric ones, remains relevant in research [29].

The use of HEL reduces collateral damage, which has a direct positive social and ethical impact. In the context of deploying systems on military platforms (land, sea, air), ensuring structural integrity and integrity is crucial, including assessing the corrosion protection effect on joints in aircraft structures [30]. Environmental aspects include the energy efficiency of laser systems and the management of waste generated during production processes. Energy conservation and optimization of cooling systems (thermal regulation) are crucial, as evidenced by research on selecting optimal windows and their orientation in an energy-related context [31]. Furthermore, in the context of a circular economy, research on electrostatic polymer separation [32] may have implications for waste management generated in advanced military production. Finally, logistical security is essential. The design of transport systems, such as crates for carrying hazardous materials (e.g., detonators [33]), is crucial to minimizing social and environmental risks. Similarly, automation, such as in heavy-lifting systems [34], reduces the risk of accidents when handling large, heavy laser components. All of these elements, supported by advanced safety analysis methods such as smoothed bootstrap for system tuning [35], contribute to the integrated and safe implementation of laser technology.

Summary

Laser technology plays a crucial role in modern military defense, enhancing precision targeting, neutralizing threats, and ensuring secure communications. In addition to common applications of lasers for rangefinding and communication, other applications of laser technology are constantly being developed. Directed Energy Weapons use focused laser beams to quickly disable drones, missiles, and vehicles, offering cost-effective alternatives to traditional ammunition. Laser systems enhance defense capabilities by enabling rapid responses to fast-moving threats and minimizing collateral damage. Despite challenges such as power requirements and environmental limitations, laser technology is increasingly integral to military platforms on land, sea, and air, shaping future warfare and defense strategies. There have been numerous advances in shipborne and land based high-energy laser weapon systems developed by leading military powers. All these systems deliver enhanced power, precision, and versatility, marking a transition toward cost-effective, rapid-response solutions in contemporary defense strategies.

References

[1] Information on: https://www.defenseadvancement.com/suppliers/military-lidar

[2] Information on: https://www.baesystems.com/en/product/geiger-mode-lidar-camera

[3] Information on https://www.transparencymarketresearch.com/europe-military-free-space-optics-communication-market.html

[4] Information on https://www.militaryaerospace.com/power/article/55140201/laser-weapons-high-power-microwaves-high-energy-weapons

[5] O. Steinvall, Laser dazzling: an overview, Technologies for Optical Countermeasures XIX, 12738 (2023) 1-6. https://doi.org/10.1117/12.2678416

[6] Information on https://www.armyrecognition.com/news/navy-news/2025/us-navy-launches-songbow-laser-weapon-project-to-counter-drones-without-running-out-of-missiles

[7] Information on https://www.gov.uk/government/news/advanced-future-military-laser-achieves-uk-first

[8] Information on https://turdef.com/article/tuebitak-bilgem-s-isik-laser-weapon-passes-acceptance-tests

[9] A. Szczotok et al., Stereological characterization of γ' phase precipitation in CMSX-6 monocrystalline nickel-base superalloy, Materials Characterization 60 (2009) 1114-1119. https://doi.org/10.1016/j.matchar.2009.01.023

[10] D. Klimecka-Tatar et al., The kinetics of Ti-1Al-1Mn alloy thermal oxidation and charcteristic of oxide layer, Archives of Metallurgy and Materials 60 (2015) 735-738. https://doi.org/10.1515/amm-2015-0199

[11] K. Jagielska-Wiaderek et al., Corrosion resistance depth profiles of nitrided layers on austenitic stainless steel produced at elevated temperatures, Archives of Metallurgy and Materials 54 (2009) 115-120.

[12] J.E. Takosoglu, P.A. Laski, Intelligent positioning system of electro-pneumatic servo-drive, Annals of DAAAM and Proc. International DAAAM Symposium (2011) 1641-1642.

[13] I. Krzysztofik et al., Selected methods of control of the scanning and tracking gyroscope system mounted on a combat vehicle, Annual Reviews in Control 44 (2017) 173-182. https://doi.org/10.1016/j.arcontrol.2016.10.003

[14] A. Dudek, C. Kolan, Assessments of shrinkage degree in bioceramic sinters HA+ZrO2, Solid State Phenomena 165 (2010) 25-30. https://doi.org/10.4028/www.scientific.net/SSP.165.25

[15] A. Dudek, Microstructure and properties of the composites: Hydroxyapatite with addition of zirconia phase, Journal of Engineering Materials and Technology 133 (2011) art. 021006. https://doi.org/10.1115/1.4003104

[16] J. Pietraszek et al., Modeling of errors counting system for PCB soldered in the wave soldering technology, Advanced Materials Research 874 (2014) 139-143. https://doi.org/10.4028/www.scientific.net/AMR.874.139

[17] P. Szataniak et al., HSLA steels - Comparison of cutting techniques, METAL 2014 - 23rd Int. Conf. Metall. Mater. (2014) 778-783.

[18] J. Pietraszek et al., Challenges for the DOE methodology related to the introduction of Industry 4.0, Prod. Eng. Arch. 26 (2020) 190-194. https://doi.org/10.30657/pea.2020.26.33

[19] S. Borkowski et al., The use of 3x3 matrix to evaluation of ribbed wire manufacturing technology, METAL 2012 – 21st Int. Conf. Metall. Mater., (2012) 1722-1728.

[20] J. Pietraszek, Fuzzy regression compared to classical experimental design in the case of flywheel assembly, Lecture Notes in Computer Science 7267 LNAI (2012) 310-317. https://doi.org/10.1007/978-3-642-29347-4_36

[21] J. Pietraszek, The modified sequential-binary approach for fuzzy operations on correlated assessments, Lecture Notes in Computer Science 7894 LNAI (2013) 353-364. https://doi.org/10.1007/978-3-642-38658-9_32

[22] R. Dwornicka, J. Pietraszek, The outline of the expert system for the design of experiment, Prod. Eng. Arch. 20 (2018) 43-48. https://doi.org/10.30657/pea.2018.20.09

[23] J. Pietraszek et al., The estimation of accuracy for the neural network approximation in the case of sintered metal properties, Studies in Computational Intelligence 513 (2014) 125-134. https://doi.org/10.1007/978-3-319-01787-7_12

[24] J. Pietraszek et al., The parametric RSM model with higher order terms for the meat tumbler machine process, Solid State Phenomena 235 (2015) 37-44. https://doi.org/10.4028/www.scientific.net/SSP.235.37

[25] J. Pietraszek, A. Goroshko, The heuristic approach to the selection of experimental design, model and valid pre-processing transformation of DoE outcome, Advanced Materials Research 874 (2014) 145-149. https://doi.org/10.4028/www.scientific.net/AMR.874.145

[26] P.J. Romanowicz et al., Using the Effect of Compression Stress in Fatigue Analysis of the Roller Bearing for Bimodal Stress Histories, Materials 15 (2022) art. 196. https://doi.org/10.3390/ma15010196

[27] J. Pietraszek, E. Skrzypczak-Pietraszek, The uncertainty and robustness of the principal component analysis as a tool for the dimensionality reduction, Solid State Phenomena 235 (2015) 1-8. https://doi.org/10.4028/www.scientific.net/SSP.235.1

[28] J. Pietraszek, A. Gadek-Moszczak, The smooth bootstrap approach to the distribution of a shape in the ferritic stainless steel AISI 434L powders, Solid State Phenomena 197 (2013) 162-167. https://doi.org/10.4028/www.scientific.net/SSP.197.162

[29] E. Skrzypczak-Pietraszek, J. Pietraszek, Chemical profile and seasonal variation of phenolic acid content in bastard balm (Melittis melissophyllum L., Lamiaceae), Journal of Pharmaceutical and Biomedical Analysis 66 (2012) 154-161. https://doi.org/10.1016/j.jpba.2012.03.037

[30] A. Dudek et al., Effect of corrosion protection method on properties of RSW and RFSSW lap joints applied in production of thin-walled aerostructures, Materials 13 (2020) art. 1841. https://doi.org/10.3390/MA13081841

[31] J.M. Djoković et al., Selection of the Optimal Window Type and Orientation for the Two Cities in Serbia and One in Slovakia, Energies 15 (2022) art. 323. https://doi.org/10.3390/en15010323

[32] M. Zenkiewicz et al., Electrostatic separation of binary mixtures of some biodegradable polymers and poly(vinyl chloride) or poly(ethylene terephthalate), Polimery/Polymers 61 (2016) 835-843. https://doi.org/10.14314/polimery.2016.835

[33] P. Krysiak et al., The design and field tests of a box for “ERG” electric blasting caps, Lect. Note Mech. Eng. (2017) 285-292. https://doi.org/10.1007/978-3-319-50938-9_29

[34] E. Lisowski, G. Filo, Automated heavy load lifting and moving system using pneumatic cushions, Autom. Constr. 50 (2015) 91-101. https://doi.org/10.1016/j.autcon.2014.12.004

[35] R. Dwornicka et al., The smoothed bootstrap fine-tuning, System Safety: Human - Technical Facility - Environment 1 (2019) 716-723. https://doi.org/10.2478/czoto-2019-0091

Terotechnology XIV | Materials Research Forum LLC
Materials Research Proceedings 62 (2026) 211-216 | https://doi.org/10.21741/9781644904015-27

Evaluation of Manufacturing Pipe Elements in Selected Experimental Metal Forming Tests

MIŁEK Tomasz[1,a*]

[1]Kielce University of Technology, Faculty of Mechatronics and Mechanical Engineering, Al. 1000-lecia Państwa Polskiego 7, 25-314 Kielce, Poland

[a]tmatm@tu.kielce.pl

Keywords: Metal Forming, Necking, Bulging, Rolling, Hydroforming, Hydromechanical Bulge Forming

Abstract. This paper presents a review of experimental research results evaluating the feasibility of manufacturing various components from copper and steel pipes using selected metal forming methods, a such as: necking, bulging, rolling and hydroforming process. The studies were conducted for different relative wall thicknesses s_0/D (where s_0 denotes tube thickness and D is outer diameter of pipe). Experimental results from cross-rolling steel pipe reducers using internal shaping tools and hydroforming of axisymmetric copper components (valve balls) are presented. These results are supplemented by summarized results from the necking and bulging of steel reducers, as well as hydroforming of copper cross-joints and various one- and two-sided steel connections.

Introduction

Pipe components such as reducers and tube connections (including straight and skewed T-pipes and cross-joints) are widely used in the energy, chemical, machinery and automotive industries. Copper connectors are also used in plumbing, heating, gas, and sanitary installations [1]. Reducers as pipeline elements can be manufactured using various technologies, e.g., by rolling from sheet metal and then welding or turning from bars (due to the high material consumption of the process, it is more likely to be used for small dimensions). In metal forming, reducers are obtained by methods involving the reduction or increase of the diameter of the semi-finished product – pipe. In the first case, they are made by necking in a die or by crimping the pipes using steam-air hammers, presses, and swaging machines, or in small-scale production, they can be rolled by using segments with an internal shaping surface on a *WPM-120* profile-eccentric rolling mill [1-2]. By increasing the diameter or cross-section of a pipe segment, reducers are obtained through metal forming through bulging [1-3]. Classic bulging methods use tools with a steel or rubber punch; with a liquid instead of a punch (hydraulic pressing), and with expanding segments spread apart by a wedge-shaped mandrel. Specific methods for expanding pipe elements include the electrodynamic and electromagnetic methods [1,3]. However, these are only used for non-ferrous metals (such as copper, aluminium, and their alloys) because they are good conductors of electricity. The electrodynamic method is used to shape pipes with diameters exceeding 30 mm during expansion and 10 mm during compression [1,3]. Pipe connectors such as T-pipes, cross-joints (straight and skewed) and other one- and two-sided elements are made by welding, casting and metal forming methods, where, in addition to typical methods such as die forging and pulling out the stub pipes, the hydroforming technology is used [4,5].

The aim of the research in this paper was to assess the feasibility of manufacturing various components from copper and steel pipes using selected metal forming methods. Experimental results of cross-rolling steel pipe reducers using internal shaping tools and hydroforming of axisymmetric copper components (valve balls) were presented. These results were supplemented by results for the necking and bulging of steel reducers, and on the hydroforming of copper cross-joints and various one- and two-sided steel connectors.

Necking and One-Sided, and Two-Sided Bulge Forming of Reducing Tubes
The necking process involves locally reducing the pipe diameter by forcing it into a conical die cavity. The diagram of this process and scheme of laboratory tooling used in investigations is shown in Figure 1. Based on experimental tests, which used pipe sections with an outer diameter of $D_0=38mm$ and wall thicknesses of $g_0=2.5mm$ and $4\ mm$ (which corresponded relative thickness $g_0/D=0.66$ and 0.105, respectively) made of E235N steel (EN 10027) and $g_0=2.8\ mm$ ($g_0/D = 0.074$) made of P255 steel, it was found that manufacturing conical cones in the compression zone is possible while maintaining high surface quality and minimal ovalization. A die with a calibration hole diameter of $d_m=25\ mm$ and a working cone angle of $\alpha = 30°$ was used for the tests. The sample length before deformation was 55 mm. The necking factor of the manufactured cones was K = 1.52. An example of a steel reducer with a conical shape in the compression zone made of E235N steel ($g_0/D=0.66$) is shown in Figure 2.

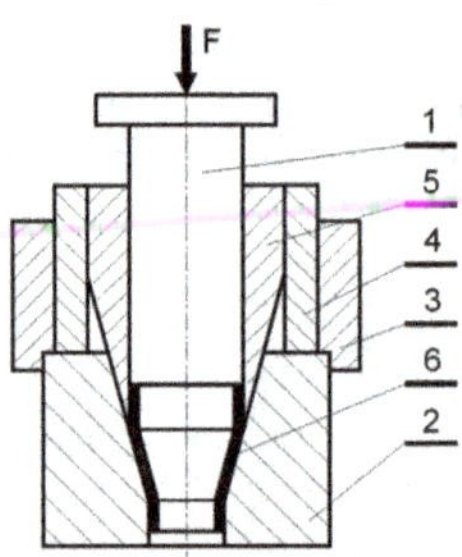

Fig. 1. Laboratory press forming-die for necking process, where: 1-punch, 2-die, 3- holder, 4-reduction sleeve, 5-pilot sleeve, 6-sample

Fig. 2. Reducer made of E235N pipe with a conical shape in the compression zone (opening angle 30^0) at relative thickness $g_0/D=0.66$ and necking factor $K=1.52$

For reducers with a small cylindrical portion (d=25 mm), the best results were obtained for P255 steel pipes ($g_0=2.9\ mm$). The resulting reducers were characterized by the best surface quality and ovalization in the range of $A=0.24\text{-}0.69\%$. The worst results were obtained when necking pipes with a thickness of $g_0=4\ mm$ made of *E235N* steel. When necking reducers with a cylindrical portion, the resulting shapes in this zone deviated from the intended die shape (large concavities, uneven faces of the cylindrical portion of the reducer). The ovalization of the necked cylindrical portions of the reducers ranged from 0.96-3.6%.

The bulging of reducers involves increasing the diameter or cross-section of pipe sections. The scope of the bulging tests conducted [3] included classical unilateral expansion of pipe reducers in the so-called open die (without the use of a cylindrical clamp on the reducer part - Figure 3a), as well as simultaneous (bilateral) expansion of both ends of the pipe section according to the scheme shown in Figure 3b. In the case of unilateral expansion tests, they were performed for steel pipe sections (E235N) with relative initial pipe thicknesses $s_0/D=0.05;\ 0.07;\ 0.10$; in terms of evaluation of pressure forces and changes in geometric parameters of the expanded reducers for different flare angles $\alpha=30^0;\ 36^0;\ 45^0;\ 60^0$. On this basis, it was found, among other things, that the pressure force during bulging process increases with the relative ratios $\Delta l/l_0$, s_0/D and the flare angle α, and its maximum value was obtained at the highest ratio $s_0/D=0.10$ and angle

Materials Research Proceedings 62 (2026) 211-216
https://doi.org/10.21741/9781644904015-27

$\alpha=60^0$. Moreover, the nature of changes in the wall thickness distribution of reducers bulged unilaterally with a displacement of $\Delta l=20\ mm$ ($\Delta l/l_0=0.36$) at different α and s_0/D is very similar. In the cylindrical part of the reducer, a slight wall thickening occurs. In the expanded part, the wall thickness decreases compared to the initial thickness s_0. The maximum wall thinning was found at the front of the conical part of the reducer. Based on the comparison of the results of the tests of bilateral and unilateral expansion of steel pipe reducers with the relative initial pipe wall thickness $s_0/D=0.07$; and different flare angles α, α_1, α_2 amounting to 30^0; 45^0; 60^0 (Fig. 3a and 3b) it was found that there is a possibility of bilateral expansion of steel reducers not only with identical ($\alpha_1=\alpha_2=30^0$; $\alpha_1=\alpha_2=45^0$; $\alpha_1=\alpha_2=60^0$) but also with different opening angles ($\alpha_1=60^0$ and $\alpha_2=30^0$; $\alpha_1=60^0$ and $\alpha_2=45^0$; $\alpha_1=45^0$ and $\alpha_2=30^0$) of punches, as evidenced by successfully conducted tests with their relative displacement up to $\Delta l/l_0=0.75$. Both in the case of bilateral and unilateral bulging, the pressure force during expansion of the reducers increases with the increase of the relative displacement $\Delta l/l_0$ and the punch opening angle (for the same $\alpha_1=\alpha_2$ in bilateral expansion and α in unilateral expansion). For the same angles $\alpha=\alpha_1=\alpha_2$ and relative displacements $\Delta l/l_0=0.36$ and $\Delta l/l_0=0.54$, the pressure forces during unilateral bulging are higher than the values obtained for bilateral bulging (an increase of about *17-25%*). Examples of single- and double-sided reducers made of E235N steel pipes with a diameter of $D=30mm$ and a wall thickness of $s_0=2mm$ ($s_0/D=0.07$) are shown in Figure 4.

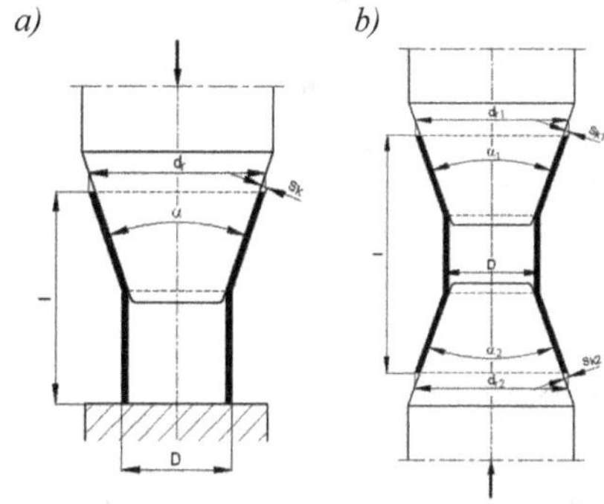

Fig. 3. Schemes of bulge forming of reducing pipes: a) one-sided, b) two-sided [3]

Fig. 4 One-sided and two-sided reducers made from E235N steel with 30mm diameter and wall thickness $s_0=2mm$ ($s_0/D=0.07$)) after bulging [3].

Reducing Pipes on a WPM-120 Profile-Eccentric Rolling Mill

In this method, the process of reducing the diameter of a given length of pipe occurs through successive movements of the rolling segments with their internal shaping surfaces and axial shifts of the material. The rolling tools move parallel on circular paths and always occupy a symmetrical position relative to the centre of the system – the axis of the workpiece. They are mounted in two jaws, each driven by two eccentric shafts. The pipe diameter is reduced as the tool rolls along the internal surface of the tools in the first phase of the cycle. In the second phase, the tool loses contact with the material, which is used to axially shift the material by the feed rate. A simplified diagram of this method is shown in Figure 5. The tubes intended for rolling on the *WPM-120* were drawn on a mandrel and were made of *ASTM A513 (MT 1020)* steel with a carbon content of *0.2081%*, magnesium *0.3-0.6%*, sulphur *0.006393%*, and phosphorus max. *0.05%*. Carbon and sulphur composition analysis was performed using a *LECO CS 230* analyser. The tubes had an outer diameter of $D=34\ mm$ and a wall thickness of $s_o=3.2\ mm$, corresponding to a relative wall thickness of $s_0/D=0.09$. The sample length was $l_0=265$ *mm*, which resulted from the need to clamp them in the mill feeder during rolling. Rolling dies with an internal shaping surface designed to reduce the diameters of shafts and pipes were used. The dies had a diameter of *26 mm* and a working part angle of *30°*. An eccentricity of $e=1.161$ *mm* was assumed during rolling. After rolling, the obtained reducers had a diameter of $d=24.3$ *mm* and a reduced part length of approximately *85* mm. The elongation factor for the reduced part of the pipes was $\lambda=1.958$, and the compression factor $K=1.4$. A successfully rolled sample of *ASTM A513 (MT 1020)* steel for the parameters specified above is shown in Figure 6.

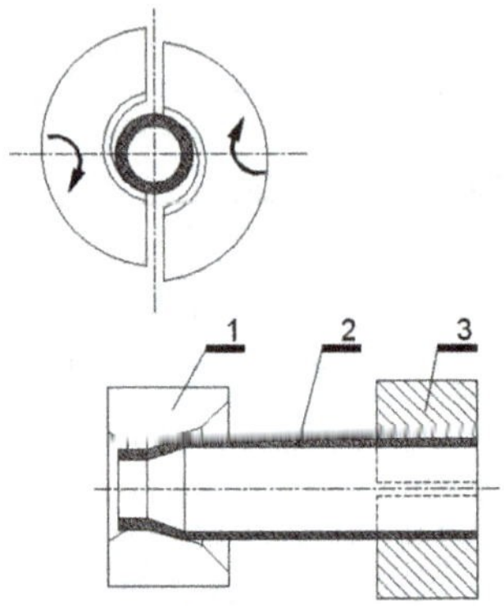

Fig. 5. Schematic diagram of the rolling on the WPM-120 rolling mill, where: 1 – tool with an internal shaping surface, 2 – sample, 3 – mill feeder holder

Fig. 6. Steel reducer rolled on the WPM-120 rolling mill.

Hydroforming

Hydromechanical bulge forming (hydroforming) tests [5] of copper pipes with a relative wall thickness of $s_0/D=0.05$ allowed us to determine the technological suitability of this method for equal and reducing cross-joints with $d/D=0.9$ and $d/D=0.8$. For the determined ranges of working pressures and pipe upsetting ratios $\Delta l/lo=(0.1\text{-}0.45)$, an accurate representation of the die insert pattern was obtained with a transition radius of the pipe part into the stub $R=5$ *mm*. An example of an equal copper X-branch component is shown in Figure 7a. The possibility of manufacturing steel T-pipes with a one-sided (Fig. 7b) and two-sided (Fig. 7c) branch with a pipe diameter of $D=32$ *mm* and a stub diameter of $d=29$ *mm* and an angle of inclination between the body and the branch of $\alpha=70^0$ was also tested [5].

a) b) c)

Fig. 7. Pipe elements after hydroforming tests: a) equal copper cross-joint with $d/D=1$ and $s_0/D=0.05$ [5]; b) one-sided and c) two-sided steel reducing tees with $s_0/D=0.05$; $d/D=0.9$ and $\Delta l/l_0=0.44$ made from P265TR1 steel [3]

Materials Research Proceedings 62 (2026) 211-216 https://doi.org/10.21741/9781644904015-27

Based on these results, it was found, among other things, that at a constant upset ratio *$\Delta l/l_0=0.32$* and a similar pressure change, the length of the stub pipes *h* in one-sided T-pipes increases slightly with the initial length of the pipe l_o, while there are no significant differences in the obtained pressure force curves. At a constant ratio $\Delta l/l_0=0.44$, hydroforming of connectors with a double branch requires the use of higher pressure than for single-sided connectors, while the obtained forces are lower for double-sided connectors. The nature of the change in connector thickness is the same, despite the increase in the upset ratio from $\Delta l/l_0=0.32$ na $\Delta l/l_0=0.44$ [3].

Hydroforming of axisymmetric elements, shaped similarly to valve balls, made of copper pipes was performed according to the diagram in Fig. 8. This was performed for copper pipes with a diameter of *$D=22$ mm* and a wall thickness of *$s_0 = 1$ mm*, corresponding to a relative wall thickness of *$s_0/D= 0.045$*. Samples were prepared as sections with a length of *$l_0=115$ mm*. Successfully formed samples were expanded by changes in force and fluid pressure as shown in Figure 9a. An example of an hydroformed axisymmetric element is shown in Figure 9b.The maximum value of force at a relative ratio of *$\Delta l/l_0=0.06$* was *83.69 kN*, and the pressure was *60 MPa*. Exceeding the pressure by approximately 25% resulted in the bursting on spherical cup. For successfully hydromechanically bulged elements, the *h/d* ratio was 0.67 (where *h* denotes height and *d* is diameter of spherical cup), while the bulge coefficient *$k=1.36$* (where *$k=d/D$*) [5]. The obtained *k* coefficient is higher than the maximum bulge coefficient for copper pipes (k=1.33) using only liquid pressure (without upsetting force) given in the literature [4].

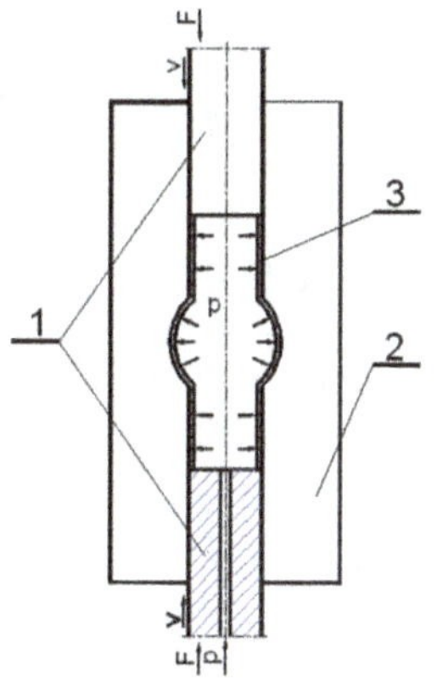

a)

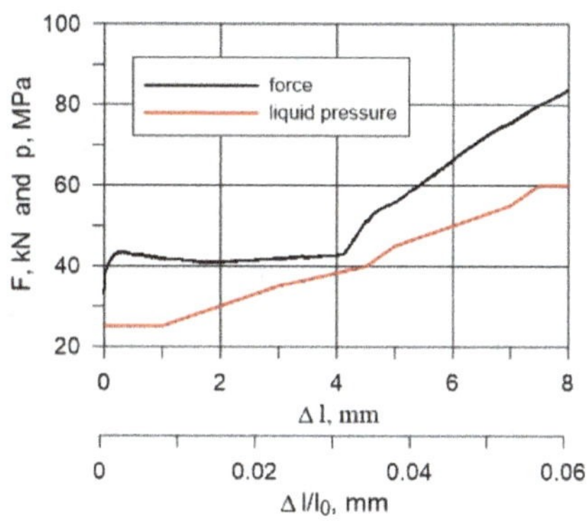

b)

Fig. 8. Hydroforming of axisymmetric components, where: 1-punches, 2-die, 3-hydroformed component [5]

Fig. 8. Changes in force and liquid pressure (a) for hydroforming of axisymmetric components made for copper (b) at $s_0/D=0.04$ and $\Delta l/l_0=0.06$

Summary

The paper synthetically describes the results of the author's many years of experimental research in the field of selected metal forming processes for shaping components from tubes. On their basis, it was found that:

1. Necking of reducers from tubes with a conical shape of the die (at the flare angle α=30°) from *E235N* steel (at relative pipe wall thicknesses $g_0/D=0.66$ and *0.105*) and P255 steel ($g_0/D=0.074$) is possible for the factor K=1.52 while maintaining their high surface quality and low ovalization.
2. Forming of reducers by cross-rolling with tools with an internal shaping surface of ASTM A513 (MT 1020) steel pipes at $s_0/D=0.09$, flare angle $\alpha=30°$ and factor K=1.4; with a significant reduced part is possible with appropriately selected eccentricity and pipe feed of rolling mill.
3. The bulging method can be used to manufacture steel reducers both by one-sided forming of pipes (at $s_0/D= 0.05; 0.07; 0.10$) and by two-sided forming ($s_0/D=0.07$) for different angles of dies $\alpha = 30^0; 45^0; 60^0$.
4. In the tests of hydroforming process of copper pipes with a relative wall thickness of $s_0/D=0.05$, the technological suitability of this method was found for equal and reducing cross-joints with $d/D=0.9$ and $d/D=0.8$, for the determined ranges of working pressures and pipe upsetting ratios $\Delta l/lo=(0.1\text{-}0.45)$. Similarly, in the case of steel skewed T-pipes with a single-sided and double-sided branch with an angle between the body and the branch of $\alpha=70^0$.
5. It is possible to hydroforming axisymmetric valve ball-shaped parts from copper tubes with $s_0/D=0.04$ for a fixed change in pressure and force as a function of displacement. The resulting bulging coefficient is greater than the maximum value for the liquid pressure expansion method.

References

[1] Z. Gronostajski et al., Recent development trends in metal forming, Archives of Civil and Mechanical Engineering 19/3 (2019) 898-941. https://doi.org/10.1016/j.acme.2019.04.005

[2] J. Pacanowski, Metody wykonywania wytłoczek kołowo-symetrycznych, in: Projektowanie procesów ciągnienia wytłoczek kołowo-symetrycznych i konstrukcji tłoczników, t.1. Metody i zasady ciągnienia wytłoczek kołowo-symetrycznych, Wydawnictwo Politechniki Świętokrzyskiej, Kielce, 2018, 82-173.

[3] T. Miłek, Experimental investigations into one-sided and two-sided bulge forming of steel reducing tubes, Steel Research International Special edition (2012) 543-546.

[4] J. Chałupczak, A. Długosz, Hydromechanical bulge forming of axisymmetric components, Mechanik 4 (1986) 167-168.

[5] Miłek T., The analysis of microhardness variations of hydroformed P265TR1 steel axisymmetric elements from tubes, Archives of Foundry Engineering 20/3 (2020) 74-78. https://doi.org/10.24425/afe.2020.133333

Terotechnology XIV | Materials Research Forum LLC
Materials Research Proceedings 62 (2026) 217-221 | https://doi.org/10.21741/9781644904015-28

The Possibility of Laser Cladding Usage for the Regeneration of External Surfaces of Cast Iron Engine Cylinder Liners

Piotr SĘK[1, a*]

[1]Kielce University of Technology, Faculty of Mechatronics and Mechanical Engineering, Al. 1000-lecia Państwa Polskiego 7, 25-314 Kielce, Poland

[a]psek@tu.kielce.pl

Keywords: Laser Cladding with Powders, Machine Parts Regeneration, Cylinder Liners

Abstract. The aim of the work is to determine the possibility of using laser powder coating for the regeneration of cast iron engine cylinder liners and to analyze the entire process: selection of cladding parameters, laboratory analysis: hardness, microscopic and profilometric. The disadvantages of this treatment are costs, long and difficult technological process, required stress relief treatment.The advantage is the possibility of applying layer surfacing as a buffer, and the possibility of transforming the used cylinder liner into new ones with satisfactory parameters.

Introduction

Cylinder liners are made of cast iron because it is durable, has excellent wear resistance, and naturally retains oil. The material's porosity and graphite content help to maintain the lubricating oil film between the piston rings and the cylinder wall, which reduces friction and wear. Additionally, cast iron is effective at damping engine vibrations and can be produced at a lower cost compared to some other materials. The key reasons for using cast iron are: self-lubricating properties - Ccast iron has a micro-structure containing graphite, which helps hold lubricating oil, reducing wear on the piston rings and the liner itself, wear resistance - it is a hard and durable material that can withstand the high pressures and temperatures of engine operation, vibration damping - this material is good at absorbing vibrations, which leads to a quieter and smoother running engine, cost effectiveness - often cheaper to produce than materials like certain steel alloys, making it a more economical choice for many applications, porosity: - the porous nature of cast iron helps to retain oil, which is crucial for the lubrication of the piston rings [1,2]. Lower coefficient of thermal expansion allows smaller bore fit clearance and/or more open tolerances for piston and bore hole diameter. Can allow more stability in bore or allow a shorter skirt. Cast iron has one major disadvantage - is over 2–1/2 times heavier than aluminum. The piston is constantly being accelerated and deaccelerated from zero to maximum velocity and back to zero in a linear direction for every crankshaft rotation. This causes more stress on connecting rods and the crankshaft. Cast iron also adds to the overall weight of an engine [1,2]

Laser Cladding of Cast Iron

Laser resurfacing regenerates cast iron cylinder liners by improving surface properties like hardness and wear resistance through localized heating and melting or by creating wear-resistant coatings. This process can create oil-retaining micro-holes and achieve a more durable surface than the original material, leading to increased power output and reduced oil consumption, as explained in [3]. Laser cladding, also known as Laser Metal Deposition (LMD), is a relatively simple process that creates a new layer of material applied in the form of powder or wire on the surface of a base element. Therefore, this method is increasingly used to apply coatings to iron alloys and other metals. Laser cladding involves melting the surface of the base material with a laser beam while simultaneously applying an alloying powder or wire. The materials used to produce the powders for this process depend primarily on the mechanical properties of the desired

Materials Research Proceedings 62 (2026) 217-221 https://doi.org/10.21741/9781644904015-28

layer, but the most common are cobalt-based and nickel-based powders. Using this method, we can also produce composite surface layers in which the layer created using metallic powder or wire serves only as a matrix, and carbides or oxides can be used as reinforcing particles. Using laser cladding of this type of material, it is possible to quickly and thoroughly regenerate worn machine parts, as well as to create both single- and multi-layer coatings for parts operating in extremely difficult conditions. The laser cladding method allows for the introduction of filler material directly under the action of a laser beam. This technology requires the use of an inert shielding gas, which prevents oxidation of the fed filler material and the surface of the base material. Thanks to the use of modern 5-axis laser machining centers or numerically controlled manipulators using computer-aided CAD/CAM systems, the filler material can be precisely applied to create a coating with high surface quality or to reproduce a complex product shape. Proper selection of processing parameters, such as laser power, beam diameter, feed speed, powder feed rate, and gas feed rate, allows for the creation of a weld with distinct mechanical properties and a microstructure distinct from the surface of the base material. The use of appropriate materials in laser hardfacing technology significantly improves the operational properties of the welded part, particularly its resistance to frictional wear. Because the clad is heated and solidified at high speed in an inert gas atmosphere, high metallurgical purity of the surface and a fine-grained metal structure of the newly formed layer are ensured. Rapid cooling of the still-liquid layer results in greater surface hardness compared to other hardfacing methods while maintaining an identical chemical composition. The advantage of high-power density in the laser beam limits the heat-affected zone, as well as the stresses and deformations acting on the component [4,5].

Materials and Parameters

Metco 41C powder was used for the cladding process. This austenitic powder consists primarily of nickel and chromium. Coatings produced with this powder exhibit good wear and corrosion resistance. They are light in color and are used for the regeneration of worn or mismatched parts. They are characterized by work hardening. The appropriate molybdenum content ensures high resistance to pitting and crevice corrosion, as well as high tensile strength at high temperatures and impact strength. Fig.1 shows its microstructure and grain size.

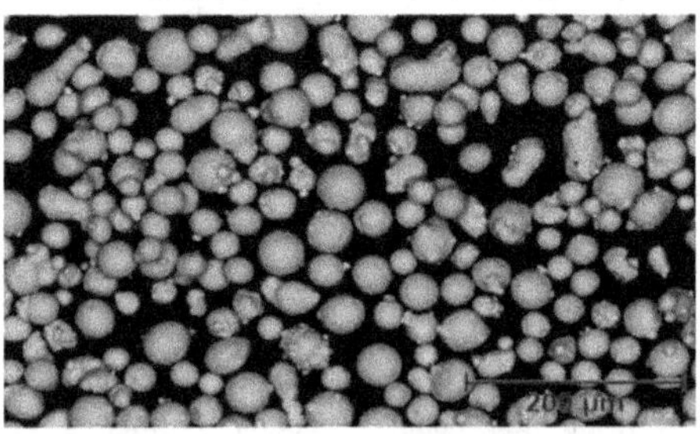

Fig. 1. Microstructure and grain size of Metco 41C powder.

During laser cladding, the filler material and the substrate material are simultaneously melted, mixing to form a new layer of material that is firmly bonded to the substrate. The power and diameter of the laser beam represent the beam's power density. In laser cladding, in most cases, this density ranges from 102 to 106 W/cm^2 wich density was used during experiment. The workstation where the experiments were conducted is equipped with a TRUMPF LASERCELL 1005 laser. Two sleeves with a diameter of 79.5mm, a length of 152mm, and a wall width of 2mm were used for the tests. To simulate mechanical damage to the outer surface of the sleeves, 1mm-deep cuts were introduced. The irregularity of the cut edges is intentional to accurately reflect the actual damage to the components in use as shown in Fig.2.

Fig. 2. Method of making cuts on the test sleeve.

Laser Cladding of Engine Sleeves

The experiment was divided into two parts. In the first, the weld deposits were applied to the sleeves without any pretreatment. In the second, the sleeve was preheated to a temperature of approximately 330°C and then, after the weld deposits were applied, it was cooled with the furnace turned off for approximately 2 hours to prevent cold cracks.

During experimental work on cladding without prior heating, many disturbing results were observed, depending on the surfacing process parameters such as laser power, beam feed speed, gas flow rate or powder flow rate, as shown in Fig.3. Numerous spalling and surface irregularities of the clad surface were observed due to insufficient fusion of the filler material, cracks extending into the liner's base material, insufficient filling of the liner cavity due to insufficient powder flow, visible transverse cracks in the clad, numerous spatters, and irregular clad edges. Cracks resulting from the cladding process on unheated material disqualify such a liner because they would cause engine compression pressure drops during operation and allow coolant from the water jacket to enter the cylinder. A linear analysis of the elemental distribution at the weld-liner junction was conducted. Using a filler material with a different chemical composition than the liner material allows for precise determination of the degree of mixing of elements such as chromium and nickel. Analyzing the linear distributions of elements in the sample, it can be seen that the nature of the connection is diffusive and not, as might be expected, adhesive.

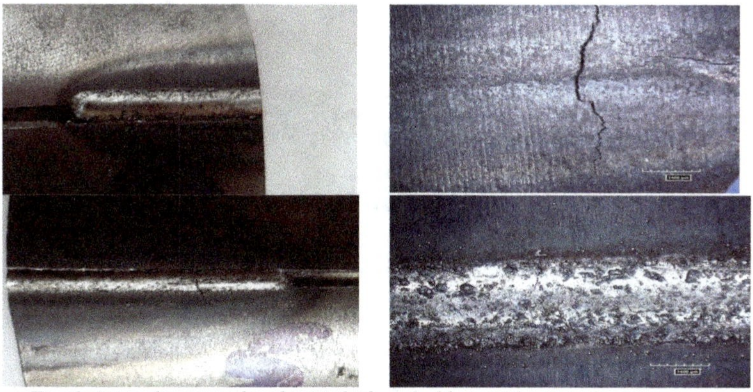

Fig. 3. Disturbing results obtained after cladding without preheating.

In tests for depositing the weld overlay onto a sleeve heated to 330°C, the parameters were selected similarly to those that achieved the best results in the first experiment. This allowed for the entire kerf to be filled in just four beam passes. The clad has even edges and perfectly fills the entire groove width. To reduce the heat-affected zone, the weld deposition process was divided into two cycles of two beam passes each. As expected, the heat-affected zone was almost halved compared to the test without preheating. When analyzing the linear distribution of elements in the clad overlay applied to the heated material, a sharp increase in chromium and nickel counts can be observed over a distance of approximately 50 μm. This is due to the adhesive nature of the weld bond with the base material. This is another advantage of welds deposited onto the heated sleeve surface. No cracks were observed as shown on Fig.4.

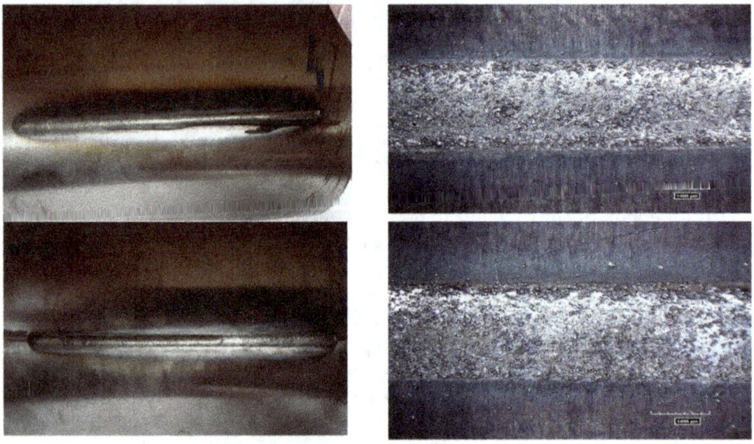

Fig. 4. Results of laser cladding with preheating.

Examination of the clads created by preheating the base material using an optical profilometer confirms the correct clad shape. Only minor surface imperfections, not exceeding 30μm, are visible, as shown in Figure 5.

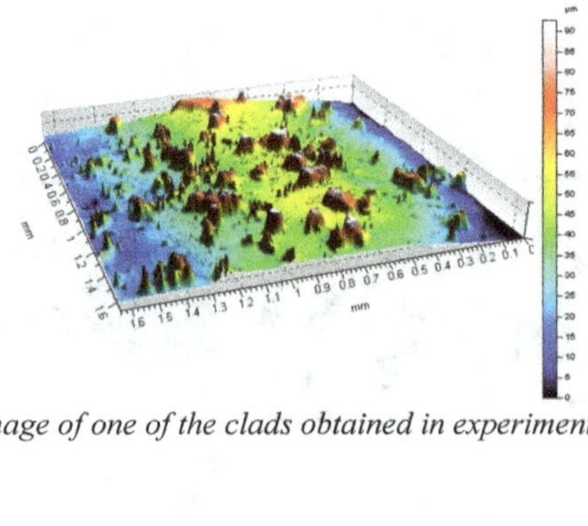

Fig. 5. Isometric image of one of the clads obtained in experiment with preheating.

Summary

The use of laser cladding technology for the regeneration of the external surfaces of engine liners is possible; however, it presents numerous challenges. The cast iron from which the liners are made tends to develop cold cracks during the cooling of the material after the cladding process. The small wall thickness—approximately 1 mm after damage—required extremely precise parameter selection.

In most trials without preheating, cold cracking occurred both in the cladding material and in the base material of the liner. Moreover, when the laser power exceeded 2 kW, full penetration of the material occurred. This caused damage to the honed structure of the inner surface of the liner, resulting in a loss of perfect roundness. The shape of the best-quality clad bead, as examined using a profilometer, was satisfactory, and the amount of spatter from the filler material remained within acceptable limits. Trials conducted with preheating yielded significantly better results. Although cold cracks were still present, they were smaller and occurred far less frequently. In two of the tests, the inner side of the liner remained unaffected, and no thermal changes were observed in the material structure. Analysis of the linear distribution of elements within the cladding layer showed satisfactory adhesion to the substrate. Considering the results of this study, it can be concluded that the regeneration of external surfaces of engine liners using laser cladding is quite difficult but feasible. However, a question arises as to whether this process is economically viable. Liners for mass-produced engines are relatively inexpensive to purchase. This technology may prove useful in the case of vintage engines, for which replacement parts are no longer manufactured, or in large-displacement engines, where the cost of producing a new liner significantly exceeds that of its regeneration.

References

[1] A. Posmyk, S. Witaszek, Ocena właściwości tribologicznych skojarzenia żeliwo-kompozyt w warunkach eksploatacji silnika spalinowego, Tribologia, 39 (2008) 281-288.

[2] T. Burakowski, T. Wierzchoń, Surface engineering of metals – principle, equipment, technology, CRC Press, Boca Raton, 1999.

[3] S.A. Alves et al., Enhanced tribological performance of cylinder liners made of cast aluminum alloy with high silicon content through plasma electrolytic oxidation, Surface and Coatings Technology 433 (2022) art. 128146. https://doi.org/10.1016/j.surfcoat.2022.128146

[4] A. Klimpel, Technologie laserowe w spawalnictwie, Wydawnictwo Politechniki Śląskiej, Gliwice, 2011.

[5] B. Zabala et al., Friction and wear of a piston ring/cylinder liner at the top dead centre: Experimental study and modelling, Tribology International 106 (2017) 23-33. https://doi.org/10.1016/j.triboint.2016.10.005

Materials Research Proceedings 62 (2026) 222-227 https://doi.org/10.21741/9781644904015-29

Machine Learning Methods and Hybrid Models in Supply Chain Forecasting

Marek LEWIŃSKI[1,2,a *], Katarzyna A. MŁYNARCZYK[2,b]

[1]Cracow University of Technology, Al. Jana Pawła II 37, 31-864 Kraków, Poland

[2]GE HealthCare, ul. Życzkowskiego 20, 31-864 Kraków, Poland

[a]mlewinski@pk.edu.pl, [b]katarzyna.mlynarczyk025@outlook.com

Keywords: Machine Learning, Forecasting, Supply Chain Forecasting, Deep Learning, Transformers, Hybrid Models

Abstract. Supply chain forecasting is a broad domain embracing number of topics like demand planning, pricing, supply and service prediction. It is also an integral part of many processes that those affect. This paper describes machine learning methods used in forecasting like regressions, random forest, XGBoost but also showcases the newest contributions to deep learning architectures used for forecasting like RNNs, LSTMs as well as state-of-art Transformer models created for long-term forecasting: TimeGPT and PatchTST. Paper also describes another robust approach for time series forecasting - hybrid approaches that combine both statistical and machine learning models alongside their applications.

Introduction

In current data-rich world, classical statistical forecasting methods while still strong and widely utilized often do not have enough predictive power to model very complex patterns, cannot be easily extended to incorporate additional exogenous information and features, and are limited by the assumptions that govern their internal structure [1]. Machine learning models allow forecasters and supply chain planners to perform more informed decision making – (1) by incorporating both external indicators and internal features, (2) utilizing cross-learning between multiple products (i.e. recognizing patterns governing the demand of one product, and applying them to newly introduced products), and (3) gaining insight into the models' reasoning using interpretable machine learning methods, such as LIME (*Local Interpretable Model-Agnostic Explanations*) and Shapley values, that provide information on the relative importance and the impact of particular feature value. Machine learning models prove to be more scalable than classical statistical models, enable for faster experimentation due to more structured tuning process and shorter fitting time (especially when utilizing *global forecasting models*, which even combined with hyperparameter tuning requires only a fraction of time required by *local forecasting* approach usually utilized by statistical methods), and work seamlessly with different approaches – hierarchical forecasting, probabilistic forecasting, real-time and hybrid modelling [1]. The aim of this article is to provide an overview of machine learning and deep learning models utilized for forecasting, combining them with statistical methods in powerful hybrid architectures, and show their applicability to contemporary supply chain planning.

Machine Learning Models in Forecasting Applications

Classical statistical forecasting models, such as SARIMA, incorporate the characteristics of time series data into their structure—for example, by assuming an ARMA (autoregressive/moving average) process [2]. Due to these assumptions, the forecasted values can be computed directly using the model's mathematical formulation. In contrast, machine learning models do not offer a straightforward method for calculating the next value. Instead, standard machine learning models

Materials Research Proceedings 62 (2026) 222-227 https://doi.org/10.21741/9781644904015-29

approximate a parametric function under the assumption that data points are i.i.d. (independent and identically distributed). This assumption is invalid in forecasting because [3]:

- Individual data points depend on previous observations (e.g., through autoregression).
- The underlying distribution may change over time due to the heteroskedastic or non-stationary nature of the data-generating process.

To overcome this limitation, we can either encode the temporal characteristics of the data into a tabular format suitable for supervised learning, using techniques like lag features and rolling statistics [1, 2], or employ sequential models such as Recurrent Neural Networks (RNNs), Long Short-Term Memory networks (LSTMs), or Transformer architectures, which are designed to capture temporal dependencies directly.

From the standard machine learning perspective, the most commonly utilized models are linear regression (with optional regularization terms) and ensemble modeling [4]:

- *Linear regression* – models the target variable as a weighted linear combination of previous time steps as in Eq. 1. This approach is often utilized due to its interpretability – the weight associated with a particular variable corresponds to its relative importance, ease of use and native support of multivariable forecasting (exogenous features can be added as a regular variables). Eq.2. presents the case with time-aligned exogenous features (i.e. sharing the indexing set).

$$\hat{y_t} = \beta_h y_{t-1} + \beta_{n-1} y_{t-2} + \cdots + \beta_1 y_{t-h} + \beta_0 = \sum_{i=1}^{h} \beta_i y_{t-h+i} + \beta_0. \quad (1)$$

$$\hat{y_t} = \sum_{i=1}^{h} \beta_i y_{t-h+i} + \sum_{i=1}^{h} \alpha_i x_{t-h+i} + \beta_0 \quad (2)$$

- *LASSO regression* – by addition of $\mathcal{L}_1$ regularization term to the linear regression model objective function, as shown in Eq. 3. (with X representing generalized features, i.e. including temporal features), a sparse model can be obtained, i.e. a model with fewer non-zero coefficients assigned to the variables deemed significant by the optimization algorithm. This model enables automated feature selection (based on their statistical importance, not necessarily their business meaning or significance) and results in a simpler, interpretable model with respect to Ockham's Razor principle.

$$\min_{\beta} \frac{1}{2n_{samples}} ||X\beta - y||_2^2 + \alpha ||\beta||_1 \quad (3)$$

- *Random forest* – ensemble built from multiple decision trees that utilizes bagging (*bootstrap aggregation*) and random subspaces method (i.e. each tree is trained on a different subset of *features*, not only different samples – incorporating this approach reduces the chances of the model to overfit). Random forest models allow for estimation of relative feature importance – statistically, features that frequently appear near the top of decision trees are considered more influential, as they contribute more to reducing impurity during splits (as measured by MDI – *Mean Decrease in Impurity*). Additionally, random forests can be easily extended with additional temporal and exogenous features.
- *XGBoost* – boosting-based method with decision trees built sequentially, where each tree is trained to minimize the residual errors of its predecessors [5]. XGBoost excels in modelling complex, non-linear relationships with reasonably high predictive accuracy, efficient training and inference speed (due to parallelization) while remaining robust against overfitting. Similarly to random forests, XGBoost is scalable and can incorporate exogenous features.

Materials Research Proceedings 62 (2026) 222-227 https://doi.org/10.21741/9781644904015-29

Deep Learning Models

Identifying very complex patterns over a long horizon, with multi-seasonal effects (i.e. daily, weekly, quarterly etc. seasonal patterns) and non-linear relationships among multiple variables might not be feasible or efficient using classical machine learning algorithms. In such cases the deep learning forecast might achieve better results. Deep learning algorithms for forecasting can be divided into three categories:

- based on recurrent neural networks (RNN, LSTM, GRU), which are designed to capture temporal dependencies through recursive edges and hidden states.
- utilizing feed-forward networks that try to model the time series as tabular data
- contemporary transformer-based architectures (which are discussed later in this paper) that leverage the attention mechanism to model long-horizon temporal dependencies.

On the other hand, neural networks tend to struggle with data irregularities, sparsity and by their nature are more challenging in terms of interpretability. Below we describe several representatives of the aforementioned classes of DL models:

- RNN (Recurrent Neural Network) – in contrast to classical FFN (*Feed-Forward NN*) it provides its own output at time *t* as the input for step *t+1*. Because of that, the network can the patterns not only from direct input, but also from a temporal context (i.e. what had happened in the previous step). As a side effect of their design, they suffer from *exploding gradients problem* which limits their ability to model long-term relationships.
- LSTM/GRU units – *Long-Short Term Memory / Gated Recurrent Units* – consists of internal state that tries to model the long-term relationships – this internal state is "buffered" from the direct backpropagation of values; thus, it reduces the exploding/vanishing gradients problem while improving the network's ability to recognize patterns over a longer horizon [6].
- N-BEATS - *Neural Basis Expansion Analysis for Time Series Forecasting* – in contrast to RNN networks, it processes the sequence using sliding window approach with feed-forward layers. Model structure is divided into stacks, each of which learns different time series characteristic – trend, seasonality etc. Each stack consists of multiple blocks that implement learning on residuals to improve accuracy. N-BEATS model is interpretable in the sense that it is possible to extract specific components of the forecast from the stacks [7].
- N-HiTS - *Neural Hierarchical Interpolation for Time Series Forecasting* – this architecture has been proposed to fight the volatility of the forecasts over a long horizon. It is done by a mechanism called *hierarchical interpolation*, which decomposes the input signal using multiple sampling rates (e.g. yearly, quarterly, monthly) to identify the underlying structure and then reconstruct it back to form the prediction – the approach is similar to wavelet decomposition. N-HiTS model excels with complex patterns and long forecasting range [8].

Hybrid Approaches

Real world data contains complex patterns that can be both linear and nonlinear at the same time. Using statistical approach allows the capture of simple patterns, while machine learning methods have application in more intricate non-linearities. When dealing with multivariate time series more complex tools are needed. Therefore, alongside statistical and machine learning methods emerged another option combining those two approaches. Hybrid modeling can join both different types of method (e.g. statistical and machine learning) but also the same types of models (statistical and statistical). The aim of hybrid modeling is to join advantages of used methods to achieve more accurate and robust results. Statistical models can be used to capture the linear aspects of time series, then machine learning methods are being trained to handle nonlinearities. Combining models' outputs can increase the forecast accuracy significantly. This claim is backed up by multiple hybrid models outperforming pure statistical and machine learning models in M4 competition [9].

There are two types of hybrid modeling: residual learning and model-as-a-feature. The pipelines of both are shown on Fig.1. In the first approach first model is trained on dataset. After the training the values of fitted model are retrieved. Next, fitted values are used for each record to calculate residuals by subtracting the fitted value from corresponding historical value. After that, final prediction is made by combining outputs of both models. This approach was used to combine predictive power of ARIMA and ANN for short-term forecasting series. ARIMA model was designed to capture linear dependencies while deep Artificial Neural Network was designed to capture nonlinearities [10]. The same approach was used in [11], with ARIMA model capturing linear patterns and XGBoost model handling nonlinear residuals.

Model as a feature is an approach that is based on using fitted values, parameters or learned features as another input along the historical values for the model applying concept of meta-learning. This method helps provide second model with a cleaner signal, propagates information and redacts noise [12].

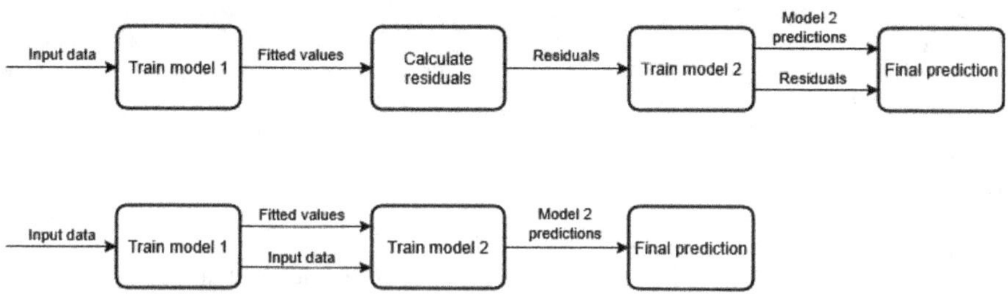

Fig. 1. *Comparison of pipelines of hybrid modeling approaches shown on example of two models for: a) Residual approach and b) Model as a feature approach.*

Combining models can increase accuracy, but there is a trade-off in interpretability. Although studies show that using hybrid models can be effective, using more models and adding complexity to the structure can be ineffective and attribute to higher bias and overfitting [13].

Advances with Generative AI – Transformer Models

Although deep learning models might be effective at predicting the multivariate time series, they can have less generalization capabilities. Recently another deep learning models family emerged offering better efficiency and scalability – transformers. These models are mainly used in natural language processing and text generation due to its capabilities of processing multiple sequential inputs simultaneously. Transformer architecture introduces many mechanisms that can mitigate RNNs and LSTMs weaknesses. Embedding, self-attention and encoder-decoder structures boost predictive power of models while increasing efficiency. Application of self-attention mechanism helps models learn both short-term and long-term patterns and exceed the performance of other deep learning architectures [14]. Below we describe state-of-the-art transformer models used for time series forecasting:

- TimeGPT – first foundation model for time series forecasting introduced by Garza et al. (2024). TimeGPT is a pre-trained architecture on diverse datasets designed to be used across various domains to detect anomalies and produce accurate predictions applying transfer learning principle. Thanks to that it is enabled for zero-shot inference – being able to forecast without prior training [15].
- PatchTST – transformer model that introduces channel independent patch architecture. The data points are aggregated to subseries-level patches to improve model's understanding of different data points correlations and semantic information. Additionally, every channel that

appears in multivariate data is tokenized independently. Those mechanisms result in improved long-term forecasting [16].

There is a pending scholar discussion regarding the use of transformers in time series forecasting suggesting that the usage of self-attention mechanism and tokenizing sub-series leads to temporal information loss. This can especially affect accuracy of long-term series forecasting (LTSF) [17].

Applications in Supply Chain Planning

Use of machine learning and deep learning methods within supply chain domains is becoming increasingly significant. As the COVID-19 pandemic and global instability impacted most of global supply chains, practitioners and planners highlighted the need to utilize more data for planning and forecasting, especially macroeconomic factors, leading indicators and endogenous data. This trend is visible in This trend is visible in Douaioui et al. (2024) – a bibliometric review that shows clearly the growing interest in ML-based forecasting solutions, with the most significant raise after 2020 [18]. The most common themes (motor themes) were *deep learning*, *risk management* and *demand forecasting*. Also, the study highlights the accuracy improvements of machine learning models, especially when combined with other models into hybrid solutions. As the authors highlight, the use of machine learning and artificial intelligence tends to increase the competitive advantage of many firms, with the most prominent application being the demand forecasting. In another article, Babai et al. (2025) identify the inventory management as the second most common application of ML techniques, with transportation, production, and procurement being next in terms of interest [19]. This article also highlights an interesting finding – growing interest in *reinforcement learning* for automated decision making in SC applications. Deep learning and artificial neural networks are prominently utilized in other parts of the SCOR model too – spanning from forecasting supplies, production capabilities, predictive maintenance, service part requirements, and returns [19]. All forecasting and machine learning-powered tasks are heavily reliant on data quality and require interpretability for better informed decision making and regulatory compliance. Haque (2025) highlights this as one of the core issues with machine learning systems adoption in supply chain management [20]. Additional use-cases that are highlighted in this article are: (1) the alignment of production scheduling to demand forecasts, especially in push systems, and (2) delivery time prediction for higher customer satisfaction and supply chain predictability [20].

Summary

Presented paper explains the benefits of machine learning forecasting over traditional statistical methods. It contains a description and an overview of several most popular machine- and deep learning models utilized in supply chain forecasting area. Additionally, hybrid approaches that consist of multiple phases of modelling are presented, highlighting that a combination of multiple models and approaches can yield better results when compared to the individual models. Contemporary forecasting methods that can incorporate multiple exogenous features, real-time data and comprehensive feature engineering found their applicability in various domains of supply chain management, particularly within demand forecasting.

References

[1] M. Lewiński, S.A. Lariane, Statistical Forecasting Methods and Machine Learning Models in Hierarchical Forecasting for Supply Chain Applications, Materials Research Proceedings 45 (2024) 286-295. https://doi.org/10.21741/9781644903315-33

[2] F. Petropoulos et al., Forecasting: theory and practice, International Journal of Forecasting 38 (2022) 705-871. https://doi.org/10.1016/j.ijforecast.2021.11.001

[3] R.J. Hyndman, G. Athanasopoulos, Forecasting: principles and practice, 3rd Ed., OTexts, Melbourne, Australia, 2021. http://OTexts.com/fpp3. Accessed on 2025-07-31.

[4] R.P. Masini, M.C. Medeiros, Machine Learning Advances for Time Series Forecasting, arXiv:2012.12802 [econ.EM], https://doi.org/10.48550/arXiv.2012.12802

[5] T. Chen, C. Guestrin, XGBoost: A Scalable Tree Boosting System, in: Proc. 22nd ACM SIGKDD Int. Conf. Knowl. Discovery and Data Mining (KDD '16). Association for Computing Machinery, New York, NY, USA, 2016, 785–794. https://doi.org/10.1145/2939672.2939785

[6] S. Hochreiter, J. Schmidhuber, Long Short-Term Memory. Neural Comput. 9 (1997) 1735-1780. https://doi.org/10.1162/neco.1997.9.8.1735

[7] B.N. Oreshkin et al., N-BEATS: Neural basis expansion analysis for interpretable time series forecasting, arXiv:1905.10437 [cs.LG], https://doi.org/10.48550/arXiv.1905.10437

[8] C. Challu et al., N-HiTS: Neural Hierarchical Interpolation for Time Series Forecasting, arXiv:2201.12886 [cs.LG], https://doi.org/10.48550/arXiv.2201.12886

[9] S. Makridakis et al., The M4 Competition: Results, findings, conclusion and way forward, Int. J. Forecasting 34 (2018) 802-808. https://doi.org/10.1016/j.ijforecast.2018.06.001

[10] Ü.Ç. Büyükşahin, Ş. Ertekin, Improving forecasting accuracy of time series data using a new ARIMA-ANN hybrid method and empirical mode decomposition, Neurocomputing 361 (2019) 151-163. https://doi.org/10.1016/j.neucom.2019.05.099

[11] P. Li, J.-S. Zhang, A New Hybrid Method for China's Energy Supply Security Forecasting Based on ARIMA and XGBoost. Energies, Vol. 11, 2018, https://doi.org/10.3390/en11071687.

[12] W. Tingyu, L. Wenyang, X. Jun, Supply chain sales forecasting based on lightGBM and LSTM combination model, Industrial Management & Data Systems 120 (2020) 265-279. https://doi.org/10.1108/IMDS-03-2019-0170.

[13] X. Wang et al., Forecast combinations: An over 50-year review, Int. J. Forecasting 39 (2023) 1518-1547. https://doi.org/10.1016/j.ijforecast.2022.11.005

[14] J. Kim et al. A comprehensive survey of deep learning for time series forecasting: architectural diversity and open challenges, Artif. Intell. Rev. 58 (2025) art.216. https://doi.org/10.1007/s10462-025-11223-9

[15] A. Garza et al., TimeGPT-1, arXiv:2310.03589 [cs.LG], https://doi.org/10.48550/arXiv.2310.03589

[16] Y. Nie et al., A Time Series is Worth 64 Words: Long-term Forecasting with Transformers, arXiv:2211.14730 [cs.LG], https://doi.org/10.48550/arXiv.2211.14730.

[17] A. Zeng et al., Are Transformers Effective for Time Series Forecasting?, arXiv:2205.13504 [cs.AI], https://doi.org/10.48550/arXiv.2205.13504.

[18] K. Douaioui et al., Machine Learning and Deep Learning Models for Demand Forecasting in Supply Chain Management: A Critical Review. Appl. Syst. Innov. 7 (2024) art.93. https://doi.org/10.3390/asi7050093

[19] M.Z. Babai et al., On the use of machine learning in supply chain management: a systematic review, IMA J. Manag. Mathematics 36 (2025) 21-49. https://doi.org/10.1093/imaman/dpae029

[20] S.R. Haque, Machine Learning Applications in End-To-End Supply Chain Management: A Comprehensive Review, Int. J. Sci. Res. Manag. 13 (2025) 2226-2241. https://doi.org/10.18535/ijsrm/v13i06.ec03

Materials Research Proceedings 62 (2026) 228-236 https://doi.org/10.21741/9781644904015-30

Analysis of Key Performance Indicators for High-Pressure Die Casting Process in the Context of Industry 4.0

PACANA Andrzej[1,a *], CZERWIŃSKA Karolina[2,b] and Renata DWORNICKA[3,c]

[1]Rzeszow University of Technology, The Faculty of Mechanical Engineering and Aeronautics al. Powstańców Warszawy 8, 35-959 Rzeszów, Poland, ORCID: 0000-0003-1121-6352

[2]Rzeszow University of Technology, The Faculty of Mechanical Engineering and Aeronautics al. Powstańców Warszawy 8, 35-959 Rzeszów, Poland, ORCID: 000-0003-2150-0963

[3]Cracow University of Technology, Faculty of Mechanical Engineering, ul. Warszawska 24, 31-155 Kraków, Poland, ORCID: 0000-0002-2979-1614

[a]app@prz.edu.pl, [b]k.czerwinska@prz.edu.pl, [c]renata.dwornicka@pk.edu.pl

Keywords: Process Optimization, Data Analysis, Process Monitoring, Smart Manufacturing, Quality Improvement, Real-Time Monitoring, Mechanical Engineering

Abstract. The aim of the study was to create a model enabling the identification of key performance indicators (KPIs) for the die casting process in aluminum industry companies, as well as their implementation and visualization using Industry 4.0 tools. The model's assumptions were based on the idea of smart manufacturing. The proposed indicators concerned the key stages of the die casting process. The model integrates indicators' data from sensors and control systems, enabling real-time monitoring and data-driven decision-making. The presented KPI classification and evaluation procedure supports process optimization, improvement of energy efficiency, and the quality of aluminum alloy products. The implementation of digital KPI monitoring tools (visualization dashboards) increases resource efficiency, process stability, and supports sustainable manufacturing goals in the foundry industry. The model can be used by engineers, Industry 4.0 specialists, and management staff in the analysis and improvement of processes and data-driven decision-making.

Introduction

In the face of ongoing technological change and growing competition, modern manufacturing companies are forced to continuously optimize their production processes, minimize costs, and improve product quality. Implementing Industry 4.0 solutions allows for increased operational efficiency and competitiveness [1-3]. The foundry industry is one of the key sectors of the metallurgical industry. It plays a crucial role in supply chains for the engineering, automotive, energy, and aviation industries [4]. In recent decades, the development of this industry has been linked to the implementation of modern production control methods, process automation, and general technological progress [5, 6]. Contemporary foundries are increasingly implementing digital technologies such as automation, robotization, computer simulations, and real-time monitoring systems, which improve the productivity, quality, and efficiency of production processes [7,8]. At the same time, despite the visible signs of ongoing technological transformation, the foundry sector is still considered conservative and relatively slow to adapt to innovation compared to other industrial sectors [9]. This is the result of several interrelated technical, economic, and organizational reasons, among which the following can be distinguished: high investment costs with a long payback period, the complexity of the technological processes involved, difficult environmental conditions, and a low level of integration of production systems [10]. The conditions for the operation of the foundry industry result from the considerable complexity of technological processes and the need to ensure stable material and thermal conditions while striving to ensure adequate product quality and reduce environmental impact. Each stage of the production of cast products, from the preparation of the alloy, through the mold filling process, to cooling and final processing, requires precise monitoring of process parameters [11-13]. This necessity favors the implementation of modern monitoring and management systems based on the principles of Industry 4.0. The integration of automation, robotization, data analytics, and intelligent control systems not only contributes to improving quality and efficiency but also supports the implementation of the idea of sustainable manufacturing, which combines technological, economic, and environmental aspects in order to implement the principles of responsible and sustainable manufacturing [14,15].

The considerable complexity and dynamic nature of the die casting process mean that its effective management requires the use of accurate and measurable tools for monitoring and evaluating efficiency. The high sensitivity of the casting process to deviations in technological parameters (e.g., melt temperature, injection pressure, filling and cooling speed) means that even relatively small deviations can result in an increase in the frequency of casting defects and an increase in the proportion of material lost in the process [16]. In this respect, the implementation of key performance indicators (KPIs) becomes essential for conscious process monitoring, early identification of irregularities, and detection of areas requiring optimization. The use of an adequate set of KPIs allows production data to be transformed into constructive information supporting accurate decision-making [17]. The aim of the study was to create a model enabling the identification of key performance indicators for the die casting process in modern aluminum industry enterprises, as well as their implementation and visualization using Industry 4.0 tools, such as navigation and visual screens for real-time process monitoring.

Methodology of Research

Implementing appropriate KPIs in a manufacturing company is a systematic process involving several key stages. Fig.1 schematically illustrates the stages of implementing the model for identifying and implementing KPIs for the die casting process.

Table 1. Characteristics of key performance indicators (KPIs) in the die casting process

Stage of the process	Name	Indicator	Expected value	Description / Meaning
Melting and alloy preparation	Metal Temperature Stability Index	$MTSI = \frac{\sigma T}{T_{avg}} \cdot 100\%$ where: σT – standard deviation of metal temperature in a given period [°C]; T_{avg} – average metal temperature during the same period [°C]	metal temperature: – in the crucible: ≤2-3% – in the feed channel: ≤1-2%	It indicates the stability of the liquid metal temperature before pouring.
Preparing the mold	Lubricant Coverage Efficiency	$LCE = \frac{A_{covered}}{A_{total}} \cdot 100\%$ where: $A_{covered}$ – mold surface evenly coated with release agent; A_{total} – total mold area	≥95%	It demonstrates the uniformity of the release agent coverage on the mould.
Injection (filling the chamber, filling the mould)	Injection Pressure Effectiveness	$IPE = \frac{P_{avg}}{P_{nom}} \cdot 100\%$ where: P_{avg} – average injection pressure during the cycle; P_{nom} – nominal pressure recommended for casting	95-105%	It demonstrates the efficiency and precision of the injection process.

Table 1. Characteristics of key performance indicators (KPIs) in the die casting process

Stage of the process	**Name**	**Indicator**	**Expected value**	**Description / Meaning**
Pressure and solidification	Porosity Index	$$PI = \frac{V_{por}}{V_{casting}} \cdot 100\%$$ where: V_{por} – pore volume in casting; $V_{casting}$ – total casting volume	≤ 2%	It indicates the level of porosity in the casting volume.
Ejection and mould opening	Ejection Efficiency	$$EE = \frac{N_{success}}{N_{totalc}} \cdot 100\%$$ where: $N_{success}$ - number of cycles in which the casting was correctly ejected from the mould (without sticking, damage or scratches); N_{total} - total number of production cycles (all discharges in a given period)	≥98%	It demonstrates the efficiency and stability of the process of pushing castings out of the mould.
Cutting and cleaning	Casting Yield	$$CY = \frac{M_{casting}}{M_{casting} + M_{sprue}} \cdot 100\%$$ where: $M_{casting}$ – mass of the finished, usable casting; M_{sprue} – the mass of sprues, gates and other elements of the gating system that are not part of the finished casting	70-85%	It demonstrates the efficiency of metal utilisation in the process.
Maintaining form and press	Die Life Utilization	$$DLU = \frac{N_{cyclesu}}{N_{cyclesm}} \cdot 100\%$$ where: $N_{cyclesu}$ – number of cycles performed on a given mould; $N_{cyclesm}$ – maximum predictable number of mould cycles before replacement/regeneration is required	≤80% before regeneration	It indicates the degree of utilisation of the casting mould during its life cycle.
Quality control	Defect Rate	$$DR = \frac{N_{defective}}{N_{produced}} \cdot 100\%$$ where: $N_{defective}$ – number of castings that do not meet quality requirements; $N_{produced}$ – number of castings produced in a given period	≤1-3%	It demonstrates the quality, stability and maturity of the process.

Materials Research Proceedings 62 (2026) 228-236 | https://doi.org/10.21741/9781644904015-30

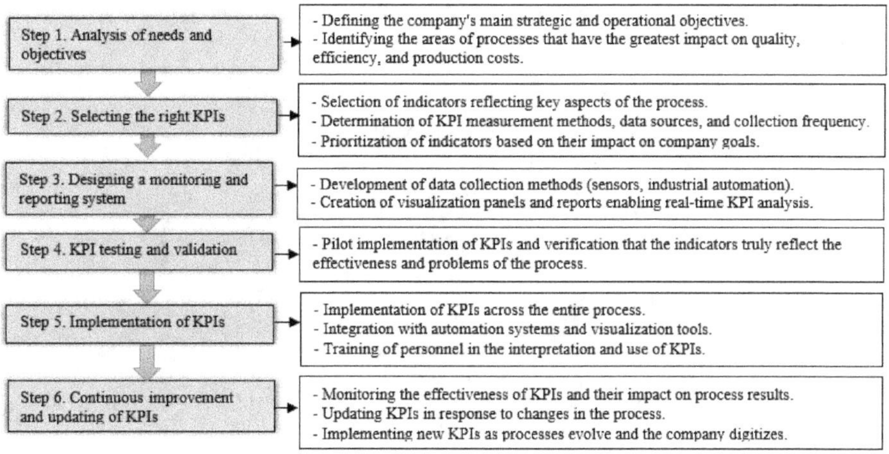

Fig. 1. Diagram of the model for identifying and implementing KPIs for the pressure casting process. Source: own study.

The adopted model for identifying and implementing KPIs is cyclical. This means that the indicators are not a fixed set. With the development of technology, changes in production processes, or the emergence of new strategic goals, it becomes necessary to regularly modify and adapt them to the changing needs and goals of the company. The iterative approach to KPIs is part of the concept of continuous process improvement, thus forming the basis for building a growing company capable of efficiently adapting to changing industrial conditions.

Results

The model was tested at a foundry located in south-eastern Poland. The company offers light alloy products for industries such as robotics, energy, automotive, aviation, and railways. The tests were carried out in the first quarter of 2025.

Based on the literature on the subject of the casting process and the use of KPIs in industrial environments, a potential set of KPIs was created [18-23]. Meetings with management staff were also organized to identify available information and determine the degree of implementation of Industry 4.0 solutions. The KPI selection process included a survey assessing the importance of individual indicators on a Likert scale (where 1 meant the least important and 5 the most important). The results were then sorted according to average ratings, and the most relevant ones were verified with the participation of representatives of the management and executive staff. Of the 55 surveys sent out, 54 were considered valuable, reflecting a very high response rate. The respondents were selected in cooperation with the management and the HR department, which allowed for the involvement of participants who were most important for the implementation of the study. The selected KPIs are characterized in Table 1. Their significant number resulted from the lack of maturity in the conscious management of the die casting process. Once the expected level of quality has been achieved and the process stabilized, the number of indicators will be reduced.

The presented KPIs enable monitoring of the efficiency, quality and safety of the die casting process at every stage – from melting the alloy, through injection and solidification, to ejection and quality control. A visualisation dashboard (Figure 2) has been developed for the indicators in Table 1.

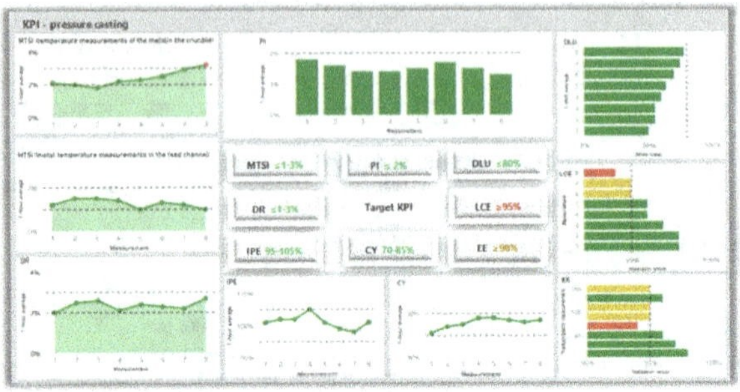

Fig. 2. A visualization dashboard developed for a set of KPIs for the pressure casting process. (source: own study)

The development of a visualization dashboard for KPIs in the die casting process and its implementation enables transparent, real-time monitoring of the process, identification of deviations from established standards, and response to identified quality issues. Supervision using a visualization dashboard facilitates the analysis of causes and effects (e.g., the correlation between unstable melt temperature and the porosity of cast products) and supports decision-making at the operational and strategic levels (e.g., optimization of injection parameters or improvement of energy efficiency in the melting and casting process). In addition, real-time visualization of key aspects of the process improves internal communication, automates the reporting process, and enables easy examination of trends and the effects of implemented improvements, which supports the development of a culture of continuous improvement.

Economic and Technological Impact in Other Industries

Creating a "digital twin" using, among others, the DOE methodology [23,24] supported by heuristics [25] and an expert system [26] is one of the key elements of Industry 4.0 [27,28], as it allows for free modeling of the behavior of a physical system [29] even in the case of incomplete information [30,31] requiring intensive data imputation [32,33]. This is effective both in the case of foundries [34,35], forges [36], coating technologies [37,38] and welding [39]. Creating such a complete model naturally allows for process optimization and implementation of the "lean management" concept [40,41] with the support of advanced analytical methods [42,43], while maintaining or improving quality [44-46] and reducing waste [47].

Summary

The aim of the study was to create a model enabling the identification of key performance indicators for the die casting process in modern aluminum industry enterprises, as well as their implementation and visualization using Industry 4.0 tools, such as navigation and visual screens for real-time process monitoring.

Positive verification of the developed model confirmed its substantive correctness and consistent structure. The study proposed a comprehensive methodology for assessing the correctness of the die casting process through a dedicated set of key performance indicators (KPIs). In addition, a design for visualization panels linked to KPIs was developed, enabling continuous real-time transmission of information, which ensured a high level of transparency and clarity of the data presented.

The developed KPI identification model for the die casting process can be successfully implemented in aluminum industry companies. The parameters displayed in real time and the collected data support activities aimed at improvement, loss reduction, and further implementation of modern digital solutions in line with the Industry 4.0 concept. Future research will focus on expanding the set of KPIs to include indicators related to energy efficiency and loss minimization, and then on adapting the developed KPI model to foundries with varying production scales and technologies.

References

[1] TY. Chai, SY. Cheng, Intelligent operational decision-making in industrial process: development and prospects, Engineering 52 (2025) 40-52. https://doi.org/10.1016/j.eng.2025.08.010

[2] XJ. Liu et al., Financial technology, financing constraints, and future industrial development, Finance Research Letters 73 (2025) art.106671. https://doi.org/10.1016/j.frl.2024.106671

[3] M. Korzynski, A. Pacana, Centreless burnishing and influence of its parameters on machining effects. J. Mater. Proces. Technol. 210 (2010) 1217-1223. https://doi.org/10.1016/j.jmatprotec.2010.03.008

[4] C. Kolmasiak, Decarbonization of production systems in foundries, Archives Of Foundry Engineering 24 (2024) 104-109. https://doi.org/10.24425/afe.2024.149276

[5] P. King et al., Hands-on digital tools for metalcasting engineering education in foundry science, Int. J. Metalcast. 19 (2025) 2531-2543. https://doi.org/10.1007/s40962-024-01488-1

[6] A. Pacana, K. Czerwińska, Model of diagnosing and searching for incompatibilities in aluminium castings, Materials 14 (2021) art. 6497. https://doi.org/10.3390/ma14216497

[7] A. Pacana, A., K. Czerwińska, Improving the quality level in the automotive industry. Prod. Eng. Arch. 26 (2020) 162-166. https://doi.org/10.30657/pea.2020.26.29

[8] V.B. Bhagwat et al., An overview of machine learning applications in metal casting industries, Arch. Metall. Mater. 69 (2024) 1577-1584. https://doi.org/10.24425/amm.2024.151428

[9] CM. Rosca et al., Artificial intelligence of things infrastructure for quality control in cast manufacturing environments shedding light on industry changes, Applied Sciences 15 (2025) art. 2068. https://doi.org/10.3390/app15042068

[10] K. Czerwińska, A. Pacana, Analysis of the maturity of process monitoring in manufacturing companies, Prod. Eng. Arch. 28 (2022) 246-251. https://doi.org/10.30657/pea.2022.28.30

[11] Y.V. Bazhin et al., Influence of temperature regime of the combined process of casting and rolling of strips from high-alloy aluminium alloys, Non-Ferrous Metals 2 (2024) 45-51. https://doi.org/10.17580/nfm.2024.02.07

[12] A.I. Pronin et al., Study of the dynamics of processing parameters when turning an aluminomatrix cast composite workblank, J. Machin. Manuf. Rel. 53 (2024) 322-327. https://doi.org/10.1134/S1052618824700286

[13] K. Sabri, M. Gaceb, Smart casting: vision-driven defect detection for high-precision manufacturing, Int. J. Adv. Manuf. Technol. 140 (2025) 1871-1894. https://doi.org/10.1007/s00170-025-16390-1

[14] W.P. Liu et al., Simulation-assisted multi-process integrated optimization for greentelligent aluminum casting, Applied Energy 336 (2023) art. 120831. https://doi.org/10.1016/j.apenergy.2023.120831

[14] A. Pacana et al., Decision support in the production of packaging films by cost-quality analysis. Przemysł Chemiczny 95 (2016) 1042-1044, https://doi.org/10.15199/62.2016.5.31

[15] J. Yang et al., Data extension-based analysis and application selection of process-composition-properties of die casting aluminum alloy, Eng. App. Artif. Intell. 133 (2024) art. 108514. https://doi.org/10.1016/j.engappai.2024.108514

[16] H.Y. Hu et al., An intelligent indicator for warning of short-term high CO2 damage to garlic scapes, Packaging Technol. Sci. 36 (2023) 557-567. https://doi.org/10.1002/pts.2727

[17] M. Ruberti, Environmental performance and trends of the world's semiconductor foundry industry, Journal of Industrial Ecology 28 (2024) 1183-1197.

[18] J. Haraldsson et al., Taxonomy, saving potentials and key performance indicators for energy end-use and greenhouse gas emissions in the aluminium industry and aluminium casting foundries, Energies 14 (2021) art. 123571. https://doi.org/10.3390/en14123571

[19] A. Pacana et al., Effect of selected factors of the production process of stretch film for its resistance to puncture. Przemysl Chemiczny 93 (2014) 2263-2264

[20] T.C. Uyan et al., Industry 4.0 foundry data management and supervised machine learning in low-pressure die casting quality improvement, International Journal of Metalcasting 17 (2023) 414-429. https://doi.org/10.1007/s40962-022-00783-z

[21] K.E.T. Chen et al., Empirical study of foundry efficiency improvement based on data-driven techniques, Processes 9 (2021) art. 1083. https://doi.org/10.3390/pr9071083

[22] J. Ruzbarsky, S. Gaspar, Analysis of selected production parameters for the quality of pressure castings as a tool to increase competitiveness, Applied Sciences 13 (2023) art. 8098. https://doi.org/10.3390/app13148098

[23] A. Pacana et al., Effect of selected factors of the production process of stretch film for its resistance to puncture. Przemysł Chemiczny 93 (2014) 2263-2264.

[23] J. Pietraszek, A. Szczotok and E. Kocyłowska, Factorial approach to assessment of GPU computational efficiency in surrogate models, Advanced Materials Research 874 (2014) 157-162. https://doi.org/10.4028/www.scientific.net/AMR.874.157

[24] J. Pietraszek, N. Radek and A.V. Goroshko, Challenges for the DOE methodology related to the introduction of Industry 4.0, Production Engineering Archives 26 (2020) 190-194. https://doi.org/10.30657/pea.2020.26.33

[25] J. Pietraszek, A. Goroshko, The heuristic approach to the selection of experimental design, model and valid pre-processing transformation of DoE outcome, Advanced Materials Research 874 (2014) 145-149. https://doi.org/10.4028/www.scientific.net/AMR.874.145

[26] R. Dwornicka, J. Pietraszek, The outline of the expert system for the design of experiment, Production Engineering Archives 20 (2018) 43-48. https://doi.org/10.30657/pea.2018.20.09

[27] R. Ulewicz, B. Krstić and M. Ingaldi, Mining Industry 4.0 – Opportunities and Barriers, Acta Montanistica Slovaca 27 (2022) 291-305. https://doi.org/10.46544/AMS.v2i2.02

[28] R. Ulewicz, K. Czerwińska and A. Pacana, A Rank Model of Casting Non-Conformity Detection Methods in the Context of Industry 4.0, Materials 16 (2023) art. 723. https://doi.org/10.3390/ma16020723

[29] J. Pietraszek, A. Gadek-Moszczak and T. Toruński, Modeling of errors counting system for PCB soldered in the wave soldering technology, Advanced Materials Research 874 (2014) 139-143. https://doi.org/10.4028/www.scientific.net/AMR.874.139

[30] J. Pietraszek, Fuzzy regression compared to classical experimental design in the case of flywheel assembly, Lecture Notes in Computer Science 7267 LNAI (2012) 310-317. https://doi.org/10.1007/978-3-642-29347-4_36

[31] J. Pietraszek, The modified sequential-binary approach for fuzzy operations on correlated assessments, Lecture Notes in Computer Science 7894 LNAI (2013) 353-364. https://doi.org/10.1007/978-3-642-38658-9_32

[32] J. Pietraszek, A. Gadek-Moszczak, The smooth bootstrap approach to the distribution of a shape in the ferritic stainless steel AISI 434L powders, Solid State Phenomena 197 (2013) 162-167. https://doi.org/10.4028/www.scientific.net/SSP.197.162

[33] A. Szczotok et al., The bootstrap analysis of one-way ANOVA stability in the case of the ceramic shell mould of airfoil blade casting, Solid State Phenomena 235 (2015) 24-30. https://doi.org/10.4028/www.scientific.net/SSP.235.24

[34] R. Ulewicz, Practical application of quality tools in the cast iron foundry, Manufacturing Technology 14 (2014) 104-111.

[35] A. Szczotok, J. Nawrocki and J. Pietraszek, The Impact of the Thickness of the Ceramic Shell Mould on the ($\gamma + \gamma'$) Eutectic in the IN713C Superalloy Airfoil Blade Casting, Archives of Metallurgy and Materials 62 (2017) 587-593. https://doi.org/10.1515/amm-2017-0087

[36] R. Dindorf, J. Takosoglu and P. Wos, Prediction of the parameters and the hot open die elongation forging process on an 80 MN hydraulic press, Open Engineering 11 (2021) 528-534. https://doi.org/10.1515/eng-2021-0056

[37] N. Radek, E. Wajs and M. Luchka, The WC-Co electrospark alloying coatings modified by laser treatment, Powder Metallurgy and Metal Ceramics 47 (2008) 197-201. https://doi.org/10.1007/s11106-008-9005-7

[38] Ł.J. Orman et al., Application of laser treatment technology for boiling heat transfer augmentation, Production Engineering Archives 30 (2024) 259-265. https://doi.org/10.30657/pea.2024.30.25

[39] N. Radek et al., Properties of Steel Welded with CO2 Laser, Lecture Notes in Mechanical Engineering (2020) 571-580. https://doi.org/10.1007/978-3-030-33146-7_65

[40] R. Ulewicz, M. Ulewicz, Problems in the Implementation of the Lean Concept in the Construction Industries, Lecture Notes in Civil Engineering 47 (2020) 495-500. https://doi.org/10.1007/978-3-030-27011-7_63

[41] R. Ulewicz, D. Kleszcz and M. Ulewicz, Implementation of Lean Instruments in Ceramics Industries, Management Systems in Production Engineering 29 (2021) 203-207. https://doi.org/10.2478/mspe-2021-0025

[42] A. Szczotok, S. Roskosz, New possibilities of light microscopy research resulting from digital recording of images, Materials Science- Poland 23 (2005) 559-565.

[43] E. Skrzypczak-Pietraszek, J. Pietraszek, Chemical profile and seasonal variation of phenolic acid content in bastard balm (Melittis melissophyllum L., Lamiaceae), Journal of Pharmaceutical and Biomedical Analysis 66 (2012) 154-161. https://doi.org/10.1016/j.jpba.2012.03.037

[44] R. Ulewicz, Outsorcing quality control in the automotive industry, MATEC Web of Conferences 183 (2018) 0-0. https://doi.org/10.1051/matecconf/201818303001

[45] D. Siwiec, R. Dwornicka and A. Pacana, Improving the non-destructive test by initiating the quality management techniques on an example of the turbine nozzle outlet, Materials Research Proceedings 17 (2020) 16-22. https://doi.org/10.21741/9781644901038-3

[46] A. Pacana, D. Siwiec, Analysis of the possibility of used of the quality management techniques with non-destructive testing, Tehnicki Vjesnik 28 (2021) 45-51. https://doi.org/10.17559/TV-20190714075651

[47] E. Radzyminska-Lenarcik, M. Ulewicz, Polymer inclusion membranes (PIMs) doped with alkylimidazole and their application in the separation of non-ferrous metal ions, Polymers 11 (2019) art. 1780. https://doi.org/10.3390/polym11111780

Terotechnology XIV Materials Research Forum LLC
Materials Research Proceedings 62 (2026) 237-244 https://doi.org/10.21741/9781644904015-31

Machine Learning-Driven Cost Forecasting in Automotive Production Engineering: Towards Sustainable and Intelligent Manufacturing

Dorota KLIMECKA-TATAR[1,a *] and Muhammad Taimur HAMZA[2,b]

[1]Częstochowa University of Technology, Faculty of Management, Częstochowa, Poland

[a]dorota.klimecka-tatar@pcz.pl, [b]muhammad.hamza@pcz.pl

Keywords: Machine Learning, Cost Prediction, Production Engineering, Automotive Production, Sustainability, Industry 4.0

Abstract. Accurate forecasting of manufacturing costs remains a central challenge for the automotive sector, where volatile supply chains, dynamic energy prices, and sustainability pressures continuously reshape production systems. This paper proposes a machine-learning-based framework for cost prediction in automotive production engineering, integrating historical pricing data, supplier performance, and material indices. Several regression models were trained and compared, including Linear Regression, Decision Tree, Random Forest, Gradient Boosting, and XGBoost. The XGBoost algorithm achieved the best performance (MAE = 34.49, R^2 = 0.90), confirming its robustness and interpretability for complex industrial datasets. Furthermore, the study explores stochastic and hybrid modeling techniques, combining deterministic ML prediction with uncertainty quantification to improve reliability under variable market conditions. The proposed approach demonstrates the potential of machine learning to support sustainable and intelligent manufacturing, enhancing both economic efficiency and resource responsibility.

Introduction

Contemporary automotive manufacturing operates in an environment of unprecedented complexity, where rapid technological progress coincides with strict sustainability and cost-efficiency requirements. The ability to accurately forecast production costs is crucial for making informed design decisions, effective procurement planning, and accurate profitability assessments. Traditional cost-estimation techniques, such as rule-based calculations or expert judgment, tend to lack flexibility and scalability in data-intensive, highly variable contexts.

Machine learning (ML) offers a transformative alternative by exploiting large volumes of production data to discover latent correlations between process parameters and cost structures. Through data-driven modeling, ML systems can adapt dynamically to variations in raw-material prices, supplier reliability, and production cycles, creating predictive systems that evolve in parallel with the factory itself. Within the broader Industry 4.0 paradigm, such intelligent forecasting mechanisms represent a vital step toward the digitalization and sustainability of manufacturing operations [1-10]. In the context of the automotive supply chain, the impact of supplier quality on costs is confirmed by case studies of outsourcing quality control and CSL2 activities [11].

Cost Engineering and Production Systems in the Automotive Industry

Cost engineering in the automotive sector combines technical, economic, and managerial perspectives to ensure efficient product design and optimal allocation of manufacturing resources. Automotive products are composed of thousands of interdependent elements, and each design or material change can alter the overall cost structure. Therefore, identifying and quantifying key cost drivers, including materials, tooling, labor, logistics, and energy consumption, is fundamental to sustainable competitiveness. The growing use of electric mobility, composite materials, and lightweight structures has further diversified the parameters affecting production cost variability [1-5].

Within production engineering, cost management has shifted from static, experience-based estimation to data-driven predictive modeling. The digitalization of manufacturing under Industry 4.0, enabled by cyber-physical systems, IIoT, and digital twins, allows real-time monitoring of machines, materials, and processes, forming a basis for analytical cost forecasting [2,12]. Such integration transforms cost engineering into a dynamic, feedback-oriented system capable of adapting to supply chain and energy market volatility [3].

An essential tool in this transformation is Value Stream Mapping (VSM), which visualizes process and information flows to identify inefficiencies. When extended to VSM 4.0, it integrates digital sensors and data analytics, enabling quantitative assessment of cost drivers and waste sources [4]. The rise of machine learning (ML) further enhances forecasting accuracy, as ensemble algorithms such as Random Forest, Gradient Boosting, and XGBoost effectively capture nonlinear relationships and complex feature interactions. By converting operational data into actionable insight, ML supports adaptive planning and precise cost control within production engineering [6-10].

Finally, cost forecasting must align with sustainability goals. Data-driven analytics facilitate energy-efficient scheduling, waste reduction, and responsible resource management, aligning with ISO 14001 and circular economy principles. Thus, intelligent cost engineering becomes a strategic instrument for eco-efficiency and long-term industrial resilience [3].

Methodology

The empirical study utilized an anonymized dataset derived from historical procurement data of automotive components. The input variables included supplier quotations, component categories, historical price trends, inventory levels, and exchange rate indicators. The data underwent normalization, imputation of missing values, and categorical encoding to ensure statistical consistency. The sample was divided into training (80%) and testing (20%) subsets. Five regression models were implemented using Python and Scikit-learn:

- Linear Regression – baseline model for linear dependency;
- Decision Tree – interpretable non-linear partitioning;
- Random Forest – ensemble averaging to reduce variance;
- Gradient Boosting – sequential additive optimization;
- XGBoost – advanced boosting with regularization and shrinkage.

Performance was evaluated using Mean Absolute Error (MAE), Root Mean Squared Error (RMSE), and the coefficient of determination (R^2). A dedicated desktop application was created using PyQt5, providing an interactive interface for data input, model selection, and visualization of prediction outputs. This implementation demonstrates the practical feasibility of ML-based forecasting tools for engineering environments.

Results and Discussion

The comparative results of the regression algorithms are summarized in Table 1, which presents the main performance indicators: Mean Absolute Error (MAE), Root Mean Squared Error (RMSE), and the coefficient of determination (R^2). These metrics jointly characterize both the predictive precision and the robustness of each model. The Linear Regression model, serving as the baseline, exhibited the weakest predictive performance (MAE = 128.12, $R^2 = 0.62$). This result indicates that the cost data exhibit non-linear dependencies and interactions between features (e.g., supplier variability, price dynamics, and inventory levels) that linear models cannot adequately capture. Consequently, Linear Regression systematically underestimated cost peaks and overestimated low-value observations, revealing a bias toward the mean.

Table 1. Performance comparison of hybrid models for cost prediction (based on [9, 10, 16, 17])

Model	**MAE**	**RMSE**	**R^2**
Linear Regression	128.12	176.88	0.62
Decision Tree	92.45	121.60	0.75
Random Forest	48.90	68.21	0.87
Gradient Boosting	43.10	59.73	0.89
XGBoost	**34.49**	**51.14**	**0.90**

The Linear Regression model, serving as the baseline, exhibited the weakest predictive performance (MAE = 128.12, $R^2 = 0.62$). This result indicates that the cost data exhibit non-linear dependencies and interactions between features (e.g., supplier variability, price dynamics, and inventory levels) that linear models cannot adequately capture. Consequently, Linear Regression systematically underestimated cost peaks and overestimated low-value observations, revealing a bias toward the mean. The Decision Tree model showed substantial improvement (MAE = 92.45, $R^2 = 0.75$), primarily due to its ability to partition the feature space hierarchically. However, the model's performance was still limited by high variance, a typical issue in single-tree architectures that leads to overfitting on specific price segments or supplier categories.

Ensemble-based algorithms, particularly Random Forest and Gradient Boosting, delivered markedly higher accuracy ($R^2 = 0.87$ and 0.89, respectively). The Random Forest model benefited from averaging across multiple decorrelated trees, which stabilized predictions and reduced the variance observed in the Decision Tree approach. Gradient Boosting further enhanced performance by sequentially correcting residual errors, demonstrating its capability to learn complex, multi-dimensional cost structures in the automotive dataset. Its RMSE of 59.73 suggests a strong alignment between predicted and actual cost values, making it suitable for most industrial forecasting applications. The XGBoost algorithm achieved the best performance, with MAE = 34.49, RMSE = 51.14, and $R^2 = 0.90$. This indicates that the model explains 90% of the total cost variance, leaving only 10% unexplained by the selected features. The improvement of nearly 73% in MAE compared with the baseline linear regression confirms that regularized gradient boosting effectively handles both non-linearity and noise. Moreover, XGBoost's inherent L1/L2 regularization mechanisms prevent overfitting, ensuring model generalization across unseen data.

Residual analysis revealed that the XGBoost model exhibited a symmetric, zero-centered error distribution, with a narrow standard deviation ($\sigma \approx 50$), indicating stable predictions across the cost range. The Random Forest model, though competitive, showed slightly wider dispersion, reflecting sensitivity to small input perturbations.

Correlation analysis between actual and predicted costs yielded a correlation coefficient of $r = 0.95$ for XGBoost, confirming an almost linear relationship between the forecasted and observed values. Gradient Boosting followed closely with $r = 0.93$, whereas Linear Regression achieved only $r = 0.78$, further demonstrating the superiority of non-linear ensemble techniques. Feature importance analysis, derived from the XGBoost model, identified historical pricing trends, supplier type, and inventory level as the most influential predictors of cost variability. Historical price data accounted for nearly 70 % of the model's total explanatory power, underscoring the temporal dependency of cost evolution. Supplier attributes contributed approximately 20%, reflecting the impact of market heterogeneity and procurement terms, while inventory-related features represented about 10%, indicating moderate sensitivity to stock levels and supply continuity. This is consistent with the results showing that DDMRP parameterization by Deep RL improves inventory levels and on-time delivery [13].

The progression from linear to ensemble models illustrates a clear non-linear learning curve in model accuracy. While simple regressors are computationally efficient, they overlook high-order

Materials Research Proceedings 62 (2026) 237-244 https://doi.org/10.21741/9781644904015-31

interactions that are typical of manufacturing datasets. Conversely, boosting-based algorithms, particularly XGBoost, strike a balance between complexity and interpretability, providing a robust trade-off between precision and generalization.

Visualization of XGBoost Performance

The graphical representation in Fig. 1 illustrates the relationship between the predicted and actual cost values obtained from the XGBoost model. The data points are closely aligned along the diagonal reference line, indicating a strong linear correlation (r = 0.95) and confirming the model's capacity to replicate observed cost behavior with high fidelity. The residual dispersion around the line remains minimal, demonstrating low prediction error and consistent performance across the test dataset.

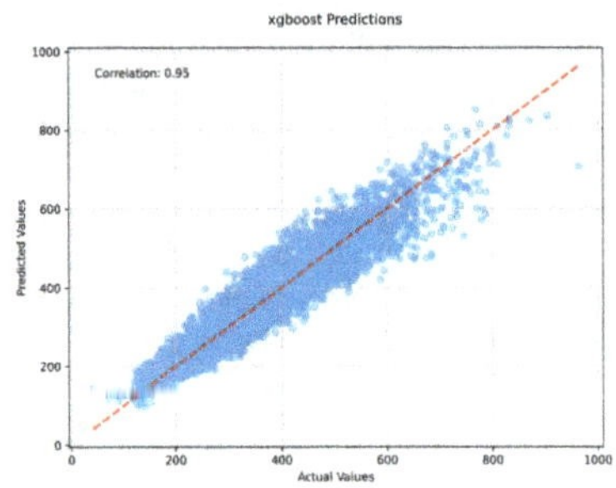

Fig 1. Detailed Results of XG-Boost

From a production-engineering perspective, these findings confirm that machine learning enables granular cost visibility at the component level, facilitating proactive budget control and supplier management. The predictive performance achieved by XGBoost positions it as a viable foundation for real-time cost monitoring systems integrated into digital manufacturing environments and ERP infrastructures [14,15].

The proposed ML-driven forecasting framework contributes to sustainable production engineering in multiple dimensions:

- Resource optimization: Precise cost predictions support material-use efficiency.
- Energy management: Data-driven scheduling reduces idle machine time and energy waste.
- Circular economy: Better cost visibility promotes component reuse and remanufacturing feasibility.
- Strategic agility: Rapid forecasting enables real-time responses to supplier disruptions.

Consequently, machine learning serves not merely as a computational tool but as a strategic enabler of intelligent, low-impact manufacturing systems consistent with Industry 4.0 and emerging Industry 5.0 paradigms.

Economic and Technological Impact

Cost forecasting is inherently burdened with high uncertainty and requires the use of appropriate approaches supporting both classical statistical methods [18,19] and explicit fuzzy approaches in the machine [20-22] and process [23] areas. Complex cost models form the basis for management primarily in logistics [24,25], as well as in the machinery [26] and wood [27] industries, which is reflected in the quality assurance systems implemented in these sectors [28-30]. This combined approach to cost requirements, both from the production and operational perspectives, was the impetus for the widespread implementation of modifications to the technological features of the surface layer [31-33] using special coatings such as carbonyl [34], DLC [35], and ESD [36, 37] finished by laser micromachining [38]. To implement such technologies as cost-effective as possible, it is necessary to utilize DOE experiment optimization techniques [39, 40] supported by complex process models [41] and resampling methods [42].

Summary

The study demonstrates that machine learning, particularly the XGBoost algorithm, markedly enhances the accuracy and robustness of cost forecasting in automotive production engineering. XGBoost achieved the best performance (MAE = 34.49; R^2 = 0.90), effectively capturing the non-linear relationships among key process variables, such as historical prices, supplier variability, and inventory levels. Its built-in regularization ensured stable, generalizable results, making it a reliable tool for cost management in smart manufacturing.

Practically, the proposed data-driven framework enables real-time cost estimation, enhances supplier transparency, and supports resource optimization in line with the principles of Lean and Industry 4.0. Its interpretability, based on feature-importance analysis, facilitates integration with enterprise systems, reinforcing strategic cost control and sustainability. Moreover, combining deterministic ML models with stochastic or optimization components can better capture uncertainty from market and supply fluctuations. These developments reflect the Industry 5.0 vision, where intelligent systems and human expertise converge to achieve sustainable and resilient manufacturing. Future research should focus on:

- Integrating the developed ML framework with digital twins and ERP/MES systems to enable real-time cost tracking;
- Applying explainable AI (XAI) tools such as SHAP or LIME to enhance model transparency;
- Extending the dataset to include electric-vehicle and battery-module manufacturing;
- Developing online learning mechanisms for continuous model updating in dynamic market conditions.

In conclusion, this work demonstrates that the synergy between machine learning, production engineering, and sustainability principles provides a powerful foundation for the next generation of intelligent manufacturing systems, bridging the gap between technological capability and responsible industrial transformation.

References

[1] D. Klimecka-Tatar, M. Obrecht, Eco-design and sustainability in automotive production systems, *Prod. Eng. Arch.* 26 (2020) 131–137. https://doi.org/10.30657/pea.2020.26.25

[2] H. Lasi, P. Fettke, H.-G. Kemper, T. Feld, M. Hoffmann, Industry 4.0, *Bus. Inf. Syst. Eng.* 6 (2014) 239–242. https://doi.org/10.1007/s12599-014-0334-4

[3] S. Kamble, A. Gunasekaran, S. Gawankar, Industry 4.0 and lean manufacturing practices for sustainable organisational performance in Indian manufacturing companies, *Int. J. Prod. Res.* 56 (2018) 2979–2998. https://doi.org/10.1080/00207543.2018.1444806

[4] B. Tjahjono, C. Esplugues, E. Ares, G. Pelaez, What does Industry 4.0 mean to Value Stream Mapping?, *Procedia Manuf.* 51 (2020) 149–156. https://doi.org/10.1016/j.promfg.2020.10.021

[5] T. Stock, G. Seliger, Opportunities of sustainable manufacturing in Industry 4.0, *Procedia CIRP* 40 (2016) 536–541. https://doi.org/10.1016/j.procir.2016.01.129

[6] J. Bokrantz, A. Skoogh, C. Berlin, J. Stahre, Smart Maintenance: an empirically grounded conceptualization, *Int. J. Prod. Econ.* 223 (2019) 107534. https://doi.org/10.1016/j.ijpe.2019.107534

[7] C. Hennebold, K. Klöpfer, P. Lettenbauer, M. Huber, Machine learning-based cost prediction for product development in mechanical engineering, *Procedia CIRP* 107 (2022) 264–269. https://doi.org/10.1016/j.procir.2022.04.044

[8] P. Afonso, V. Vyas, A. Antunes, S. Silva, B. Bret, A stochastic framework for product costing in advanced manufacturing, *Mathematics* 9 (2021) 2238. https://doi.org/10.3390/math9182238

[9] A.A. ForouzeshNejad, F. Arabikhan, S. Aheleroff, Optimizing Project Time and Cost Prediction Using a Hybrid XGBoost and Simulated Annealing Algorithm, *Machines* 12 (2024) 867. https://doi.org/10.3390/machines12120867

[10] Y. Li, C. Stasinakis, W.M. Yeo, A hybrid XGBoost–MLP model for credit risk assessment on digital supply chain finance, *Forecasting* 4 (2022) 184–207. https://doi.org/10.3390/forecast4010012

[11] R. Ulewicz, Outsourcing quality control in the automotive industry, *MATEC Web of Conferences* (2018), https://doi.org//matecconf/201818303001

[12] N. Wagner, M. Ingaldi, Two faces of automation: cobots and autonomous mobile robots in action, *Prod. Eng. Arch.* 31(3) (2025), 276-290, https://doi.org/10.30657/pea.2025.31.27

[13] M. El Marzougui, N. Messaoudi, W. Dachry, B. Bensassi. A Model for Decision-Making to Parameterizing Demand-Driven Material Requirement Planning Using Deep Reinforcement Learning. *Prod. Eng. Arch.*, 30(3) (2024), 377–393. https://doi.org/10.30657/pea.2024.30.37

[14] H. L. Yarfi, N. Motaki, M., Derrhi, I. Lahlou, Towards a Decision-Making Framework for Successful ERP Project Implementation: A Qualitative Study. *Prod. Eng. Arch.*, 31(1) (2025) , 91–105. https://doi.org/10.30657/pea.2025.31.9

[15] D. Cetindamar, R. Phaal, Technology Management in the Age of Digital Technologies, *IEEE Trans. Eng. Manag.* 70 (2023) 2507–2515. https://doi.org/10.1109/TEM.2021.3101196

[16] M. Abdel-Jaber, N. Makhoul, M. Abdel-Jaber, R. Beale, Cost prediction for product development using hybrid deep learning model: a meta-heuristic model, *Multimed. Tools Appl.* 84 (2025) 33307–33330. https://doi.org/10.1007/s11042-024-20437-y

[17] D.-J. Pang, Hybrid Machine Learning Model Performance in IT Project Cost and Duration Prediction, *Adv. Sci. Technol. Eng. Syst. J.* 8 (2023) 108–115. https://doi.org/10.25046/aj080212

[18] J. Pietraszek, A. Gadek-Moszczak and T. Toruński, Modeling of errors counting system for PCB soldered in the wave soldering technology, Advanced Materials Research 874 (2014) 139-143. https://doi.org/10.4028/www.scientific.net/AMR.874.139

[19] R. Dwornicka et al., Fuzzy Statistics-Aided Inference in Experimental Design, World Congress in Computational Mechanics and ECCOMAS Congress (2024). https://doi.org/10.23967/wccm.2024.131

[20] J. Pietraszek, Fuzzy regression compared to classical experimental design in the case of flywheel assembly, Lecture Notes in Computer Science 7267 LNAI (2012) 310-317. https://doi.org/10.1007/978-3-642-29347-4_36

[21] J. Pietraszek, The modified sequential-binary approach for fuzzy operations on correlated assessments, Lecture Notes in Computer Science 7894 LNAI (2013) 353-364. https://doi.org/10.1007/978-3-642-38658-9_32

[22] G. Filo, A Review of Fuzzy Logic Method Development in Hydraulic and Pneumatic Systems, Energies 16 (2023) art. 7584. https://doi.org/10.3390/en16227584

[23] A. Pacana, D. Siwiec, Method of Fuzzy Analysis of Qualitative-Environmental Threat in Improving Products and Processes (Fuzzy QE-FMEA), Materials 16 (2023) art. 1651. https://doi.org/10.3390/ma16041651

[24] M. Nowicka-Skowron, R. Ulewicz, Quality management in logistics processes in metal branch, METAL 2015 - 24th Int. Conf. Metall. Mater. (2015) 1707-1712.

[25] R. Ulewicz et al., Logistic controlling processes and quality issues in a cast iron foundry, Materials Research Proceedings 17 (2020) 65-71. https://doi.org/10.21741/9781644901038-10

[26] D. Siwiec et al., Improving the non-destructive test by initiating the quality management techniques on an example of the turbine nozzle outlet, Materials Research Proceedings 17 (2020) 16-22. https://doi.org/10.21741/9781644901038-3

[27] R. Ulewicz, Quality management system operation in the woodworking industry, The Path Forward for Wood Products: A Global Perspective – Proceedings of Scientific Papers (2016) 51-56.

[28] R. Ulewicz, Outsorcing quality control in the automotive industry, MATEC Web of Conf. 183 (2018). https://doi.org/10.1051/matecconf/201818303001

[29] R. Ulewicz, F. Novy and R. Dwornicka, Quality and work safety in metal foundry, METAL 2020 - 29th Int. Conf. Metall. Mater. (2020) 1287-1293. https://doi.org/10.37904/metal.2020.3649

[30] A. Pacana, K. Czerwinska and R. Dwornicka, Analysis of quality control efficiency in the automotive industry, Transportation Research Procedia 55 (2021) 691-698. https://doi.org/10.1016/j.trpro.2021.07.037

[31] N. Radek et al., The influence of plasma cutting parameters on the geometric structure of cut surfaces, Materials Research Proceedings 17 (2020) 132-137. https://doi.org/10.21741/9781644901038-20

[32] D. Nowakowski et al., Application of machine learning in the analysis of surface quality - the detection the surface layer damage of the vehicle body, METAL 2021 - 30th Int. Conf. Metall. Mater. (2021) 864-869. https://doi.org/10.37904/metal.2021.4210

[33] A. Dudek et al., Laser Surface Alloying of Sintered Stainless Steel, Materials 15 (2022) art. 6061. https://doi.org/10.3390/ma15176061

[34] W. Przybył et al., Microwave absorption properties of carbonyl iron-based paint coatings for military applications, Defence Technology 22 (2023) 1-9. https://doi.org/10.1016/j.dt.2022.06.013

[35] N. Radek et al., Operational properties of DLC coatings and their potential application, METAL 2022 – 31st Int. Conf. Metall. Mater., (2022) 531-536. https://doi.org/10.37904/metal.2022.4491

[36] N. Radek et al., The effect of laser treatment on operational properties of ESD coatings, METAL 2021 – 30th Int. Conf. Metall. Mater. (2021) 876-882. https://doi.org/10.37904/metal.2021.4212

[37] N. Radek et al., Technology and applications of ESD coatings before and after laser processing, METAL 2023 – 32nd Int. Conf. Metall. Mater. (2024) 500-505. https://doi.org/10.37904/metal.2023.4727

[38] A. Kalinowski et al., Laser surface texturing: characteristics and applications, System Safety: Human - Technical Facility - Environment 5 (2023) 240-248. https://doi.org/10.2478/czoto-2023-0026

[39] J. Pietraszek, A. Goroshko, The heuristic approach to the selection of experimental design, model and valid pre-processing transformation of DoE outcome, Advanced Materials Research 874 (2014) 145-149. https://doi.org/10.4028/www.scientific.net/AMR.874.145

[40] J. Pietraszek, N. Radek and A.V. Goroshko, Challenges for the DOE methodology related to the introduction of Industry 4.0, Production Engineering Archives 26 (2020) 190-194. https://doi.org/10.30657/pea.2020.26.33

[41] Ł.J. Orman et al., Comparative Analysis of Indoor Environmental Quality and Self-Reported Productivity in Intelligent and Traditional Buildings, Energies 16 (2023) art. 6663. https://doi.org/10.3390/en16186663

[42] J. Pietraszek, A. Gadek-Moszczak, The smooth bootstrap approach to the distribution of a shape in the ferritic stainless steel AISI 434L powders, Solid State Phenomena 197 (2013) 162-167. https://doi.org/10.4028/www.scientific.net/SSP.197.162

Terotechnology XIV | Materials Research Forum LLC
Materials Research Proceedings 62 (2026) 245-252 | https://doi.org/10.21741/9781644904015-32

WCAG 2.1/2.2: An Optimisation Technology for Digital Asset Lifecycle Management

Konrad WISOWSKI[1,a *], Paweł LEMPA[1,b] and Grzegorz FILO[1,c]

[1]Cracow University of Technology, Faculty of Mechanical Engineering, Jana Pawła II Ave 37, 31-864 Cracow, Poland

[a]konrad.wisowski@pk.edu.pl, [b]pawel.lempa@pk.edu.pl, [c]grzegorz.filo@pk.edu.pl

Keywords: WCAG, Web Accessibility, Terotechnology, Web Design Optimization, Accessibility Evaluation, Universal Design, Inclusive Web Development Accessibility Compliance

Abstract. The Web Content Accessibility Guidelines (WCAG) 2.1 and 2.2 are considered critical frameworks for optimizing digital assets under the terotechnology paradigm, managing the development and design lifecycle on the web. This paper examines the role of WCAG standards as a structured framework for coordinating the entire lifecycle of web-based digital assets, encompassing everything from conception to maintenance. The benefits and challenges of WCAG-based optimization are examined through existing implementation practices, evaluation approaches, and new technologies. The results show that although WCAG compliance increases inclusivity, usability, and legal compliance, the complexity of the assessment, technical implementation, and continuing maintenance needs remain significant hurdles.

Introduction

Terotechnology, the holistic management of physical assets from their point of origin through to their end-point use, also increasingly applies to digital assets found in web environments. The Web Content Accessibility Guidelines (WCAG) 2.1 and 2.2 represent a set of principles that focus on all aspects of the accessibility lifecycle, from project inception and design to construction, maintenance, and ultimately, retirement and migration [1].

With more than one billion people worldwide living with disabilities, the design of an accessible Web is now a matter of not only ethics but also business importance and legal obligation [2]. WCAG standards ensure that these are well-defined and measurable standards for web-based digital assets, helping organizations ensure that these assets remain usable, maintainable, and valuable throughout their lifecycle. This article focuses specifically on web design, web development, and web accessibility, analyzing the use of WCAG 2.1/2.2 as an optimization technology that lies at the nexus of digital asset lifecycle management processes, including web accessibility.

WCAG as a Lifecycle Optimization Framework

WCAG is a lifecycle optimization tool designed to support a holistic optimization framework for digital asset lifecycles, specifically helping to optimize all phases of a real-world asset lifecycle. This transition from WCAG 2.0 to 2.1 and 2.2 is a best practice in lifecycle management [3]. The following four distinct stages comprise this migration process: an assessment of current accessibility status, strategic planning for implementation, systematic execution of needed changes, and thorough testing for compliance. Design systems are critical to providing the enabling environment necessary for such widespread digital transition. Successful migration involves cross-functional collaboration, including designers, developers, content creators, and, crucially, end users with disabilities at every step. This cross-functional approach enables accessibility to be integrated from the outset, avoiding the need for retrofitting later in the lifecycle.

Materials Research Proceedings 62 (2026) 245-252	https://doi.org/10.21741/9781644904015-32

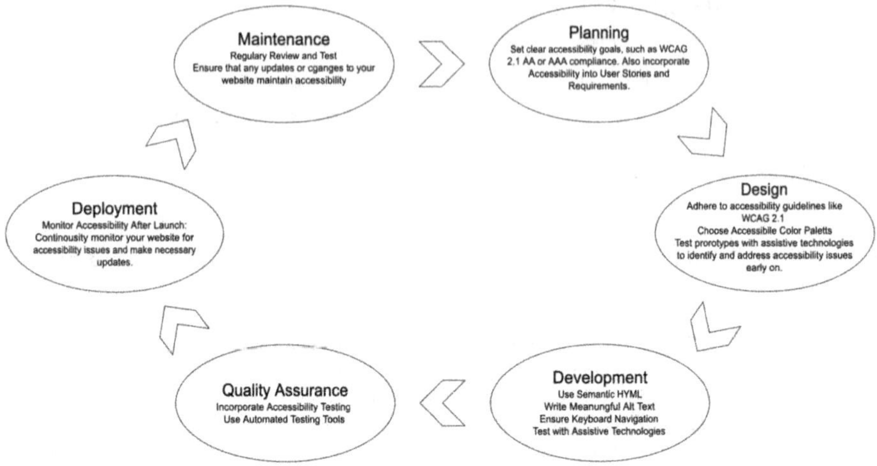

Figure 1. WCAG 2.1/2.2 Lifecycle Optimization Framework. Continuous assessment, planning, execution, testing, and maintenance cycle for web-based digital assets accessibility compliance.

Lifecycle management of digital assets requires constant maintenance and optimization. WCAG compliance is not a one-time event; it is something that the organization must continually monitor and reevaluate as standards change, content is added or modified, and technology advances. Organizations should also ensure that systematic monitoring, utilizing both automated and manual assessment systems, is implemented to ensure compliance at every stage of the asset lifecycle. The need for the maintenance phase is also more challenging when websites develop, and complexity and the number of contents increase. While automated evaluation tools can promptly surface technical violations, they cover only some aspects of WCAG success metrics. While trained evaluators may be able to provide manual assessments for lifecycle management, these methods require more time, significant financial investment, and specialized expertise.

Advantages of WCAG-Based Optimization

Compliance with WCAG increases the pool of users who can access digital assets more easily. Accessibility features are valuable for users with not only permanent disabilities but also for those with temporary disabilities, impairments, situational limitations, or age-related changes in physical function. This is especially true for the expanding elderly population of society, which will benefit from more accessible design changes in terms of color contrast, textual layout, and navigation design [4]. That growth in coverage adds business value. Companies that meet WCAG standards can reach a wider demographic, enhance customer value, and demonstrate good corporate citizenship behavior. From a lifecycle perspective, building accessibility into the design phase at the outset of development is more economical than retrofitting inaccessible assets later in their lifecycle.

The universal design principles of WCAG create design improvements that benefit not only users with disabilities [5] but also those without. Semantic HTML structures make content more maintainable and easier to update as we move from the asset lifecycle to the development phase. The ARIA (Accessible Rich Internet Applications) roles and attributes enhance content organization and meaning, providing an advantage that benefits assistive computing, as well as search engine optimization. Keyboard navigation alternatives enhance usability for all users, including those with motor impairments. If we use alternative text for images, then text access will improve comprehension for users with low bandwidth and facilitate indexing of content. These integrated advantages demonstrate that optimized accessibility enhances digital asset quality and sustainability throughout the lifecycle.

For any organization, compliance with WCAG allows meeting the legal requirements of different accessibility regulations and laws worldwide. Through a terotechnology lens, compliance minimizes operational risk throughout the life of a digital asset. Non-compliance comes with heavy charges for lawsuits, mandatory remediation (such as that imposed by regulatory authorities), and damage to reputation – all of which have a negative impact on the overall cost of ownership of digital assets. Implementing WCAG early in the initial stages of development helps to minimize these lifecycle risks and costs. Companies that embed accessibility into their standard development processes experience improved long-term asset performance and reduced total lifecycle costs.

Accessible website design is very similar to search engine best practices. Semantic HTML formatting and heading hierarchies, descriptive links, and alternative text for images are all ways to improve visibility to search engines. This alignment of accessibility and SEO serves as a reminder of how the WCAG optimization process adds value directly by analyzing online assets from multiple dimensions at every stage of the business lifecycle.

Difficulties and Implementation Barriers

A growing body of research indicates that WCAG evaluation presents real challenges – even for those trained to do so. According to experiments conducted among computer science students, 65% of the WCAG success criteria do not achieve 80% agreement among evaluators, indicating considerable subjectivity and difficulty in interpretation [6]. This evaluation difficulty is a significant problem for lifecycle management: an unreliable compliance status will always be present throughout the lifetime, as inconsistent ratings are issued. The majority of WCAG criteria (86%) cannot be effectively solved using automatic testing tools and require time-consuming manual testing. Over one-third of the success criteria are difficult to assess despite using manual methods. These evaluation challenges create bottlenecks during the testing phase of the asset lifecycle and raise the overall cost associated with WCAG compliance over time.

The ongoing WCAG implementation challenges stem from widespread knowledge gaps in the web development and designer communities to date. Numerous professionals have not received formal training on accessible design principles and are unfamiliar with the assistive technologies used by people with disabilities. This knowledge gap permeates every stage of the digital asset lifecycle, from concept development to maintenance. Some common technical challenges include the lack of color contrast between text and background elements, incomplete keyboard navigation support, improper use or absence of ARIA attributes, and limited alternative text for non-decorative images. This technical weakness in many places also reflects a lack of knowledge about accessibility rather than technological deficiencies.

Few schools, health authorities, and commercial websites are fully compliant with the WCAG. Even when library websites, government portals, and e-learning platforms have been prioritized due to their crucial role for people with disabilities, accessibility barriers still exist [7, 8]. This partial compliance reflects broader issues in digital asset lifecycle management. Frequently, organizations choose to deploy software quickly instead of accessibility testing. The WCAG compliance is treated as an add-on rather than a fundamental core requirement at every stage of the development.

Most accessibility assessments are largely performed using automated tools, with minimal real-world user testing involving individuals with disabilities [9]. Even if it effectively identifies the technical violation, automated tools cannot determine subjective aspects, such as the content's clarity, logical flow, and alternative text descriptions. Such over-reliance on automated testing leads to shortcomings in quality assurance throughout the asset lifecycle. User testing with disabled audiences reveals deep insights that automated tools are unable to document. However, hiring heterogeneous user groups, thorough testing, and incorporating feedback take time and resources that organizations do not always allocate evenly over the life cycle of an asset.

Making the change from current digital assets to newer versions of WCAG is difficult. Legacy systems may not have the necessary technology to support modern accessibility, which can require significant refactoring or a total redesign. Organizations will need to weigh up the need for enhanced accessibility against the expense of major system upgrades and the consequences of such modifications. The migration process is complex, as it requires planning to maintain services while implementing accessibility enhancements. This complexity underscores the need to design accessibility into the initial stages of a system's development, rather than retrofitting inaccessible systems later in their lifecycle.

Tools and Techniques for WCAG Optimization

Proper use of semantic HTML elements provides inherent accessibility benefits by conveying content structure and meaning. ARIA roles, states, and properties supplement HTML semantics where native elements fall short, most notably in complex interactive applications. We cannot focus on these foundational techniques after the initial design phase has finished. Automated tools and manual tests ensure text/background color contrast ratios are sufficient. Those tools are especially crucial in e-learning and content-rich websites, where the readability of text itself is significantly related to the usability of such information for consumers with visual impairments or those exposed to demanding content [10, 11]. Comprehensive switching of keyboard navigation alternatives means that all interactive components are available independently of pointing devices. This comes with visible focus indicators, logical tab ordering, and keyboard shortcuts for complex interfaces. Creating effective alternative text for images requires consideration of both context and purpose. Decorative images must be annotated as such, and informative images should have descriptions that are sufficiently informative to convey nearly the same information as the visual content.

```html
<!-- Violation: Missing label for form input -->
<!-- WCAG 2.1 Level A - Success Criterion 3.3.2: Labels or Instructions -->
<input type="email" id="userEmail" placeholder="Enter your email">
```

```html
<!-- Compliant: Proper label association -->
<!-- WCAG 2.1 Level A - Success Criterion 3.3.2: Labels or Instructions -->
<label for="userEmail">Email Address:</label>
<input type="email" id="userEmail" placeholder="Enter your email">
```

Figure 2. Form Input Labels Comparison

```css
/* Violation: Insufficient color contrast ratio */
/* WCAG 2.1 Level AA - Success Criterion 1.4.3: Contrast (Minimum) */
.button-primary {
  background-color: #cccccc;
  color: #999999; /* Contrast ratio: 2.2:1 (fails WCAG AA) */
  padding: 10px 20px;
}
```

```css
/* Compliant: Adequate color contrast ratio */
/* WCAG 2.1 Level AA - Success Criterion 1.4.3: Contrast (Minimum) */
.button-primary {
  background-color: #0066cc;
  color: #ffffff; /* Contrast ratio: 7.5:1 (passes WCAG AA) */
  padding: 10px 20px;
}
```

Figure 3. Color Contrast Comparison

```html
<!-- Violation: Missing alternative text -->
<!-- WCAG 2.1 Level A - Success Criterion 1.1.1: Non-text Content -->
<img src="quarterly-report-chart.png">
```

```html
<!-- Compliant: Descriptive alternative text provided -->
<!-- WCAG 2.1 Level A - Success Criterion 1.1.1: Non-text Content -->
<img src="quarterly-report-chart.png"
alt="Bar chart showing quarterly revenue growth Q1 to Q4 2024, with 15% increase">
```

Figure 4. Image Alternative Text Comparison

```html
<!-- Violation: Button without accessible name -->
<!-- WCAG 2.1 Level A - Success Criterion 4.1.2: Name, Role, Value -->
<button class="icon-btn">
  <img src="icons/delete.svg">
</button>
```

```html
<!-- Compliant: Button with proper ARIA label -->
<!-- WCAG 2.1 Level A - Success Criterion 4.1.2: Name, Role, Value -->
<button class="icon-btn" aria-label="Delete item">
  <img src="icons/delete.svg" alt="" role="presentation">
</button>
```

Figure 5. Button Accessible Name Comparison

Materials Research Proceedings 62 (2026) 245-252 https://doi.org/10.21741/9781644904015-32

```html
<!-- Violation: Non-descriptive link text -->
<!-- WCAG 2.1 Level A - Success Criterion 2.4.4: Link Purpose (In Context) -->
<p>To download the annual report, <a href="report2024.pdf">click here</a>.</p>
```

```html
<!-- Compliant: Descriptive link text -->
<!-- WCAG 2.1 Level A - Success Criterion 2.4.4: Link Purpose (In Context) -->
<p>
  <a href="report2024.pdf">Download Annual Financial Report 2024 (PDF, 2.5MB)</a>
</p>
```

Figure 6. Link Text Description Comparison

```html
<!-- Violation: Improper heading hierarchy -->
<!-- WCAG 2.1 Level A - Success Criterion 1.3.1: Info and Relations -->
<h1>Main Page Title</h1>
<h3>Section Content</h3>
<h2>Subsection Title</h2>
```

```html
<!-- Compliant: Proper heading hierarchy -->
<!-- WCAG 2.1 Level A - Success Criterion 1.3.1: Info and Relations
<h1>Main Page Title</h1>
<h2>Section Content</h2>
<h3>Subsection Title</h3>
```

Figure 7. Heading Hierarchy Comparison

```html
<!-- Violation: Custom interactive element without keyboard support -->
<!-- WCAG 2.1 Level A - Success Criterion 2.1.1: Keyboard -->
<div class="dropdown" onclick="toggleMenu()">
  Menu Options
</div>
```

```html
<!-- Compliant: Keyboard accessible interactive element -->
<!-- WCAG 2.1 Level A - Success Criterion 2.1.1: Keyboard -->
<button class="dropdown"
        onclick="toggleMenu()"
        onkeypress="handleKeyPress(event)"
        aria-expanded="false"
        aria-haspopup="true">
  Menu Options
</button>
```

Figure 8. Keyboard Navigation Comparison

These automated accessibility-testing tools enable quick initial screening for technical WCAG violations. Integrating tools into continuous integration pipelines helps creators identify and address accessibility issues as they arise throughout development [12]. For the same reason, organizations can also benefit from understanding that only a part of the WCAG success criteria are addressed through automated tools, and these are by no means a replacement for manual evaluation of WCAG and user testing. However, full manual evaluation by trained accessibility specialists remains critical for a comprehensive WCAG assessment. Manual testing can detect problems that automated tools miss, such as navigation logic issues, content clarity, or the effectiveness of accessibility resources. A manual audit schedule should be implemented periodically throughout the digital asset lifecycle in organizations.

The best way to evaluate real-world accessibility is to include users with disabilities in the testing process. User feedback indicates the implementation effectiveness of accessibility features and pinpoints obstacles that technical testing can bypass. Organizations should establish ongoing relationships with a diverse range of user groups throughout the asset lifecycle to facilitate continuous accessibility improvement.

Emerging Optimization Technologies

Generative artificial intelligence is a potentially powerful new tool for WCAG optimization [13]. AI systems can automate specific accessibility evaluations, recognize accessibility violation patterns, recommend remediation strategies, and even create accessible code implementations. These features could ease the resources required to test and maintain accessibility throughout the lifecycle of a digital asset. AI can also provide real-time training and guidance to developers, addressing knowledge gaps that currently impede WCAG implementation. The maturation of these technologies may yield more consistent accessibility compliance with less manual effort.

Research is ongoing to develop better evaluation techniques that address existing limitations in WCAG assessment. Integrating automated tools with manual expert review and user testing holds significant promise for more robust and reliable accessibility assessments. Developing standardized evaluation protocols with clear interpretation guidelines could mitigate the subjectivity currently plaguing WCAG assessment. Modern web development frameworks are increasingly incorporating accessibility features as standard components rather than optional add-ons. These frameworks support accessible development practices from the beginning of the project, embedding WCAG compliance in the fundamental development lifecycle rather than viewing it as a separate concern.

Legal Frameworks and Compliance Mechanisms

Established legal systems provide mechanisms for individuals to seek redress when digital assets fail to meet accessibility requirements [14]. These complaint procedures assess the effectiveness of organizational accessibility measures and can result in mandatory remediation requirements. From a lifecycle management perspective, these legal frameworks create accountability structures that incentivize proactive WCAG compliance. Organizations must understand relevant legal requirements and establish internal processes to efficiently respond to accessibility complaints. Proactive accessibility management throughout the digital asset lifecycle reduces the likelihood of formal complaints and associated legal risks.

WCAG standards serve as the foundation for accessibility requirements in numerous national and international regulations. This global alignment makes WCAG compliance essential for organizations operating across multiple jurisdictions. Lifecycle management strategies must account for varying legal requirements while maintaining consistent accessibility standards across all digital assets.

Impact on Users with Disabilities

Research demonstrates that mobile applications with robust accessibility features significantly enhance independence for users with visual impairments and other disabilities. User feedback consistently rates applications highly when they follow WCAG and related mobile accessibility standards [15]. This positive impact underscores the importance of applying accessibility principles across all digital platforms throughout their development and operational lifecycles.

The systematic collection and integration of user feedback from people with disabilities enhance accessibility outcomes throughout the digital asset lifecycle. Organizations that establish ongoing feedback mechanisms can identify and address accessibility barriers more effectively than those relying solely on technical testing. This user-centered approach aligns with both terotechnology principles and best practices for accessibility.

Taking the needs of people with disabilities into account in website design also significantly impacts software interfaces, including industrial ones, which increasingly take the form of web applications. Such changes significantly impact both educational services [16] and communication with industrial users [17, 18], where inconsistency and ambiguity in communication [19] can lead to substantial losses or even pose a security risk. Modifications that facilitate communication are also important in service-based services, where the communication context determines the speed of response and mutual understanding of needs [20, 21].

Conclusions and Future Directions

WCAG 2.1 and 2.2 function as comprehensive optimization technologies for digital asset lifecycle management in web design and development. These standards provide a systematic framework for ensuring that web-based assets remain accessible, usable, and valuable throughout their operational life. The advantages of WCAG-based optimization extend beyond disability accommodation to encompass broader usability improvements, legal compliance, enhanced search engine visibility, and expanded market reach.

However, significant challenges persist in implementing and evaluating WCAG. The complexity and subjectivity of WCAG assessment, widespread knowledge gaps among developers, resource-intensive manual testing requirements, and difficulties in migrating legacy systems all impede comprehensive accessibility optimization. These barriers affect all phases of the digital asset lifecycle, from initial development through ongoing maintenance.

Addressing these challenges requires multifaceted approaches that combine improved training and education for development professionals. These enhanced evaluation methodologies reduce subjectivity and improve efficiency by integrating accessibility considerations from project inception rather than as retrofit activities. They also involve systematic user testing with people with disabilities throughout the development lifecycle and the adoption of emerging technologies, such as AI-driven accessibility tools.

Organizations that successfully integrate WCAG principles into their digital asset lifecycle management practices realize substantial benefits in inclusivity, usability, legal compliance, and overall asset quality. As accessibility standards continue to evolve and supporting technologies advance, the potential for more efficient and effective WCAG optimization will grow. The key to success lies in treating accessibility not as a compliance checkbox but as a fundamental quality attribute embedded throughout the entire digital asset lifecycle—from initial conception through ultimate retirement or migration.

Future research should focus on developing more objective and efficient WCAG evaluation methodologies, creating better training resources for development professionals, exploring AI applications in accessibility testing and remediation, and establishing best practices for accessibility integration in agile development environments. By advancing these areas, the web development community can make WCAG optimization more achievable and sustainable as a core component of digital asset lifecycle management.

References

[1] T. Berners-Lee, Weaving the Web: The Original Design and Ultimate Destiny of the World Wide Web, Harper Business (2000).

[2] P. Acosta-Vargas et al., Accessibility in Native Mobile Applications for Users with Disabilities: A Scoping Review, Applied Sciences 11 (2021) art.5707. https://doi.org/10.3390/app11125707

[3] H. Shah, Advancing Web Accessibility: A Guide to Transitioning Design Systems from WCAG 2.0 to WCAG 2.1, Computer Science & Information Technology (CS & IT) – CSCP 2023 (2023) 233-245. https://doi.org/10.5121/csit.2023.132218

[4] M. Li, L. Peng, Research on Inclusive Analysis and Accessibility Optimization of Healthcare Service Websites Based on the Needs of Elderly Users, in: Proc. Int. Conf. Accessibility and Design, IOS Press, 384 (2024) 156-165. https://doi.org/10.3233/FAIA240018

[5] L. Patan, Designing Inclusive Web Applications with Accessibility in Mind, Int. J. for Multid. Research (IJFMR) 6 (2024) 1-13. https://doi.org/10.36948/ijfmr.2024.v06i05.29091

[6] F. Iniesto, C. Rodrigo, The use of WCAG and automatic tools by computer science students: a case study evaluating MOOC accessibility, Journal of Universal Computer Science 30 (2024) 85-105. https://doi.org/10.3897/jucs.101704

[7] S. Dongaonkar et al., Content Accessibility Evaluation of Government Website using WCAG (Web Content Accessibility Guidelines), Proc. Second Int. Conf. Res. Intell. Comp. Eng. (2017) 27-31. https://doi.org/10.15439/2017R79

[8] M.M. Campoverde et al., Empirical Studies on Web Accessibility of Educational Websites: A Systematic Literature Review, IEEE Access 8 (2020) 91676-91700. https://doi.org/10.1109/ACCESS.2020.2994288

[9] S. Panda, R. Chakravarty, Evaluating the Web Accessibility of IIT Libraries: A Study of Web Content Accessibility Guidelines, Performance Measurement and Metrics 21 (2020) 121-145. https://doi.org/10.1108/PMM-02-2020-0011

[10] K. Dziedzic et al., Analysis of Compliance with WCAG Guidelines Regarding Contrast Implementation in an E-Learning Quiz, Journal of Computer Sciences Institute 31 (2024) 103-108. https://doi.org/10.35784/jcsi.6034

[11] I. Maulana, Web Accessibility: Designing User-Friendly Websites for Individuals with Visual Impairments, Golden Ratio of Data in Summary 5 (2025) 14-19. https://doi.org/10.52970/grdis.v5i1.896

[12] N. Palani, The Web Accessibility Project: Development and Testing Best Practices, Auerbach Publications, New York (2022). https://doi.org/10.1201/9781003299431

[13] V. Vemuri, Advancing Web Accessibility: Leveraging Generative AI for Enhanced WCAG Compliance, Journal of Artificial Intelligence & Cloud Computing 3 (2024) 1-6. https://doi.org/10.47363/JAICC/2024(3)E261

[14] A. Cebera, Complaint Procedure for Failure to Ensure Accessibility for People with Special Needs, Acta Universitatis Lodziensis. Folia Iuridica, Special Issue (2023) 41-51. https://doi.org/10.18778/0208-6069.S.2023.04

[15] B. Christy, A. Pillai, User feedback on usefulness and accessibility features of mobile applications by people with visual impairment, Indian Journal of Ophthalmology 69 (2021) 555-558. https://doi.org/10.4103/ijo.IJO_1042_20

[16] R. Ulewicz, Application of servqual method for evaluation of quality of educational services at the university of higher education, Polish Journal of Management Studies 9 (2014) 254-264.

[17] D. Siwiec et al., Improving the non-destructive test by initiating the quality management techniques on an example of the turbine nozzle outlet, Materials Research Proceedings 17 (2020) 16-22. https://doi.org/10.21741/9781644901038-3

[18] A. Pacana, D. Siwiec, Analysis of the possibility of used of the quality management techniques with non-destructive testing, Tehnicki Vjesnik 28 (2021) 45-51. https://doi.org/10.17559/TV-20190714075651

[19] A. Pacana, K. Czerwińska, Analysis of the causes of control panel inconsistencies in the gravitational casting process by means of quality management instruments, Production Engineering Archives 25 (2019) 12-16. https://doi.org/10.30657/pea.2019.25.03

[20] K. Kowalik, D. Klimecka-Tatar, The process approach to service quality management, Production Engineering Archives 18 (2018) 31-34. https://doi.org/10.30657/pea.2018.18.05

[21] M. Ingaldi, Overview of the main methods of service quality analysis, Production Engineering Archives 18 (2018) 54-59. https://doi.org/10.30657/pea.2018.18.10

Terotechnology XIV Materials Research Forum LLC
Materials Research Proceedings 62 (2026) 253-261 https://doi.org/10.21741/9781644904015-33

Identification of Military Vehicles Using Vision Systems and AI Methods

Piotr CHOCHÓŁ[a], Piotr WITKOWSKI[b], Wojciech ŁAKOMIEC[c], Norbert KOT[d], Jakub GÓRSKI[e], Michał KLUZEK[f] and Paweł A. ŁASKI[g]*

Kielce University of Technology, Faculty of Mechatronics and Mechanical Engineering, Department of Automation and Robotics, Al. 1000-lecia Państwa Polskiego 7, 25-314 Kielce, Poland

[a,b,c,d,e,f]robotyka@tu.kielce.pl, [g]pawell@tu.kielce.pl

Keywords: Military Vehicles, Neural Network Model, Identifying Graphical

Abstract. This paper presents a method for identifying graphical equivalents of military vehicles. A YOLO11L neural network model, an enlarged version of the YOLO11 network, was used. YOLO models are algorithms that use the RGB color palette for object detection. LabelStudio software was used to train the neural network for image labeling. To properly train the artificial neural network for vehicle detection, approximately 3,000 images were used [1-3], each containing over 12,000 silhouettes of allied Polish/American and enemy Russian military vehicle models. The model was trained at various standard resolutions: 640x640 and 1024x1024. Because the computing power available on personal computers was insufficient for model training and time was a key factor, Google servers on the Google Colab platform were utilized, leveraging A100 GPUs. Achieved around 7-15 fps for object detection using laptops with RTX 3060 graphics cards. Further research is planned on modifying the network models and training the final network model [3-5].

Introduction

Since the emergence of the first civilizations, humans have been engaged in conflict with one another. Military advantage has also depended on accurate identification of friends and foes. Currently, this has become particularly important in applications involving the autonomous destruction of military facilities and vehicles using drones. Therefore, one effective method for gaining an advantage over the enemy in offensive operations is the accurate identification of vehicle silhouettes based on their unique visual characteristics, similar to what happens in soldiers' brains. Collecting a sufficiently large set of vehicle image samples enables the training and deployment of artificial neural networks to accurately identify vehicles.

Results and Discussion

Fig. 1 presents a set of example military vehicle silhouettes, including both wheeled and aircraft types. The set contains fifteen black vehicle silhouettes. Training the artificial neural network required collecting several thousand vehicle samples under various lighting conditions and angles, as well as on different backgrounds and at different size scales. Subsequently, the appropriate assignment of the labeled objects was performed. The set of graphics prepared in this way constituted the training dataset.

Fig.2 provides a statistical overview of the training dataset, showing the number of instances for each vehicle class and the characteristics of the object distribution within the images. The bar graph in the upper left corner shows the level of class balance, revealing which vehicle models, such as the AH-64 Apache or Ka-52, are most represented, which is crucial for understanding what the model focused on during training the neural network. Other visualizations, including a heatmap and a scatterplot, indicate that most military vehicle silhouettes are located in the central part of the frames (left image xy positions in the image) and are characterized by varying sizes, which allows the neural network model to correctly train, detecting both close-up and distant objects (in this case, the objects were mostly small or medium-sized relative to the image size).

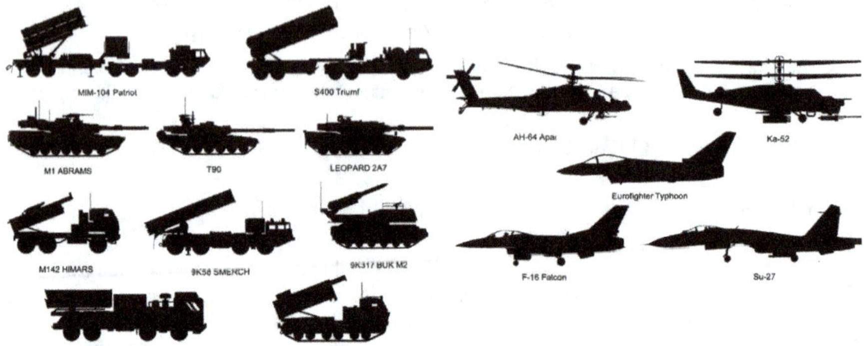

Fig. 1. Graphical models of military objects for identification during missions

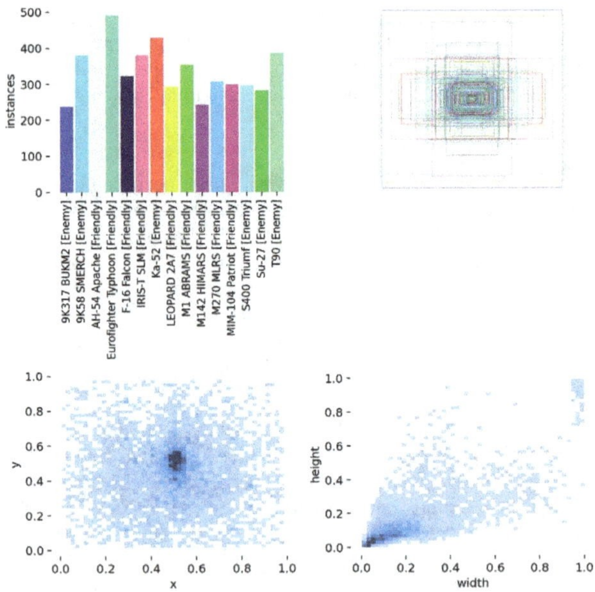

Fig. 2. Statistical overview of the dataset used for training

A label correlogram is an advanced analytical visualization that examines the interrelationship between an object's center position on the X and Y axes and its dimensions (width and height) across the entire dataset (Fig.3). A grid of histograms and scatter plots allows for assessing the consistency of geometric data, showing, for example, whether objects at the edges of images have different proportions than those in the center, which could affect detection errors. In this case, the graph confirms that the training data is naturally distributed and the model has had the opportunity to learn from a variety of silhouette examples, which promotes better generalization and performance in real-world combat environments.

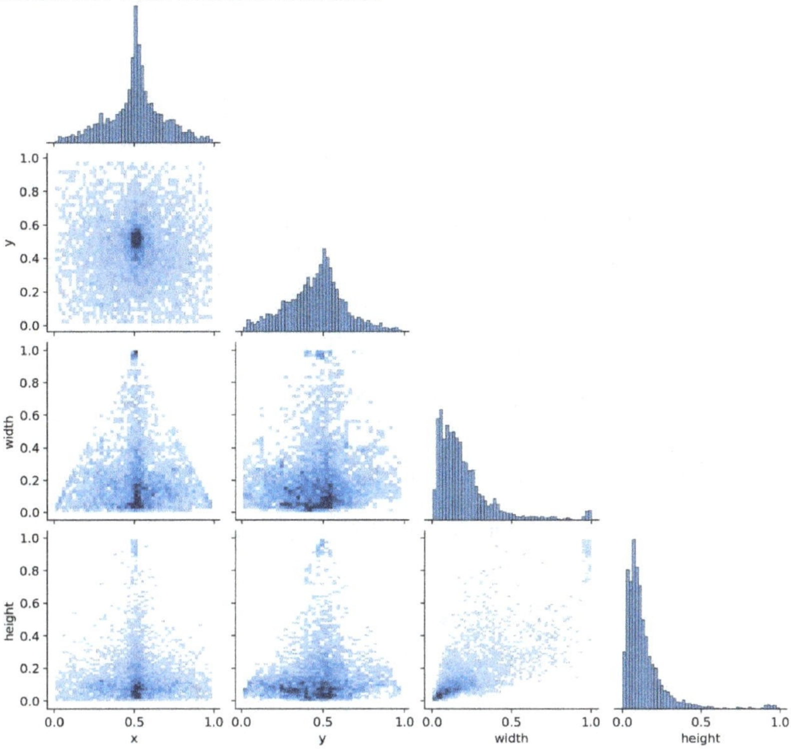

Fig. 3. Grid of histograms and scatter plots of geometric data coherence – a graphical interpretation showing objects located at the edges of photos and having different proportions than those placed centrally.

The F1-Confidence curve plotted in Figure 4 illustrates the harmonic mean of the model's precision and sensitivity as a function of the detection confidence threshold. Determining this function is essential to determining the optimal operating point of the parent application. The blue line representing the average for all object classes reaches its maximum score of 0.94 at a confidence threshold of 0.601, indicating that setting this parameter provides the optimal balance between detecting all vehicles and minimizing false alarms. The remaining thinner lines show that some vehicles, such as the IRIS-T SLM, are easier for the trained neural network model to unambiguously classify, while others require more precise tuning of the sensitivity threshold.

Fig.5 presents a quantitative error matrix of "mistakes," which complements the percentage version and shows the exact number of objects from the validation set that the model assigned to specific categories. This graph reveals not only theoretical performance but also the actual scale of testing, for example, how many T90 tanks or Su-27 aircraft were correctly identified relative to the number of misses. This ensures that high percentages are supported by a robust number of test samples and not simply due to a small number of a given class in the dataset.

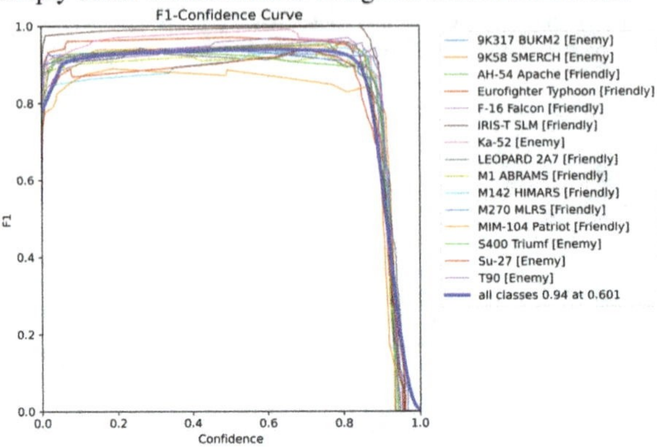

Fig. 4. Grid of histograms and scatter

Fig.5 presents a quantitative error matrix of "mistakes," which complements the percentage version and shows the exact number of objects from the validation set that the model assigned to specific categories. This graph reveals not only theoretical performance but also the actual scale of testing, for example, how many T90 tanks or Su-27 aircraft were correctly identified relative to the number of misses. This ensures that high percentages are supported by a robust number of test samples and not simply due to a small number of a given class in the dataset.

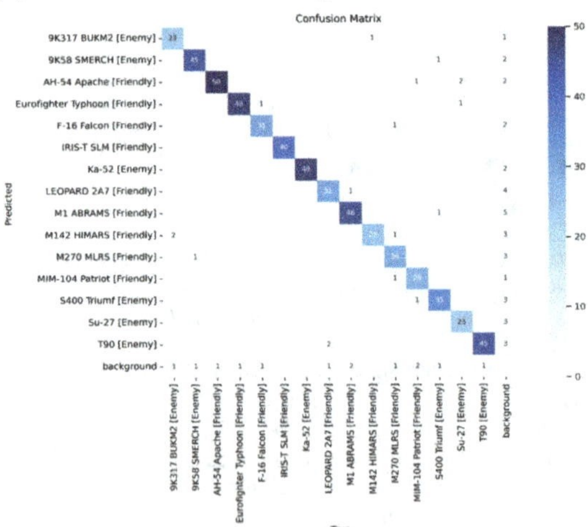

Fig. 5. Error matrix

Fig. 6 summarizes the training graphs, which document the neural network's learning process over 150 epochs, demonstrating a dynamic decline in the loss function and a systematic increase in key performance metrics. The left side of the graph illustrates how the model made fewer and fewer errors with each iteration in accurately enclosing and naming objects, while the right side shows increasing precision and the mAP index, which stabilizes towards the end of the training process. The flattening of the graphs in the final phase suggests that the model has reached its maximum potential for this dataset and has not overtrained, guaranteeing its stable performance on new, previously unseen battlefield images.

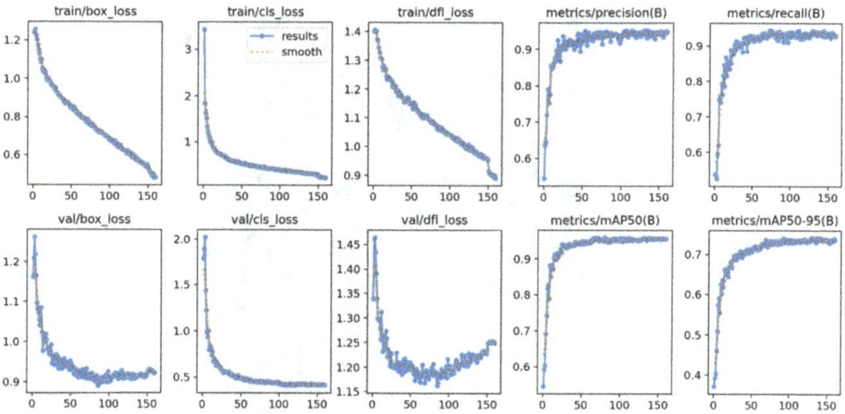

Fig. 6. The training graphs document the neural network's learning process over 150 epochs.

To verify the accuracy of automatic object detection, field tests were conducted. Vehicle silhouettes were placed against various backgrounds, including partially hidden ones. The tests were performed under various lighting conditions and at different distances from the targets. The vast majority of objects were detected correctly and unambiguously, which is particularly important for the application of a neural network trained in this way. Figures 7 and 8 present a sample set of photos from the field tests.

Fig. 7. Example sets of graphics used when training a neural network model

Fig. 8. Correct identification of aircraft and mobile robot control console during operation

Technological, Economic, Environmental, and Social Impact

The introduction of advanced neural network-based vision systems (YOLO11L) for rapid identification of military vehicles represents a significant step forward in military and security technology. The primary technological impact is centered on automation and precision, particularly in the context of mini-drone usage. Such vision systems enable autonomous mobile robot manipulators [6], which are crucial for unmanned combat platforms. The effective operation of AI systems is closely linked to the computational efficiency of the hardware; analysis of GPU computational efficiency directly translates into the speed and cost-effectiveness of model training [7]. Furthermore, the development and prototyping of actuators, such as proportional valves with piezoelectric actuators [8] and angular position control systems utilizing pneumatic artificial muscles [9], are crucial for advanced military robotics and the precise control of combat platforms. This process also requires the use of advanced engineering techniques, including laser forming technology [10] and the construction of device prototypes on processor-based platforms, e.g., the Delta parallel robot [11]. This entire development aligns with the broader challenges of the design of experiments (DOE) methodology in relation to the transformation towards Industry 4.0 [12].

The economic aspect of implementing AI systems is primarily related to optimizing and minimizing operating costs, as well as increasing the efficiency of equipment production. The use of fuzzy regression in production processes, such as flywheel assembly [13], as well as a modified sequential-binary approach for fuzzy operations [14], enables more precise quality control and reduces material and time waste. Methods such as ex-ante risk assessment using interval analysis [15] help manage uncertainty and financial project risk. Furthermore, in the context of mass-producing components, ensuring high reliability and durability is crucial. To this end, static load analysis of key components, such as slewing bearings [16], and fatigue failure of components [17] are analyzed, which translates into a longer product lifecycle and lower maintenance costs. In the context of Industry 4.0 and its associated barriers [18], models such as the Rank Model for

Materials Research Proceedings 62 (2026) 253-261 https://doi.org/10.21741/9781644904015-33

detecting casting inconsistencies [19] contribute to faster and more cost-effective quality control in the supply chain.

The environmental impact of military technologies (and the systems supporting them) is often overlooked, but it is significant, especially in the context of increased operational and production efficiency. This environmental impact is indirectly related to materials engineering and process optimization. For example, studying the stability of shape distribution in stainless steel powders [20] has applications in improving powder technology, which can lead to more material-efficient and less energy-intensive production. Although direct references are rare, optimization of energy systems and heat recovery [21] or the selection of optimal window types in construction [22] demonstrate a global trend toward sustainability [23], which may also apply to military infrastructure. Process control, such as non-parametric analyses [24], enables a deeper understanding of physical systems. Data dimensionality reduction, such as PCA [25], in process analysis also contributes to more efficient use of computational resources, reducing energy consumption.

The social aspect is inextricably linked to issues of safety, responsibility, and the impact of advanced technology on life and health. Military materials technologies directly impact the safety of our own and allied forces. The use of advanced materials, such as bioceramic coatings in medicine [26] or composites with zirconium phase additions [27], demonstrates the application of advanced materials engineering in the context of life and health. More directly, safety is impacted by field research and testing of components, such as containers for electric detonators [28]. Furthermore, the use of specialized coatings, such as fire-resistant [29] or anti-graffiti [30] coatings on railway rolling stock (which can be used in military transport), increases public safety and the durability of infrastructure. In the context of reliability, nonparametric uncertainty assessment in process analysis [31-33] is crucial for the reliability of critical systems.

Summary

The article describes a method for automatically identifying military vehicle silhouettes (both allied and enemy) using a vision system and the YOLOv11L neural network – an enlarged version of the YOLOv11 model. The aim is to quickly and unambiguously distinguish vehicles on the battlefield (e.g., for use with strike drones) based on their graphical silhouettes. The LabelStudio software was used for object annotation, and the network was trained in Google Colab on an A100 GPU due to the limited computational power of standard computers. Approximately 3,000 images were used, containing a total of over 12,000 silhouettes of fifteen types of vehicles (Polish/Allied and Russian), under various lighting conditions, on different backgrounds, and at different scales. Analysis of the dataset statistics (plots on pp. 2–3) confirms good class balance and natural object distribution in the images. The F1–Confidence curve (Fig. 4) indicates that the optimal trade-off between recall and precision was achieved at a confidence threshold of 0.601 (F1 $\approx$ score 0.94). The confusion matrix (Fig. 5) indicates high effectiveness in recognizing specific vehicle types, provided a sufficient number of validation examples is available. The training progresses over 150 epochs (Fig. 6), demonstrating a stable improvement of metrics and the absence of overfitting. Field tests with real-world scenes – including partially occluded objects and varying illumination (Figs. 7–8) – confirmed that most vehicles are detected correctly at a speed of approximately 7–15 frames per second (fps) on a laptop with an RTX 3060. Further work is planned on modifying the architecture and the network training procedure.

References

[1] T. Hiippala, Recognizing military vehicles in social media images using deep learning, 2017 IEEE International Conference on Intelligence and Security Informatics: Security and Big Data, ISI 2017. https://doi.org/10.1109/ISI.2017.8004875

[2] D. Legendre, J. Vankka, Military Vehicle Recognition with Different Image Machine Learning Techniques, Communications in Computer and Information Science 1283 (2020) 220-242. https://doi.org/10.1007/978-3-030-59506-7_19

[3] A. Sachenko et al., Real-Time Military Vehicle Classification via Convolutional Neural Networks, Modern Data Science Technologies (MoDaST 2025) Doctoral Consortium, CEUR Workshop Proceedings 4005 (2025) 239-251.

[4] H. Liu et al., A Military Object Detection Model of UAV Reconnaissance Image and Feature Visualization, Applied Sciences 12 (2022) art.12236. https://doi.org/10.3390/app122312236

[5] H. Surrisyad, W. Wahyono, A Fast Military Object Recognition using Extreme Learning Approach on CNN, International Journal of Advanced Computer Science and Applications 11.12 (2020). https://doi.org/10.14569/IJACSA.2020.0111227

[6] A. Annusewicz-Mistal et al., Autonomous Manipulator of a Mobile Robot Based on a Vision System, Applied Sciences (Switzerland) 13 (2023) art. 439. https://doi.org/10.3390/app13010439

[7] J. Pietraszek et al., Factorial approach to assessment of GPU computational efficiency in surrogate models, Advanced Materials Research 874 (2014) 157-162. https://doi.org/10.4028/www.scientific.net/AMR.874.157

[8] P.A. Laski, Proportional valve with a piezoelectric actuator, EPJ Web of Conferences 143 (2017) art. 02064. https://doi.org/10.1051/epjconf/201714302064

[9] J. Takosoglu, Angular position control system of pneumatic artificial muscles, Open Engineering 10 (2020) 681-687. https://doi.org/10.1515/eng-2020-0077

[10] N. Radek et al., Technology of laser forming, Manufacturing Technology 14 (2014) 428-431.

[11] P.A. Laski, M. Smykowski, Using a development platform with an stm32 processor to prototype an inexpensive 4-dof delta parallel robot, Sensors 21 (2021) art. 7962. https://doi.org/10.3390/s21237962

[12] J. Pietraszek et al., Challenges for the DOE methodology related to the introduction of Industry 4.0, Prod. Eng. Arch. 26 (2020) 190-194. https://doi.org/10.30657/pea.2020.26.33

[13] J. Pietraszek, Fuzzy regression compared to classical experimental design in the case of flywheel assembly, Lecture Notes in Computer Science 7267 LNAI (2012) 310-317. https://doi.org/10.1007/978-3-642-29347-4_36

[14] J. Pietraszek, The modified sequential-binary approach for fuzzy operations on correlated assessments, Lecture Notes in Computer Science 7894 LNAI (2013) 353-364. https://doi.org/10.1007/978-3-642-38658-9_32

[15] E. Kozien, M.S. Kozien, The EX ante risk assessment in the project in interval analysis description, Materials Research Proceedings 17 (2020) 31-35. https://doi.org/10.21741/9781644901038-5

[16] M. Krynke, R. Ulewicz, Analysis of the influence of slewing bearing mounting on their static load capacity, Transp. Res. Proc. 40 (2019) 745-750. https://doi.org/10.1016/j.trpro.2019.07.105

[17] R. Ulewicz et al., The investigation of the fatigue failure of passenger carriage draw-hook, Engineering Failure Analysis 104 (2019) 609-616. https://doi.org/10.1016/j.engfailanal.2019.06.036

[18] R. Ulewicz et al., Mining Industry 4.0 – Opportunities and Barriers, Acta Montanistica Slovaca 27 (2022) 291-305. https://doi.org/10.46544/AMS.v2i2.02

[19] R. Ulewicz et al., A Rank Model of Casting Non-Conformity Detection Methods in the Context of Industry 4.0, Materials 16 (2023) art. 723. https://doi.org/10.3390/ma16020723

[20] J. Pietraszek, A. Gadek-Moszczak, The smooth bootstrap approach to the distribution of a shape in the ferritic stainless steel AISI 434L powders, Solid State Phenomena 197 (2013) 162-167. https://doi.org/10.4028/www.scientific.net/SSP.197.162

[21] N. Fialko et al., Environmental aspects of heat recovery systems of boiler plants, E3S Web of Conf. 100 (2019) art. 00015. https://doi.org/10.1051/e3sconf/201910000015

[22] J.M. Djoković et al., Selection of the Optimal Window Type and Orientation for the Two Cities in Serbia and One in Slovakia, Energies 15 (2022) art. 323. https://doi.org/10.3390/en15010323

[23] A. Pacana et al., Influence of natural aggregate crushing process on crushing strength index, Sustainability 13 (2021) art. 8353. https://doi.org/10.3390/su13158353

[24] E. Skrzypczak-Pietraszek, J. Pietraszek, Chemical profile and seasonal variation of phenolic acid content in bastard balm (Melittis melissophyllum L., Lamiaceae), Journal of Pharmaceutical and Biomedical Analysis 66 (2012) 154-161. https://doi.org/10.1016/j.jpba.2012.03.037

[25] J. Pietraszek, E. Skrzypczak-Pietraszek, The uncertainty and robustness of the principal component analysis as a tool for the dimensionality reduction, Solid State Phenomena 235 (2015) 1-8. https://doi.org/10.4028/www.scientific.net/SSP.235.1

[26] A. Dudek, Investigations of microstructure and properties in bioceramic coatings used in medicine, Archives of Metallurgy and Materials 56 (2011) 135-140. https://doi.org/10.2478/v10172-011-0015-y

[27] A. Dudek, Microstructure and properties of the composites: Hydroxyapatite with addition of zirconia phase, Journal of Engineering Materials and Technology 133 (2011) art. 021006. https://doi.org/10.1115/1.4003104

[28] P. Krysiak et al., The design and field tests of a box for "ERG" electric blasting caps, Lecture Notes in Mechanical Engineering (2017) 285-292. https://doi.org/10.1007/978-3-319-50938-9_29

[29] N. Radek, R. Dwornicka, Fire properties of intumescent coating systems for the rolling stock, Communications – Scientific Letters of the University of Žilina 22 (2020) 90-96. https://doi.org/10.26552/com.C.2020.4.90-96

[30] N. Radek et al., Technology and application of anti-graffiti coating systems for rolling stock, METAL 2019 – 28th Int. Conf. Metall. Mater. (2019) 1127-1132.

[31] J. Pietraszek, Fuzzy regression compared to classical experimental design in the case of flywheel assembly, Lecture Notes in Computer Science 7267 LNAI (2012) 310-317. https://doi.org/10.1007/978-3-642-29347-4_36

[32] A. Szczotok, R. Przeliorz, Phase transformations in CMSX-4 nickel-base superalloy, IOP Conf. Series: Mater. Sci. Eng. 35 (2012) art. 012005. https://doi.org/10.1088/1757-899X/35/1/012005

[33] J. Pietraszek et al., Non-parametric assessment of the uncertainty in the analysis of the airfoil blade traces, METAL 2017 – 26th Int. Conf. Metall. Mater. (2017) 1412-1418.

An Analysis of the Implementation of Robotic Process Automation in Shipping Logistics. A Case Study

PACANA Andrzej[1,a *], PILCH Milosz[2,b] and Anna G. PIETRASZEK[3,c]

[1]Rzeszow University of Technology, The Faculty of Mechanical Engineering and Aeronautics al. Powstańców Warszawy 8, 35-959 Rzeszów, Poland, ORCID: 0000-0003-1121-6352

[2]Safran Aircraft Engines, ul.Południowa 23, 39-120 Sedziszow Małopolski, Poland, ORCID: 0009-0003-2306-3215

[3]Cracow University of Economics, ul. Rakowicka 27, 31-510 Cracow, Poland, ORCID: 0009-0001-2557-2422

[a]app@prz.edu.pl, [b]8pilchu@gmail.com, [c]237233@student.uek.krakow.pl

Keywords: Robotic Process Automation, RPA, Automation, Sales Logistics, Logistics 4.0, Operational Efficiency, Digital Transformation

Abstract. This article presents the results of a case study on the implementation of robotic process automation (RPA) in sales logistics and the shipping of finished products to customers at a company in the aerospace and defense industries. The RPA implementation encompassed the automation of ERP (Enterprise Resource Planning) tasks, including issuing finished products from stock, generating Certificates of Conformity (CofC), and archiving and printing the generated certificates. The study was conducted to assess the impact of RPA on process efficiency and to evaluate the involvement of logistics staff during weekly shipments [1-3]. An analysis of data extracted from the ERP system revealed that although the use of RPA did not shorten the time required to sell and validate a single finished product set, it nonetheless significantly reduced the amount of human labor involved in the process. Based on historical data, before the RPA implementation, logistics employees processed a certain average number of sets per day while also performing other duties. After implementation, finished-product sets could be processed automatically one after another throughout the day without interruption, which enabled nearly twice as many sets to be sold in a single day, while also eliminating the need for employee involvement in monotonous tasks related to sales and certificate generation. The findings confirm that implementing RPA in sales and shipping logistics significantly contributed to increased productivity and the effective use of human resources. The example analyzed constitutes an important element of enterprises' digital transformation and Industry 4.0.

Introduction

Industry 4.0 has brought the integration of the physical world and machines with the Internet and accompanying information technologies. People and machines can communicate with one another independently. Information flows both vertically (between machines and the company's IT department) and horizontally (between machines involved in the production process) [4]. Ongoing digitization – where successive production and management processes are transitioned into the digital sphere through information technologies such as cloud computing, artificial intelligence, data analytics, and so-called big data – enables deeper monitoring and optimization of these processes. In turn, further robotization enables collaboration between robots and humans within cognitive systems, combining human creativity and flexibility with the precision and speed of robotic work [5].

Rapid technological development and steadily increasing competitive pressure compel enterprises to invest in innovative technologies such as robotic process automation – not only as tools to streamline work, but also as components of strategic development. With the ever-growing complexity of logistics processes and a shortage of qualified personnel, solutions that enable the automation of repetitive administrative activities related to these processes become crucial. The topic of RPA is highly important and timely in the era of digital transformation [6]. RPA enables the automation of repetitive and monotonous tasks that humans previously performed through software bots (robots). Once programmed, they can operate across the various IT systems and applications used in an enterprise. These robots can take over tasks previously performed by people, such as extracting content from documents and forms, opening websites with account login capabilities, transferring information from one software application to another, opening emails and their attachments, using computations and workflows, and reading from and writing to databases [7]. The impetus for implementing robotic process automation in the company under study was the increasing volume of customer orders, which resulted in a higher workload for logistics personnel. In subsequent years, this growth would have necessitated hiring an additional person to process shipments. Conversely, the process of shipping the finished product to the customer was an ideal candidate for RPA due to its stability, full standardization, and the absence of any need for subjective judgment – the finished product must be compliant to be shipped. The conclusions drawn from this analysis may contribute to the more effective implementation of RPA in other processes.

RPA in Logistics

Robotic process automation is a technology that enables the automation of tasks performed by humans in computer applications and systems by emulating human interactions with the user interface and orchestrating workflows. It complements classical BPM (Business Process Management) and BPA (Business Process Automation), as well as API (Application Programming Interface) and EDI (Electronic Data Interchange) system integration, and workstation scripts, allowing for deployment without modifying application code or changing the system. Three variants of RPA are distinguished: a robot operating under user supervision (the attended mode), where robots share a workstation with the user; robots operating autonomously on servers or in virtual environments (the unattended mode), managed by schedules and events; and robots that combine RPA with elements of artificial intelligence and machine learning to support decision-making. The architecture of RPA typically consists of an execution robot, an orchestrator (responsible for scheduling, queues, and task distribution), a log repository, and a connector layer (e.g., for email, a browser, or ERP, WMS (Warehouse Management System), and TMS (Transportation Management System) systems), and a security module (logs, passwords, and access control) [8,9].

In logistics domains, three integration patterns of robotic process automation predominate: UI-driven (User Interface) scenarios, in which the robot operates solely through the user interfaces of ERP/WMS/TMS systems; a hybrid approach, which combines user-interface automation with data access (selected APIs, SQL (Structured Query Language) queries to the database, exports of CSV (Comma-Separated Values) and XML (Extensible Markup Language) files); and event-driven scenarios, in which the robot responds to the occurrence of process events such as the appearance of a file or report, an order, or a carrier status update [10]. From the perspective of logistics management, RPA aligns with the concept of lean digital operations, which involves improving information flow in IT systems and administration by eliminating waste. Moreover, RPA serves as a bridge between traditional ERP systems and modern AI-based solutions, including cognitive automation, enabling enterprises to advance to a higher level of digital transformation [11].

The use of robotic process automation in logistics enables the automation of key transport-related processes. Robots prove effective in procurement and inventory management processes by pulling data from multiple sources and comparing it against customer expectations – while simultaneously using reports to optimize inventory levels in real-time. Planning and tracking shipments are supported by extracting information from email and carrier delivery schedules posted on their websites, thereby relieving employees of these activities. Invoice management involves utilizing robots to track overdue payments, analyze key information contained in invoices, and automate payment execution. By delegating these tasks to robots, the risk of errors is reduced – errors that could otherwise necessitate re-entering data, manually adding data to an invoice, or issuing a correction. RPA also greatly simplifies the creation of reports on order statuses, payment information, and transport fleet data. In conjunction with artificial intelligence, robots can acquire key data and generate ready-to-use reports based on it, minimizing manual work [12]. Based on the cited analyses, a research gap has been identified in the lack of descriptions of practical RPA implementations in the logistics domain, which is highly significant from the perspective of many strategic indicators. Examining such an implementation may be useful in this enterprise and, in the future, in many others pursuing continuous improvement.

Study of the Robotic Automation of a Logistics Process

The study was conducted at a company operating in the aerospace and defense industries, located within the Subcarpathian Aviation Valley. This valley brings together aerospace firms, research and development centers, and associated educational and training facilities. It is a large, global enterprise employing over 300 people at its plant located in southeastern Poland. Due to the nature of the industry, the company pursues continuous improvement, driving numerous innovative initiatives, including those in logistics. One such initiative is the implementation of robotic process automation. Shipping activities are organized weekly, and the customer requires a Certificate of Conformity to accompany the finished product. Production has the characteristics of mass production. As part of the RPA implementation, a robot was developed to perform actions within the ERP system and utilize computer software for validating the finished product and printing certificates. The robot runs on a dedicated computer located in the logistics office, where it can be started by various employees responsible for shipments and act on their behalf after they log in to the machine. The robot operates according to a hybrid pattern, combining user-interface automation for ERP systems without prior modifications with access to selected services via API integration. The robot was designed to perform the following process. Before being issued from the system, the finished product undergoes an additional compliance check in the ERP system. After this, the set is finally transferred to the finished-product warehouse and is ready for shipment. Sets are consolidated in the system's finished-product warehouse in larger quantities until the full safety stock is reached (or replenished), after which the RPA robot is started. The robot issues finished products from the system according to the customer's order and then generates a certificate of conformance. The logistics employee can initiate any number of certificates for printing via the robot while individual sets are being processed. The only parts of the process still performed manually by the logistics employee are: the final check in the ERP system; completing carrier waybills with carrier data, also in the ERP system; printing those waybills and delivering them to the warehouse together with the printed certificates, which must be attached to the pallets with the finished products. The time spent on these tasks represents a small portion of the sales and shipping process, which – due to robotic process automation – has been automated by approximately 80–90%.

Analysis of RPA Implementation Results

The data used in the study covers the number of processed finished-product sets for two analogous periods: three months before the implementation of robotic process automation and three months after its implementation. Only full working days were considered, which always consist of work on the first shift only, as well as the average number of sets processed on those days. The time to handle the weekly shipment is the period during which sets are processed, when the logistics employee also performs other duties and tasks or attends meetings. Due to the confidentiality of company data, the results are presented using relative indices, with the baseline state before the RPA implementation set to 100 units. Although the implementation of RPA did not shorten the time needed to process a single finished-product set (the robot mimics human work), it nevertheless significantly reduced human labor input, improving daily throughput. The impact of the RPA implementation on the shipping process is presented in Table 1.

Table 1. The Impact of the RPA Implementation on the Shipping Process

KPI	How measured?	Before RPA	After RPA	Change
Daily throughput	Number of sets per day (index)	100	192	+92%
Time to handle the weekly shipment	Calendar days per shipment (index)	100	52	-48%
Estimated work time for 1 set	Employee active time (index)	100	17	-83%

Analyzing the results in Table 1, it can be observed that, thanks to the use of RPA, the time a logistics employee must devote to manual tasks related to processing shipments has been reduced to a minimum, essentially boiling down to starting the robot and arranging transportation. Moreover, the shipping process itself can be started later than before. In contrast, under manual processing, the logistics employee would have had no time for any other activities (having to "work like a robot," performing highly repetitive tasks). As a result, the production department has more time to deliver the finished product to the warehouse for shipment, resulting in a highly efficient and cohesive process. Ultimately, because the robot can work continuously, daily throughput has nearly doubled compared to the logistics employee's previous manual work. The results confirm that implementing RPA in the company under study increases throughput and efficiency without the need to hire additional staff. This is relevant considering long-term forecasts by aircraft manufacturers, which indicate that the global passenger and cargo fleet is expected to grow significantly in the future. Both Europe's Airbus and U.S.-based Boeing project an increase in the global fleet to around 50,000 aircraft by 2044, which, according to Airbus, represents a 99% increase and, according to Boeing, about an 80% increase relative to 2025 levels [13-15]. This will contribute to increased demand for logistics and production processes, and the use of RPA will enable maintaining operational efficiency under rising workloads.

Discussion

Given the complexity of the process and the use of not only the ERP system but also additional applications to certify the customer's finished-product sets, it was necessary to assess the feasibility of integrating the interfaces of these applications with robotic process automation tools. Although, in theory, once the robot takes control of the keyboard and mouse and optical character recognition (OCR) is applied, it is possible to run the process without application programming interfaces (APIs), in practice, their integration is the better solution. This shortens the robot's execution time and reduces the risk of an error that would interrupt its operation.

Because robotic process automation streamlines day-to-day work, enterprises become more profitable, more flexible, and better prepared for change. Employee satisfaction, engagement, and productivity increase when they are relieved of repetitive and monotonous tasks in the shipping process. The implementation of RPA in the company under study was a sound decision due to:

1. A stable sales and shipping process in which there is no need for the logistics employee to intervene to decide whether a given finished product set is suitable for shipment. When a set appears in the ERP system's finished-product warehouse, it is compliant and ready to ship. Therefore, the robot can issue and certify set after set without third-party involvement.

2. High repetitiveness and monotony of the sales and shipping process, during which the employee merely clicked through screens and sent emails to certify sets. This could lead to employee fatigue and result in errors.
3. Time freed up, which allowed the logistics employee to focus on duties they perform best and value most—innovation, collaboration with other departments, and interaction with the customer—securing material for production and focusing on delivering the finished product on time. The enterprise benefited from increased satisfaction and productivity.
4. A first step toward digital transformation, aimed at the potential deployment of artificial intelligence assistants. Combining RPA with artificial intelligence—in the form of machine-learning models, natural-language processing, and character and image recognition—can radically enhance robots' ability to handle processes that require understanding documents, interpreting on-screen content, and executing employees' instructions

The next natural step for the company under review is to implement RPA in other logistics processes and departments. This involves selecting appropriate processes characterized by high repeatability, system stability, and a low exception rate at the point of data entry. Risk analysis and return on investment (ROI) assessment will also be necessary.

Economic and Technological Impact

The automation of robotic processes discussed in the article is a valuable guide not only for existing robotic implementations [16-18], but also for the machinery industry, where there are areas requiring high precision and, at the same time, posing a potential environmental hazard to workers. These include processes related to the production of special anti-wear coatings [19-21], including DLC [22,23] and ESD [24-26], as well as fire-retardant coatings [27]. Appropriate optimization of the mentioned processes, requiring the use of appropriate DOE techniques [28] together with appropriate methodological support in the case of small data sets [29,30] or incomplete/uncertain data [31,32], as well as the use of appropriately advanced analytical methods [33,34], should significantly improve process management [35,36]. As a result, there should be a significant improvement in occupational health and safety, but also in the quality and durability of products [37], and thus a reduction in the environmental burden of production and post-operational waste [38,39].

Summary

Robotic process automation is a technology that enables the automation of tasks performed by employees within an enterprise's computer applications and systems. Implementing RPA is not an easy undertaking. Therefore, this article analyzes a case study of RPA implementation in the area of sales logistics at a company in the aerospace and defense sectors, which belongs to the Aviation Valley industry association in Poland. The study examines the sales process in which shipments are organized on a weekly basis, and the customer requires a Certificate of Conformity to accompany the finished product. It also characterizes the post-implementation operating model, under which, during the sales process, the logistics employee has been relieved of most activities, with their work reduced to launching the robot and handling the shipment waybills. An analysis was conducted of the results from two analogous periods: three months before the implementation of RPA and three months after its implementation. The data used for the analysis comes directly from the company's ERP system. Due to the confidentiality of the results, they are presented using baseline indices for the pre-implementation state. Although the time required to process a single finished-product set did not change, the employee's labor input decreased dramatically, and the process itself was automated to over 80%. Daily throughput nearly doubled, which shortened the overall lead time for the shipping process, allowing it to start later and providing greater operational flexibility for the production department to deliver the finished product to the warehouse. In the discussion section, the practical aspects of the robot's operation are outlined,

Materials Research Proceedings 62 (2026) 262-269 https://doi.org/10.21741/9781644904015-34

along with a preference for integrating systems and applications at the API level over relying solely on the UI – thereby reducing the risk of errors and interruptions to the robot's work. The advantages of implementing robotic process automation are highlighted, particularly the absence of third-party involvement during the robot's operation and the increased satisfaction of logistics employees, who were able to devote the time saved through implementation to other tasks. The direction the company should pursue is also outlined, treating the automation of the shipping process as a prelude to further implementations in other processes, including those supporting production. The results obtained can also be applied to implement RPA in other enterprises and organizations.

References

[1] A. Pacana et al., Improving the Process of Product Design in a Phase of Life Cycle Assessment (LCA). Processes 11 (2023) art.2579. https://doi.org/10.3390/pr11092579

[2] A. Pacana, D. Siwiec, Method of Determining Sequence Actions of Products Improvement. Materials 15 (2022) art.6321. https://doi.org/10.3390/ma15186321

[3] D. Siwiec, A. Pacana, A New Model Supporting Stability Quality of Materials and Industrial Products. Materials 15 (2022) art. 4440. https://doi.org/10.3390/ma15134440

[4] B. Siuta-Tokarska, Przemysł 4.0 i sztuczna inteligencja: szansa czy zagrożenie dla realizacji koncepcji zrównoważonego i trwałego rozwoju?, Nierówności społeczne a wzrost gospodarczy 65 (2021) 7-26. http://repozytorium.ur.edu.pl/handle/item/6274

[5] W. Furmarek, Najważniejsze idee czwartej rewolucji przemysłowej (Industry 4.0), Dydaktyka Informatyki T. 13 (2018) 55-63. http://repozytorium.ur.edu.pl/handle/item/3932

[6] L. Ivančić et al., Robotic process automation: systematic literature review, Lecture Notes in Bus. Inf. Proc. 361 (2019) 280-295. https://doi.org/10.1007/978-3-030-30429-4_19

[7] W.M.P. van der Aalst et al., Robotic Process Automation, Bus. Inf. Syst. Eng. 60 (2018) 269-272. https://doi.org/10.1007/s12599-018-0542-4

[8] A. Pacana et al., Decision support in the production of packaging films by cost-quality analysis. Przemysł Chemiczny 95 (2016) 1042-1044. https://doi.org/10.15199/62.2016.5.31

[9] T. Taulli, The Robotic Process Automation Handbook: A Guide to Implementing RPA Systems, Apress, Berkeley, 2020.

[10] J. Krakau et al., Robotic process automation in logistics: Implementation model and factors of success, in: C. Jahn, W. Kersten, C.M. Ringle (Eds.), Adapting to the Future: Maritime and City Logistics in the Context of Digitalization and Sustainability. Proc. Hamburg Int. Conf. Logistics (HICL) 32 (2021) 219-256. https://doi.org/10.15480/882.4005

[11] C. Flechsig, The impact of intelligent process automation on purchasing and supply management – initial insights from a multiple case study, Lecture Notes in Logistics (2021) 67-89. https://doi.org/10.1007/978-3-030-85843-8_5

[12] S. Khan et al., Application of robotic process automation (RPA) for supply chain management, smart transportation and logistics, International Journal of Health Sciences 6 (2022) 11051-11063. https://doi.org/10.53730/ijhs.v6ns3.8554

[13] Airbus Global Market Forecast 2025-2044, Aviation industry outlook, (2025), https://www.airbus.com/en/products-services/commercial-aircraft/global-market-forecast [access date: 28-10-2025]

Materials Research Forum LLC
https://doi.org/10.21741/9781644904015-34

[14] A. Pacana et al., Analysis of the incompatibility of the product with fluorescent method, Metalurgija 58 (2019) 337-340.

[15] Boeing. Commercial Market Outlook 2025–2044, (2025), https://www.boeing.com/commercial/market/commercial-market-outlook [access date: 28-10-2025]

[16] P. Wos et al., Bricklaying robot lifting and levelling system, Communications – Sci. Letters Univ. Žilina 23 (2021) B257-B264. https://doi.org/10.26552/COM.C.2021.4.B257-B264

[17] P.A. Laski, M. Smykowski, Using a development platform with an STM32 processor to prototype an inexpensive 4-dof delta parallel robot, Sensors 21 (2021) art. 7962. https://doi.org/10.3390/s21237962

[18] A. Annusewicz-Mistal et al., Autonomous Manipulator of a Mobile Robot Based on a Vision System, Applied Sciences 13 (2023) art. 439. https://doi.org/10.3390/app13010439

[19] W. Zórawski et al., Plasma-sprayed composite coatings with reduced friction coefficient, Surf. Coat. Technol. 202 (2008) 4578-4582. https://doi.org/10.1016/j.surfcoat.2008.04.026

[20] L. Adamczyk et al., Electrochemical preparation of composite coatings of 3,4-etylenodioxythiophene (EDOT) and 4-(pyrrole-1-yl) benzoic acid (PyBA) with heteropolyanions, Mater. Chem. Phys. 144 (2014) 418-424. https://doi.org/10.1016/j.matchemphys.2014.01.012

[21] M. Scendo et al., Influence of laser power on the corrosive resistance of WC-Cu coating, Surf. Coat. Technol. 259 (2014) 401-407. https://doi.org/10.1016/j.surfcoat.2014.10.062

[22] A. Kalinowski et al., The effect of laser infrared radiation on the surface modification of DLC coatings, AIP Conf. Proc. 3130 (2024) art. 020014. https://doi.org/10.1063/5.0204985

[23] A. Kalinowski et al., The effect of micro-machining with a single-mode and multi-mode fiber laser on the surface of DLC coatings, METAL 2024 – 33rd Int. Conf. Metall. Mater. (2024) 328-333. https://doi.org/10.37904/metal.2024.4952

[24] M. Scendo et al., Influence of laser treatment on the corrosive resistance of WC-Cu coating produced by electrospark deposition, Int. J. Electrochem. Sci. 8 (2013) 9264-9277.

[25] N. Radek et al., The impact of laser processing on the performance properties of electro-spark coatings, World Congress in Computational Mechanics and ECCOMAS Congress 1000 (2021). https://doi.org/10.23967/wccm-eccomas.2020.336

[26] N. Radek et al., Technology and applications of ESD coatings before and after laser processing, METAL 2023 – 32nd Int. Conf. Metall. Mater. (2024) 500-505. https://doi.org/10.37904/metal.2023.4727

[27] N. Radek, R. Dwornicka, Fire properties of intumescent coating systems for the rolling stock, Communications – Sci. Letters Univ. Žilina 22 (2020) 90-96. https://doi.org/10.26552/com.C.2020.4.90-96

[28] J. Pietraszek et al., Challenges for the DOE methodology related to the introduction of Industry 4.0, Prod. Eng. Arch. 26 (2020) 190-194. https://doi.org/10.30657/pea.2020.26.33

[29] J. Pietraszek, A. Gadek-Moszczak, The smooth bootstrap approach to the distribution of a shape in the ferritic stainless steel AISI 434L powders, Solid State Phenomena 197 (2013) 162-167. https://doi.org/10.4028/www.scientific.net/SSP.197.162

[30] A. Gądek-Moszczak et al., The bootstrap approach to the comparison of two methods applied to the evaluation of the growth index in the analysis of the digital X-ray image of a bone

regenerate, Studies in Comp. Intell. 572 (2015) 127-136. https://doi.org/10.1007/978-3-319-10774-5_12

[31] J. Pietraszek, Fuzzy regression compared to classical experimental design in the case of flywheel assembly, Lecture Notes in Computer Science 7267 LNAI (2012) 310-317. https://doi.org/10.1007/978-3-642-29347-4_36

[32] J. Pietraszek, The modified sequential-binary approach for fuzzy operations on correlated assessments, Lecture Notes in Computer Science 7894 LNAI (2013) 353-364. https://doi.org/10.1007/978-3-642-38658-9_32

[33] A. Gądek-Moszczak et al., Nano X-ray Tomography Application for Quantitative Surface Layer Geometry Analysis after Laser Beam Modification, Materials 15 (2022) art. 5935. https://doi.org/10.3390/ma15175935

[34] Ł.J. Orman et al., Comparative Analysis of Indoor Environmental Quality and Self-Reported Productivity in Intelligent and Traditional Buildings, Energies 16 (2023) art. 6663. https://doi.org/10.3390/en16186663

[35] D. Siwiec et al., Improving the non-destructive test by initiating the quality management techniques on an example of the turbine nozzle outlet, Materials Research Proceedings 17 (2020) 16-22. https://doi.org/10.21741/9781644901038-3

[36] R. Ulewicz et al., Logistic controlling processes and quality issues in a cast iron foundry, Materials Research Proceedings 17 (2020) 65-71. https://doi.org/10.21741/9781644901038-10

[37] T. Lipiński, R. Ulewicz, The effect of the impurities spaces on the quality of structural steel working at variable loads, Open Engineering 11 (2021) 233-238. https://doi.org/10.1515/eng-2021-0024

[38] A. Pacana, D. Siwiec, Method of Fuzzy Analysis of Qualitative-Environmental Threat in Improving Products and Processes (Fuzzy QE-FMEA), Materials 16 (2023) art. 1651. https://doi.org/10.3390/ma16041651

[39] R. Ulewicz et al., Sustainable Vehicle Design Considering Quality Level and Life Cycle Environmental Assessment (LCA), Energies 16 (2023) art. 8122. https://doi.org/10.3390/en16248122

Functional Materials for 24/7 Autonomy: EMC, Infrared Signature, and Environmental Durability

Ernest SZAJNA[1,a], Agata DUDEK[2,b *], Maciej WIELGOSZ[3,c], Yaroslav HIRNYAK[4,d], Konrad SZAJNA[1,e]

[1]Czestochowa University of Technology, Czestochowa, Poland

[2]WEA TechLab sp. z o.o., Dąbrowa Górnicza, Poland

[3]AGH University of Krakow, Kraków, Poland, Academic Computer Centre CYFRONET AGH, Kraków, Poland

[4]Skyfrr ApS, DK-5270 Odense N, Danmark

[a]e.szajna@weatechlab.com, [b]agata.dudek@pcz.pl, [c]wielgosz@agh.edu.pl, [d]yaro@skyfrr.one, [e]k.szajna@iseraph.dev

Keywords: VTOL UAVs, Electromagnetic Compatibility (EMC), RF Materials, Radomes, Thermal Management, Infrared (IR) Signature, Environmental Durability, Reliability

Abstract. This paper distills materials requirements for around-the-clock autonomy in VTOL UAVs across three domains: EMC and communications, thermal and infrared management, and environmental durability. We map system functions, such as GNSS-independent navigation, anti-jamming capability, and mesh networking, to material and integration choices. These include conductive and absorbing shielding with low-loss RF laminates, radomes with controlled permittivity and low loss tangent in the operating band with uniform thickness to limit pattern distortion, low impedance heat paths with emissivity engineered coatings, and structures and barrier systems that preserve mechanical and dielectric properties after exposure to moisture, ultraviolet radiation, and thermal cycling. We describe trade-offs between shielding and antenna efficiency, emissivity and cooling, and mass versus resistance to erosion and fatigue, and codify the corresponding selection criteria. A representative layered reference architecture (RA) serves as the foundation for the analysis, providing a framework for the co-design case study. We close with an outlook on an AI-tunable multimodal camouflage layer that extends the same co-design principles.

Introduction

An autonomous system is a closed-loop arrangement comprising perception (sensor data acquisition), localization and navigation, decision-making (planning, collision avoidance, and coordination), and execution (control of propulsion and actuators). The operator defines objectives and constraints, while the system maintains operational continuity independently. In practice, this applies to VTOL-class drones capable of route planning, swarm operation, long-range communications, and continued mission execution with limited access to GNSS.

The new generation of such solutions is built on a layered architecture that integrates mechanics and aerodynamics, electronics and radio communications, navigation with autonomy algorithms, networking and security layers, and HMI. This arrangement enables 24/7 operation and long-duration, cooperative missions of multiple unmanned vehicles in varied operating conditions. The software-defined paradigm remains essential, but critical functions, such as cooperative navigation and reliable long-range communications, as well as a hardware root of trust, require specialized hardware. Whether functional layers operate stably depends on material properties and the selected manufacturing technologies, ranging from the load-bearing structure to enclosures and antenna radomes, as well as RF path elements and the heat-removal system.

Material and manufacturing choices set the limits of achievable performance and determine the architecture of parts and assemblies. Priority is given to high-stiffness, low-mass composites (CFRP with lay-ups tailored to load paths and lightweight cores), additive manufacturing of structural elements (including continuous fiber reinforcement with a focus on isotropy and vibration resistance), shape-memory materials for adaptive elements, thermal solutions that stabilize temperature and infrared signature, and materials for RF paths, both for shielding and absorption of interference and for building low-loss transmission lines and radomes with controlled dielectric permittivity.

From these areas, follow the requirements at the part and subassembly levels: low mass with high stiffness and vibration damping, dimensional and thermal stability, adequate thermal conductivity, controlled IR emissivity, electromagnetic properties that ensure shielding or transmission of signals, and environmental durability with maintainability.

The importance of these issues is increasing in today's higher-risk environment. Ongoing armed conflicts, deliberate disruption and spoofing of navigation, jamming of communications, operation without infrastructure, and extreme weather phenomena all demand higher hardware resilience. In such conditions, mission success is determined not only by algorithms. Equally important is the system's materials foundation—the ability to maintain communications, control signatures, and preserve mechanical and electrical parameters after environmental loads.

In the sections that follow, requirements in EMC, IR, and durability are mapped directly to material decisions adopted in a representative layered reference architecture (RA), used here solely as a neutral case study [1-10].

Reference Architecture for 24/7 Autonomy: Design Assumptions and Programmability

We define a system-level reference architecture (RA) for next-generation autonomous robotics. It enables continuous, multi-domain operation by fleets of unmanned vehicles, singly or in swarms. The RA spans the full stack: structures and aerodynamics, power and propulsion, custom electronics, RF front-ends and sensors, navigation, perception and autonomy, distributed networking and timing, cryptography and cybersecurity, and human–machine interfaces. It fixes module boundaries, data, RF interfaces, and performance budgets required for EMC, thermal management, and environmental durability.

Fig. 1. Reference architecture functional stack and collaboration scale

The reference architecture is layered, with higher-level functions built upon lower-layer functions, starting from the hardware. Changes span the entire technology stack, encompassing mechanical and aerodynamic structures, radio electronics, navigation and autonomy algorithms, networking and security, as well as human–machine interfaces. The core remains a software-

Materials Research Proceedings 62 (2026) 270-276
https://doi.org/10.21741/9781644904015-35

defined approach, extended to hardware elements as well. Some critical functions—such as cooperative navigation, long-range communications, and a hardware root of trust—require dedicated modules; software alone is not sufficient. The layer structure and function allocation are shown in Fig. 2.

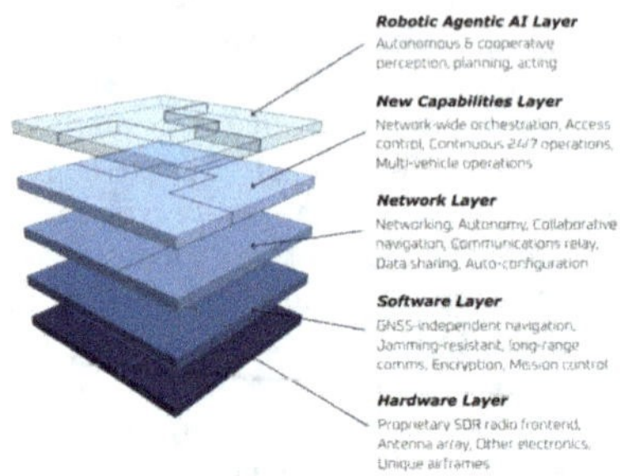

Fig. 2. Layered structure of the reference system

Materials are part of the programmable architecture: the stability of the hardware's electromagnetic and thermal parameters conditions the performance of navigation, communications, and autonomy. System operating requirements translate directly into requirements for the materials used in parts and assemblies. In communications and networking, it is crucial to suppress electromagnetic interference while keeping losses low in high-frequency paths. This drives the use of shielding and absorbing materials, low-loss laminates, and ceramics in RF sections, as well as radomes with controlled permittivity and low loss. For 24/7 navigation and autonomy, the airframe needs low mass with high stiffness, resistance to vibration and fatigue, dimensional stability, and control of the infrared signature. Proper selection of materials and manufacturing processes is a prerequisite for achieving the required performance of the entire system.

EMC and Communications: Materials and RF Path Integration

EMC and communications design in unmanned systems requires the concurrent selection of materials, geometry, and the antenna/RF layout. Already at the packaging stage, return-current paths, grounding points, and zones with elevated dielectric requirements are planned, ensuring that, after mechanical integration, the link budget is preserved, crosstalk and unintended emissions are minimized, and antenna matching remains stable across the full environmental envelope. Enclosures and internal bulkheads must form a continuous shield with high surface conductivity and uniform thickness. Shielding continuity is ensured by properly prepared contact surfaces and conductive gaskets at joints and access panels. RF paths rely on dielectrics with controlled permittivity and low loss. In practice, laminates based on PTFE or LCP are used, as well as ceramic LTCC modules.

The antenna radome should be as "radio-transparent" as possible, meaning it should attenuate the signal as little as possible in the operating bands and exhibit a constant, uniform permittivity so as not to detune the antenna or distort its radiation pattern. The most common choices are GFRP (glass-fiber-reinforced polymer) laminates or QFRP (quartz-fiber-reinforced polymer). Where mass is critical, sandwich constructions with lightweight cores are applied. Key attributes include uniform wall thickness, smooth surface finish, and low water uptake, as variations in thickness and moisture increase losses and distort radiation characteristics. No high-loss coatings are applied in the radiating area, for example, coatings with carbon or metallic fillers. If protection is required, dielectric protective coatings are used with a low loss tangent in the operating band and a permittivity close to that of the radome substrate, with tightly controlled coating thickness. Surfaces exposed to rain and sand are protected with flexible polyurethane erosion-resistant coatings or replaceable leading-edge strips, and hydrophobic fluoropolymer topcoats are used to reduce water film formation [1].

Thermal Management and Infrared (IR) Signature

Thermal and IR signature design aims to maintain the power system, RF paths, and sensors at stable temperatures during 24/7 operation, thereby preventing degradation of antenna impedance matching or radiation patterns. The heat-removal path should be short with low thermal impedance. In-plane spreading is achieved with thin, high-conductivity inserts, for example, pyrolytic graphite sheets (PGS), and transport to load-bearing elements is provided by vapor chambers or heat pipes. Thermal interfaces are filled with TIMs of low thermal resistance. Effectiveness depends on contact quality, specifically flat and smooth surfaces, and a proper clamping force, as even micrometer-scale air gaps significantly increase interfacial thermal resistance. Near RF transmission lines (microstrip, CPW, stripline), one uses dielectrics with low loss and stable permittivity. Cooling hardware is routed outside the near-field region with the required clearances to prevent changes in impedance, detuning of circuits, or increased losses. The IR signature is shaped by selecting surface emissivity and panel geometry. When thermal masking is the priority, low-emissivity coatings are used, and, if necessary, directional radiation shields are employed to further enhance the effectiveness. When radiative cooling dominates, high-emissivity coatings are applied that remain stable after temperature and humidity cycling.

Above antennas and in RF "windows," exclusion zones are defined for coatings with elevated dielectric loss. If protection is required, dielectric protective coatings are used with a low loss tangent in the operating band and a permittivity close to that of the substrate. Coating thickness is maintained within 50–150 µm with a tight tolerance to prevent changes in impedance matching or antenna radiation patterns. Surfaces exposed to rain and abrasive particles are protected with flexible polyurethane erosion-resistant systems or replaceable leading-edge strips. The formation of thin water films is limited by hydrophobic fluoropolymer topcoats, which exhibit high contact angles and low water uptake. The geometry of the antenna radome is designed to shed water rapidly from the active area [2-5].

Environmental Durability and Reliability

The environmental durability and reliability of a VTOL unmanned platform are properties of the entire structural–electronic system. They encompass the airframe structural material, composite–metal interfaces, electronics enclosures, antenna radomes, and the applied coatings and seals. The system must retain its mechanical, electrical, and dielectric parameters after exposure to moisture and salt fog, dust and sand, UV radiation, thermal cycling, vibration, shock, and erosion. The primary load-bearing structure is most often made from CFRP, GFRP, or hybrid laminates, with localized use of light alloys. Interfaces between CFRP and aluminum or titanium require electrical isolation – typically achieved through glass interlayers, insulating layers, and dielectric bushings in holes – to limit stray currents and galvanic corrosion. Metallic parts are protected with barrier coating systems and adhesion-promoting primers resistant to UV and salt fog. Electronics enclosures are designed to be sealed, made from low-absorption materials, with pressure-equalization membranes and controlled water-drain paths. PCBs and connectors are protected with conformal coatings selected for the environmental profile, with clearly defined exclusion zones above antennas and tuned elements. Adhesive joints are chosen for coefficient-of-thermal-expansion compatibility and appropriate adhesive properties, which reduces stress concentrations under thermal cycling and vibration. Leading edges and inlets are protected with flexible, erosion-

Materials Research Proceedings 62 (2026) 270-276 https://doi.org/10.21741/9781644904015-35

resistant coatings or replaceable strips that take the wear. In impact-prone areas, parts are made from fiber-reinforced PEEK or PEI. The antenna radome should maintain stable permittivity and a low loss tangent after exposure to moisture and thermal cycling. This requires materials with low water uptake and uniform wall thickness, as well as dielectric protective coatings that do not introduce significant loss within the antenna's operating band and have permittivity close to that of the radome substrate [6-8].

Machine-Learning-Driven Intelligent Camouflage for UAVs with AI-Tunable Material

This section presents a materials-to-algorithms approach for intelligent camouflage on VTOL-class drones, in which onboard AI tunes surface and structural materials to reshape the multimodal signature in the thermal infrared, acoustic, visual, and RF domains according to the current threat and mission goals. The approach extends the selection and integration rules established for EMC and communications, infrared signature control, and environmental durability, by adding reconfigurable materials and learning-based control strategies. We treat this as another software-defined layer, co-designed with networking, autonomy, and the SDR and RF subsystems. In operations with multiple UAVs, camouflage states can be coordinated across the swarm through the same semantic mapping and software-defined architecture.

The layer actuates tunable surface layers and panels. Graphene and electrochromic films, as well as electrically tunable infrared modulators, provide fast control of emissivity. VO_2 thermochromic multilayers enable simultaneous camouflage in the visible and infrared domains. Hybrid visible and infrared modules combine emissivity tuning with a thin thermoelectric backplane to produce rapid changes in apparent temperature. Recent devices report wide emissivity ranges and field-relevant spans of apparent temperature, which supports the feasibility of lightweight UAV skins. On the RF side, reconfigurable metasurfaces, known as intelligent reflecting surfaces, enable software-defined control of absorption, reflection, and phase. Deep reinforcement learning controllers have been demonstrated for rapid beam and phase optimization in dynamic scenes. A compact sensing, decision, and actuation loop closes the control loop: skin thermistors and infrared pixels, together with a radar warning receiver and software-defined radio features, feed a detection risk estimator. Model predictive control stabilizes tightly constrained thermal loops, while reinforcement learning explores metasurface phase maps subject to power and electromagnetic compatibility limits [9].

To reshape the noise signature, thin acoustic metamaterial liners, such as microperforated or metalined ducts, target blade pass tones. At the same time, machine learning-aided active noise control addresses environment-dependent broadband components. Recent aeroacoustics studies on drones and eVTOLs report multidecibel reductions, along with design guidelines that can be applied to small UAVs [10]. Camouflage across these domains can be configured to degrade the performance of AI-based detection and tracking models.

The issues discussed in this article are also relevant to other areas of military applications. They provide significant contributions to both radar signature reduction [11] and general camouflage [12]. Wear resistance is also important in the production of special DLC [13] and ESD [14,15] coatings, including those subject to laser micromachining [16]. These are technological issues strongly supported by the classic DOE methodology [17,18].

Summary

This paper presents material domains critical for 24/7 autonomy, including EMC and communications, thermal management, and infrared signature, as well as environmental durability. It shows how stability of dielectric parameters, thermal conductivity, and environmental resistance translates into reliable system functions in navigation, communications, and autonomy. The neutral reference architecture (RA) serves as a case study in which the co-design of materials, electronics, and algorithms sustains system operation under interference, harsh weather, and infrastructure-denied conditions. This integrated approach is what separates a demonstrator from a system ready for continuous service.

References

[1] T. Dahms et al., Measurement of Dielectric Properties of Thin Materials for Radomes Using Waveguide Cavities. Telecom 5 (2024) 706-722. https://doi.org/10.3390/telecom5030035

[2] F. Zhou et al., A novel ultra-thin vapor chamber with radial-gradient hierarchical wick for high-power electronics cooling, International Journal of Thermal Sciences 183 (2023) art.107896. https://doi.org/10.1016/j.ijthermalsci.2022.107896

[3] Y-G. Lv et al., Review on thermal management technologies for electronics in spacecraft environment, Energy Storage and Saving 3 (2024) 153-189. https://doi.org/10.1016/j.enss.2024.03.001

[4] J. Wei et al. Durable superhydrophobic coatings for prevention of rain attenuation of 5G/weather radomes. Nature Commun. 14 (2023) art.2862. https://doi.org/10.1038/s41467-023-38678-0

[5] NASA Small Spacecraft Systems Virtual Institute, Small Spacecraft Technology State-of-the-Art: Thermal Control (Chapter 7), 2024 Edition, NASA, Feb 2025. URL: nasa.gov/smallsat-institute/sst-soa/thermal-control (access date: 11.10.2025)

[6] R.K. Bheema et al., A review on recent progress in polymer composites for effective electromagnetic interference shielding properties – structures, process, and sustainability approaches, Nanoscale Advances 6 (2024) 5773-5802. https://doi.org/10.1039/D4NA00572D.

[7] I. Haider et al., Environmental Aging of Reinforced Polymer Composite Radomes: Dielectric Performance over Wideband and Moisture Effects, Frontiers in Materials 11 (2024) art. 1427541. https://doi.org/10.3389/fmats.2024.1427541

[8] N. Osa-uwagboe et al., Review on Mechanical Performance of Fibre-Reinforced Polymers under Moisture Ingress for Maritime Applications, Applied Composite Materials 31 (2024) 1991-2018. https://doi.org/10.1007/s10443-024-10247-8

[9] Y. Huang et al. Adaptive visible-infrared camouflage with wide-range radiation control for extreme ambient temperatures. PhotoniX 6 (2025) art.25. https://doi.org/10.1186/s43074-025-00184-5

[10] H. Xiao et al. Broadband sound absorption of subwavelength porous meta-liner. Frontiers in Materials 9 (2022) art.845597. https://doi.org/10.3389/fmats.2022.845597

[11] W. Przybył et al., Microwave absorption properties of carbonyl iron-based paint coatings for military applications, Defence Technology 22 (2023) 1-9. https://doi.org/10.1016/j.dt.2022.06.013

[12] M. Michalski et al., The Influence of Operational Exposure on Changes in Parameters of Effective Camouflage of Coatings Used in Military Technology, Advances in Science and Technology Research Journal 17 (2023) 182-196. https://doi.org/10.12913/22998624/156940

[13] N. Radek et al., Operational properties of DLC coatings and their potential application, METAL 2022 – 31st Int. Conf. Metall. Mater., (2022) 531-536. https://doi.org/10.37904/metal.2022.4491

[14] N. Radek et al., The effect of laser treatment on operational properties of ESD coatings, METAL 2021 – 30th Int. Conf. Metall. Mater. (2021) 876-882. https://doi.org/10.37904/metal.2021.4212

[15] N. Radek et al., Technology and applications of ESD coatings before and after laser processing, METAL 2024 – 33rd Int. Conf. Metall. Mater. (2024) 500-505. https://doi.org/10.37904/metal.2023.4727

[16] A. Kalinowski et al., Laser surface texturing: characteristics and applications, System Safety: Human - Technical Facility - Environment 5 (2023) 240-248. https://doi.org/10.2478/czoto-2023-0026

[17] J. Pietraszek, A. Goroshko, The heuristic approach to the selection of experimental design, model and valid pre-processing transformation of DoE outcome, Advanced Materials Research 874 (2014) 145-149. https://doi.org/10.4028/www.scientific.net/AMR.874.145

[18] J. Pietraszek et al., Challenges for the DOE methodology related to the introduction of Industry 4.0, Prod. Eng. Arch. 26 (2020) 190-194. https://doi.org/10.30657/pea.2020.26.33

Materials Research Proceedings 62 (2026) 277-284 | https://doi.org/10.21741/9781644904015-36

Innovative Technical Solutions for Engineer Ordnance and Equipment

Piotr KRYSIAK[1,a*], Marcin SZCZEPANIAK[1,b] and Jacek PARTYKA[1,c]

[1]Military Institute of Engineering Technology, Obornicka 136, 50-961 Wrocław, Poland

[a]krysiak@witi.wroc.pl, [b]szczepaniak@witi.wroc.pl, [c]partyka@witi.wroc.pl

Keywords: EFP Charge, Anti-Tank Mine, Engineer Ordnance, Fragmentation Mines, Detection and Localization

Abstract. The paper presents the latest advancements in the design and manufacturing of munitions that can be utilized both in peacetime and wartime. In the former, they can serve as a protective measure for strategic facilities and the state border. At the same time, in the latter, they are designed to neutralize lightly armored vehicles and personnel (fragmentation charges) and to destroy armored vehicles (off-route mines employing explosive-formed projectiles, or EFPs). The study showcases two systems: controlled explosive charges and guided anti-tank mines. These are modern systems operating automatically, utilizing the latest available technologies.

Introduction

Over recent decades, the world has witnessed armed conflicts of both symmetric and asymmetric nature, as well as stabilization and peacekeeping missions, in many of which the Polish Armed Forces have participated. This has provided the military and the broader military science community with substantial data for analysis, resulting in experiences that have been leveraged to develop modern military equipment in the realm of ordnance. World wars, local conflicts (e.g., Afghanistan, the Balkans, Iraq), and the current conflict in Ukraine have confirmed that mines play a significant role as one of the most effective means of land warfare. Mines are most used in defensive operations in the form of minefields, minefield sectors, mine groups, and individual mines. They significantly impact the durability and activity of the defense, thereby restricting and impeding the enemy's maneuverability. The objective of combatant emplacing minefields is to maximally delay the enemy's actions and inflict the highest possible casualties with minimal expenditure of their own forces and weapon systems [1]. The use of minefields in defensive operations represents one of the many ways to employ combat engineers as a "combat capability multiplier" for combined arms formations [2]. In summary, the use of minefields significantly enhances the combat effectiveness of the combined arms units that include combat engineers, which is of great importance on the modern air-land battlefield.

New Weapon Systems

The Controlled Explosive Charges System is designed for constructing engineering barriers that align with the concept of an intelligent battlefield. It is a munition system functioning as an anti-personnel mine; however, under the constraints of the Ottawa Convention, it is intended for use by tactical-level military engineering units.

The system comprises the following effector elements:

1. Narrow-Area Controlled Explosive Charges (Fig. 1a),
2. Wide-Area Controlled Explosive Charges (Fig. 1b),
3. Controlled Fragmentation Grenade Launchers (Fig. 1c).

Materials Research Proceedings 62 (2026) 277-284 https://doi.org/10.21741/9781644904015-36

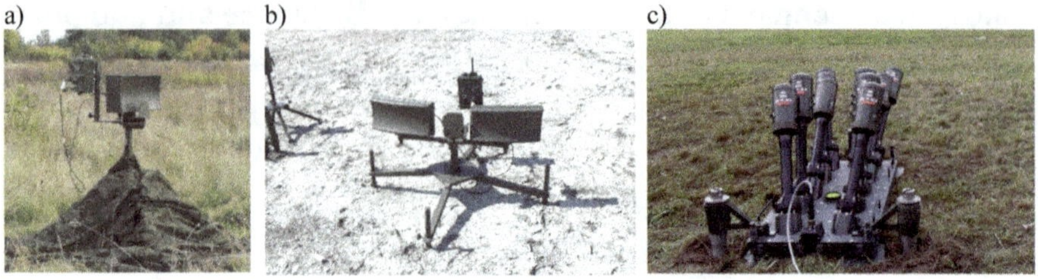

Fig. 1. Engagement system components

Each controlled explosive charge within the system is in communication with the operator control unit, which displays its status and operational functions.

The system operator can observe the monitored sector of the battlefield and can control the action of each explosive charge, including the capability to disarm them. The system is capable of operating:
- continuously for a minimum period of 15 days;
- in field conditions;
- during both day and night;
- under various weather conditions;
- exclusively in a supervised/manual mode.

The narrow-effect controlled charge is designed for the controlled engagement of armored vehicles by damaging their observation and sighting devices, lights, and windows, as well as for protecting anti-tank minefields, strategic facilities, and borders. The narrow-effect charge is an element of the engineering barrier construction system, falling within the area of counter-mobility and enhancing troop survivability. It is constructed using an aluminum-composite body filled with steel balls and a 60/40 explosive mixture of RDX/TNT, weighing 17 kg. The average density of projectiles per unit area for this munition is approximately 18 frag./m^2. The width of the engagement sector at a distance of 100 meters is 20 meters.

The controlled wide-effect charge is constructed with a double aluminum casing, each filled with steel balls and 6 kg of RDX/TNT 60/40 explosive. The average number of projectiles impacting a target per unit area for this munition type is approximately 45 frag./m^2. The width of the engagement sector at a distance of 30 m is 30 m.

The controlled fragmentation grenade launcher possesses 10 barrels, onto which fragmentation grenades and the guidance system are mounted. Upon receiving a signal from the controller, the grenades are launched sequentially from the launcher and engage targets within the upper hemisphere at ranges of 70 to 100 m. Each grenade contains a propellant charge located in the base and a 0.3 kg RDX/TNT 60/40 explosive charge with fragmentation elements in the warhead. The average number of projectiles impacting a target per unit area for the grenade is approximately 15 frag./m^2.

The system also includes a warning zone kit (Fig. 2a), a denial zone kit (Fig. 2b), and a detection and localization system, including the system controller (Fig. 2c). These system components support its proper operation in detecting and identifying intruders and significantly aid operators in the decision-making process regarding system and munition deployment.

Fig. 2. System components

Guided anti-tank mine system [3, 4] aligns with the development trend of anti-tank mines, creating an "intelligent" minefield. This type of technologically advanced barrier is designed to detect a moving object using appropriate sensors, then determine whether it is a suitable target, track its trajectory, and destroy it from a significant distance at the appropriate moment.

In this type, the payload of mine consists of an Explosively Formed Projectile (EFP), traveling at a velocity of approximately 2000 m/s and capable of penetrating 100 mm of armor. The projectile strikes one of the least armored areas, specifically the side of the vehicle at the level of the running gear and below the turret (within a height range of 0.5-1.2 m). The impact energy and its effect on the vehicle are significantly greater than those of anti-tank artillery shells and missiles. The charge is most commonly constructed using classical materials such as steel and copper [5-9], but modern solutions are also based on composite materials [10-15]. During the development of the charge, numerous numerical and experimental studies were conducted, proving the effectiveness of the generated projectile. Selected results of the numerical simulations are presented in Fig. 3.

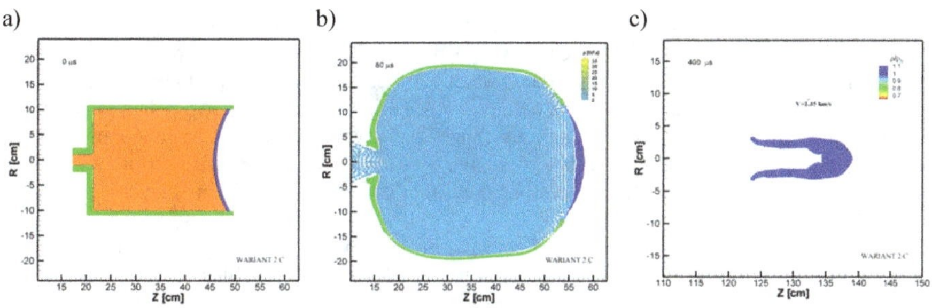

Fig. 3. Numerical analysis of EFP

Figure 4 presents selected frames from the high-speed footage recorded during testing using a Phantom V711 high-speed camera.

Materials Research Proceedings 62 (2026) 277-284 https://doi.org/10.21741/9781644904015-36

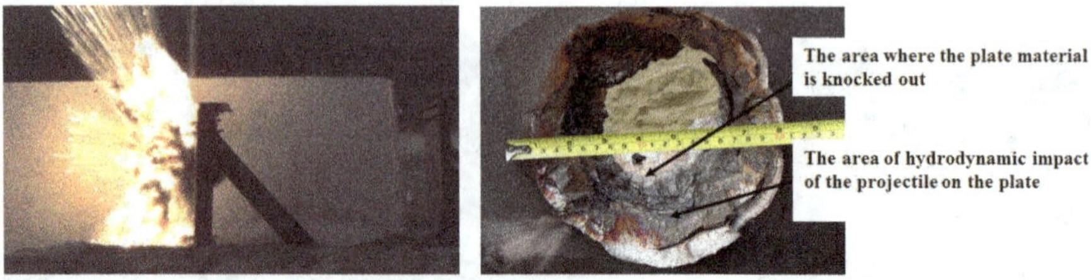

Fig. 3. Field test results of EFP charge

a) b)

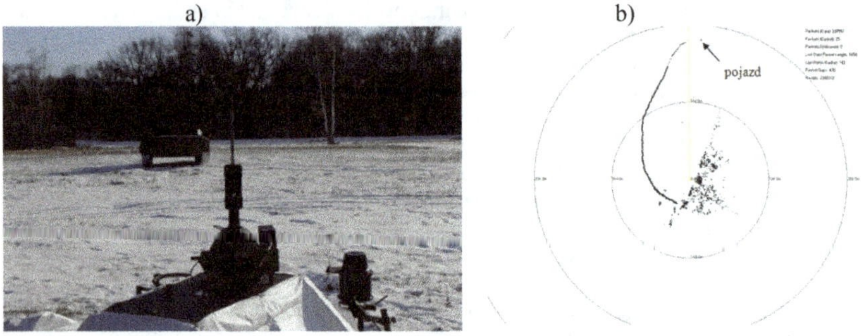

Fig. 4. The mine during field tests (a) with an armored vehicle within its engagement range, and a view of the vehicle's trajectory (b) on the controller screen

By design, Guided Anti-Tank Mines operate entirely in automatic mode, resulting in high effectiveness against armored vehicles at a significantly lower cost compared to, for example, guided anti-tank missiles. Optionally, the operator can decide the moment of charge detonation.

The structure of a minefield constructed from Guided Anti-Tank Mines means that the application of classical clearance methods (such as mechanical mine-clearance or explosive clearance using line charges) does not guarantee a safe passage through the minefield. This is because combat vehicles attempting to cross the minefield via a cleared lane can still be engaged by adjacent mines.

The Guided Anti-Tank Mine System represents the next evolutionary stage of engineering barriers, aligning with the concept of the intelligent battlefield, and enables:

- The emplacement of minefields or mine groups.
- Automatic targeting of individual mines onto distant targets.
- Omnidirectional operation, i.e., destroying or neutralizing targets located at significant distances (up to 80 m), regardless of their position relative to the mine.
- Monitoring the operational status of all the mines within the minefield.
- Remote control over the status of all the mines within the minefield.
- The capability to deactivate an emplaced minefield and dismantle the mines.
- The ability to re-emplace the minefield in a different location using the same components.
- Increased minefield resistance against all types of clearance (manual, mechanical, explosive).
- Operation in all climatic conditions, day and night.

Technological Impact

The war in Ukraine demonstrated that, regardless of the complexity of precision weapons, the availability of cheap, mass-produced ammunition is crucial for combat effectiveness. A key

Materials Research Proceedings 62 (2026) 277-284 https://doi.org/10.21741/9781644904015-36

element is the development of well-managed production processes [16-18] that deliver high-quality products [19,20] within an acceptable cost range [21,22]. Due to the specific nature of mines, it is essential to protect them against corrosion [23-25], which can be achieved through the use of appropriate special coatings [26-28]. An additional element supporting quality [29] is the use of appropriate materials [30,31], which may be modified in the surface layer [32,33] through laser micromachining [34,35]. This increases durability both in storage and later, after deployment in mined areas. The use of advanced materials and technologies requires optimization support, which can be provided by the DOE methodology [36,37]. The assessment of unspecified conditions for a specific "operation" can be modeled along the lines of Taguchi's "robust design" using fuzzy descriptions [38, 39].

Summary

Fragmentation mines equipped with effectors in the form of balls or cylinders form the basis for protecting strategic areas, safeguarding national borders, and can serve as a protective element for more strategically significant minefields. Minefields composed of modern, highly automated omnidirectional mines enable effective defense.

As demonstrated by the tests, the system's parameters (both functional, such as penetration ability, and operating, like omnidirectional operation ability) are superior to those of classic off-route mines (e.g., the Polish MPB mine), which were their predecessors. The sensor systems detect the target, determine its direction, and destroy it if it is located within a distance of 80 m or closer. The ordnance is an EFP moving at a speed of approximately. 2,500 m/s and capable of penetrating 100 mm of armor. The projectile hits one of the weakest armored parts, i.e., the vehicle's side at the drive system height and below the turret (within the height range of 0.4–1.2 m). The energy of impact that affects the vehicle is much greater than anti-tank gun shells and missiles. The system's ordnance and sensors operate automatically by design, which provides high effectiveness of armored vehicle engagement at approximately 80%. The operator may control the moment of charge detonation. The minefield's structure is such that the use of classic demining methods (mechanical, electronic, or explosive demining using line charges) does not guarantee that a clear breach in the minefield can be achieved. This stems from the fact that combat vehicles attempting to cross the minefield along the demined section can still be engaged by neighboring mines. The system's design allows deactivation of the entire minefield or parts of it, e.g., during the retreat of friendly troops or to allow demining vehicles, recon vehicles, or the first group of combat vehicles to pass through the minefield, and subsequently, specific mines or the entire field to be reactivated. The system enables the formation of a 900 x 300 m minefield. Defending a similar area with classic anti-tank mines would require using approximately. 1.2 thousand (influence or scatterable mines) up to 4 thousand (contact fuse mines) pieces of ordnance. The relatively low number of mines in the system results in a lower minefield price compared to fields of similar size laid with classic or scatterable anti-tank mines.

References

[1] Z. Szewczuk, T. Wysocki, Efektywność zapór minowych, Wyd. Przegląd Wojsk Lądowych 1988/10.

[2] S. Bornhoft, Saperzy jako mnożnik możliwości bojowych wojsk, Wyd. Wojskowy Przegląd Zagraniczny Nr 1 (155), Warszawa 1984.

[3] M. Szczepaniak et al., Armored Fighting Vehicle Destruction System. Problems Of Mechatronics. Armament, Aviation, Safety Engineering 10 (2019) 135-142. https://doi.org/10.5604/01.3001.0013.0802

[4] J. Śliwiński, T. Wysocki, Efektywność min przeciwburtowych, Konferencja Naukowo-Techniczna „Problemy rozwoju produkcji i eksploatacji uzbrojenia", Rynia 1996.

[5] A. Wojewódka, T. Witkowski, Metodyka modelowania procesu formowania strumienia kumulacyjnego wydłużonych ładunków materiałów wybuchowych, Chemik 65 (2011) 28-35.

[6] Ł. Czubaty et al., Dynamika oddziaływania na przegrodę pocisku formowanego wybuchowo, IX Konferencja Naukowo-Techniczna „Inżynieria wojskowa, problemy i perspektywy", 2014, 451-460.

[7] R. Panowicz et al., Wpływ wybranych typów uszkodzeń/deformacji ładunków kumulacyjnych na ich efektywność, Zeszyty Naukowe WSOWL 2 (172) (2014) 124-140. https://doi.org/10.5604/17318157.1127102

[8] L.-K. Hao et. al., Damage of a large-scale reinforced concrete wall caused by an explosively formed projectile (EFP), Defence Technology 28 (2023) 280-297. https://doi.org/10.1016/j.dt.2022.11.003

[9] B. Zygmunt et al., The Conception of the Powder Liners Technology for EFP Charges, Problemy Mechatroniki Uzbrojenie, Lotnictwo, Inżynieria Bezpieczeństwa 5(2) (16) (2014) 63-76.

[10] P. Krysiak et al., Identification of strains in a multilayer composite pipe. J. Sci. Mil. Acad. Land Forces 49 (2017) 272-282. https://doi.org/10.5604/01.3001.0010.7233

[11] P. Krysiak et al., Strength Testing and Ring Stiffness Testing of Underground Composite Pressure Pipes, Mater. Res. Proc. 17 (2020) 191-202. https://doi.org/10.21741/9781644901038-29

[12] P. Krysiak et al., Theoretical and Experimental Analysis of Inter-Layer Stresses in Filament-Wound Cylindrical Composite Structures. Materials 14 (2021) art. 7037. https://doi.org/10.3390/ma14227037

[13] A. Błachut, P. Krysiak. Analysis of mechanical properties of epoxy resins used in composite high-pressure tank. Interdisciplinary Journal of Engineering Sciences 4 (2016) 50-55.

[14] P. Krysiak et al., Strength testing of a composite mounting frame for a multi-sensor detection system, Materials Research Proceedings 17 (2020) 165-170, https://doi.org/10.21741/9781644901038-25

[15] P. Krysiak et al., Influence of fiber type and the layer thickness on the stress distribution in composite pipe, Interdisciplinary Journal of Engineering Sciences 2 (2014) 17-20.

[16] S. Borkowski et al., The use of 3x3 matrix to evaluation of ribbed wire manufacturing technology, METAL 2012 – 21st Int. Conf. Metall. Mater., (2012) 1722-1728.

[17] M. Ingaldi, S.T. Dziuba, Modernity evaluation of the machines used during production process of metal products, METAL 2015 – 24th Int. Conf. Metall. Mater. (2015) 1908-1914.

[18] A. Pacana et al., Decision support in the production of packaging films by cost-quality analysis, Przemysl Chemiczny 95 (2016) 1042-1044. https://doi.org/10.15199/62.2016.5.31

[19] D. Siwiec et al., Quality analysis of the mechanical lever and improvement of non-destructive testing process, System Safety: Human - Technical Facility - Environment 2 (2020) 299-306. https://doi.org/10.2478/czoto-2020-0037

[20] D. Siwiec et al., Concept of a model to predict the qualitative-cost level considering customers' expectations, Polish Journal of Management Studies 26 (2022) 330-340. https://doi.org/10.17512/pjms.2022.26.2.20

[21] E. Kozień, Assessment of technical risk in maintenance and improvement of a manufacturing process, Open Eng. 10 (2020) 658-664. https://doi.org/10.1515/eng-2020-0047

[22] D. Siwiec et al., Analysis of aluminum can by eco-costs and life cycle assessments (LCA), METAL 2024 – 33rd Int. Conf. Metall. Mater., (2024) 627-632. https://doi.org/10.37904/metal.2024.4975

[23] D. Klimecka-Tatar et al., The effect of powder particle biencapsulation with Ni-P layer on local corrosion of bonded Nd-(Fe, Co)-B magnetic material, Archives of Metallurgy and Materials 60 (2015) 154-157. https://doi.org/10.1515/amm-2015-0024

[24] A. Bartkowska et al., Microstructure, microhardness, corrosion resistance and chemical composition of Mo, B and Mo-B coatings produced using laser processing, Materials 13 (2020) art. 3249. https://doi.org/10.3390/MA13153249

[25] T. Lipiński, J. Pietraszek, Corrosion of the S235JR Carbon Steel after Normalizing and Overheating Annealing in 2.5% Sulphuric Acid at Room Temperature, Materials Research Proceedings 24 (2022) 102-108. https://doi.org/10.21741/9781644902059-16

[26] N. Radek et al., Electrospark alloying of carbon steel with WC-Co-Al2O3: Deposition technique and coating properties, Advanced Materials Research 874 (2014) 101-106. https://doi.org/10.4028/www.scientific.net/AMR.874.101

[27] L. Adamczyk et al., Electrochemical preparation of composite coatings of 3,4-etylenodioxythiophene (EDOT) and 4-(pyrrole-1-yl) benzoic acid (PyBA) with heteropolyanions, Materials Chemistry and Physics 144 (2014) 418-424. https://doi.org/10.1016/j.matchemphys.2014.01.012

[28] N. Radek et al., The effect of laser beam processing on the properties of WC-Co coatings deposited on steel, Materials 14 (2021) art. 538. https://doi.org/10.3390/ma14030538

[29] A. Pacana et al., Method Supporting Improving Products in Terms of Qualitative-Environmental, Materials Research Proceedings 34 (2023) 226-236. https://doi.org/10.21741/9781644902691-27

[30] T. Lipiński, P. Szabracki, Mechanical properties of AlSi9Mg alloy with a sodium modifier, Solid State Phenomena 223 (2015) 78-86. https://doi.org/10.4028/www.scientific.net/SSP.223.78

[31] T. Lipiński, A. Wach, Influence of inclusions on bending fatigue strength coefficient the medium carbon steel melted in an electric furnace, Production Engineering Archives 26 (2020) 88-91. https://doi.org/10.30657/pea.2020.26.18

[32] A. Wrońska, A. Dudek, Characteristics of surface layer of sintered stainless steels after remelting using GTAW method, Archives of Civil and Mechanical Engineering 14 (2014) 425-432. https://doi.org/10.1016/j.acme.2013.10.011

[33] N. Radek et al., The influence of plasma cutting parameters on the geometric structure of cut surfaces, Materials Research Proceedings 17 (2020) 132-137. https://doi.org/10.21741/9781644901038-20

[34] M. Kawalec et al., Laser assisted machining of aluminium composite reinforced by SiC particle, ICALEO 2008 – 27th Int. Congr. App. Lasers and Electro-Optics (2008) 895-900. https://doi.org/10.2351/1.5061278

[35] M. Kukliński et al., Influence of microstructure and chemical composition on microhardness and wear properties of laser borided monel 400, Materials 13 (2020) art. 5757. https://doi.org/10.3390/ma13245757

[36] J. Pietraszek, A. Gadek-Moszczak, The smooth bootstrap approach to the distribution of a shape in the ferritic stainless steel AISI 434L powders, Solid State Phenomena 197 (2013) 162-167. https://doi.org/10.4028/www.scientific.net/SSP.197.162

[37] J. Pietraszek et al., Challenges for the DOE methodology related to the introduction of Industry 4.0, Prod. Eng. Arch. 26 (2020) 190-194. https://doi.org/10.30657/pea.2020.26.33

[38] J. Pietraszek, Fuzzy regression compared to classical experimental design in the case of flywheel assembly, Lecture Notes in Computer Science 7267 LNAI (2012) 310-317. https://doi.org/10.1007/978-3-642-29347-4_36

[39] J. Pietraszek, The modified sequential-binary approach for fuzzy operations on correlated assessments, Lecture Notes in Computer Science 7894 LNAI (2013) 353-364. https://doi.org/10.1007/978-3-642-38658-9_32

Design and Construction of a Mobile Robot for the URC and ERC Competitions

Krzysztof BORKOWSKI[a], Gabriel BRACHA[b], Dawid S. PIETRALA[c], Hubert WIŚNIEWSKI[d], Anna ANNUSEWICZ-MIŚTAL[e], Szczepan KOSTECKI[f], Jakub PAJĄK[g], Adam SIEROŃ[h], Michał WZOREK[i], Paweł A. ŁASKI[j]*, Piotr CHOCHÓŁ[k], Piotr WITKOWSKI[l], Wojciech ŁAKOMIEC[m], Norbert KOT[n], Jakub GÓRSKI[o] and Michał KLUZEK[p],

Kielce University of Technology, Faculty of Mechatronics and Mechanical Engineering, Department of Automation and Robotics, Al. 1000-lecia Państwa Polskiego 7, 25-314 Kielce, Poland

[a]kborkowski@tu.kielce.pl, [b]gbracha@tu.kielce.pl, [c]dpietrala @tu.kielce.pl, [d]hubert.w@tu.kielce.pl, [e]aannusewicz@tu.kielce.pl, [f]skostecki@tu.kielce.pl, [g]jpajak@tu.kielce.pl, [h]asieron@tu.kielce.pl, [i]mwzorek@tu.kielce.pl, [j]pawell@tu.kielce.pl, [k,l,m,n,o,p]robotyka@tu.kielce.pl

Keywords: Mars Rover, Robotic Missions, Designing Mobile Robots

Abstract. This article presents the components of a mobile robot prepared for the international URC and ERC "Mars rovers" competitions. The URC and ERC competitions are the largest robotics and space events, targeting the worlds of science and business, the new technologies sector, and the general public. The European Rover Challenge, like the University Rover Challenge, is part of the Rover Challenge Series – a league of the most prestigious competitions for robotic mobile vehicles. The core of the competition is the international Mars robot competition, which attracts students, graduates, and academics from universities around the world. Teams of around a dozen people work for twelve months to develop a fully functional, next-generation Mars rover. Each team includes specialists in mechatronics, automation, robotics, autonomy, communications, and navigation.

Introduction

To prepare the robot for the competition, we utilized laboratories equipped with workstations featuring SolidWorks, nTopology, and MATLAB software. Furthermore, the robot's components were manufactured in a rapid prototyping laboratory using FDM printing techniques and a CNC milling machine. We also utilized resources available at the Faculty of Mechatronics and Mechanical Engineering, including an industrial SLS printer, laser cutters, and conventional metalworking machines. These resources allowed us to produce the mechanical components in-house. The robot's chassis is constructed in a truss-like structure and is made of PA6G polyamide sheets. The wheel assemblies were optimized to minimize weight while maintaining strength. The main drive wheels are powered by high-performance BLDC motors, controlled by a motherboard that contains an STM32H7 microprocessor and actuator modules. The structure features a multi-link suspension that allows it to climb objects up to the height of the wheels. Figure 1 presents the concept of preparation, creation, and validation during the creation and preparation of the robot for the competition.

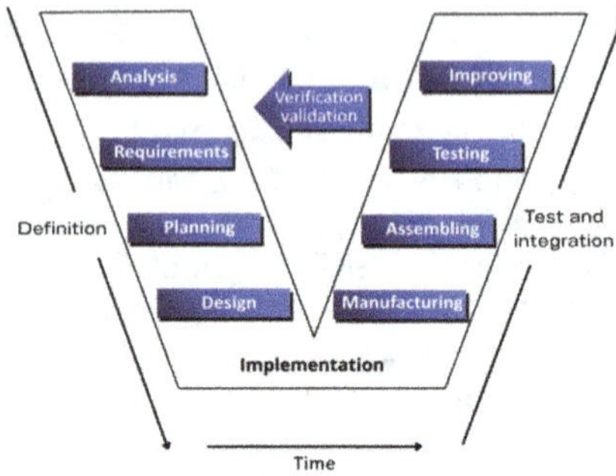

Fig. 1. Engineering, integration, and testing plan

Robot Control System

One of the robot's most important systems is the control system. A distributed control system was employed, featuring an STM32 main processor, with additional subsystems controlling the wheels and the 6-degree-of-freedom arm, both of which were also managed by STM32 processors. Furthermore, for tasks requiring autonomous execution, an image analysis system based on the Nvidia Jetson development platform and camera arrays was used. Basic communication utilizes the 2.4 GHz and 5.8 GHz frequency bands. Fig.2 graphically illustrates the signal flow and the basic robot control systems and subsystems.

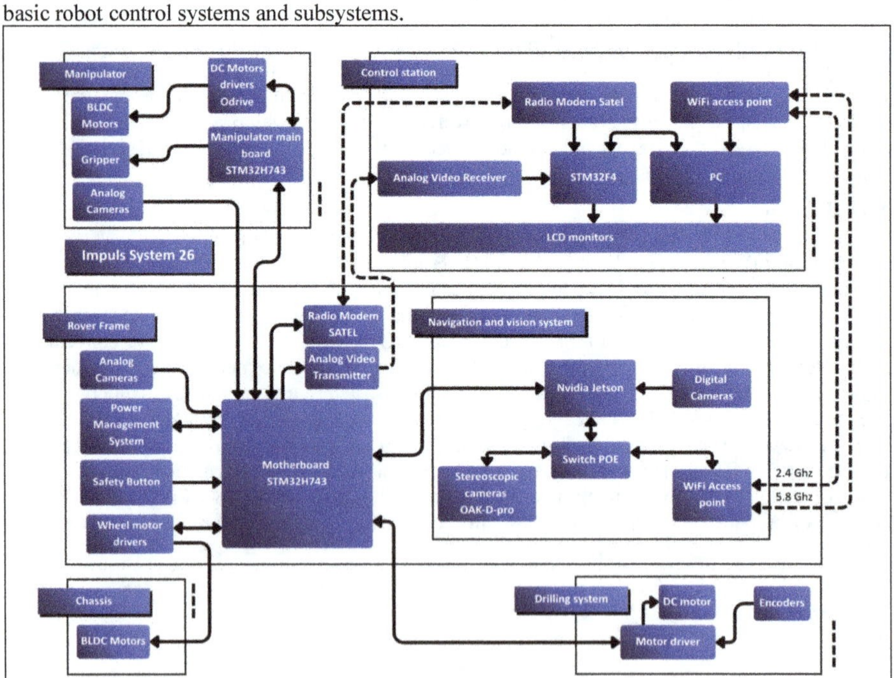

Fig. 2. System architecture

Fig.3 shows the new and modified components for the URC competition.

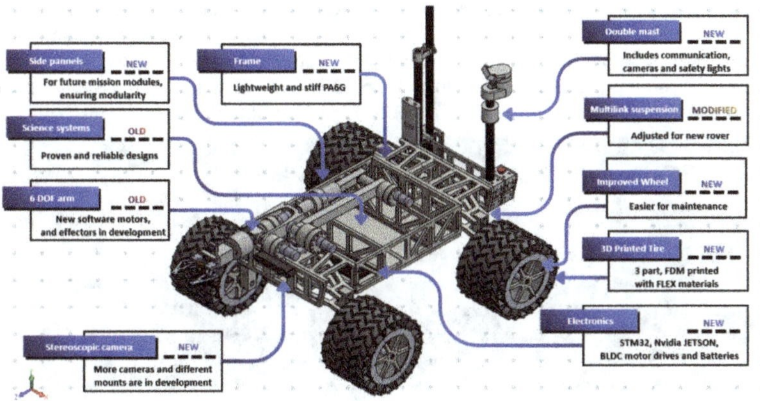

Fig. 3. Impuls M3 Rover, isometric view

Results and Discussion

To reduce the weight of the designed structure, FEM studies and topological optimization of selected robotic arm components, designed for additive 3D printing, were conducted. The NTopology environment was used for this purpose. The first step was to convert the solid model into a surface model, and then to determine the fixation surfaces, force application areas, and areas that remained invariant during the optimization process. The next step in preparing the model for optimization was to introduce the FEM mesh. To achieve this, a series of steps were performed: generating a surface mesh (Mesh from CAD body), improving and correcting the mesh quality (Remesh Surface), converting the surface mesh to a volume mesh (Volume Mesh), and finally determining the FEM mesh (FE Volume Mesh). This resulted in a correct and acceptable parametric mesh, which was used in subsequent optimization steps employing the FEM method. In each case, the goal of the optimization was to minimize the Structural Compliance Response parameter, which is the inverse of stiffness. This means that the algorithm, regardless of the constraints applied, aimed to maintain the highest possible stiffness of the structure (Fig.4). For the presented case, the initial mass of the element was 1.2 kg, which was then reduced to 0.7 kg after the optimization process, which indicates almost a 40% mass reduction without losing the initial strength and stiffness.

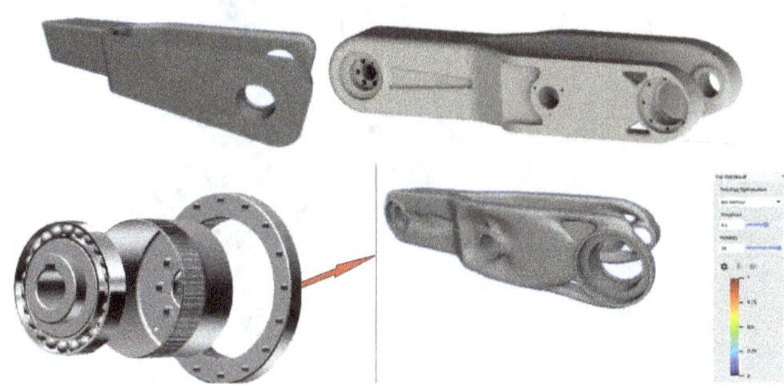

Fig. 4. Design of the manipulator arm elements based on the results of topological optimization

As mentioned, a significant portion of the robot's structural components are made of plastics, both through additive manufacturing and CNC machining. According to the URC and ERC competition regulations, the robot must meet numerous criteria, including maximum weight, battery life, power, and radio frequency bands, as well as robotic arm payload. During the competition, the robot is adapted to subsequent tasks and missions. One mission involves searching for an astronaut and delivering items such as tools or a water bottle to their vicinity (Fig.5b).

Fig. 5. A view of the robot prepared for the "Mars Rovers" competition and a view of the robot performing the task.

When performing a hardware maintenance mission (Fig.6a), the robot performs tasks designed to simulate the repairability of complex equipment. The robot turns components on and off on the service panel, enters appropriate sequences and alphanumeric characters on the keyboard, and inserts and removes objects from the service panel module.

Fig. 6. View of the robot during a servicing mission.

Two of the most challenging tasks during the competition include scientific and autonomous navigation missions. During the scientific mission, activities include collecting appropriate samples from the surface and from a depth of approximately 150 mm, detecting potential "life" within them, and detecting the presence of basic elements N, K, and P, which may indicate the existence of "biological life." Furthermore, the robot's tasks include autonomously searching for and collecting technical elements (Fig.7b). During the navigation mission, the robot moves independently and autonomously between GPS points indicated on the map, informing the base of its position each time. The greatest challenge lies in the fact that the robot moves through an area containing hills, depressions, and rock formations. Fig.7a shows the robot during the navigation mission.

Fig. 7. View of the robot during its navigation and science missions.

During all missions, the robot is controlled remotely. The operator has access to a set of levers and monitors that display the robot's parameters in real time. Images from multiple cameras are also displayed (Fig. 8). Only during autonomous operations can the operator not use the panel and act as a passive observer, supervising the task. The designed operator panel enables remote task execution, allowing the operator to be in a safe location while the robot completes the assigned tasks and accomplishes the mission.

Fig. 8. Operator's view while performing a mission during a competition.

Technological, Economic, Environmental, and Social Impact

The design and construction of the IMPULS M3 mobile rover, intended for the international URC and ERC competitions, constitute a comprehensive engineering undertaking that exerts a significant and multidimensional impact in four key areas: technological, economic, environmental, and social.

The construction of the Mars rover is a project driven by advanced technologies, reflecting a holistic approach to mechatronic systems design. A key aspect of the design process was the structural optimization of components. Weight reduction, achieved through FEM analysis and topology optimization, is fundamental to improving the rover's operational and energy performance. The use of advanced engineering methods, such as structural and thermal process optimization, is crucial to increasing reliability and efficiency [4]. To ensure precision and reliability, advanced drive control systems are essential, including intelligent electro-pneumatic servo drive positioning systems [5] and proportional valves based on piezoelectric actuators [6].

Manipulator design requires in-depth knowledge of robotics and kinematics, especially in the context of parallel robots with high degrees of freedom [7]. In the context of competition tasks requiring high payloads, research into automatic systems for lifting and moving heavy loads [8] is crucial, as well as leveling and lifting systems based on electro-hydraulic technology [9] or construction robot mechanics [10]. The control and monitoring of the rover's precise movements and its manipulator are continually being developed, including advanced methods such as fractional-order control (FOC) in pneumatic servo drives [11] and control systems based on pneumatic artificial muscles [12, 13].

The development of the rover as a research and educational platform also relies on advanced engineering and manufacturing methodologies. Design of Experiments (DoE) methodology is essential for modeling and analyzing the significance of parameters in regression models [14], as well as for addressing the new challenges of Industry 4.0 [15]. In the context of component production, rapid prototyping processes, such as pneumatic valves [16], are essential. A holistic approach to designing advanced systems, similar to the design of scientific instrumentation [17], is essential. Ultimately, component reliability and durability require advanced quality control and defect identification, which is particularly important in the context of casting defect detection methods within the Industry 4.0 paradigm [18]. To ensure autonomy, the development of manipulation requires computer vision-based systems [19].

The IMPULS M3 rover project is a complex undertaking with economic implications focused on cost efficiency, process optimization, and risk management. High energy and operational efficiency, achieved through weight optimization, directly translates into reduced operating costs and increased project competitiveness. The design process itself is optimized through the analysis and selection of technological and production strategies, as exemplified by the production of complex components [20]. Furthermore, project effectiveness depends on the implementation of effective production planning and control, including the use of logical process chains [21]. In managing such a complex project, the use of ex-ante risk assessment methods is crucial, as they enable the identification of potential problems at early stages and minimize losses [22]. The team's success in the competition also translates into the innovation capacity and technological position of the unit (university or company), which can be measured through indicator analysis [23].

Although the rover's design is primarily focused on space exploration, it also contributes to the development of engineering methods that support sustainable development, particularly by minimizing resource and energy consumption. The emphasis on reducing energy consumption [24] and optimizing the drive mass are direct steps towards improving efficiency, which has analogies with the pursuit of minimizing pollution in traditional industrial processes [25]. Increasing environmental awareness also requires product life cycle assessment, which is crucial in modern engineering and applies to rover components, e.g., in the context of LCA analysis of wheels [26]. Overall, the project aligns with the broader context of sustainable development and promotes the Circular Economy in technology and engineering [27].

Summary

The article describes selected elements of the process of designing and building the mobile robot "IMPULS M3 Rover", intended for participation in the international URC and ERC Mars rover competitions. An important part of the work was the mass optimization of selected manipulator components using FEM analysis methods and topology optimization. This made it possible to reduce the mass of one of the key elements by about 40% while maintaining the required stiffness and strength, which directly translated into improved robot parameters such as payload capacity, energy consumption, and drive system safety. The typical tasks that the rover must perform in competition are discussed. The article demonstrates the comprehensive nature of the approach to Mars rover design, encompassing the initial concept, mechanical structure, electronics, software, structural optimization, and practical verification in real-world competition tasks. Currently, the robot complies with the URC and ERC regulations, serving as a research and educational platform that can be further developed in subsequent iterations of the project.

References

[1] A. Zakręcki, B. Gaczorek, A. Kania, K. Berent, Project of Mars rover wheels based on structural infill using additive manufacturing, Mechanik, R. 91, nr 3, 244–248, 2018.

[2] D. Snider, M. Mirvish, M. Barcis, V. Aksoy Tezer, University Rover Challenge: Tutorials and Team Survey, in: A. Koubaa (ed.), Robot Operating System (ROS): The Complete Reference (Volume 3), Studies in Computational Intelligence, vol. 778, Springer, Cham, 2019, 315–370.

[3] B. Vinod, P. Gokul, S. Umashankar, A. Vishal, D. Gokul, Design and Analysis of Mars Rover, NVEO – Natural Volatiles & Essential Oils, vol. 8, iss. 5, 2620–2631, 2021.

[4] P. Duda, R. Dwornicka, Optimization of heating and cooling operations of steam gate valve, Structural and Multidisciplinary Optimization 40 (2010) 529-535. https://doi.org/10.1007/s00158-009-0370-8

[5] J.E. Takosoglu, P.A. Laski, Intelligent positioning system of electro-pneumatic servo-drive, Annals of DAAAM and Proceedings of the International DAAAM Symposium (2011) 1641-1642.

[6] P.A. Laski, Proportional valve with a piezoelectric actuator, EPJ Web of Conferences 143 (2017) art. 02064. https://doi.org/10.1051/epjconf/201714302064

[7] A.M. Annusewicz, P.A. Łaski, Design and kinematic analysis of the parallel robot 4-DoF SCARA, Advances in Intelligent Systems and Computing 743 (2018) 518-527. https://doi.org/10.1007/978-3-319-77179-3_49

[8] E. Lisowski, G. Filo, Automated heavy load lifting and moving system using pneumatic cushions, Automation in Construction 50 (2015) 91-101. https://doi.org/10.1016/j.autcon.2014.12.004

[9] P. Wos et al., The Electro-Hydraulic Lifting and Leveling System for the Bricklaying Robot, Lecture Notes in Mechanical Engineering 24 (2021) 216-227. https://doi.org/10.1007/978-3-030-59509-8_19

[10] P. Wos et al., Bricklaying robot lifting and levelling system, Communications - Scientific Letters of the University of Žilina 23 (2021) B257-B264. https://doi.org/10.26552/COM.C.2021.4.B257-B264

[11] P.A. Laski, Fractional-order feedback control of a pneumatic servo-drive, Bulletin of the Polish Academy of Sciences: Technical Sciences 67 (2019) 53-59. https://doi.org/10.24425/bpas.2018.126120

[12] J. Takosoglu, Angular position control system of pneumatic artificial muscles, Open Engineering 10 (2020) 681-687. https://doi.org/10.1515/eng-2020-0077

[13] D.S. Pietrala, P.A. Laski, Design and Control of a Pneumatic Muscle Servo Drive Containing Its Own Pneumatic Muscles, App. Sci. 12 (2022) art. 11024. https://doi.org/10.3390/app122111024

[14] J. Pietraszek, L. Wojnar, The bootstrap approach to the statistical significance of parameters in RSM model, ECCOMAS Congress 2016 – Proc. 7th Europ. Congr. Comput. Methods in App. Sci. Eng. 1 (2016) 2003-2009. https://doi.org/10.7712/100016.1937.9138

[15] J. Pietraszek et al., Challenges for the DOE methodology related to the introduction of Industry 4.0, Production Engineering Archives 26 (2020) 190-194. https://doi.org/10.30657/pea.2020.26.33

[16] S. Blasiak et al., Rapid prototyping of pneumatic directional control valves, Polymers 13 (2021) art. 1458. https://doi.org/10.3390/polym13091458

[17] J. Smyrski et al., Design of the forward straw tube tracker for the PANDA experiment, Journal of Instrumentation 12 (2017) art. C06032. https://doi.org/10.1088/1748-0221/12/06/C06032

[18] R. Ulewicz et al., A Rank Model of Casting Non-Conformity Detection Methods in the Context of Industry 4.0, Materials 16 (2023) art. 723. https://doi.org/10.3390/ma16020723

[19] A. Annusewicz-Mistal et al., Autonomous Manipulator of a Mobile Robot Based on a Vision System, Applied Sciences (Switzerland) 13 (2023) art. 439. https://doi.org/10.3390/app13010439

[20] M. Konstanciak, Analysis of technological strategies on the example of the production of the tramway wheels, Archives of Materials Science and Engineering 57 (2012) 69-74

[21] R. Ulewicz et al., Implementation of logic flow in planning and production control, Management and Production Engineering Review 7 (2016) 89-94. https://doi.org/10.1515/mper-2016-0010

[22] E. Kozień, M.S. Kozień, Ex-ante risk estimation in the production project, System Safety: Human - Technical Facility - Environment 1 (2019) 708-715. https://doi.org/10.2478/czoto-2019-0090

[23] A. Pacana, K. Czerwińska, Indicator analysis of the technological position of a manufacturing company, Prod. Eng. Arch. 29 (2023) 162-167. https://doi.org/10.30657/pea.2023.29.19

[24] G. Filo et al., Pressure loss reduction in an innovative directional poppet control valve, Energies 13 (2020) art. 3149. https://doi.org/10.3390/en13123149

[25] R. Dwornicka, The impact of the power plant unit start-up scheme on the pollution load, Adv. Mater. Res. 874 (2014) 63-69. https://doi.org/10.4028/www.scientific.net/AMR.874.63

[26] A. Pacana et al., Life Cycle Assessment (LCA) of Truck Steel Wheels, Materials Research Proceedings 45 (2024) 213-222. https://doi.org/10.21741/9781644903315-25

[27] E. Skrzypczak-Pietraszek, A.G. Pietraszek, Phytoremediation and Allelopathy as an Element of Sustainable Circular Economy to Prevent Environmental Pollution, Materials Research Proceedings 45 (2024) 140-145. https://doi.org/10.21741/9781644903315-17

Materials Research Proceedings 62 (2026) 293-301 | https://doi.org/10.21741/9781644904015-38

Capabilities to Assess Object Size Using Electromagnetic Metal Detectors

Wiesław MADEJ[1,a *], Paweł OTRĘBA[1,b] and Piotr KRYSIAK[1,c]

[1]Military Institute of Engineering Technology, Obornicka 136 Str.,50-961 Wrocław, Poland

[a]madej@witi.wroc.pl, [b]potreba@witi.wroc.pl, [c]krysiak@witi.wroc.pl

Keywords: Metal Detector, Electromagnetic Induction, Induction Coil, Mine

Abstract. Metal detectors operating on the principle of electromagnetic induction are widely used in civilian industry and military technology. Based on scientific literature, a review of the fundamental types of electromagnetic detectors is presented, along with the test results of a pulse induction metal detector regarding its capability to assess the size and distance of detected objects.

Introduction

The detection of metals or metal-containing objects is widely employed in various civilian industries, including technological processes and archaeology. These systems are utilized to ensure safety and protect human life. Such devices are commonly used in the food and pharmaceutical industries. In daily life, we frequently encounter metal detectors during passenger and baggage screening in public transportation, as well as at mass events, as these devices are used to detect weapons, ammunition, and dangerous metallic objects.

Metal detectors have found broad application in military technology due to the significant challenge posed by detecting landmines and hazardous items. Since clearing areas of mines after armed conflicts represents a major technical, humanitarian, and economic challenge, electromagnetic metal detectors play a crucial role in the process of detecting and locating landmines.

The development of metal detectors focuses on enhancing effectiveness, minimizing weight, ensuring operational stability, and simplifying their operation. In response to the evolution of mine designs towards minimizing metal content, detectors must be upgraded to detect minimal quantities of metal. Such advanced designs necessitate the implementation of microprocessor systems for processing acquired signals to identify the detected object. The aim is to achieve identification that allows for the determination of the detected material (distinguishing between metals, ferromagnetic, and diamagnetic materials), the object's shape, its size, and its burial depth.

As armed conflicts cover vast territories, the safe detection and removal of hazardous ordnance can take many years. To increase the speed of demining operations, multi-section mine detectors mounted on vehicles are being designed, enabling the scanning of large areas in a short time [1,2].

This article focuses on Pulse Induction (PI) technology, which, due to its properties, demonstrates significant effectiveness in detecting objects at considerable depths in various environments, including soils with high mineral content.

Operating Principle of an Electromagnetic Metal Detector

Electromagnetic metal detectors operate on the principle of electromagnetic induction. The relationship between the signal generated at the detector's output and the presence of metallic objects in its vicinity is described by Ampère's and Faraday's laws [3]. For simplicity, the device is considered to consist of a search head containing one or more coils carrying a time-varying electric current. This generates an electromagnetic field, the distribution of which is disturbed by metallic elements within its area of influence, such as a metal object buried beneath the ground surface [4]. Metallic objects near the detector's coil affect the electromagnetic permeability or alter

Materials Research Proceedings 62 (2026) 293-301
https://doi.org/10.21741/9781644904015-38

the magnetic field distribution due to the secondary magnetic field created by eddy currents generated within the object [5]. The effect influences the output signal level of the detector's sensing coil, which is then subjected to further processing in the receiver circuit. There are three commonly used MD design techniques, which will be described below [6]:
- Beat Frequency Oscillator (BFO),
- Very Low Frequency (VLF),
- Pulse Induction (PI).

Beat Frequency Oscillator

This is one of the simplest methods, utilizing a single coil. The operating principle of the Beat Frequency Oscillator (BFO) detector relies on utilizing two oscillators. The first one serves as a frequency reference, while the second is coupled with the detector's search coil [7]. When brought near a metallic element, the inductance of the coil winding changes, thereby disturbing the frequency of the measurement oscillator. The result of this proximity to metal is a change in frequency within the measurement circuit, which indicates metal detection.

The primary advantages of BFO detectors are their simple construction and the ability to distinguish between detected ferromagnetic metals (which cause an increase in coil inductance) and non-ferromagnetic metals (which cause a decrease in coil inductance). Their disadvantages include a short detection range, a lack of immunity to electromagnetic interference, and poor thermal stability.

Very Low Frequency

Most modern frequency-domain metal detectors indeed use separate transmitting and receiving circuits. They typically operate within the Very Low Frequency range of the spectrum, usually from a few kilohertz up to several tens of kilohertz (1 Hz up to 50 kHz).

The transmitter coil generates an alternating electromagnetic field within the detector's range. The receiver coil is positioned to detect changes in the induced voltage caused by the presence of a metallic element, which manifests as a change in the mutual inductance between the transmitter and receiver coils. The relative arrangement of the coils is of particular importance; they must have the lowest possible mutual inductance in the absence of a metal object within the detection area [8].

The measurement system consists of a resonant transmitter that generates a sinusoidal signal, which is supplied to the transmitter probe. Subsequently, the signal from the receiver probe is amplified and processed, with changes in both its amplitude and phase relative to the transmitter signal being analyzed. The received signal can be represented based on two parameters [9].

The first parameter Vr denotes the component of the received signal that is in phase with the transmitted signal, referred to as the in-phase or resistive component. The second parameter represents the component of the received signal that is phase-shifted by 90 degrees relative to the transmitted signal, known as the quadrature-phase or reactive component. By analyzing the measurements of these two parameters, the control unit can classify the detected object based on the metal alloy from which it is composed.

Devices based on VLF technology have largely eliminated the problems associated with BFO designs, achieving thermal stability and an increased detection range. However, in VLF-based systems, a key challenge becomes covering a large detection area simultaneously. The simultaneous operation of multiple coils leads to mutual interference, which degrades operational performance. Increasing the coil size to a very large dimension, for instance, 1m x 1m, results in a loss of stability. While a larger coil size increases the detection range, it also reduces the detector's sensitivity.

Materials Research Proceedings 62 (2026) 293-301
https://doi.org/10.21741/9781644904015-38

Pulse Induction

Time domain metal detectors work as instruments that operate by sending pulses of current through a coil (with a typical repetition rate of about 1 kHz), while ensuring that the current switch-off time is in the range of a few microseconds (For this purpose, switching transistors are used) [9]. In this solution, the coil functions as both a signal transmitter and receiver. The operating principle involves applying a brief pulse to the coil by closing a switch, which initiates current flow. Subsequently, the switch is opened, inducing a large voltage spike (Figure 1a). The cycle repeats itself. The eddy currents induced in a conductive target attempt to restore the magnetic field that has just vanished [10]. As a result of these eddy currents, a secondary magnetic field is generated, producing a small voltage in the coil that decays at the same rate. A Time-Domain metal detector essentially detects the rate at which the momentarily created magnetic field collapses, slowing down in the presence of metal within the coil field. This time depends predominantly on the target's size, conductivity, and permeability [11].

Figure 1b presents the voltage decay time in the coil. The solid line indicates the voltage waveform in the absence of a metallic element near the coil. In contrast, the added dashed line shows how the voltage waveform changes upon detection of a metallic object.

Pulse Induction (PI) detectors feature a simple design, offer a long detection range, and, above all, facilitate the integration of multiple coils operating in tandem. The use of several coils enables a large detection area to be achieved. However, the most significant disadvantages of this type of detector are its high power consumption and the inability to discriminate the type of metal from which the detected object is made.

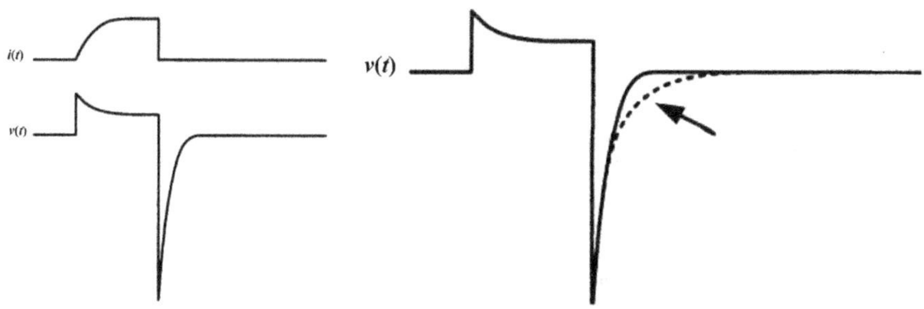

Figure 1. *a) Current and voltage waveform in the PI detector coil; b) change in the voltage waveform in the PI detector coil due to the presence of metallic objects in the coil's vicinity [5]*

The deployment of PI detectors for mine detection requires ensuring an appropriate coil power supply method, so that the generated magnetic field does not activate magnetic fuses used in anti-tank mines.

Experimental Study

When searching for hazardous objects buried in the ground, the ability to classify whether a detected metal object could be a potential hazard is highly desirable. Electromagnetic detector technology does not allow for such classification with a high degree of reliability. Only BFO (Beat Frequency Oscillation) and VLF (Very Low Frequency) detectors allow for limited discrimination between detected metal types. However, as mentioned earlier, BFO and VLF detectors have several drawbacks and are not suitable for constructing multi-channel detectors intended for searching large areas. In this case, Pulse Induction (PI) detectors perform best, and for the tests, the Plate Mine Detector (Polish acronym: PWM), developed by WITI based on this technology, was used.

A single PWM panel is constructed with four measurement channels, each coil operating simultaneously. The use of several such panels enables the detection of objects over a large area in a short time. During the research, a PWM consisting of three panels was utilized. Tests of the PWM detector were conducted to determine parameters potentially useful for the further classification of detected objects.

The size of an object detected by a PI detector can be determined based on the ground area over which the detection signal occurs. In the case of multi-channel detectors like the PWM, the simultaneous appearance of detection signals on adjacent channels indicates the presence of a large object. Example output signal waveforms for the four channels are presented in Fig. 2.

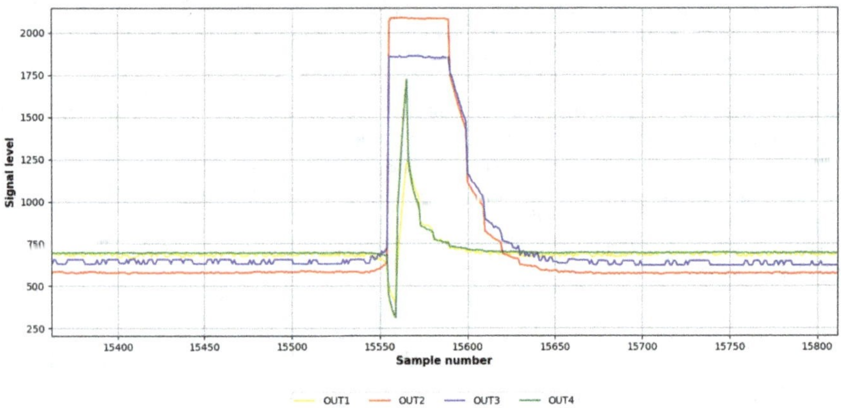

Figure 2. *Waveforms of the OUT1 to OUT4 signals from the PWM detector during the detection of a large object*

Based on the surface area of the coils used in the detector and considering their displacement, the cross-sectional area of the detected object can be estimated.

By determining the object's size on this basis and analyzing the signal level, the distance of the object from the detector coil can be estimated. As part of the tests conducted by the authors on the PWM detector, the relationship between the output signal level and distance was determined. The tests were performed for a steel plate object with a diameter of 300 mm and a thickness of 4 mm, placed on the ground surface at a distance (x) ranging from 0.2 m to 0.6 m in 0.05 m increments. At a distance of 0.2 m, the maximum output signal was obtained for the detector's minimum sensitivity. For each position, 3 to 5 measurements were taken. The average values of the normalized output signal (Sn) and its range (ΔSn) as a function of distance (x) are presented in the graph in Fig.4.

Materials Research Proceedings 62 (2026) 293-301 https://doi.org/10.21741/9781644904015-38

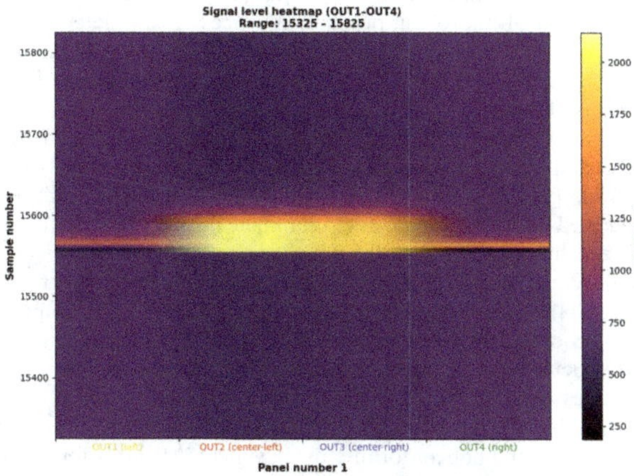

Figure 3. *Surface plot of the OUT1 to OUT4 signals from the PWM detector during the detection of a large object*

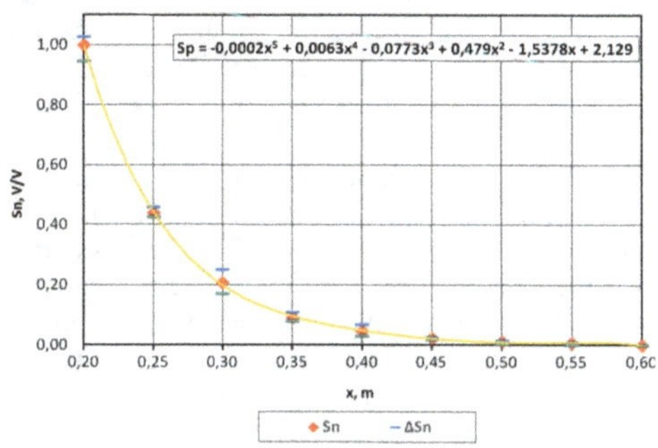

Figure 4*. Normalised output signal as a function of distance from the object*

Based on the obtained results, a relationship was determined between the output signal and distance in the form of a 5th-order polynomial (1).

$$Sp = -0.0002x^5 + 0.0063x^4 - 0.0773x^3 + 0.479x^2 - 1.5378x + 2.129 \quad (1)$$

Such a relationship indicates rapid signal decay with increasing distance, creating significant challenges in achieving the desired selectivity for the aforementioned purposes.

Technological and Economic Impact

Research into the ability to assess the size and distance of detected objects using electromagnetic metal detectors, particularly pulse induction detectors, has significant technological and economic implications in various industrial and military sectors. Precisely determining object characteristics (size, shape, distance) increases the informational value of detection, which translates into operational efficiency and safety. This is contributing to a new level of quality control and non-destructive testing (NDT). Improved detection and assessment of the size of metallic contaminants (in the food and recycling industries) or defects in metal components (bearings) is crucial for quality control. The ability to precisely determine object properties (cracks, inclusions) using NDT is mentioned in the context of improving production processes and defect analysis [12,13]. The use of advanced analysis methods, such as interval analysis in risk assessment, can also be applied to the interpretation of detector data, increasing its reliability [14]. This significantly improves the quality of final products [15,16].

The development of detectors that provide more data than just a binary "detected/not detected" signal aligns with the challenges and goals of Industry 4.0, which require advanced real-time data monitoring and analysis [17]. Precise positioning and control systems, such as pneumatic servo systems, can be integrated with metal detectors to automate inspection or sorting processes [18,19], as well as precision machining [20]. Data from higher-resolution detectors can be used to verify and optimize metallurgical and manufacturing processes, including heat treatment (such as induction hardening [21]) or cutting [22], as well as welding [23,24]. Improved input control (material purity) and process control (tool position tracking) contribute to enhanced mechanical properties and microstructure of materials [25-27], which in turn impact corrosion resistance [28].

A more accurate assessment of the size and location of a detected object minimizes the risk of misidentification (including product nonconformity), which is a key element of quality management, including risk and nonconformity management. Methods for identifying the causes of nonconformity and quantitative risk assessment can support the use of detector data to minimize losses and improve safety [29-31]. Precise detection and size identification of metal objects in waste streams, such as those in recycling, enable more efficient sorting, which directly impacts environmental costs (eco-costs) and the overall product life-cycle assessment (LCA), as demonstrated by analyses of aluminum cans [32]. Eliminating defective products early reduces material and energy losses, resulting in cost reductions and improved resource utilization.

Improved control over production processes, including a logical flow in production planning and control, as well as the use of Lean Production tools [33], such as Value Stream Mapping (VSM) [34], enables increased organizational efficiency. Detectors capable of assessing object size can support these processes by providing reliable input data, thus supporting logistics optimization and production planning. The use of advanced measurement techniques, including statistical methods such as the smooth bootstrap [35] and fuzzy regression [36, 37], in materials research and process evaluation contributes to innovation and increased competitiveness. Research results on metal detectors provide valuable input for such analyses, thereby improving the quality and reliability of final products and increasing competitiveness.

Developing the ability of electromagnetic metal detectors to assess object size is a significant step toward moving from simple detection to advanced object characterization, which is essential in the context of increasing quality and efficiency requirements in modern industry.

Summary

The analysis of the technology used in electromagnetic metal detectors and the test results obtained for the PWM detector indicate that while it is possible to assess the size of detected objects and estimate their burial depth, this requires a multi-parameter analysis of the signals received from the detector. However, the results obtained can be subject to significant errors due to the strong dependence of the signals on the object's size and distance. Classification of object types based solely on metal detector data is not feasible; therefore, the use of other detection methods is necessary. The utilization of GPR (Ground Penetrating Radar) and the joint analysis of data from both types of detectors will enhance the reliability of classifying detected objects [38,39].

References

[1] M. Szczepaniak et al., Wieloczujnikowy system wykrywania przedmiotów niebezpiecznych – Konferencja Naukowo-Techniczna WITI, Kościerzyna, 2010.

[2] M. Szczepaniak et al., Urządzenie inżynieryjne do rozpoznania i rozminowania dla kołowych pojazdów opancerzonych – projekt, obliczenia, wykonanie, Projektowanie i Konstrukcje Inżynierskie 7/8 (2012). https://konstrukcjeinzynierskie.pl/urzadzenia-inzynieryjne-do-rozpoznania-i-rozminowania-dla-kolowych-pojazdow-opancerzonych/

[3] D. Halliday, R. Resnick, J. Walker, Fundamentals of Physics. Part 3. Wiley, Hoboken, 2007.

[4] S. Tumanski, Induction Coil Sensor – a Review, Meas. Sci. Technol. 18 (2007) R31. https://doi.org/10.1088/0957-0233/18/3/R01

[5] G. Overton, C. Moreland, Inside the Metal Detector, Geotech Press, 2012.

[6] C. Garrett, Totally Revised Modern Metal Detectors, 2nd Ed., Ram Publishing Company, Dallas, Texas, 1995.

[7] A.M. Jawad et al., Design of a Beat Frequency Oscillator Metal Detector, IOSR Journal of Electronics and Communication Engineering 9 (2014) 56-62. https://doi.org/10.9790/2834-09215662.

[8] L.J. Carter, Z. Tang, Effects of Placement Errors on Performance of VLF Metal Detector Heads, Int. J. Smart Sensing Intell. Syst. 4 (2011) 675-685. https://doi.org/10.21307/IJSSIS-2017-462

[8] C. Bruschini, A Multidisciplinary Analysis of Frequency Domain Metal Detectors, Ph.D. Thesis, Vrije Universiteit Brussel, Brussels, Belgium, 2002.

[9] S.F. Minhas et al., Deep learning-based classification of anti-personnel mines and sub-gram metal content in mineralized soil (DL-MMD), Scientific Reports 14 (2024) art.10830. https://doi.org/10.1038/s41598-024-60592-8

[10] B. Candy, Metal Detector Basics and Theory, Minelab Pty Ltd., 2017

[11] B. Kim et al., Pulse-induction metal detector with time-domain bucking circuit for landmine detection, Electronics Letters 51 (2015) 159-160. https://doi.org/10.1049/el.2014.3895

[12] D. Siwiec et al., Improving the Non-Destructive Testing Process of the Outer Bearing Ring, Materials Research Proceedings 24 (2022) 174-180. https://doi.org/10.21741/9781644902059-26

[13] A. Pacana et al., Method of precise determination of the main causes of the problems on example of the pin connecting discs in an engine gear, Transportation Research Procedia 55 (2021) 683-690. https://doi.org/10.1016/j.trpro.2021.07.036

[14] E. Kozien, M.S. Kozien, The EX ante risk assessment in the project in interval analysis description, Materials Research Proceedings 17 (2020) 31-35. https://doi.org/10.21741/9781644901038-5

[15] E. Radzyminska-Lenarcik, M. Ulewicz, Selective Transport of Cu(II) across a Polymer Inclusion Membrane with 1-Alkylimidazole from Nitrate Solutions, Separation Science and Technology (Philadelphia) 47 (2012) 1113-1118. https://doi.org/10.1080/01496395.2011.646378

[16] M. Zenkiewicz et al., Electrostatic separation of binary mixtures of some biodegradable polymers and poly(vinyl chloride) or poly(ethylene terephthalate), Polimery/Polymers 61 (2016) 835-843. https://doi.org/10.14314/polimery.2016.835

[17] J. Pietraszek et al., Challenges for the DOE methodology related to the introduction of Industry 4.0, Prod. Eng. Arch. 26 (2020) 190-194. https://doi.org/10.30657/pea.2020.26.33

[18] J.E. Takosoglu, P.A. Laski, Intelligent positioning system of electro-pneumatic servo-drive, Annals of DAAAM and Proc. Int. DAAAM Symposium (2011) 1641-1642.

[19] D.S. Pietrala, P.A. Laski, Design and Control of a Pneumatic Muscle Servo Drive Containing Its Own Pneumatic Muscles, Applied Sciences 12 (2022) art. 11024. https://doi.org/10.3390/app122111024

[20] Ł.J. Orman et al., Application of laser treatment technology for boiling heat transfer augmentation, Prod. Eng. Arch. 30 (2024) 259-265. https://doi.org/10.30657/pea.2024.30.25

[21] A. Szczotok et al., Effect of the induction hardening on microstructures of the selected steels, METAL 2018 – 27th Int. Conf. Metall. Mater. (2018) 1264-1269.

[22] P. Szataniak et al., HSLA steels - Comparison of cutting techniques, METAL 2014 – 23rd Int. Conf. Metall. Mater. (2014) 778-783.

[23] I. Miletić et al., Analysis of selected properties of welded joints of the HSLA Steels, Materials 13 (2020) art. 1301. https://doi.org/10.3390/ma13061301

[24] N. Radek et al., Properties of Steel Welded with CO2 Laser, Lecture Notes in Mechanical Engineering (2020) 571-580. https://doi.org/10.1007/978-3-030-33146-7_65

[25] D. Siwiec et al., An analysis of the influence double normalization process on mechanical properties and microstructure 38MnVS6 steel, METAL 2021 - 30th Anniversary Int. Conf. Metall. Mater. (2021) 579-584 https://doi.org/10.37904/metal.2021.4162

[26] D. Siwiec et al., Improving the process of achieving required microstructure and mechanical properties of 38MnVS6 steel, METAL 2020 – 29th Int. Conf. Metall. Mater. (2020) 591-596. https://doi.org/10.37904/metal.2020.3525

[27] T. Lipinski, Modification of Al-11% Si alloy with Cl-based modifier, Manufacturing Technology 15 (2015) 581-587.

[28] T. Lipinski, J. Pietraszek, Influence of animal slurry on carbon C35 steel with different microstructure at room temperature, Engineering for Rural Development 21 (2022) 344-350. https://doi.org/10.22616/ERDev.2022.21.TF115

[29] R. Dwornicka et al., Development of a Hybrid Method for Identifying the Causes of Product Incompatibility in Metallurgical Manufacturing, Materials Research Proceedings 34 (2023) 199-206. https://doi.org/10.21741/9781644902691-24

[30] E. Kozień, Quantitative Method of Ex-Ante Project Risk Assessment Based on Checklist Method, Management and Production Engineering Review 15 (2024) 0-0. https://doi.org/10.24425/mper.2024.153123

[31] C. Karolina et al., Improvement of the production process with the use of selected KPIs, System Safety: Human - Technical Facility - Environment 2 (2020) 307-315. https://doi.org/10.2478/czoto-2020-0038

[32] D. Siwiec et al., Analysis of aluminum can by eco-costs and life cycle assessments (LCA), METAL 2024 – Int. Conf. Metall. Mater. (2024) 627-632. https://doi.org/10.37904/metal.2024.4975

[33] R. Ulewicz et al., Implementation of logic flow in planning and production control, Management and Production Engineering Review 7 (2016) 89-94. https://doi.org/10.1515/mper-2016-0010

[34] D. Klimecka-Tatar, Value Stream Mapping as Lean Production tool to improve the production process organization - Case study in packaging manufacturing, Prod. Eng. Arch. 17 (2017) 40-44. https://doi.org/10.30657/pea.2017.17.09

[35] J. Pietraszek, A. Gadek-Moszczak, The smooth bootstrap approach to the distribution of a shape in the ferritic stainless steel AISI 434L powders, Solid State Phenomena 197 (2013) 162-167. https://doi.org/10.4028/www.scientific.net/SSP.197.162

[36] J. Pietraszek, Fuzzy regression compared to classical experimental design in the case of flywheel assembly, Lecture Notes in Computer Science 7267 LNAI (2012) 310-317. https://doi.org/10.1007/978-3-642-29347-4_36

[37] J. Pietraszek, The modified sequential-binary approach for fuzzy operations on correlated assessments, Lecture Notes in Computer Science 7894 LNAI (2013) 353-364. https://doi.org/10.1007/978-3-642-38658-9_32

[38] J. Ishikawa et al., Experimental design for test and evaluation of anti-personnel landmine detection based on vehicle-mounted GPR systems, in: Proc. SPIE 5794, Detection and Remediation Technologies for Mines and Minelike Targets X (2005). https://doi.org/10.1117/12.603149

[39] R. Pavlik et al., System concept for GPR reflectance measurement method in time and frequency domain, VOP-026 Sternberk s.p., VTUO Brno Division, 2007.

Materials Research Proceedings 62 (2026) 302-310 https://doi.org/10.21741/9781644904015-39

Effect of Bearing Type on the Wear and Surface Condition of Shafts under Long-Term Service

Hubert KUŹDUB[1,a *] and Joanna BOROWIECKA-JAMROZEK[2,b]

[1]MESKO S.A., ul. Legionów 122, 26-110 Skarżysko-Kamienna, Poland

[2]Kielce University of Technology, Faculty of Mechatronics and Mechanical Engineering, Al.Tysiąclecia PP 7, 25-314 Kielce, Poland

[a]h.kuzdub@gmail.com, [b]jamrozek@tu.kielce.pl

Keywords: Needle Roller Bearing, Bronze Bushing, Shaft Wear, Surface Roughness, Tribology, Bearing Durability, Machine Reliability

Abstract. This study presents a comparative analysis of two shafts on which material feeder discs were mounted. In the first configuration, a bronze sleeve bushing was used, whereas in the second one, two needle roller bearings (SKF, 40×47×16 mm) were applied. Both shafts operated under identical service conditions on equivalent machines positioned side by side, in a two-shift system over a period of five years, corresponding to approximately 15,000 operating hours. Both maintenance and service of the systems were carried out in accordance with the manufacturer's recommendations, which ensured comparable operating conditions. The conducted study aimed to determine the differences in wear, reliability, and maintenance costs between the shaft equipped with a bronze bushing and the one fitted with needle roller bearings, as illustrated by the condition of the components shown in Fig.1. The obtained results may serve as a basis for optimizing design solutions in similar drive systems used in industrial applications [1-3].

Introduction

In machine operation processes, a crucial factor determining durability and reliability is the proper selection of bearing and sliding elements within rotating assemblies. The type of bearing applied affects not only the resistance to motion but also the efficiency of load transmission and the wear resistance during prolonged operation. In industrial practice, various design solutions are often compared in terms of their operational durability and maintenance costs.

The optimization of bearing systems has a direct impact on the energy efficiency, service life, and reliability of machines used across multiple industrial sectors. Proper selection between sliding and rolling bearings is particularly important in applications where continuous operation, high rotational speeds, or variable loads are present. Such conditions are common in process equipment employed in the chemical, food processing, and defense industries, as well as in transportation and manufacturing systems where shafts, rollers, and feeders are key structural components.

Accurate identification of wear mechanisms and quantitative assessment of dimensional changes enable engineers to optimize the design of bearing assemblies, minimize friction losses, and extend maintenance intervals. By analyzing the wear behavior of different bearing types under real operating conditions, it becomes possible to formulate guidelines for material selection, lubrication strategies, and surface treatments tailored to specific working environments.

The outcomes of such comparative studies are of high practical significance, as they support predictive maintenance planning and reliability-based design approaches in industrial machinery. Moreover, the implementation of optimized bearing solutions contributes to reducing energy consumption, improving operational safety, and ensuring long-term stability of production systems. Therefore, the research into the tribological performance of sliding and rolling bearings

Materials Research Proceedings 62 (2026) 302-310 https://doi.org/10.21741/9781644904015-39

under comparable service conditions provides a valuable basis for engineering decision-making and technological advancement in modern mechanical design [4].

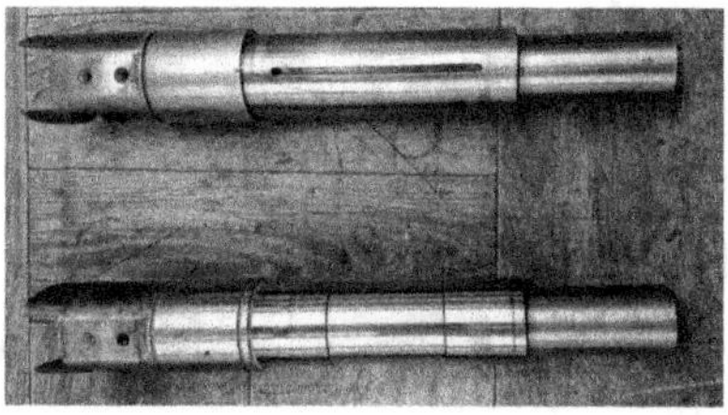

Fig. 1. Shafts selected for wear analysis.

Materials and Bearing Components

The shafts used in the tested assemblies were made of high-quality non-alloy structural steel grade C45, commonly applied in machine construction due to its favorable balance between strength and machinability. C45 steel contains approximately 0.45% carbon, which enables effective heat treatment to enhance its mechanical strength and wear resistance.

After machining, the components underwent heat treatment, consisting of quenching and tempering, to achieve a hardness of approximately 30 HRC. These parameters of final heat treatment ensure a microstructure with good fatigue strength while maintaining adequate core ductility.

A hardness range of 28–32 HRC allows for effective transmission of dynamic loads and minimizes plastic deformation in the contact zones with bearing elements.

Following heat treatment, the shaft surfaces were subjected to black oxide coating (alkaline oxidation). This treatment imparts corrosion-resistant properties to the surfaces and enhances their resistance to environmental factors, including moisture and oil mist. Additionally, the iron oxide layer formed during blackening reduces the coefficient of friction. It enhances the retention of the lubricating film on the surface, thereby improving the tribological conditions of cooperating components. As a result, the combination of heat treatment and black oxide finishing yielded a material with balanced mechanical properties, high fatigue life, and increased resistance to both abrasive and corrosive wear [7].

In the first of the analyzed configurations, a sliding bearing was used in the form of a bushing made of tin bronze grade B101. This alloy is a copper-based material with approximately 10% tin content, characterized by high wear resistance, good thermal conductivity, and anti-adhesive properties, which prevent seizure during operation under limited lubrication conditions. The inner surface of the bushing was machined with a figure-eight pattern of lubrication grooves spaced at 90 mm, as well as a single circumferential groove with a radius of R = 2 mm, ensuring uniform distribution of lubricant along the entire contact surface [2,3,7].

In the second configuration, rolling bearings were used in the form of two SKF needle roller bearings (manufacturer's catalog designation) with dimensions of 40×47×16 mm. Bearings of this type are distinguished by their high load-carrying capacity, combined with compact dimensions, allowing for the efficient transmission of radial loads in applications with limited installation space. The use of small-diameter rollers provides favorable pressure distribution and a low coefficient of friction compared to sliding bearing solutions. SKF needle roller bearings are also characterized by high manufacturing precision and long operational life, which makes them a preferred choice in systems subjected to intensive operation and variable loading conditions [5].

Methodology and Results of Shaft Diameter Measurements

Diameter measurements of the working surfaces of the shafts after long-term operation were performed using a Sylvac SCAN F60LT optical measuring system, designed for non-contact measurement of rotationally symmetrical components. The device enables high-precision measurements with excellent repeatability, eliminating the errors associated with contact pressure that are typical of tactile measurement methods. For each shaft, measurements were taken in several control cross-sections, and the arithmetic mean of the obtained values was calculated.

Shaft Operating with the Bronze Bushing

For the shaft operating with the sliding bronze bushing, the nominal journal diameter was 40f7 consistent with the original design geometry presented in Fig.2. After five years of operation, the average measured diameter was 39.461 mm, based on five independent measurements taken in different cross-sections, as shown in Fig.3. The resulting difference of approximately 0.54 mm relative to the nominal dimension indicates a significant wear of the working surface.

The reduction in journal diameter is the result of long-term sliding friction acting within the contact zone between the shaft and the bronze bushing. Despite regular lubrication, the steel material experienced gradual wear during service, particularly in areas where the lubricating film was periodically disrupted. Additionally, the presence of contaminants in the working environment likely contributed to microscopic scratches and surface erosion, which consequently increased the operational clearance and led to a gradual reduction in journal diameter [3-6].

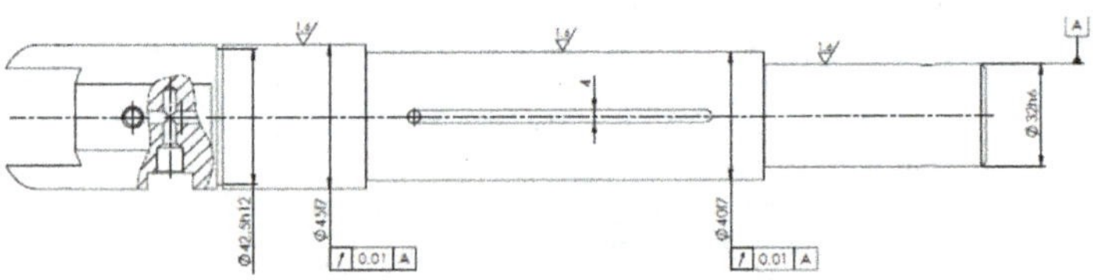

Fig. 2. Technical drawing of the shaft cooperating with the bronze sleeve bearing.

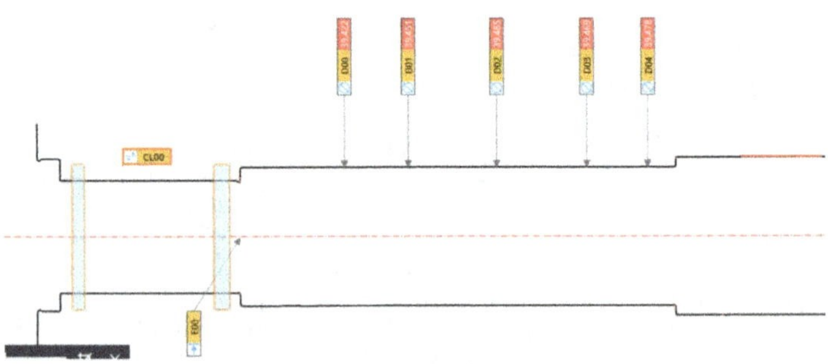

Fig. 3. Measurement results of the shaft cooperating with the bronze sleeve bearing.

Shaft Operating with Needle Roller Bearings

For the shaft fitted with two needle roller bearings, the nominal journal diameter was 40k6, consistent with the original design geometry presented in Fig.4. After the operational period, six control measurements were conducted, yielding an arithmetic mean value of 40.007 mm, as presented in Fig.5. The obtained result indicates a minimal deviation from the nominal dimension, remaining well within the tolerance limits of the specified fit.

The maintenance of the journal diameter at an almost unchanged level confirms the favorable operating conditions of the rolling bearing arrangement. The presence of a rolling element layer between the shaft surface and the bearing raceway prevents direct material contact, thereby minimizing adhesive and abrasive wear phenomena. Moreover, needle roller bearings provide stable shaft guidance with a low coefficient of friction, which significantly reduces wear on the mating surfaces compared to sliding bearing configurations [5].

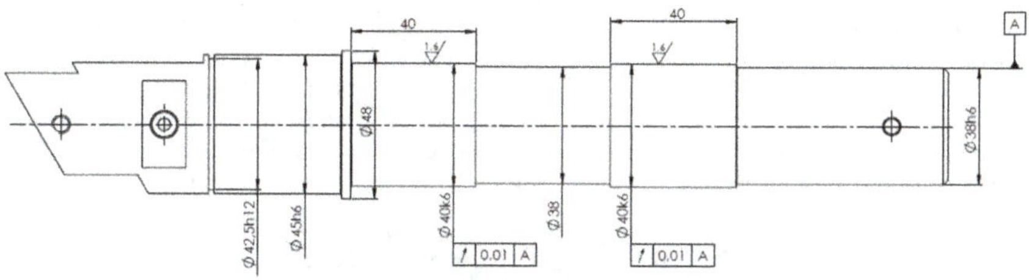

Fig. 4. Technical drawing of the shaft cooperating with needle roller bearings.

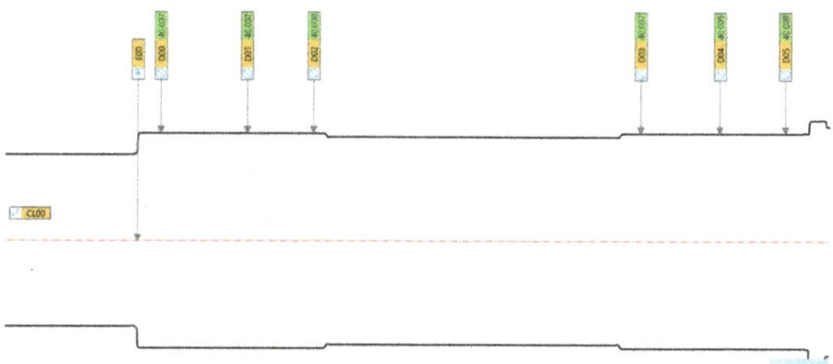

Fig. 5. Measurement results of the shaft cooperating with needle roller bearings.

Surface Roughness Analysis and Evaluation of Shaft Surfaces after Operation

The condition of the working surfaces of the shafts after the operational period was evaluated based on surface roughness measurements and microscopic observations. Initially, the journal surfaces were finished to a roughness value of Ra 1.6 μm, in accordance with the design requirements for surfaces cooperating with bushings and bearings.

After the operation period and subsequent measurements, significant differences in surface roughness parameters were observed between the analyzed components. For the shaft operating with the bronze bushing, the obtained Ra values were 0.85, 1.07, 0.59, and 0.41 μm. In contrast, for the shaft operating with needle roller bearings, the Ra values were 0.15, 0.11, 0.22, and 0.17 μm.

The decrease in Ra values in both cases indicates the occurrence of the so-called operational smoothing process, which takes place during prolonged contact between cooperating surfaces. As a result of friction, surface asperities are gradually sheared off, leading to a reduction in roughness amplitude and an improvement in the smoothness of the working surface.

In the case of the shaft operating with needle roller bearings, the obtained Ra values, in the range of 0.1–0.2 μm, correspond to roughness grades N4–N5, which are typical of precisely ground or polished surfaces. Such low roughness values indicate very favorable rolling conditions and minimal wear effects on the shaft surface.

For the shaft operating with the bronze bushing, the measured Ra values (0.4–1.1 μm) correspond to roughness grades N6–N7, characteristic of ground or lapped surfaces. Slightly higher Ra values in this case result from the nature of sliding contact, where direct metal-to-metal interaction occurs, accompanied by intensified asperity deformation under load. Despite the presence of a lubricating film, its periodic disruption and the presence of fine contaminants could have contributed to local increases in roughness and the formation of micro-scratches [2,3,4].

For the shaft operating with needle roller bearings, at a magnification corresponding to an area of approximately 1 mm², fine, parallel surface scratches were observed, as illustrated in Fig.6. These marks result from the natural rolling contact process, where minor irregularities on the raceway and needles leave shallow traces on the surface. The absence of visible point defects or pits indicates proper lubrication and minimal fatigue-related wear of the material.

Fig. 6. Surface topography of the shaft cooperating with needle roller bearings.

In contrast, for the shaft operating with the bronze bushing, surface analysis at a magnification corresponding to an area of approximately 75 μm² revealed the presence of deep and wide abrasive scratches, as shown in Fig.7. These scratches were caused by the intensive action of contaminant particles entering the friction zone, either with the lubricating oil or through minor system leaks. The presence of hard particles (e.g., metallic dust or oxide particles) acted as an abrasive medium, locally removing material from the shaft surface. This phenomenon is typical for sliding bearings operating in environments with elevated particulate contamination and represents one of the main factors accelerating surface wear.

Fig. 7. Surface topography of the shaft cooperating with a bronze sliding bushing.

Materials Research Proceedings 62 (2026) 302-310 | https://doi.org/10.21741/9781644904015-39

Comparative Analysis

The analysis of shaft diameter measurements, surface roughness, and surface topography after the operational period revealed clear differences in wear between the bronze sliding bushing and needle roller bearing solutions.

For the shaft operating with the bronze bushing, a significant loss of nominal diameter was observed (reduction of approximately 0.54 mm) along with the presence of deep and wide abrasive scratches. These effects resulted from prolonged sliding contact and the influence of contaminant particles. The phenomena indicate intensive surface material wear and gradual deterioration of tribological performance. Despite proper lubrication, the sliding bearing exhibited limited resistance to environmental factors and local overheating in the friction zone.

In contrast, the shaft with needle roller bearings maintained an almost unchanged journal diameter (a difference of approximately 0.007 mm) and very low surface roughness (Ra: 0.11-0.22 µm). The working surface exhibited minor, regularly aligned scratches typical of rolling contact, with no signs of adhesive wear. These results indicate that rolling bearings provided significantly more favorable operating conditions, reducing friction and wear on the shaft material [3].

From a tribological perspective, it can be concluded that needle roller bearing arrangements enable a substantial extension of the shaft–bearing system's service life while minimizing energy losses due to friction. Conversely, although sliding bearings are simple in design and low-cost to manufacture, they require careful maintenance of lubricant cleanliness and system sealing, as even minor contaminants can significantly accelerate surface wear [4].

Technological, Economic, Environmental, and Social Impact

The results of the comparative study, demonstrating the significant superiority of needle roller bearings over plain bearings in terms of dimensional stability and shaft surface quality, have direct and broad implications for industrial engineering.

The confirmed tenfold increase in the durability of a shaft mated to a rolling bearing is a key argument for optimizing design solutions in industrial machines, such as the material feeder systems studied. Improved reliability and extended service life are possible due to the performance properties of modern materials and coatings, which are specifically designed to enhance wear resistance [8]. Surface treatment processes, including laser processing, can further enhance the operational properties of components [9].

The aspect of surface wear, exhibiting mild rolling marks in contrast to intense abrasive scratches on the sliding surface, underscores the need for advanced analysis of the tribological behavior of materials, such as in sliding contact with composites [10]. Understanding the impact of bearing installation on static load capacity is also crucial, which directly translates into long-term system reliability [11]. Using rolling bearings enables better control of system dynamics, thereby reducing the need for complex methods to minimize vibration in machines whose design principles are still being intensively studied [12]. The durability and safety of structures, for example, in transportation systems, are crucial, as confirmed by research on the fatigue failure of fasteners [13].

In the context of developing mechanics and automation, research results on the use of rolling bearings support the design of precise positioning and drive systems, such as parallel manipulators [14] or servo drive systems with directional valves [15]. Fatigue analysis, considering complex load histories, is essential for verifying the durability of shafts using rolling bearings [16]. Designing advanced drive systems also requires multiphysics modeling that considers dynamic factors, such as rotor imbalance [17]. Replacing plain bearings with rolling bearings, which results in a more than tenfold increase in durability, directly reduces maintenance costs (LCC) and enhances economic efficiency. Analysis of technological strategies used in industry, such as wheel production [18], must consider such solutions to minimize downtime. Optimization of production and control processes is supported by modern analytical tools, including statistical methods such

Materials Research Proceedings 62 (2026) 302-310 | https://doi.org/10.21741/9781644904015-39

as bootstrap approaches for analyzing the distribution of material characteristics [19, 20] and methods based on fuzzy regression for comparing experimental results [21, 22]. Such approaches enable precise data management and decision-making in the context of implementing Industry 4.0 assumptions [23, 24], which in turn requires the implementation of flow logic in production planning [25].

Computational methods that account for data uncertainty, such as modified sequential-binary approaches for fuzzy operations [22], are also employed for precise economic evaluation. Improvements in the production process can be measured using precisely selected KPIs [26]. Furthermore, the development of expert systems to support the design of experiments (DoE) [27] enables effective technological validation of optimal solutions, reducing testing costs. Replacing sliding elements with rolling elements, which are less sensitive to contaminants in the lubricant, has a positive impact on the environment. This reduces material waste (by reducing shaft wear) and minimizes the risk of environmental pollution from faulty components. The pursuit of sustainable development [28] necessitates a comprehensive environmental analysis, encompassing assessments of phenomena related to pollutant emissions, such as the impact of start-up schedules on environmental impact [29].

The issue of contamination is also evident in studies of solids accumulation on steel surfaces during biomass co-firing [30], which is analogous to the harsh operating conditions of material feeders. The environmental aspect also encompasses the broader context of closed-loop management and phytoremediation as pollution prevention measures [31], which justifies the selection of components with long service lives and low failure rates. Research results on the electrostatic separation of biodegradable polymers [32] indicate progress in waste management in industry. Increasing the reliability and durability of industrial machine shafts through the use of needle bearings contributes to improved occupational safety and reduced risk of sudden failures. This action aligns with the goals of Corporate Social Responsibility (CSR), a concept whose importance is confirmed by the perspectives of customers and the general public [33]. The social aspect also encompasses the quality of infrastructure, the maintenance of which is expensive and time-consuming. The durability of spare parts, including those for railway rolling stock, is constantly being optimized, for example, through the use of anti-graffiti systems [34].

Conclusions

The dimensional durability of the shaft operating with needle roller bearings was over ten times greater than that of the shaft with a sliding bronze bushing, confirming the superiority of rolling bearing arrangements for long-term operation.

1. Surface roughness analysis indicates that rolling surfaces exhibit self-smoothing, whereas sliding surfaces show local degradation due to contact with contaminants.
2. Microscopic observations confirmed the different wear patterns: mild rolling marks in the needle bearings versus intense abrasive scratches in the bronze bushing.
3. In engineering practice, the results obtained can serve as a basis for optimizing the selection of bearing types in material feeder systems and other machinery under similar operating conditions, particularly in environments where lubricant cleanliness is difficult to control.
4. For future research, it is recommended to extend the analysis to include hardness measurements of surface layers and spectrometric studies of lubricant wear, which would allow a more precise determination of surface degradation mechanisms.

The comparison of both solutions confirms that the use of needle roller bearings provides significantly more favorable operating conditions for the shaft, reduces material wear, and improves the overall reliability of the drive system. These findings have direct practical applications in industry, enabling a more informed selection of bearing components in machines designed for long operational cycles.

References

[1] T. Burakowski, T. Wierzchoń, Surface engineering of metals - principle, equipment, technology, CRC Press, Boca Raton, 1999.

[2] B. Bhushan, Introduction to Tribology, 2nd Edition, John Wiley & Sons, Hoboken, 2013.

[3] M. Nosonovsky, B. Bhushan, Multiscale Friction, Lubrication, and Wear, Springer, New York, 2008.

[4] R. Radzimirski, Tribology: Friction, Wear and Surface Engineering, WNT, Warszawa, 2010.

[5] SKF Engineering Handbook, 14th Edition, SKF Group, Gothenburg, 2017.

[6] N. Radek et al., Operational tests of coating systems in military technology applications, Eksploatacja i Niezawodność 25 (2023) art.12. https://doi.org/10.17531/ein.2023.1.12

[7] J. Totten, Steel Heat Treatment Handbook, 2nd Edition, CRC Press, 2006.

[8] N. Radek, Determining the operational properties of steel beaters after electrospark deposition, Eksploatacja i Niezawodnosc 44 (2009) 10-16.

[9] N. Radek et al., The impact of laser processing on the performance properties of electro-spark coatings, World Congress in Computational Mechanics and ECCOMAS Congress 1000 (2021) 1-10. https://doi.org/10.23967/wccm-eccomas.2020.336

[10[J. Korzekwa et al., Tribological behaviour of Al2O3/inorganic fullerene-like WS2 composite layer sliding against plastic, International Journal of Surface Science and Engineering 10 (2016) 570-584. https://doi.org/10.1504/IJSURFSE.2016.081035

[11] M. Krynke, R. Ulewicz, Analysis of the influence of slewing bearing mounting on their static load capacity, Transp. Res. Proc. 40 (2019) 745-750. https://doi.org/10.1016/j.trpro.2019.07.105

[12] I. Drach et al., Design Principles of Horizontal Drum Machines with Low Vibration, Advances in Sci. Technol. Res. J. 15 (2021) 258-268. https://doi.org/10.12913/22998624/136441

[13] R. Ulewicz et al., The investigation of the fatigue failure of passenger carriage draw-hook, Engineering Failure Analysis 104 (2019) 609-616. https://doi.org/10.1016/j.engfailanal.2019.06.036

[14] J. Takosoglu et al., A Design of a 2 DoF Planar Parallel Manipulator with an Electro-Pneumatic Servo-Drive – Part 2, Energies 16 (2023) art. 2970. https://doi.org/10.3390/en16072970

[15] G. Filo et al., Pressure loss reduction in an innovative directional poppet control valve, Energies 13 (2020) art. 3149. https://doi.org/10.3390/en13123149

[16] P.J. Romanowicz et al., Using the Effect of Compression Stress in Fatigue Analysis of the Roller Bearing for Bimodal Stress Histories, Materials 15 (2022) art. 196. https://doi.org/10.3390/ma15010196

[17] A. Goroshko et al., Multiphysics Modeling of Direct-Start Induction Motor Considering Rotor Unbalance, Materials Research Proceedings 45 (2024) 267-276. https://doi.org/10.21741/9781644903315-31

[18] M. Konstanciak, Analysis of technological strategies on the example of the production of the tramway wheels, Archives of Materials Science and Engineering 57 (2012) 69-74.

[19] J. Pietraszek, A. Gadek-Moszczak, The smooth bootstrap approach to the distribution of a shape in the ferritic stainless steel AISI 434L powders, Solid State Phenomena 197 (2013) 162-167. https://doi.org/10.4028/www.scientific.net/SSP.197.162

[20] E. Skrzypczak-Pietraszek, J. Pietraszek, Chemical profile and seasonal variation of phenolic acid content in bastard balm (Melittis melissophyllum L., Lamiaceae), Journal of Pharmaceutical and Biomedical Analysis 66 (2012) 154-161. https://doi.org/10.1016/j.jpba.2012.03.037

[21] J. Pietraszek, Fuzzy regression compared to classical experimental design in the case of flywheel assembly, Lecture Notes in Computer Science 7267 LNAI (2012) 310-317. https://doi.org/10.1007/978-3-642-29347-4_36

[22] J. Pietraszek, The modified sequential-binary approach for fuzzy operations on correlated assessments, Lecture Notes in Computer Science 7894 LNAI (2013) 353-364. https://doi.org/10.1007/978-3-642-38658-9_32

[23] J. Pietraszek et al., Challenges for the DOE methodology related to the introduction of Industry 4.0, Production Engineering Archives 26 (2020) 190-194. https://doi.org/10.30657/pea.2020.26.33

[24] R. Ulewicz et al., Mining Industry 4.0 – Opportunities and Barriers, Acta Montanistica Slovaca 27 (2022) 291-305. https://doi.org/10.46544/AMS.v2i2.02

[25] R. Ulewicz et al., Implementation of logic flow in planning and production control, Management and Production Engineering Review 7 (2016) 89-94. https://doi.org/10.1515/mper-2016-0010

[26] C. Karolina et al., Improvement of the production process with the use of selected kpis, System Safety: Human - Technical Facility - Environment 2 (2020) 307-315. https://doi.org/10.2478/czoto-2020-0038

[27] R. Dwornicka, J. Pietraszek, The outline of the expert system for the design of experiment, Production Engineering Archives 20 (2018) 43-48. https://doi.org/10.30657/pea.2018.20.09

[28] A. Pacana et al., Analysis of a practical approach to the concept of sustainable development in a manufacturing company in the automotive sector, Waste Forum (2020) 151-161.

[29] R. Dwornicka, The impact of the power plant unit start-up scheme on the pollution load, Advanced Materials Research 874 (2014) 63-69. https://doi.org/10.4028/www.scientific.net/AMR.874.63

[30] M. Opydo et al., The effect of biomass co-combustion in a CFB boiler on solids accumulation on surfaces of P91 steel tube samples, Biomass and Bioenergy 85 (2016) 61-68. https://doi.org/10.1016/j.biombioe.2015.12.011

[31] E. Skrzypczak-Pietraszek, A.G. Pietraszek, Phytoremediation and Allelopathy as an Element of Sustainable Circular Economy to Prevent Environmental Pollution, Materials Research Proceedings 45 (2024) 140-145. https://doi.org/10.21741/9781644903315-17

[32] M. Zenkiewicz et al., Electrostatic separation of binary mixtures of some biodegradable polymers and poly(vinyl chloride) or poly(ethylene terephthalate), Polimery/Polymers 61 (2016) 835-843. https://doi.org/10.14314/polimery.2016.835

[33] E. Kozień, A. Kozień, The Importance of Corporate Social Responsibility from the Perspective of Customers in the Light of Own Research, Materials Research Proceedings 45 (2024) 249-257. https://doi.org/10.21741/9781644903315-29

[34] N. Radek et al., Technology and application of anti-graffiti coating systems for rolling stock, METAL 2019 – 28th Int. Conf. Metall. Mater. (2019) 1127-1132.

Analysis and Prediction of Delivery Reliability in the Retail Sector

Robert SZCZUR[1,a], Patrycja GUZANEK[1,b*] and Anna BORUCKA[2,c]

[1]Doctoral School, Military University of Technology, 2 Gen. Sylwestra Kaliskiego Street, 00-908 Warsaw, Poland

[2]Faculty of Safety, Logistics and Management, Military University of Technology, 2B Gen. Sylwestra Kaliskiego Street, 00-908 Warsaw, Poland

[a]robert.szczur@wat.edu.pl, [b]patrycja.guzanek@student.wat.edu.pl, [c]anna.borucka@wat.edu.pl

Keywords: Delivery Reliability, Retail Logistics, Delivery Completeness, Loss Rate

Abstract. This study analyzes delivery reliability in the retail sector, focusing on delivery completeness and losses over time. Using descriptive statistics, ANOVA, the Kruskal-Wallis test, and Tukey's post-hoc comparisons, the research found no significant monthly or seasonal variations. The correlation between order volume and delivery completeness was also insignificant, indicating that disruptions are incidental rather than cyclical. These results suggest that risk management should focus on identifying and eliminating one-off issues, while stable logistics processes provide a solid basis for continuous improvement in supply chain performance.

Introduction

Delivery reliability is a key component of logistics performance, directly influencing supply chain efficiency and customer satisfaction [1]. It reflects a company's ability to provide complete and timely deliveries, ensuring product availability and strengthening brand reputation [2], [3]. In the cosmetics retail sector, where demand is dynamic and shaped by trends, promotions, and new product launches, maintaining consistent deliveries is particularly challenging [4]. Even minor disruptions may cause shortages, lost sales, and reduced customer trust [5]. The aim of this study is to evaluate delivery reliability in a cosmetics retail company by analyzing two indicators: delivery completeness and delivery losses. The research investigates whether these indicators vary across months, show seasonal or cyclical patterns, and whether order volume affects delivery performance. The findings are intended to distinguish structural from incidental fluctuations and provide practical insights for logistics process improvement and risk management. Studies have shown that delivery reliability is a fundamental determinant of supply chain effectiveness [6], [7], [8]. Reliable deliveries foster stable customer relationships, reduce costs, and enhance operational flexibility [9], [10]. Incomplete or delayed deliveries, by contrast, can generate both direct losses and indirect consequences such as lower service quality and diminished competitiveness [11], [12]. Research in the retail and FMCG sectors highlights that delivery reliability depends on effective coordination of inventory, transportation, and forecasting [13], [14]. Furthermore, seasonal demand fluctuations and operational bottlenecks are commonly identified as sources of delivery variability [15], [16]. In industries such as cosmetics, the main challenge lies in distinguishing recurring seasonal effects from isolated operational disruptions [17]. Companies with mature logistics systems often maintain stable performance regardless of seasonal changes, reflecting strong process discipline and adaptability [18].

Materials and Methods

Company Description and Data Characteristics. The study used operational data from a cosmetics retail distributor operating in the beauty and FMCG sector. The company supplies regional outlets from a central warehouse, offering cosmetics products. Deliveries occur several times per week to ensure product availability and prevent stockouts. Reliability is crucial, as cosmetics are highly

Materials Research Proceedings 62 (2026) 311-317 https://doi.org/10.21741/9781644904015-40

sensitive to marketing cycles and seasonal demand. The dataset contained monthly aggregated records over several years, including ordered and delivered quantities and calculated indicators of delivery completeness and losses.

Analytical Framework. Two quantitative indicators were used to evaluate delivery reliability: delivery completeness - the ratio of delivered to ordered items (in %), representing how fully customer orders were met, delivery losses - the ratio of undelivered to ordered items (in %), representing the share of products that did not reach their destination. The analysis involved several stages: 1) descriptive statistics - to determine means, medians, dispersion (standard deviation), and quartile ranges, 2) ANOVA (*Analysis of Variance*) and Kruskal-Wallis test - to verify whether differences between months were statistically significant, 3) Tukey's post-hoc test - to locate potential pairwise differences between months, 4) Shapiro-Wilk and Levene's tests - to assess normality and equality of variances and 5) Pearson's correlation analysis - to examine the relationship between the number of ordered items and delivery completeness.

Results

Analysis of the Trend in Delivery Completeness. In the first stage of the study, the temporal trend of delivery completeness was evaluated. Delivery completeness was defined as the ratio of delivered to ordered items, expressed as a percentage. This measure allowed the assessment of how the level of delivery completeness changed across individual months and years of the study period. The results are illustrated in Fig.1.

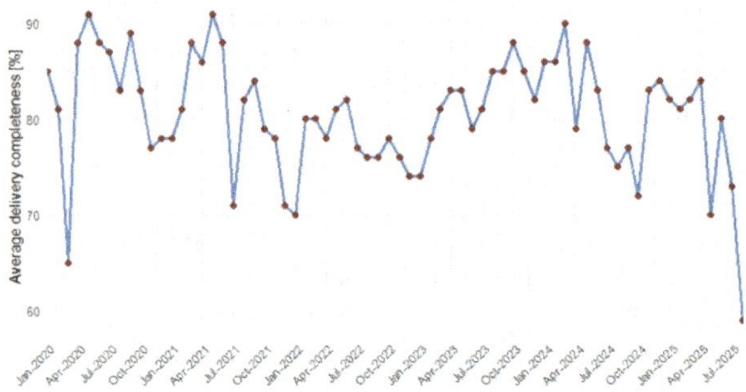

Figure 1. Trend of delivery completeness over time.

Descriptive statistics showed an average delivery completeness of 80.5% (median = 81%), indicating generally consistent performance. Monthly values ranged from 59% to 91%, with none reaching full (100%) completeness. The standard deviation (SD = 6.1) reflects moderate variability. Quartile values (Q1 = 77%, Q3 = 85%) confirm that half of the months fell within a narrow eight-point range, suggesting stable delivery operations with only minor fluctuations.

Analysis of Delivery Losses. In the next stage of the study, delivery losses were examined to evaluate the share of undelivered items in relation to the total number of items ordered. Delivery loss was defined as the ratio of undelivered to ordered items, expressed as a percentage. This measure reflects the operational inefficiency of the distribution process and represents a direct indicator of potential financial risk and customer dissatisfaction. From a managerial perspective, it highlights the proportion of orders that failed to reach retail branches on time, thus pointing to possible issues in stock availability, transportation, or warehouse coordination.

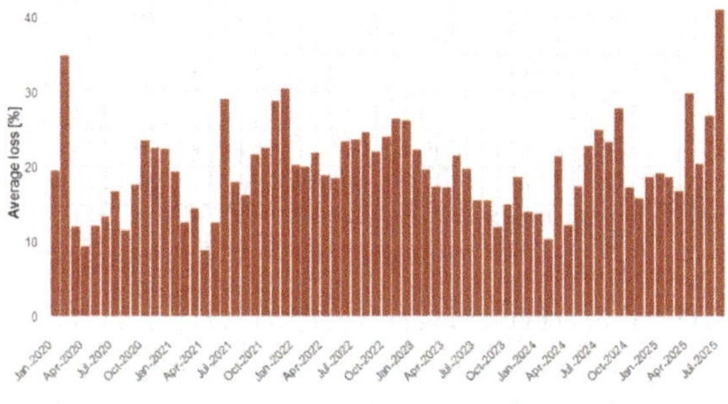

Figure 2. Trend of delivery completeness over time.

The descriptive statistics revealed that the average loss rate was 19.6%, with a median of 19.1%, confirming a stable yet concerning level of incomplete deliveries. This means that nearly one in five ordered items did not reach the retail branches - a significant indicator of supply chain inefficiency. The minimum loss recorded was 8.6%, which can be considered a benchmark month demonstrating the company's best operational performance. Conversely, the maximum loss reached 40.9%, meaning that in the worst-performing month, almost half of the ordered cosmetics were not delivered to retail outlets. The standard deviation (SD = 6.1) suggests moderate variability in losses between months. The quartile values (Q1 = 15.4%, Q3 = 22.8%) indicate that 50% of the months experienced losses within this range. This concentration implies that, although fluctuations occurred, they remained within a relatively narrow bandwidth, with occasional spikes that likely resulted from exceptional operational disturbances (Fig. 3).

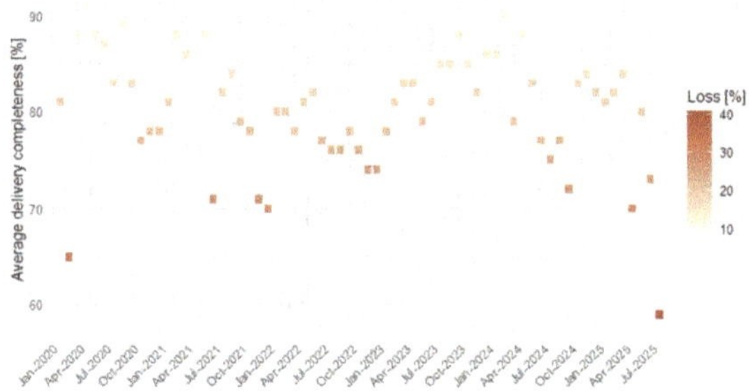

Figure 3. Delivery risk map [%].

From a business perspective, such losses pose a significant operational and financial risk. Undelivered products cause revenue loss, stockouts, and weakened customer trust-particularly critical in the cosmetics sector, where availability drives loyalty. Continuous monitoring and early detection of deviations are therefore essential to improve supply chain resilience and minimize financial and customer-related risks.

Statistical Comparison Between Months. In the next stage of the study, statistical tests were used to verify whether there were significant monthly differences in delivery completeness and loss rates. To this end, both Analysis of Variance (ANOVA) and the non-parametric Kruskal-Wallis test were applied [19], [20]. Tukey's post-hoc test was used to check for significant differences between months. Before ANOVA, data normality and variance homogeneity were verified using the Shapiro-Wilk and Levene's tests. Delivery completeness data were not normally distributed ($W = 0.94983$, $p = 0.00827$) but showed equal variances ($F = 0.8139$, $p = 0.6262$). Therefore, both ANOVA and the Kruskal-Wallis test were applied, and neither indicated significant monthly differences (ANOVA: $p = 0.6262$; Kruskal-Wallis: $p = 0.5422$). Tukey's post-hoc results confirmed this, with all p-values above 0.9.

Table 1. Tukey multiple comparisons of means for deliveries' completness.

Month	Jan	Feb	Mar	Apr	May	Jun	Jul	Aug	Sep	Oct	Nov	Dec
Jan	-	0.9999	0.9999	0.9948	0.9683	0.9897	0.9999	0.9998	0.9995	1.0000	1.0000	0.9999
Feb		-	1.0000	0.9999	0.9996	0.9999	0.9976	0.9813	1.0000	1.0000	0.9999	0.9989
Mar			-	0.9999	0.9994	0.9999	0.9984	0.9860	1.0000	1.0000	1.0000	0.9993
Apr				-	1.0000	1.0000	0.9377	0.8282	1.0000	0.9996	0.9993	0.9620
May					-	1.0000	0.8282	0.6610	0.9999	0.9950	0.9925	0.8825
Jun						-	0.9086	0.7776	1.0000	0.9990	0.9983	0.9420
Jul							-	1.0000	0.9863	0.9999	0.9999	1.0000
Aug								-	0.9420	0.9990	0.9994	1.0000
Sep									-	0.9999	0.9999	0.9924
Oct										-	1.0000	0.9999
Nov											-	0.9999

For delivery losses, a similar pattern was observed. The Shapiro-Wilk test ($W = 0.94999$, $p = 0.00842$) again indicated non-normality, while Levene's test ($F = 0.7337$, $p = 0.7018$) confirmed equal variances. Both ANOVA ($F = 0.845$, $p = 0.597$) and Kruskal-Wallis ($\chi^2 = 10.1$, $p = 0.5214$) results demonstrated no significant differences between months, and the Tukey post-hoc test (Table 2) corroborated this conclusion.

Table 2. Tukey multiple comparisons of means for deliveries' losses.

Month	Jan	Feb	Mar	Apr	May	Jun	Jul	Aug	Sep	Oct	Nov	Dec
Jan	-	0.9999	0.9999	0.9948	0.9683	0.9897	0.9999	0.9998	0.9995	1.0000	1.0000	0.9999
Feb		-	1.0000	0.9999	0.9996	0.9999	0.9976	0.9813	1.0000	1.0000	0.9999	0.9989
Mar			-	0.9999	0.9994	0.9999	0.9984	0.9860	1.0000	1.0000	1.0000	0.9993
Apr				-	1.0000	1.0000	0.9377	0.8282	1.0000	0.9996	0.9993	0.9620
May					-	1.0000	0.8282	0.6610	0.9999	0.9950	0.9925	0.8825
Jun						-	0.9086	0.7776	1.0000	0.9990	0.9983	0.9420
Jul							-	1.0000	0.9863	0.9999	0.9999	1.0000
Aug								-	0.9420	0.9990	0.9994	1.0000
Sep									-	0.9999	0.9999	0.9924
Oct										-	1.0000	0.9999
Nov											-	0.9999

Fig.4 illustrates the monthly distribution of the two analyzed indicators - delivery completeness and delivery losses. Each boxplot shows the variation in results across months, while the black dashed line represents the overall mean value for the entire observation period.

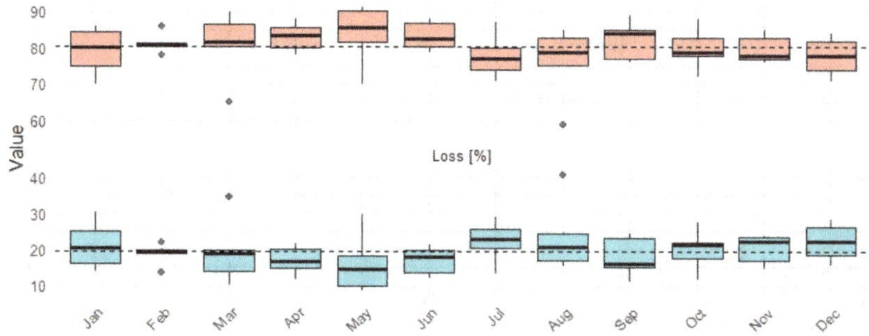

Figure 4. Monthly distribution of delivery completeness and losses.

The visualization confirms that there is no seasonality in delivery completeness or losses. Median values remain stable throughout the year, and no month shows notably higher or lower performance. This indicates that fluctuations in delivery reliability are caused by incidental rather than cyclical factors. Therefore, the company should focus on identifying and mitigating one-off disruptions instead of applying seasonal logistics adjustments.

Correlation Between Order Volume and Delivery Completeness. The relationship between order volume and delivery completeness was tested using the Pearson correlation coefficient. The result ($r = -0.10$, $p = 0.41$) showed a very weak, statistically insignificant negative correlation, indicating that order size does not affect delivery completeness.

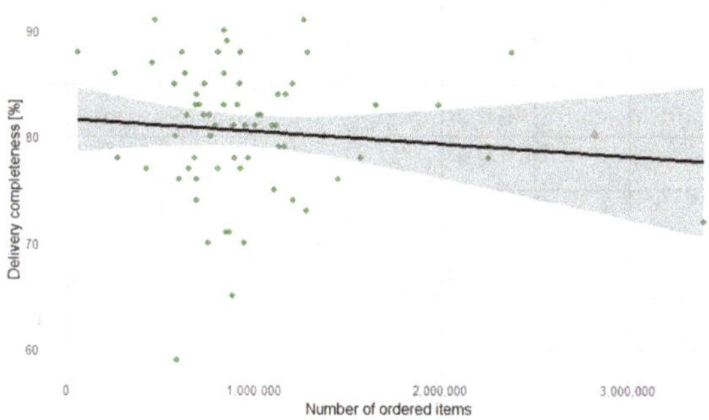

Figure 5. Relationship between the number of ordered items and delivery completeness.

The scatter plot with a fitted regression line (Figure 5) visually confirms the absence of any meaningful trend between the two variables. Data points are widely dispersed, and the regression line remains nearly flat, further supporting the conclusion that delivery reliability is not influenced by the scale of orders.

Summary
The study confirmed that delivery reliability in the cosmetics retail company is stable over time and unaffected by season or order volume. This shows that logistics processes are well-balanced and resilient to demand changes. Occasional delivery losses likely stem from isolated issues such

as supply or transport disruptions. The lack of seasonality and order-size effects highlights operational predictability, enabling efficient planning. Improvement efforts should target incidental risks and strengthen supplier coordination. Overall, the supply chain is stable and provides a solid basis for continuous improvement.

References

[1] J. Konwerski et al., Analysis of failure data for estimating availability and reliability indicators using statistical and stochastic methods, Eksploatacja i Niezawodność 28 (2025). https://doi.org/10.17531/ein/211577

[2] A. Borucka, Seasonal Methods of Demand Forecasting in the Supply Chain as Support for the Company's Sustainable Growth. Sustainability 15 (2023) art.7399. https://doi.org/10.3390/su15097399

[3] M. Grzelak et al., Application of Linear Regression for Evaluation of Production Processes Effectiveness. in: J. Machado, F. Soares, J. Trojanowska, V. Ivanov (Eds.) Innovations in Industrial Engineering. Lecture Notes in Mechanical Engineering, (2022). Springer. https://doi.org/10.1007/978-3-030-78170-5_4

[4] Y. Wang, et al., The multi-depot pickup and delivery vehicle routing problem with time windows and dynamic demands, Engineering Applications of Artificial Intelligence 139 (2025) 109700. https://doi.org/10.1016/j.engappai.2024.109700

[5] B. Wu et al., How own delivery services influence customer behavior and sales in online retail? Building trust and improving delivery quality in digital economy, Journal of Marketing 88 (2024) 131-152. https://doi.org/10.1177/0022242924123989

[6] B. Jiang et al., Impact of unreliable supplier on a supply chain with demand-generating marketing, Production and Operations Management 34 (7) (2024) 1616-1631. https://doi.org/10.1177/10591478241302786

[7] A. Abusalmaa et al., The impact of supply chain performance on financial performance: Dimensions of the SCOR model, Uncertain Supply Chain Management 12 (2024) 81-92. https://doi.org/10.5267/j.uscm.2024.4.010

[8] A. Nath, M.N. Sulthana, Exploring key performance indicators for strengthening supply chain resilience: A systematic review for strategic insights, Journal of Information Systems Engineering and Management 10 (2025) 1-12. https://doi.org/10.52783/jisem.v10i18s.2943

[9] N. Pourmohammadreza et al., Last-mile logistics with alternative delivery locations: A systematic literature review, Results in Engineering 25 (2025) 104085. https://doi.org/10.1016/j.rineng.2025.104085

[10] X. Shi et al., Is operational flexibility a viable strategy during major supply chain disruptions? Evidence from the COVID-19 pandemic, Transportation Research Part E: Logistics and Transportation Review 195 (2025) 103952. https://doi.org/10.1016/j.tre.2024.103952

[11] N. Dhone, Optimizing supply chain performance in quick commerce through strategic supplier relationships and technology integration, Academy of Marketing Studies Journal 29 (4) (2025) 1-15.

[12] S. He, Z. Ma, Service quality and price competition in crowdsourced delivery markets, International Transactions in Operational Research 31 (3) (2024). https://doi.org/10.1111/itor.13192

[13] D.G. Mogale et al., Modelling and analysing supply chain disruption: A case of online grocery retailer, Oper. Manag. Res. 16 (2023) 1901-1924. https://doi.org/10.1007/s12063-023-00405-9

[14] T. Samal, A. Ghosh, Ensemble-based predictive analytics for demand forecasting in multi-channel retailing, Expert Systems with Applications (2025) 130212. https://doi.org/10.1016/j.eswa.2025.130212

[15] J.M. Chandran, M.R.B. Khan, A strategic demand forecasting: Assessing methodologies, market volatility, and operational efficiency, Malaysian Journal of Business, Economics and Management (2024) 150-167. https://doi.org/10.56532/mjbem.v3i2.71

[16] A. Panek et al., Reliability and efficiency in technology selection in logistics facilities - multi-criteria decision support using the AHP method, Eksploatacja i Niezawodność 27 (2) (2025). https://doi.org/10.17531/ein/200684

[17] S. Kehinde, et al., Effective sales promotion: A necessary catalyst for elongating the product life cycle in a developing economy, International Journal of Advances in Social Sciences and Humanities 3 (1) (2024) 1-7. https://doi.org/10.56225/ijassh.v3i1.231

[18] X. Huang et al., Spare parts consumption prediction model for improving maintenance and operational reliability, Eksploatacja i Niezawodność (2026). https://doi.org/10.17531/ein/210721

[19] P. Gładysz et al., Reliability of Unmanned Aerial Vehicles in the Context of Selected Factors, Eksploatacja i Niezawodnosc (2026). 28(1) http://doi.org/10.17531/ein/210312

[20] T.V. Hecke, Power study of anova versus Kruskal-Wallis test. J. Stat. Manag. Syst. 15 (2012) 241-247. https://doi.org/10.1080/09720510.2012.10701623

Materials Research Proceedings 62 (2026) 318-325 | https://doi.org/10.21741/9781644904015-41

Laser Surface Modification of Reinforcing Steel to Improve Corrosion Resistance and Bond to Concrete

ULEWICZ Malgorzata[1,a*], RADEK Norbert[2,b]

[1]Czestochowa University of Technology, Dabrowskiego 69, 42-201 Czestochowa, Poland

[2]Kielce University of Technology, al. Tysiąclecia Państwa Polskiego 7, 25-314 Kielce, Poland

[a]malgorzata.ulewicz@pcz.pl, [b]norrad@tu.kielce.pl

Keywords: Laser Surface Treatment, Reinforcing Steel, Steel–Concrete Bond, Corrosion Resistance, Reinforced Concrete

Abstract. The durability of reinforced concrete structures is strongly governed by the corrosion resistance of reinforcing steel and the quality of the steel–concrete bond. Laser surface modification of steel offers a promising approach to simultaneously mitigate chloride-induced corrosion and tailor interfacial adhesion with cement paste. This paper synthesizes research on laser surface melting (LSM), laser cladding (LC), and micro-/nanotexturing (LST/LIPSS) of reinforcing and structural steels in alkaline and chloride-containing environments, highlighting their effects on microstructure, passive film stability, and steel–concrete bond performance. Key research gaps are identified, including the lack of long-term studies on bars embedded in concrete and the absence of integrated durability design models that incorporate laser-treatment effects.

Introduction

Reinforcing steel constitutes the strengthening phase in the concrete composite and is essential for compensating the material's deficiency in tensile strength, thereby ensuring load-bearing capacity, ductility, and the ability to transfer bending stresses. Under the high pH conditions of concrete pore solution, a stable passive layer forms on the steel surface, protecting it against corrosion. In construction practice, both plain and ribbed bars are used; the latter provide stronger mechanical anchorage within the concrete cover and more favorably shape the steel–concrete bond characteristics. Reinforcement is widely applied in buildings and infrastructure (including bridges and multi-storey car parks), often exposed to aggressive marine environments and de-icing salts, which underscores the importance of a durable steel–concrete interface [1]. It is estimated that global consumption of reinforcing steel currently reaches approximately 350–370 million tons per year, reflecting its key role in modern construction infrastructure [2].

Chloride-ion-induced corrosion of reinforcing steel is one of the most serious threats to the durability of reinforced concrete structures. The literature indicates that chloride ions cause local breakdown of the passive film, initiate pitting corrosion, and accelerate the degradation of reinforced concrete elements [1, 3]. Under normal conditions, the alkaline environment of the cement paste maintains steel in a passive state; however, chloride ingress-both from the surrounding environment and from contaminated concrete mix constituents–leads to loss of passive film stability and initiation of corrosion processes. Exposure of steel to aqueous NaCl solutions results in significant mass loss and material degradation even at low temperatures [4]. Importantly, chlorides are also a strong activator of corrosion in stainless steels [5], highlighting the scale of the threat when they are present in the concrete cover.

The durability of reinforced concrete structures largely depends on how the chemical composition of the cementitious matrix interacts with reinforcing steel and shapes its corrosion conditions in service. Waste used as additives or substitutes for traditional raw materials may modify not only the mechanical properties of concrete but also the pore environment in the steel cover zone. Therefore, the development of concrete with a reduced environmental footprint, aligned with circular economy principles, requires parallel assessment of strength parameters and corrosion resistance. The literature describes cement-based concrete composites incorporating biomass fly ash [6, 7], bottom and fly ash from sewage sludge incineration [8], hemp shives [9], utility ceramics [10], as well as polymer waste [11-13]. The most recent studies also use products of thermal conversion of municipal waste, such as fly ash or flue gas cleaning residues, which may contain considerable amounts of toxic metals (Zn, Cu, Pb, Ni, Cr, Ba). Recent research [14, 15] demonstrates that cement composites with satisfactory strength parameters and frost resistance can be produced. However, the presence of heavy metals requires control of their impact on aquatic and soil environments, as well as assessment of their influence on corrosion processes of reinforcing steel. This is particularly relevant for biomass and sewage sludge ashes which, through modification of microstructure, porosity, water absorption, and compressive and flexural strength, affect concrete properties, but due to their chemical composition may also influence the stability of the passive layer on the steel surface. When this layer is destabilized, corrosion is initiated, and the corrosion products are characterized by a volume several times greater than that of metallic iron. This generates tensile stresses in the concrete cover, initiates microcracking, causes concrete delamination, and degrades the steel–concrete bond [1]. Consequently, the effective load-transfer area is reduced, the steel begins to behave more brittlely, the load-bearing capacity of structural elements decreases, and their service life is shortened, resulting in substantial repair costs and the risk of premature rehabilitation.

Against this background, there is a need to seek solutions on the reinforcing steel side that could compensate for potentially adverse effects of modifying concrete composition–especially solutions that simultaneously improve steel corrosion resistance and bond performance in the steel–concrete zone. One of the most promising research directions involves laser-based techniques for surface modification of reinforcing bars, including laser texturing, laser passivation, and related methods (laser cladding, laser alloying), which enable tailoring of the micro- and nanostructure of the surface layer as well as the electrochemical properties of the steel surface.

The aim of this article is to provide a synthetic review of research on laser surface modification of reinforcing steel in the context of improving corrosion resistance and bond to concrete, with particular emphasis on the effect of laser processing on microstructure, passive layer stability, initiation of chloride-induced corrosion, and the bond–slip characteristics of reinforced concrete structures.

Mechanisms of Laser Beam Interaction with Steel

Laser beam interaction with steel depends primarily on the wavelength, power density, pulse duration, and scanning strategy. Proper selection of these parameters enables switching between different laser operating regimes, which produce distinct thermal, microstructural, and topographical effects. In the case of reinforcing steel, this is of critical importance because the microstructure and electrochemical properties of the bar surface determine both corrosion resistance in the concrete environment and the mechanical performance of the steel–concrete interface.

Numerous laser–metal interaction modes have been reported in the literature. These regimes differ in temperature range, heating and cooling rates, and interaction depth, resulting in effects ranging from modifications of residual stresses, through the formation of remelted layers, to regular micro- and nanotextures. A summary of these regimes is presented in Table 1.

From the perspective of reinforced concrete structures, the most relevant regimes are those that simultaneously: tailor surface microgeometry to enhance cement paste anchorage, stabilize the passive layer and reduce susceptibility to chloride-induced corrosion, and avoid defects that may promote corrosion initiation (microcracks, excessively deep grooves, local galvanic cells). Studies on LST, LIPSS, LSM, and LSP indicate that appropriately selected processing parameters allow concurrent control of steel topography and electrochemical behavior, opening a pathway for the deliberate design of reinforcing bar surfaces to improve the durability and load-bearing performance of the steel–concrete bond [16, 17].

Table 1. Main laser–steel interaction regimes (based on [16, 18-21])

Interaction mode	Abbreviation	Characterization
Laser cleaning	LC	Removal of rust, oxides, and contaminants from reinforcing bar surfaces
Sub-melting laser heating	-	Heating to temperatures close to melting without full material remelting.
Laser hardening	LH	Localised hardening of the surface layer due to rapid heating and self-quenching.
Laser shock peening	LSP	Introduction of compressive residual stresses via a laser-induced shock wave.
Laser surface melting	LSM	Formation of a thin remelted layer that rapidly solidifies with a modified microstructure.
Direct laser ablation	DLA	Local material removal leading to pits, grooves, or dimples.
Direct Laser Writing	DLW	Controlled beam scanning to produce regular grooves and channels
Direct laser interference patterning	DLIP	Formation of periodic microstructures resulting from interference of multiple laser beams.
Laser surface texturing/ laser induced periodic surface structures	LST / LIPSS	Generation of micro- and nanoscale surface patterns through controlled ablation and/or melting.

Classification of Laser Surface Modification Techniques for Reinforcing Steel

Increasing requirements for the durability of reinforced concrete structures, particularly those exposed to chloride ions, are driving the rapid development of surface engineering methods for reinforcing steel. Laser-based surface modification techniques are among the most intensively studied solutions because they enable simultaneous tailoring of surface topography, microstructure, electrochemical behaviour, and mechanical performance. Depending on the specific process and parameter selection, a laser beam can remove contaminants, remelt the surface, introduce compressive stresses, modify the composition of the oxide layer, and generate precise micro- and nanoscale structures [18, 22, 23].

For reinforcing steel, it is crucial that surface modifications may influence both corrosion resistance in chloride-containing environments and the steel–concrete bond, which governs the load-bearing performance of reinforced concrete elements. However, the number of studies conducted directly on reinforcing bars remains limited; therefore, an important part of current knowledge comes from investigations of reinforcing steel and structural steels tested in environments resembling service conditions of rebar in concrete (alkaline solutions, seawater, or salt spray) [24]. One of the few studies performed directly on reinforcing steel is the work of Premlall et al. [24], who applied laser surface melting (LSM) to low-carbon reinforcing bars in alkaline solutions containing sulphates and their mixtures with chlorides. The authors reported a reduction in corrosion current density and a shift of corrosion potential towards more noble values, which was attributed to microstructural refinement and stabilization of corrosion products. Although steel–concrete bond was not examined, the results suggest that LSM may help mitigate the initiation of chloride-induced corrosion of reinforcement in zones particularly vulnerable to chloride ingress. At the same time, studies on other alloys (e.g., Ti-6Al-4V) indicate that while LSM can increase hardness and alter microstructure, non-optimized parameters may deteriorate overall corrosion resistance due to martensite formation, increased roughness, and residual stresses [25]. This highlights the need to select LSM parameters for reinforcing steel in a way that minimizes unfavorable surface morphology and tensile residual stresses.

Promising results for reinforced concrete applications have also been achieved within NCHRP IDEA Project 240, which developed the concept of stainless clad rebar consisting of a metallic stainless outer layer on a low-cost carbon steel core. In the final phase of the project, cladding was applied to a semi-finished product, followed by hot rolling into reinforcing bars [26]. Accelerated macrocell corrosion tests showed at least a fourfold reduction in corrosion rate for stainless-clad bars, and in the optimized process even more than a tenfold reduction compared with conventional carbon steel reinforcement. After rolling, the cladding exhibited good ductility, enabling bending without cracking or delamination. The steel–concrete bond in pull-out tests was comparable to that of uncoated bars, and impact resistance proved clearly superior to traditional epoxy coatings. Thus, metallic cladding can be considered a realistic alternative to conventional organic coatings, provided that coating uniformity on ribbed bars, bend performance, and compatibility with the alkaline concrete environment are ensured.

Another technique relevant to durability is Laser Shock Peening (LSP), which enables effective modification of surface roughness and tribological behavior while introducing deep compressive residual stresses [20]. For high-strength E690 steel, LSP was shown to refine grains, increase compressive stresses, and improve corrosion resistance in NaCl, including a decrease in corrosion current density from ~114.5 to 5.41 μA/cm^2 [27]. Stress modelling and measurements identified characteristic "stress holes," improving understanding of how LSP parameters control the depth and distribution of compressive stresses [28]. Similar benefits–higher hardness, reduced friction coefficient, and improved fatigue behavior – have been reported for AH32 steel used in marine applications [29]. Although these studies do not directly address reinforcing bars, they suggest that LSP may be particularly useful for elements exposed to fatigue and corrosion–fatigue in marine conditions, while its influence on the mechanical steel–concrete bond mechanism remains limited because the process does not generate regular textures tailored to cementitious materials.

Against the background of LSM, metallic cladding, and LSP, especially compelling results have been reported for laser-based surface modification of stainless steels. Zheng et al. [22] applied high-field laser passivation to AISI 304 steel, producing hierarchical superhydrophobic micro/nanostructures. The authors reported up to a 100,000-fold reduction in corrosion rate in saline, acidic, and alkaline solutions and maintained ultralow corrosion rates for more than 6500 hours of exposure in NaCl. Although these findings concern stainless steels rather than conventional reinforcing steels, they strongly support the view that micro- and nanotexturing techniques such as LST/LIPSS may be the most promising when the goal is simultaneous enhancement of chloride-related corrosion resistance and modification of microgeometry relevant to steel–concrete bond. In contrast to LSP, which is primarily aimed at tailoring residual stress states and improving fatigue resistance, and metallic cladding, which focuses on barrier-type protection, LST/LIPSS can be designed to control both rib-scale topography and wettability/oxide-layer behavior at micro- and nanoscales.

Selected examples of laser texturing technologies, their documented surface and functional effects, and their potential relevance to reinforced concrete applications are presented in Table 2. The analysis of LST variants indicates that the most useful approaches for reinforcing bars include: regular microstructures enhancing mechanical anchorage of concrete, periodic LIPSS structures enabling concurrent control of wettability and oxide-layer stability, and LST textures used as sublayers to improve the effectiveness of protective coatings.

Table 2. Selected laser surface modifications and relevance to reinforcing steel

LST method/ material	Surface effect	Functional effect	Implications for reinforcing steel	Ref.
LIPSS/ on stainless steel AISI 304	Nanowavy nanostructures; minor roughness change; strongly hydrophilic surface	~2× higher interfacial shear stress vs. reference (better wetting/capillary uptake of paste)	Bond improvement can be driven mainly by wettability, even with minor roughness change	[16]
DLW / on stainless steel	Regular microchannels (~80 µm deep, ~500 µm wide)	~16× higher shear stress vs. untreated	Texture orientation is decisive; grooves perpendicular to bond forces are preferred	[16]
LST / on Al alloys	Regular micro-/nanopatterns (grooves, grids)	Improved corrosion and tribological wear resistance	Shows coupled control of topography and durability-relevant surface behavior	[30]
LST + functional coatings	Laser texture + thin coatings (epoxy, FAS-TiO_2, graphene, etc.)	Durable hydrophobicity; reduced corrosion; improved wear/fatigue	Hybrid strategy: LST as an anchoring sublayer for protective coatings	[30]
LST / on steel 316L	Dimple array (~36 µm diameter, ~3.6 µm depth)	At high texture density (75–91%): lower corrosion rate, higher Ecorr	Texture shape/density strongly governs corrosion performance	[17]
LSM/ on low-carbon steel	Thin remelted/self-quenched oxide layer	Lower icorr and corrosion rate	Surface melting alone can enhance corrosion resistance in sulphate–chloride media	[24]
Laser-cut reinforcement /on steel S355	Laser-cut plates; roughened/perforated/ribbed variants	Bond strength several-fold ↑ vs. smooth (~1 MPa → ~5–6×)	Macro-geometry (rib/hole spacing) controls mechanical interlocking; informs macro-texture design	[31]

Comparative Assessment Methodology for Laser Modification of Reinforcing Steel

To provide a concise comparison of the suitability of selected laser-based surface modification techniques for reinforcing steel, a qualitative multi-criteria expert assessment was adopted (Table 3). The procedure is based on insights from the literature on laser texturing, surface melting, cladding and shock strengthening of steels (LST/LIPSS, LSM, LC, LSP). It is not a formal optimisation algorithm (e.g., AHP, TOPSIS), but rather a structuring tool that synthesizes the current state of knowledge for applications in reinforced concrete structures. Because the number of studies performed directly on reinforcing bars embedded in concrete remains limited, the assessment also incorporates results obtained for structural and stainless steels tested in environments resembling concrete pore conditions (alkaline solutions, NaCl, seawater, salt spray). The resulting values should therefore be interpreted as a comparison of the potential of each technique rather than definitive design recommendations.

Assessment criteria were defined to reflect the key performance requirements of reinforcement in chloride-exposed RC structures:

C – effect on chloride-related corrosion resistance (0.35): the extent to which a given technique may limit corrosion initiation and propagation in the presence of Cl^- by modifying the passive film, surface-layer chemistry, or microstructure [17, 32].
A – effect on the steel–concrete bond (0.35): the potential to increase bond capacity, alter the bond–slip response and modify the failure mode; the highest improvements have been reported for LST/LIPSS [16].
F – effect on fatigue resistance (0.15): improvement of resistance to cyclic loading and corrosion–fatigue through the introduction of compressive stresses and grain refinement; this criterion particularly favors LSP [18].
I – industrial implementability (0.15): practical aspects including system/optics complexity, processing speed, integration with ribbed rebar production lines, surface preparation requirements and unit costs; higher scores were assigned to processes closer to established industrial practice [18].

Scoring and aggregation: each method was rated on a 1–5 scale for each criterion (1 – very small effect; 5 – very large effect) based on qualitative conclusions from experimental studies, cross-technique comparisons in review papers, and information on technology readiness and industrial adoption. The overall score (S) was calculated as a weighted sum.

$$S = 0.35 \cdot C + 0.35 \cdot A + 0.15 \cdot F + 0.15 \cdot I$$

Thus, the highest scores are achieved by methods that simultaneously provide strong corrosion protection, improve the bond to concrete, and exhibit at least moderate implementation potential.

Table 3. Weighted assessment of selected laser-based steel surface modification techniques (scale 1–5)

Method	C- corrosion resistance (weight 0.35)	A – steel–concrete bond (weight 0.35)	F – fatigue resistance (weight 0.15)	I – industrial implementability (weight 0.15)	Overall score
LST / LIPSS	4	5	2	2	3.75
LSM	4	2	3	3	3.00
LC	5	3	3	2	3.55
LSP	3	1	5	2	2.45

The results of the analysis (Table 3) indicate that LST/LIPSS offer the greatest potential for reinforcement in concrete, primarily due to a very strong improvement in the steel–concrete bond combined with a favorable effect on corrosion resistance. LSM and LC provide the most effective corrosion protection, but typically deliver only moderate bond enhancement, since their main function is to form a protective layer rather than to tailor surface texture. In contrast, LSP stands out in terms of fatigue resistance but has a limited effect on bonding to concrete, which lowers its overall score for conventional reinforced concrete applications. It should be emphasized that the table we present is a qualitative structuring tool reflecting the current state of scientific knowledge rather than the outcome of a formal optimization process. The ordinal numerical scores should be interpreted as a relative comparison of the potential of the individual techniques, not as a final technological recommendation. As new research results emerge–particularly those concerning conventional B500 reinforcing steel under real concrete exposure conditions–the weighting scheme and assigned values may require updating.

Conclusion

The analysis of the literature shows:

1. Laser-based surface modification of reinforcing steel has a dual potential: it can enhance resistance to chloride-induced corrosion and improve the steel–concrete bond.

2. The analysis indicates that individual laser methods exhibit varying potential in terms of corrosion protection, bond performance, and fatigue resistance.
3. The number of studies on laser-modified reinforcing steel embedded in concrete is currently very limited, which justifies the need for long-term experimental investigations and for establishing a basis for durability-oriented design of reinforced concrete structures.

References

[1] M. Ali et al., A review on chloride induced corrosion in reinforced concrete structures: lab and in situ investigation, RSC Adv. 14 (2024) 37252–37271. https://doi.org/10.1039/d4ra05506c
[2] Steel rebar market, TowardsChemandMaterials, 2025. Available at: https://www.towardschemandmaterials.com/insights/steel-rebar-market (accessed 1 November 2025).
[3] M.F. Montemor et al., Chloride-induced corrosion on reinforcing steel: from the fundamentals to the monitoring techniques, Cem. Concr. Compos. 25 (2003) 491–502. https://doi.org/10.1016/S0958-9465(02)00089-6
[4] T. Lipiński, R. Ulewicz, Degradation of R35 steel in 5% NaCl environment at 10°C, Mater. Res. Proc. 34 (2023) 77–86. https://doi.org/10.21741/9781644902691-10
[5] D. Čuchor et al., Comparison of tribological and corrosion characteristics of AISI 316Ti and AISI 430 stainless steels, Prod. Eng. Arch. 30 (2024) 565–576. https://doi.org/10.30657/pea.2024.30.52
[6] J. Jura, M. Ulewicz, The use of bottom and a mixture of bottom and fly ash from wood-sunflower biomass combustion in concrete production, Arch. Civ. Eng. 71 (2025) 19–35. https://doi.org/10.24425/ace.2025.154105
[7] J. Jura, Analysis of the impact of sludge and slag waste on basic properties of cement mortars, Syst. Saf. Hum.-Tech. Facil.-Environ. 5 (2023) 130–141. https://doi.org/10.2478/czoto-2023-0015
[8] T. Kalak et al., Modification of concrete composition doped by sewage sludge fly ash and its effect on compressive strength, Materials 16 (2023) 4043. https://doi.org/10.3390/ma16114043
[9] R. Šadzevičius et al., Research on the properties of concrete with hemp shives, Constr. Optim. Energy Poten. 12 (2023) 25–32. https://doi.org/10.17512/bozpe.2023.12.03
[10] M. Ulewicz, J. Halbiniak, Application of waste from utilitarian ceramics for production of cement mortar and concrete, Physicochem. Probl. Miner. Process. 52 (2016) 1002–1010. http://dx.doi.org/10.5277/ppmp160237
[11] A. Pietrzak, M. Ulewicz, Influence of post-consumer waste thermoplastic elastomers obtained from used car floor mats on concrete properties, Materials 16 (2023) 2231. https://doi.org/10.3390/ma1606223
[12] M. Ulewicz, A. Pietrzak, Properties and structure of concretes doped with production waste of thermoplastic elastomers from the production of car floor mats, Materials 14 (2021) 872. https://doi.org/10.3390/ma14040872
[13] W. Brachaczek et al., The influence of micronized rubber powder on the properties of deformable tile adhesives, Constr. Optim. Energy Poten. 13 (2024) 193–199. https://doi.org/10.17512/bozpe.2024.13.19
[14] A. Pietrzak et al., Application of a mixture of fly ash and solid waste from gas treatment from municipal solid waste incineration in cement mortar, Materials 18 (2025) 481. https://doi.org/10.3390/ma18030481
[15] A. Pietrzak, M. Ulewicz, The mechanical strength of ecological cement mortars based on fly ash from the combustion of municipal waste and cement kiln dust, Appl. Sci. 15 (2025) 3215. https://doi.org/10.3390/app15063215

[16] M. Pontrandolfi et al., Enhancing bonding of fresh concrete to steel through laser surface texturing, Surf. Interfaces 48 (2024) 104299. https://doi.org/10.1016/j.surfin.2024.104299
[17] J.I. Ahuir-Torres et al., The influence of texture density and free surface energy on the corrosion resistance of laser-textured 316L stainless steel, Npj Mater. Degrad. 9 (2025) 115. https://doi.org/10.1038/s41529-025-00622-6
[18] F.D. Ince, T. Özel, Laser surface texturing of materials for surface functionalization: a holistic review, Surf. Coat. Technol. 498 (2025) 131818. https://doi.org/10.1016/j.surfcoat.2025.131818
[19] J. Kusiński et al., Laser modification of the materials surface layer – a review paper, Bull. Polish Acad. Sci. 60 (2012) 711-728. https://doi.org/10.2478/v10175-012-0083-9.
[20] M. Vicen et al., Influence of the surface roughness, lubrication and grinding on tribological properties of the C35 steel shot-peened surfaces, Appl. Eng. Lett. 9 (2024) 185–194. https://doi.org/10.46793/aeletters.2024.9.4.1
[21] X. Wang et al., A review on tribological wear and corrosion resistance of surface-engineered steels, Coatings 15 (2025) 1314. https://doi.org/10.3390/coatings15111314
[22] L. Zheng et al., Significant reduction of corrosion of stainless steel by strong-field laser surface passivation, Light Sci. Appl. 14 (2025) 352. https://doi.org/10.1038/s41377-025-01952-5
[23] Ł.J. Orman et al., Application of laser treatment technology for boiling heat transfer augmentation, Prod. Eng. Arch. 30 (2024) 259–265. https://doi.org/10.30657/pea.2024.30.25
[24] K. Premlall et al., Laser surface treatment to inhibit observed corrosion of reinforcing steel in sulphate: alkaline media, Anti-Corros. Methods Mater. 58 (2011) 267–284. https://doi.org/10.1108/00035591111167749
[25] A.G. Lekatou et al., Effect of laser surface melting on the microstructure and corrosion resistance of laser powder bed fusion and wrought Ti-6Al-4V alloys, Coatings 15 (2025) 1285. https://doi.org/10.3390/coatings15111285
[26] R. McAlpine, Mechanical properties of metal expansion joint manufactured by mechanically assisted laser forming, NCHRP-IDEA Project 240 Final Report, Transportation Research Board, Washington, DC, USA, 2023. Available at: https://onlinepubs.trb.org/onlinepubs/IDEA/FinalReports/Highway/NCHRP240.pdf
[27] J. Qin et al., Laser shock peening improves the corrosion resistance of an E690 high-strength steel cladding layer, Materials 16 (2023) 5566. https://doi.org/10.3390/ma16165566
[28] Y. Cao et al., Formation mechanism and weights analysis of residual stress holes in E690 high-strength steel by laser shock peening, Coatings 12 (2022) 285. https://doi.org/10.3390/coatings12020285
[29] X. Pei et al., The influence of laser shock peening on the microstructure and mechanical properties of AH32 steel, Materials 18 (2025) 4679. https://doi.org/10.3390/ma18204679
[30] J. Zhou et al., Review of laser texturing technology for surface protection and functional regulation of aluminum alloys: wettability, anti-icing, corrosion resistance, and wear resistance, Coatings 15 (2025) 567. https://doi.org/10.3390/coatings15050567
[31] M. Takalloozadeh et al., Experimental and numerical assessment of the bond behaviour of laser-cut reinforcement, Constr. Build. Mater. 449 (2024) 137719. https://doi.org/10.1016/j.conbuildmat.2024.137719
[32] S. Zhang et al., The effects of heat treatment and surface state on the corrosion resistance of laser powder bed fusion 304L stainless steel in 3.5 wt% NaCl solution, J. Mater. Res. Technol. 29 (2024) 5620–5632. https://doi.org/10.1016/j.jmrt.2024.03.025

Energy Efficiency Management through Production Scheduling Using FlexSim Simulation and Integration of Renewable Energy Sources

Marek KRYNKE[1,a]* and Robert ULEWICZ[1,b]

[1]Department of Production Engineering and Safety, Faculty of Management, Czestochowa University of Technology, Al. Armii Krajowej 19b, 42-218 Częstochowa, Poland

[a]marek.krynke@pcz.pl, [b]robert.ulewicz@pcz.pl

Keywords: Simulation, Flexsim, Energy Management, Sustainable Production, Photovoltaic, CO_2 Emissions, Schedule Optimization

Abstract. The article presents a methodology for managing energy consumption and scheduling production processes using discrete-event simulation in FlexSim combined with the Flexon Helios energy module. The approach enables modeling of production flows, machine scheduling, and energy consumption analysis, taking into consideration renewable energy sources such as photovoltaic installations. A representative case study is presented, including energy-intensive processes such as thermal treatment and finishing operations, highlighting how scheduling tasks in alignment with available renewable energy can reduce grid energy consumption without affecting production throughput. The proposed methodology offers a practical framework suitable for various industrial processes, enabling improved energy efficiency and flexible production planning.

Introduction

Contemporary industrial enterprises face the challenge of increasing the energy efficiency of production processes while simultaneously reducing operational costs and CO_2 emissions. Energy management in manufacturing has thus become a critical component of a company's operational strategy. The literature describes numerous approaches to optimizing energy consumption, including both analytical and simulation-based methods, which enable the prediction and control of energy usage [1-3] at various stages of the production process [4,5].

Discrete Event Simulation (DES) has gained significant popularity as a tool supporting the planning and scheduling of production lines. DES allows not only the analysis of material and machine flows but also the integration of energy consumption measurements and forecasts, enabling the scheduling of energy-intensive machines according to the availability of renewable energy. Applications include both mass production lines and medium-volume industrial processes, where energy-intensive operations can be shifted to time windows with access to low-cost [6,7] or environmentally friendly energy [8,9]. Among the available simulation tools, FlexSim, in conjunction with the Flexon Helios module, enables the detailed modeling of production processes, including real-time tracking of energy consumption and the incorporation of renewable energy sources, such as photovoltaic installations [10]. This approach enables the creation of scheduling scenarios for energy-intensive machines that maximize the utilization of renewable energy while maintaining full production line performance [7,8]. This paper presents a general methodology for energy management in production processes, utilizing simulation, and illustrates it with a case study involving energy-intensive operations, such as thermal treatment and component finishing. The model enables analysis of how scheduling furnaces and finishing equipment affects energy consumption from the grid and the share of energy sourced from renewable sources. This approach demonstrates the practical application of simulation in production line design and operational planning, with a focus on energy efficiency.

Methodology

In this study, the analysis of energy management in production processes was conducted using Discrete Event Simulation (DES) in FlexSim software combined with the Flexon Helios module [11]. The model encompasses the flow of products through production stations, machine scheduling, and real-time tracking of energy consumption. In particular, the influence of renewable energy availability, such as photovoltaic (PV) installations, on the scheduling of energy-intensive operations was considered. The simulation scenarios enable the testing of various energy management strategies and the evaluation of their impact on both energy efficiency and production line performance. The model also includes a 100 kW photovoltaic farm, which supplies energy to production stations with priority. Grid electricity is used only when PV generation is insufficient to meet demand. This energy management policy maximizes the share of renewable energy in energy-intensive processes while maintaining uninterrupted production. Data on PV system performance were determined based on solar irradiation profiles obtained from the PVGIS-SARAH2 (Photovoltaic Geographical Information System) database, developed by the Joint Research Centre (JRC) of the European Commission [12]. A reference location with coordinates 45.014° N, 7.620° E (Turin, Italy) was selected, and data for June 2020 were used to represent a typical period of high PV system performance. In the simulation, a total system loss of 14% was applied, accounting for factors such as temperature losses, DC/AC conversion, module soiling, and shading [13]. The resulting power generation profile was integrated with the machine operation schedule in the model, enabling the analysis of how variable solar energy availability affects the production process. To illustrate the methodology, a simplified simulation model of a production line was prepared (Fig. 1), comprising two main stages: a hardening furnace and component finishing operations. The hardening furnace has a batch capacity of 20 components, with a heating time set to 3 hours and air cooling simulated for a minimum of 8 hours in the cooling field. Finishing operations are performed on components after heating, with processing times adapted to the type of finishing (e.g., grinding – 45 min). This simplified model preserves demonstrative functionality while limiting the number of processing units in the simulation, which is essential due to the 500-event limit in the Flexon Helios license [11].

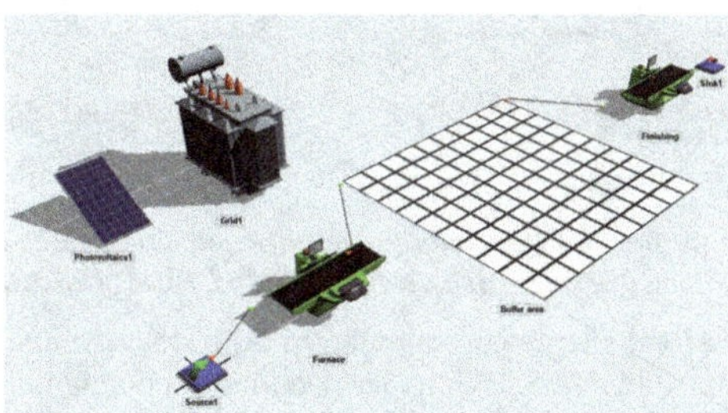

Fig. 1. Simulation model for the analyzed problem [own study]

Energy consumption parameters for both machines were defined according to their operating states (Table 1). In the second scenario, the furnace schedule was synchronized with the photovoltaic energy availability window, and the furnace was operated only between 08:00 and 14:00, corresponding to the average period of maximum PV energy generation. Finishing operations were performed in both scenarios after the minimum cooling time of the components, with an operating schedule from 08:00 to 16:00. This setup allows for the analysis of how shifting energy-intensive processes to periods of renewable energy availability affects the utilization of PV energy to meet production line demand, as well as the potential energy benefits achievable without

Materials Research Proceedings 62 (2026) 326-333 https://doi.org/10.21741/9781644904015-42

compromising throughput. Consequently, the model provides a practical example of applying computer simulation for energy management and production process scheduling, which can be readily adapted to other types of industrial processes.

Table 1. Energy consumption parameters depending on machine operating state

Operating state	Energy consumption [kW]	
	Hardening furnace	Hardening furnace
idle	0.1	0.3
processing	12.0	3.0
blocked	0.8	0.3
waiting for the operator	0.8	0.3
waiting for transport	0.5	0.2
scheduled down	0.1	0.1
setup	0.8	0.5

Results

Figure 2 presents a comparison of the power consumption profiles of the hardening furnace and the finishing station for the two scheduling scenarios. In the first case (Fig. 2a), the furnace operates continuously. In contrast, in the second case (Fig. 2b), its cycle is activated only between 11:00 and 15:00, corresponding to the period of maximum photovoltaic (PV) system output. A clear concentration of energy consumption is observed during midday hours in the second scenario, reflecting the strategy of adapting the process to renewable energy availability.

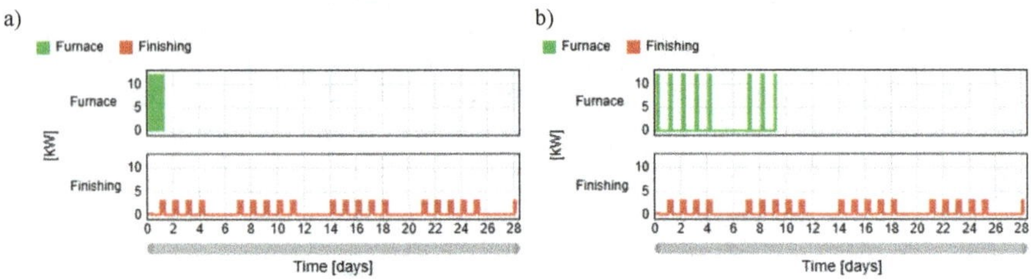

Fig. 2. Power consumption profiles of production equipment: a) furnace operating continuously, b) furnace operating between 11:00 and 15:00 [own study]

Fig. 3 summarizes the total energy consumption of both stations. Energy consumption by the hardening furnace amounted to 424.3 kWh under continuous operation and 417.8 kWh when the furnace schedule was aligned with PV availability. The energy consumption of the finishing station was similar in both scenarios (approximately 506 kWh), reflecting its constant operational cycle.

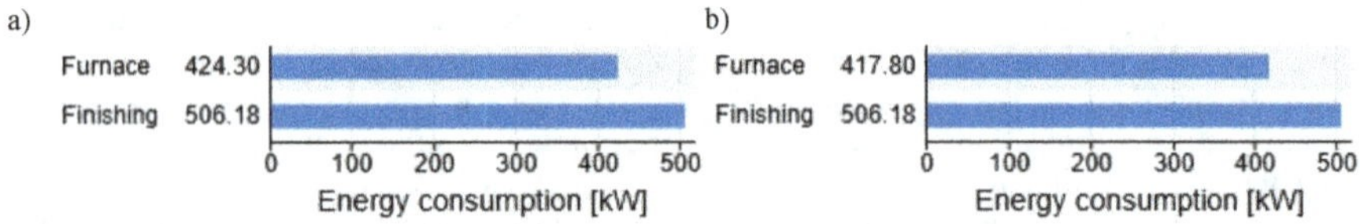

Fig.3. Energy consumption by production stations under both scheduling scenarios [own study]

Materials Research Proceedings 62 (2026) 326-333
https://doi.org/10.21741/9781644904015-42

Fig. 4 shows the CO_2 emissions resulting from equipment operation. In the continuous operation scenario, emissions from the electrical grid reached 61,967 g CO_2, whereas in the PV-synchronized scenario, they decreased to 16,706 g CO_2. This corresponds to a reduction of approximately 73%, achieved by shifting furnace operation to periods when it is powered by solar energy.

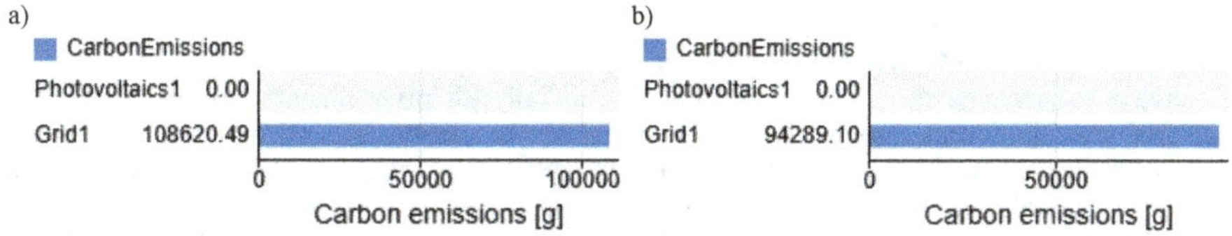

Fig. 4. CO_2 emissions depending on the energy source for the two furnace operation scenarios [own study]

Figure 5 presents the contribution of different energy sources. During continuous operation, 193.8 kWh was drawn from the grid, while the PV system supplied 736.9 kWh. With the implementation of a daily scheduling strategy, the PV share increased to 880.4 kWh, and grid consumption decreased to 50.4 kWh, resulting in a significant reduction in purchased electricity.

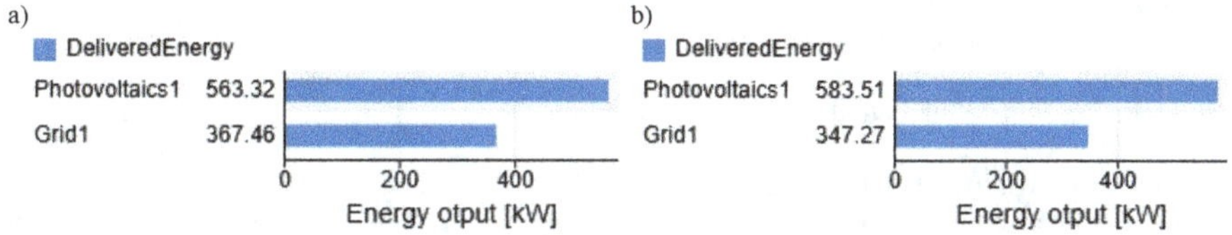

Fig. 5. Energy source distribution: PV and grid shares in the energy balance for both scenarios [own study]

Fig. 6 illustrates the temporal variability of power supplied by PV and drawn from the grid. In the PV-synchronized scenario, a better alignment of power demand with periods of peak solar energy generation is observed, confirming the effectiveness of the adopted energy management strategy.

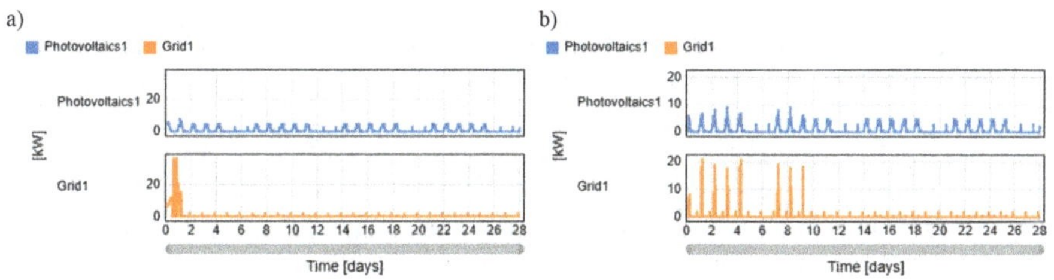

Fig. 6. Instantaneous PV generation and grid consumption profiles during the simulation [own study]

The conducted simulations demonstrated that implementing an energy management strategy based on aligning energy-intensive machine operation with periods of peak PV production leads to:

- a reduction in grid electricity consumption by over 73%,
- a decrease in CO_2 emissions by approximately 73%,

- maintenance of unchanged production throughput with minimal schedule adjustments.

The results confirm that integrating simulation tools (FlexSim and Flexon Helios) enables the effective analysis and optimization of energy consumption in production processes and supports decision-making regarding production scheduling in facilities that utilize renewable energy sources.

Economic and Technological Impact

Effective energy management [14,15] is important not only for environmental protection [16] but is also of great importance for highly energy-intensive industries [17,18]. It is also a potential source of significant savings in areas where concentrated energy streams are used in technological processes. Examples include the production of special coatings [19-21] designed to provide anti-wear and anti-corrosion properties [22,23], shaping the technological features of these coatings using laser micromachining [24,25], and the direct use of energy beams for material cutting [26,27]. The optimization of technological processes associated with this activity [28,29] strongly influences the applied simulation methods [30,31] and quality management [32,33]. On the other hand, there is a growing emphasis on the use of non-destructive methods in quality testing [34], which necessitates the use of resampling methods in modeling techniques [35] and taking into account the incompleteness of knowledge about the modeled process [36, 37].

Summary

The application of computer simulation, particularly the FlexSim Helios module, represents an effective tool for supporting energy management in production processes. This methodology enables the modeling of actual energy consumption by individual machines, accounting for different operating states and analyzing interactions with renewable energy sources, such as photovoltaic or wind installations. Consequently, it enables the quantitative assessment of the impacts of decisions regarding machine scheduling, production system configuration, or organizational changes on energy efficiency. Simulation facilitates the identification of areas with the highest potential for energy savings by:

- minimizing time spent in non-productive states, such as downtime, waiting for transport, or operator availability,
- reducing the number of start-up cycles of high-energy-consuming equipment,
- selecting optimal production buffer capacities to limit inefficient processing or storage,
- planning operations during periods of renewable energy availability, thereby reducing CO_2 emissions and energy costs.

In the presented case study, this methodology was applied to thermal treatment and finishing operations, analyzing two scheduling scenarios for the hardening furnace: continuous operation and operation synchronized with periods of maximum PV output. The results demonstrated that implementing an energy management strategy aligned with renewable energy availability can significantly reduce grid electricity consumption and CO_2 emissions without substantially affecting overall production performance.

The study confirms that integrating energy models with material flow simulation can serve as an effective tool to support decision-making in sustainable production planning and management. The methodology based on FlexSim Helios can be successfully applied not only for analyzing selected processes but also as a component of an industrial enterprise's energy planning strategy in the context of transitioning toward low-emission manufacturing.

References

[1] B. Gabrielyan et al., Energy efficiency in household sector, Production Engineering Archives 30 (2024) 136-144. https://doi.org/10.30657/pea.2024.30.13

[2] A. Wójcik-Mazur et al., Management of photovoltaic energy efficiency in smes in the light of the concept of corporate social responsibility, Polish Journal of Management Studies 24 (2021) 472-489. https://doi.org/10.17512/pjms.2021.24.1.28

[3] A. Idzikowski et al., Increasing the ecological efficiency of monocrystalline photovoltaic power plants by management their life cycle assessment, Energies 18 (2025) art. 4775. https://doi.org/10.3390/en18174775

[4] J. Bień, B. Bień, Rapid assessment of solar PV micro-system energy and greenhouse gas emissions in Poland, Production Engineering Archives 30 (2024) 326–332. https://doi.org/10.30657/pea.2024.30.32

[5] R. Ulewicz et al., Multi-Criteria Method for the Selection of Renewable Energy Sources in the Polish Industrial Sector, Energies 14 (2021) art. 2386. https://doi.org/10.3390/en14092386

[6] M. Ingaldi, R. Ulewicz, Sustainable development and technological advancements in industry 4.0: overcoming barriers in SME sector integration, Management Systems in Production Engineering 33 (2025) 144-162. https://doi.org/10.2478/mspe-2025-0015

[7] S. Pawlak, Computer simulations in planning production processes. Production Engineering Archives 31 (2025) 183-189. https://doi.org/10.30657/pea.2025.31.18

[8] J. Rosak-Szyrocka, K. Knop, Navigating the fourth industrial revolution: insights from a comprehensive bibliometric study on industry 4.0, Production Engineering Archives 30 (2024) 501-519. https://doi.org/10.30657/pea.2024.30.47

[9] M. Krynke, Analysis of the impact of effective time management on workstation efficiency using a multi-criteria optimization approach. Management Systems in Production Engineering 31 (2023) 306-311. https://doi.org/10.2478/mspe-2023-0034

[10] M. Beaverstock et al., Applied simulation modeling and analysis using FlexSim, 5th Ed. FlexSim Software Products, Inc, 2017.

[11] Flexcon Helios Tutorial. Available online: https://docs.flexsim.com/en/25.1/modules/Helios/Tutorials/BasicTutorial.html#app (accessed 2025-10-20)

[12] T. Huld et al., A new solar radiation database for estimating PV performance in Europe and Africa, Solar Energy 86 (2012) 1803-1815. https://doi.org/10.1016/j.solener.2012.03.006

[13] JRC European Commission, Photovoltaic Geographical Information System (PVGIS-SARAH2). Available online: https://re.jrc.ec.europa.eu/pvg_tools/en/ (accessed 2025-10-20).

[14] S. Maleczek et al., Tests of Acid Batteries for Hybrid Energy Storage and Buffering System—A Technical Approach, Energies 15 (2022) art. 3514. https://doi.org/10.3390/en15103514

[15] Ł.J. Orman et al., Comparative Analysis of Indoor Environmental Quality and Self-Reported Productivity in Intelligent and Traditional Buildings, Energies 16 (2023) art. 6663. https://doi.org/10.3390/en16186663

[16] A. Pacana, D. Siwiec, Method of Fuzzy Analysis of Qualitative-Environmental Threat in Improving Products and Processes (Fuzzy QE-FMEA), Materials 16 (2023) art. 1651. https://doi.org/10.3390/ma16041651

[17] O. Gumen et al., Research of thermal processes in industrial premises with energy-saving technologies of heating, Diagnostyka 18 (2017) 43-49.

[18] N. Radek et al., Formation of coatings with technologies using concentrated energy stream, Production Engineering Archives 28 (2022) 117-122. https://doi.org/10.30657/pea.2022.28.13

[19] A. Bartkowska et al., Microstructure, microhardness, corrosion resistance and chemical composition of Mo, B and Mo-B coatings produced using laser processing, Materials 13 (2020) art. 3249. https://doi.org/10.3390/MA13153249

[20] N. Radek, R. Dwornicka, Fire properties of intumescent coating systems for the rolling stock, Communications - Scientific Letters of the University of Žilina 22 (2020) 90-96. https://doi.org/10.26552/com.C.2020.4.90-96

[21] N. Radek et al., Operational properties of DLC coatings and their potential application, METAL 2022 – 31st Int. Conf. Metall. Mater., (2022) 531-536. https://doi.org/10.37904/metal.2022.4491

[22] O. Kierat et al., The effect of the corrosion medium on silane coatings deposited on titanium grade 2 and titanium alloy Ti13Nb13Zr, Materials 14 (2021) art. 6350. https://doi.org/10.3390/ma14216350

[23] T. Lipiński, J. Pietraszek, Corrosion of the S235JR Carbon Steel after Normalizing and Overheating Annealing in 2.5% Sulphuric Acid at Room Temperature, Materials Research Proceedings 24 (2022) 102-108. https://doi.org/10.21741/9781644902059-16

[24] A. Bartkowska et al., Laser processing of diffusion boronized layer produced on monel® alloy 400 – microstructure, microhardness, corrosion and wear resistance tests, Materials 14 (2021) art. 7529. https://doi.org/10.3390/ma14247529

[25] A. Kalinowski et al., Laser surface texturing: characteristics and applications, System Safety: Human - Technical Facility - Environment 5 (2023) 240-248. https://doi.org/10.2478/czoto-2023-0026

[26] N. Radek et al., The influence of plasma cutting parameters on the geometric structure of cut surfaces, Mater. Res. Proc. 17 (2020) 132-137. https://doi.org/10.21741/9781644901038-20

[27] D. Nowakowski et al., Application of machine learning in the analysis of surface quality - the detection the surface layer damage of the vehicle body, METAL 2021 – 30th Int. Conf. Metall. Mater. (2021) 864-869. https://doi.org/10.37904/metal.2021.4210

[28] J. Pietraszek, A. Goroshko, The heuristic approach to the selection of experimental design, model and valid pre-processing transformation of DoE outcome, Advanced Materials Research 874 (2014) 145-149. https://doi.org/10.4028/www.scientific.net/AMR.874.145

[29] J. Pietraszek et al., Challenges for the DOE methodology related to the introduction of Industry 4.0, Prod. Eng. Arch. 26 (2020) 190-194. https://doi.org/10.30657/pea.2020.26.33

[30] M.S. Kozień, J. Wiciak, Passive structural acoustic control of the smart plate - FEM simulation, Acta Physica Polonica A 118 (2010) 1186-1188. https://doi.org/10.12693/APhysPolA.118.1186

[31] A. Goroshko et al., Multiphysics Modeling of Direct-Start Induction Motor Considering Rotor Unbalance, Materials Research Proceedings 45 (2024) 267-276. https://doi.org/10.21741/9781644903315-31

[32] A. Pacana, K. Czerwińska, Analysis of the causes of control panel inconsistencies in the gravitational casting process by means of quality management instruments, Production Engineering Archives 25 (2019) 12-16. https://doi.org/10.30657/pea.2019.25.03

[33] A. Pacana et al., Analysis of quality control efficiency in the automotive industry, Transportation Research Procedia 55 (2021) 691-698. https://doi.org/10.1016/j.trpro.2021.07.037

[34] D. Siwiec et al., Improving the non-destructive test by initiating the quality management techniques on an example of the turbine nozzle outlet, Materials Research Proceedings 17 (2020) 16-22. https://doi.org/10.21741/9781644901038-3

[35] J. Pietraszek, A. Gadek-Moszczak, The smooth bootstrap approach to the distribution of a shape in the ferritic stainless steel aisi 434l powders, Solid State Phenomena 197 (2013) 162-167. https://doi.org/10.4028/www.scientific.net/SSP.197.162

[36] J. Pietraszek, Fuzzy regression compared to classical experimental design in the case of flywheel assembly, Lecture Notes in Computer Science 7267 LNAI (2012) 310-317. https://doi.org/10.1007/978-3-642-29347-4_36

[37] J. Pietraszek, The modified sequential-binary approach for fuzzy operations on correlated assessments, Lecture Notes in Computer Science 7894 LNAI (2013) 353-364. https://doi.org/10.1007/978-3-642-38658-9_32

Materials Research Proceedings 62 (2026) 334-342 | https://doi.org/10.21741/9781644904015-43

Effect of Corrosion S255J Steel in Birds' Slurry Water Solution

Tomasz Lipiński[1,a*] and Jacek Pietraszek[2,b]

[1]University of Warmia and Mazury in Olsztyn, The Faculty of Technical Sciences, Department of Materials and Machines Technology, ul.Oczapowskiego 11, 10-957 Olsztyn, Poland

[2]Cracow University of Technology, Faculty of Mechanical Engineering, 31-864 Kraków, ul.Warszawska 24, Poland, ORCID: 0000-0003-2851-1606

[a]tomekl@uwm.edu.pl, [b]jacek.pietraszek@mech.pk.edu.pl

Keywords: Steel, Structural Steel, Carbon Steel, Corrosion, Corrosion Rate

Abstract. Construction materials operate in various environmental conditions. Some operate in natural conditions, exposed to both the natural atmosphere and, among other things, bird droppings. These droppings, located near vines, constituted corrosive material. To ensure the homogeneity of this material and its changing chemical composition over time, it was dried and hydrated immediately before testing. The tests were conducted on typical structural steel with enhanced mechanical properties, specifically grade S355J, which is typically used for structures exposed to the type of corrosion tested. The tests were conducted on an unprotected surface by immersion in hydrated bird droppings for 384 hours, with samples analyzed every 48 hours. The results of the tests presented equations for the corrosion rate of steel in the hydrated bird droppings environment, along with their graphical representations. It was found that the corrosion rate did not change significantly during the first 144 hours, while after this soaking period, the corrosion rate of the samples changed in direct proportion to the passage of time. In the tested time range of 348 hours, no stabilization of the corrosion rate was observed in the tested corrosive environment.

Introduction

Structural steels are used in a wide range of environmental conditions. Often, operating conditions favor corrosion of the materials used [1,2]. Corrosion, depending on its type, causes gradual degradation of the structural material and a related reduction in its performance. In certain corrosive environments, steels resistant to the prevailing factors are used [3-5]. Operating conditions, economic factors, material technology, and other factors do not always allow for the use of special corrosion-resistant steels. In such cases, structural steels are used, whose properties are largely dependent on their carbon content. These steels are not predisposed to operate in aggressive environments, but such factors cannot be eliminated. Various methods are used to protect steel from corrosive factors, primarily involving sealing and isolating the surface from external factors, such as coating with a layer of paint. Natural environmental factors affecting steel include weathering, which is also influenced by various factors. Other factors that can cause corrosion of structural materials include the presence of fauna and flora in their surroundings [6-9]. Birds, natural inhabitants of all environments, also affect the condition of materials. They typically migrate and inhabit areas in large flocks. They produce droppings that can corrode structural materials. The chemical components contained in their droppings cause intense corrosion of materials. Unfortunately, testing the corrosion resistance of various groups of structural materials exposed to bird droppings is a complex process. This is due to the high heterogeneity of the droppings, which depends on the type of food consumed by the birds under their living conditions. Because these droppings contain both solid and volatile substances, and considering the possibility of subsequent chemical reactions occurring in the droppings, the chemical composition of the droppings should be considered variable over time. This opens up the possibility of conducting computer-based studies that simulate the impact of environmental and operating conditions on the properties of structural materials.

Considering the aspects briefly outlined above, it was decided to conduct corrosion rate tests on typical S355J structural steel used in critical construction structures, such as bridges, girders, machinery components, and offshore structures. These structures, due to their nature, are exposed to harsh weather conditions, particularly the effects of bird droppings. The corrosive environment was chosen to be natural bird droppings located within the vineyard. Their location, considering the birds' short digestive tract and the presence of grape seeds and skins in the droppings, suggests that grapes are the primary food source, which has a direct impact on the quality of the bird droppings.

Materials and Methods

The experiment was performed on low-carbon S355J (1.0571) steel. The real chemical composition of the tested steel is presented in Table 1.

Table 1. *Chemical composition of the S355J steel*

C	Mn	Cu	Si	P	S	N	Fe
wt. %	wt. %	wt. %	wt. %	wt. %	wt. %	wt. %	-
0.19	1.4	0.43	0.51	0.02	0.02	0.008	balance

Test samples measuring 40 x 10 x 5 mm were taken from a 5 mm thick sheet of metal, ensuring that the longer side of the sample was parallel to the rolling direction of the sheet. The samples were cut using a band saw and then ground to a roughness of no more than Ra = 40 μm. After mechanical preparation, the sample surfaces were washed with alcohol.

The corrosion medium was prepared from bird droppings collected under a grape plant. A large amount of droppings collected in one spot may indicate that they were processed grapes. The collected bird droppings were dissolved in distilled water to obtain a uniform solution. Unfortunately, this was not possible due to the presence of solid particles in the droppings, e.g., grape seeds and grape skin fragments. These particles could have deposited on the samples, limiting local access to the corrosion medium. For this reason, the suspension was filtered through a 0.32 mm square mesh strainer. The prepared solution was poured onto samples taken from the steel. These samples were placed in a flask with crushed 5 mm diameter glass rods. Every 48 hours, a sample was removed from the flask for testing. The remaining samples were poured with a new corrosion solution. Considering the variability in the chemical composition of bird droppings over time, to ensure the greatest possible repeatability of the corrosive agent, the collected bird droppings were dried, then portioned, and stored in tightly sealed containers. The corrosive media were prepared immediately before use each time. Replacing the media every 48 hours ensured the highest possible repeatability and independence from potential chemical reactions that could occur during the tests, both within the bird droppings themselves and in combination with the tested steel.

Surface roughness tests were conducted in the direction of the samples' rolling using a Diavite DH5 profilometer. The following were determined: Ra - arithmetical mean roughness value [μm], Rp - maximum roughness depth [μm], Rq - mean peak width [μm], Rt - total height of the roughness profile [μm] for time range: 0, 48, 96, 144, 192, 240, 288, 336, and 384 hours of soaking. Profile roughness parameters were analyzed according to the PN-EN 10049:2014-03 standard (Measurement of roughness average Ra and peak count RPc on metallic flat products).

The corrosion rare of the S255J steel measured in mm/year was calculated with the use of the below formula (1), measured in g/m^2 were calculated with the use the below formula (2) [10]:

Materials Research Proceedings 62 (2026) 334-342 https://doi.org/10.21741/9781644904015-43

$$r_{corm} = \frac{8760 \cdot m}{S \cdot t \cdot \varrho} \tag{1}$$

$$r_{corg} = \frac{10000 \cdot m}{S \cdot t} \tag{2}$$

where:
t – time of soaking in a birds' slurry water solution [hours],
S - surface area of the sample (the starting value was assumed) [cm^2],
m – average mass loss in solution [g],
ϱ - sample density [g/cm^3].

Results and Discussion

The relative mass loss of the tested steel is shown in Fig. 1. In the corrosion process analyses presented below (Figs. 1-3), two fits were considered. The first presented mass changes as a function of the degree most closely matching the nature of the changes, while the second was presented as a first-order function. The coefficients of determination R2 of both function descriptions (Fig. 1) are significantly higher than 0.9, indicating a good fit to the changes taking place. The mass loss of samples soaked in bird droppings during the first period of exposure to the corrosive medium (up to 48 hours) was insignificant. In subsequent soaking periods, an acceleration of mass loss was observed.

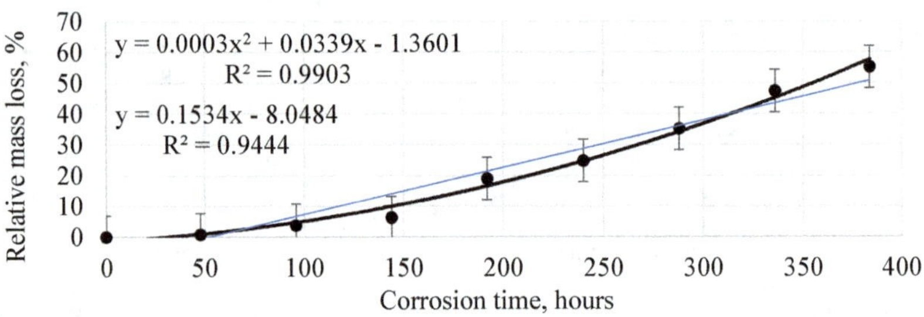

Fig. 1. Influence of soaking time of the S255J steel in the birds' slurry water solution on the relative mass loss at ambient temperature

The corrosion rate of soaking the S255J steel in birds' slurry water solution at ambient temperature, measured in mm per year, is shown in Fig. 2, and measured in grams per m^2 in Fig. 3. The corrosion rate increases gently during the initial soaking period, then accelerates, and slows down significantly in the final period of the study. This is a typical course of the corrosion process known from other studies. The slight increase in corrosion rate during the initial period is attributed to the development of surface features on samples exposed to the corrosive agent (Fig. 3 and Fig. 4). As the surface development progressed, the corrosion rate increased. After reaching a roughness level at which the progressive uniform corrosion process did not cause such a large increase in surface roughness, the corrosion rate asymptotically approached a constant value. The coefficients of determination for both the corrosion rate, expressed in mm per year and in grams per m^2, do not differ by several percent, which is not a significant difference considering the variability in corrosion.

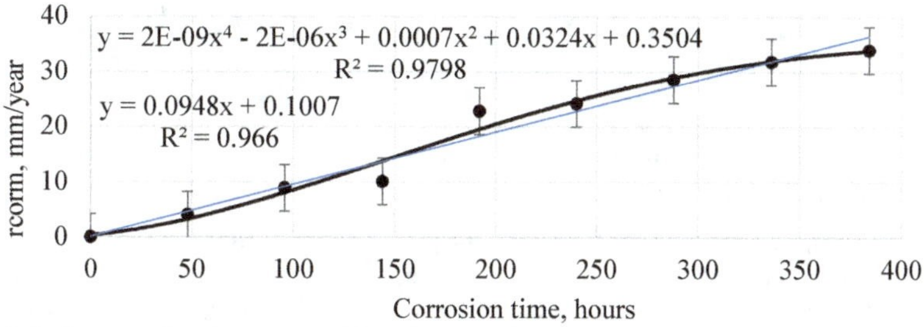

Fig. 2. Influence of soaking time of the S255J steel in birds' slurry water solution at ambient temperature on the corrosion rate measured in mm per year

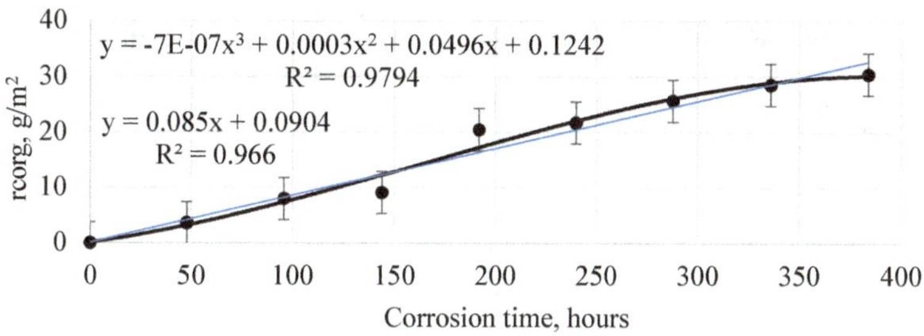

Fig. 3. Influence of soaking time of the S255J steel in birds' slurry water solution at ambient temperature on the corrosion rate measured in grams per m²

The microstructure of S355J steel in its initial state is shown in Fig. 4a. The developed surface after soaking for 336 hours in bird manure is shown in Fig. 4b. A cross-section of a sample of S355J steel soaked in bird manure for 336 hours is shown. Its roughness profile is shown in Fig. 5. This example was chosen to illustrate the surface roughness of the sample after achieving a relatively constant roughness. The steel's surface microstructure consists of light ferrite and dark pearlite (Fig. 4a). The steel's surface after soaking indicates uneven distribution of corrosion effects.

a.

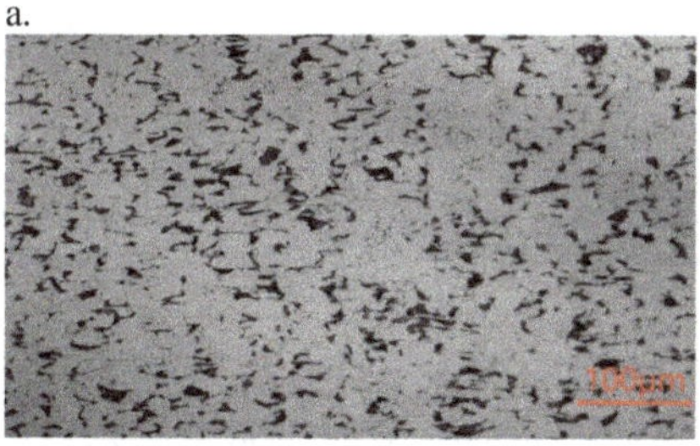

b.

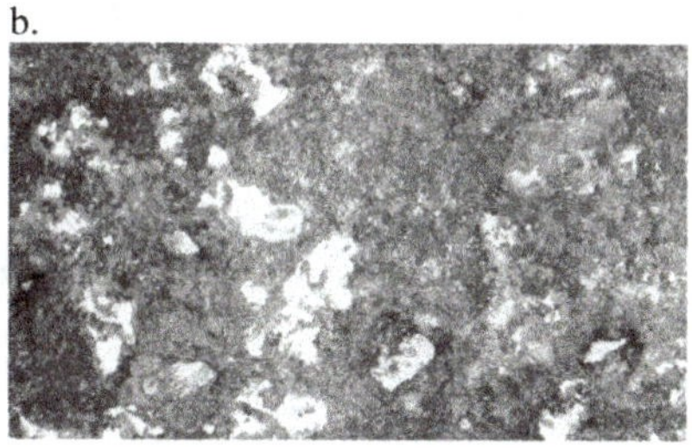

Fig. 4. Microstructure of S355J steel: a. in the initial state, b. steel surface after soaking in bird manure for 336 hours

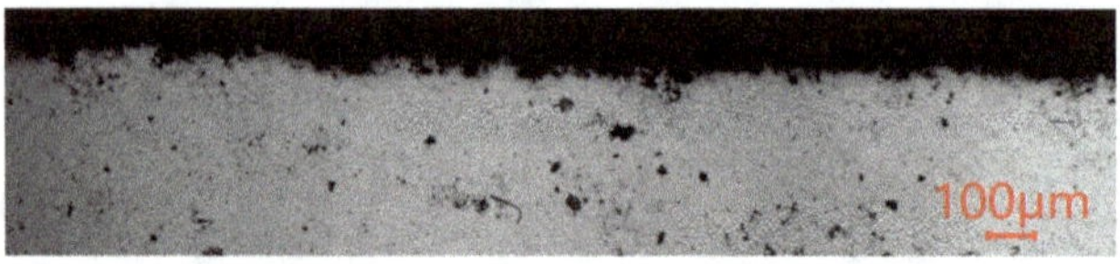

Fig. 5. Cross-section of an S355J steel sample soaked in bird manure for 336 hours

Arithmetical mean roughness value R_a and mean peak width R_q of S255J steel in birds' slurry water solution at ambient temperature for time soaked a 348 hours is presented in Fig. 7. For the first 96 hours, a gradual increase in roughness was noted, which, in conjunction with Fig. 1 to Fig. 3, indicates a low rate of corrosion development. At the beginning of the sample soaking, the corrosive environment affects the surface with a low roughness of Ra = 0.36 µm; therefore, the penetration area of the birds' slurry water solution is small. Over 96 hours, the surface development increases at a slow rate, reaching a Ra value of 1.62 µm. Further extension of the sample retention time in the corrosive solution increases the surface roughness in direct proportion to the duration of the time. For a time of 336 hours, Ra = 31.55 µm. A further extension of the soaking time to 348 hours no longer causes a significant increase in the roughness of the tested surface, with a Ra value of 32.31 µm (Fig. 7).

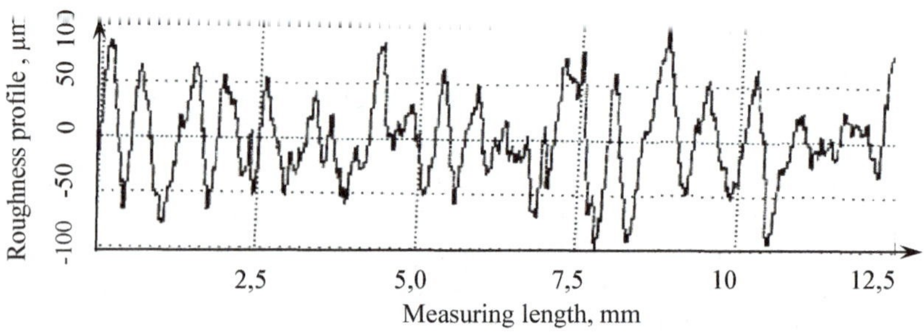

Fig. 6. Roughness profile of an S355J steel sample soaked in bird manure for 336 hours

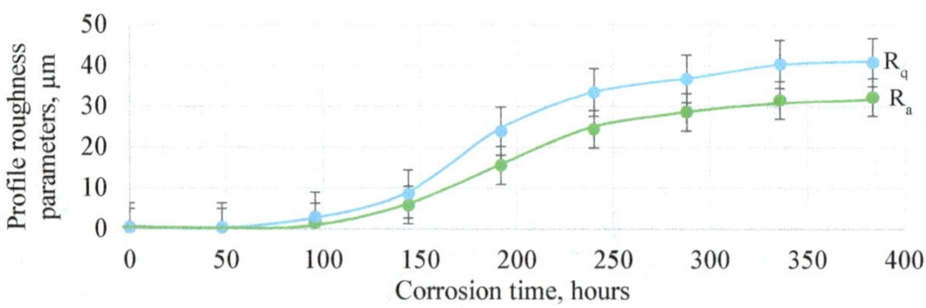

Fig. 7. Profile roughness of S255J steel in birds' slurry water solution at ambient temperature for different corrosion times: R_a - arithmetical mean roughness value [µm], R_q - mean peak width [µm]

Maximum roughness depth R_p and total height of the roughness profile R_t of S255J steel in birds' slurry water solution at ambient temperature for a time soaked of 348 hours are presented in Fig. 8.

Materials Research Proceedings 62 (2026) 334-342 https://doi.org/10.21741/9781644904015-43

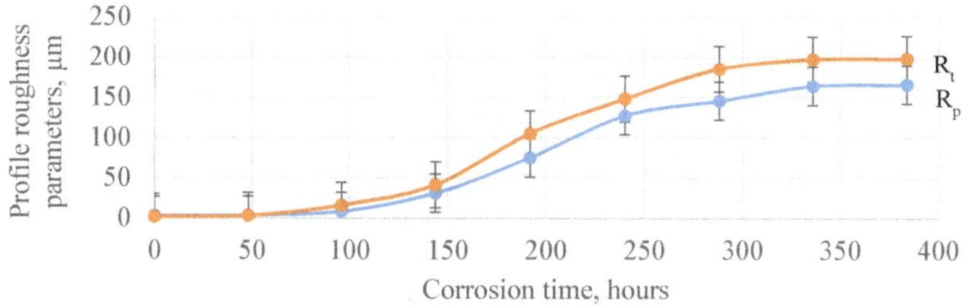

Fig. 8. Profile roughness of S255J steel in birds' slurry water solution at ambient temperature for different corrosion times: R_p - maximum roughness depth [µm], R_t - total height of the roughness profile [µm]

Their course is analogous to the arithmetic mean roughness value (Fig. 7). However, for the total height of the roughness profile, an increase was noted after exceeding 48 hours of soaking, reaching $R_t = 3.2$ µm. With a further increase in the sample soaking time, essentially both analyzed parameters increase up to 144 hours, reaching $R_t = 31.5$ µm and $R_p = 31.2$ µm. After exceeding this soaking time, both roughness parameters increase more rapidly to Rt = 185.5 µm and Rp = 145.8 µm over 288 hours of exposure to the birds' slurry water solution. A further extension of the soaking time essentially stabilizes both roughness parameters, which reach $R_t = 198.6$ µm and $R_p = 166.2$ µm for 384 hours. Comparing the presented parameters with the corrosion of comparable ferritic-pearlitic steels with a similar carbon content [11], a more intense course of corrosion was observed in the birds' slurry water solution than in the corrosion media used in other studies, i.e., NaCl and cattle slurry [12].

Technological Impact

Corrosion phenomena of metallic materials are a significant factor limiting their direct applicability and necessitating various corrosion protection methods. Appropriate preparation of materials opens up the possibility of adapting this information in the industrial field [13], particularly with respect to the tested material as a potential base for specialized coatings [14-16], as a structural material for robotic elements [17,18], and with appropriately designed control systems [19,20]. This also creates the possibility of substituting other metal materials that are more susceptible to corrosion [21-23], including specific alloys [24,25], which allows for a reduction in energy demand [26,27]. This significantly improves the company's situation by implementing the appropriate material [28], while simultaneously allowing for a substantial increase in production quality [29-31] and a notable decrease in wear [32]. Due to their resistance to environmental phenomena, suitable materials may be interesting for military purposes [33]. Regardless of this, corrosion phenomena are a significant driving force for the development of microstructure testing methods [34-36].

Conclusions

Based on the results of the conducted studies, it was concluded that:

- birds' slurry water solution at ambient temperature is a medium that causes intense corrosion of unalloyed steel;
- the first corrosion period (up to 96 hours of soaking) is characterized by a low corrosion rate with a slight increase in roughness;
- after 336 hours of soaking S355J steel in birds' slurry water solution at ambient temperature, stabilization of corrosion parameters can be expected.

References

[1] J.J. Santana Rodríguez, J.E. González González. Identification and formation of green rust 2 as an atmospheric corrosion product of carbon steel in marine atmospheres. Mater. Corros. 57 (2006) 411-417. https://doi.org/10.1002/maco.200503942

[2] E. Naveen et al., Influence of organic corrosion inhibitors on pickling corrosion behaviour of sinter-forged C45 steel and 2% Cu alloyed C45 steel. Journal of Alloys and Compounds 695 (2017) 3299-3309. https://doi.org/10.1016/j.jallcom.2016.11.133

[3] L.L. Machuca et al., Microorganisms associated with corrosion of structural steel in diverse atmospheres. Int. Biodeter. Biodegr. 114 (2016) 234-243. https://doi.org/10.1016/j.ibiod.2016.06.015

[4] M. Scendo et al., Influence of laser treatment on the corrosive resistance of WC-Cu coating produced by electrospark deposition. Int. J. Electrochem. Sci. 8 (2013) 9264-9277.

[5] A. Dudek et al., Surface remelting of 316 L+434 L sintered steel: microstructure and corrosion resistance. J. Solid State Electr. 18 (2014) 2973-2981. https://doi.org/10.1007/s10008-014-2483-2

[6] I. Hren et al., Analysis of the influence of surface condition on the corrosion behavior of alloy 2024. Manufacturing Technology 20 (2020) 603-611 https://doi.org/10.21062/MFT.2020.106

[7] J. Selejdak et al., The influence of simultaneous action of the aggressive environment and loading on strength of RC beams. MATEC Web of Conf. 183 (2018) 02002. https://doi.org/10.1051/matecconf/201818302002

[8] D.Dvorský et al., Microstructure, mechanical and corrosion properties of extruded milled magnesium powder. Manuf. Technol. 20 (2020) 708-713. https://doi.org/10.21062/MFT.2020.097

[9] T. Lipiński, J. Pietraszek. Influence of animal slurry on carbon C35 steel with different microstructure at room temperature. Engineering for Rural Development 21 (2022) 344-350. https://doi.org/10.22616/ERDev.2022.21.TF115

[10] PN-EN ISO 3651-1:2004. Determination of resistance to intergranular corrosion of corrosion-resistant steels. Part 1: Austenitic and ferritic-austenitic (duplex) corrosion-resistant steels.

[11] T. Lipiński, Corrosion Effect of 20% NaCl Solution on Basic Carbon Structural S235JR Steel. 16th International Scientific Conference Engineering for Rural Development 16 (2017) 1069-1074. https://doi.org/10.22616/ERDev2017.16.N225

[12] T. Lipinski, D. Karpisz. Effect of animal slurry on carbon structural S235JR steel at 303 K. Engineering for Rural Development 19 (2020) 1482–1487. https://doi.org/10.22616/ERDev.2020.19.TF368

[13] D. Klimecka-Tatar, R. Dwornicka, The assembly process stability assessment based on the strength parameters statistical control of complex metal products, METAL 2019 - 28th Int. Conf. Metall. Mater. (2019) 709-714.

[14] N. Radek et al., The WC-Co electrospark alloying coatings modified by laser treatment, Powder Metallurgy and Metal Ceramics 47 (2008) 197-201. https://doi.org/10.1007/s11106-008-9005-7

[15] L. Adamczyk et al., Electrochemical preparation of composite coatings of 3,4-etylenodioxythiophene (EDOT) and 4-(pyrrole-1-yl) benzoic acid (PyBA) with heteropolyanions, Materials Chemistry and Physics 144 (2014) 418-424. https://doi.org/10.1016/j.matchemphys.2014.01.012

[16] J. Korzekwa et al., Influence of the Size of Nanoparticles on the Microstructure of Oxide Coatings, Materials Science 53 (2018) 709-716. https://doi.org/10.1007/s11003-018-0127-x

[17] P. Wos et al., Bricklaying robot lifting and levelling system, Communications – Sci. Letters Univ. Žilina 23 (2021) B257-B264. https://doi.org/10.26552/COM.C.2021.4.B257-B264

[18] A. Annusewicz-Mistal et al., Autonomous Manipulator of a Mobile Robot Based on a Vision System, Applied Sciences 13 (2023) art. 439. https://doi.org/10.3390/app13010439

[19] J.E. Takosoglu, P.A. Laski, Intelligent positioning system of electro-pneumatic servo-drive, Annals of DAAAM and Proceedings of the International DAAAM Symposium (2011) 1641-1642.

[20] P.A. Laski, M. Smykowski, Using a development platform with an stm32 processor to prototype an inexpensive 4-dof delta parallel robot, Sensors 21 (2021) art. 7962. https://doi.org/10.3390/s21237962

[21] M. Scendo et al., Purine as an effective corrosion inhibitor for stainless steel in chloride acid solutions, Corrosion Reviews 30 (2012) 33-45. https://doi.org/10.1515/CORRREV-2011-0039

[22] D. Klimecka-Tatar et al., Corrosion characteristics in acid, alkaline and the ringer solution of Fe68-xCoxZr10Mo5W2B 15 metallic glasses, Journal of the Balkan Tribological Association 20 (2014) 124-130.

[23] O. Kierat et al., The effect of the corrosion medium on silane coatings deposited on titanium grade 2 and titanium alloy Ti13Nb13Zr, Materials 14 (2021) art. 6350. https://doi.org/10.3390/ma14216350

[24] A. Dudek, M. Klimas, Composites based on titanium alloy Ti-6Al-4V with an addition of inert ceramics and bioactive ceramics for medical applications fabricated by spark plasma sintering (SPS method), Materialwissenschaft und Werkstofftechnik 46 (2015) 237-247. https://doi.org/10.1002/mawe.201500334

[25] A. Dudek et al., The effect of alloying method on the structure and properties of sintered stainless steel, Arch. Metall. Mater. 62 (2017) 281-287. https://doi.org/10.1515/amm-2017-0042

[26] V. Zhelykh et al., Analysis of the processes of heat exchange on infrared heater surface, Diagnostyka 17 (2016) 81-85.

[27] J. Zwierzchowski et al., Model tests of wind turbine with a vertical axis of rotation type Lenz 2, EPJ Web of Conf. 143 (2017) art. 02150. https://doi.org/10.1051/epjconf/201714302150

[28] D. Klimecka-Tatar, M. Ingaldi, Assessment of the technological position of a selected enterprise in the metallurgical industry, Materials Research Proceedings 17 (2020) 72-78. https://doi.org/10.21741/9781644901038-11

[29] A. Pacana et al., Analysis of quality control efficiency in the automotive industry, Transportation Research Procedia 55 (2021) 691-698. https://doi.org/10.1016/j.trpro.2021.07.037

[30] D. Siwiec et al., Analysis of problems with maintaining the desired quality of gas sealant, METAL 2022 – 31st Int. Conf. Metall. Mater., (2022) 848-853. https://doi.org/10.37904/metal.2022.4524

[31] D. Siwiec et al., Improving the Quality of Products from Cast Alloy with the use of Grey Relational Analysis (GRA), Materials Research Proceedings 24 (2022) 204-211. https://doi.org/10.21741/9781644902059-30

[32] A. Pacana et al., Influence of natural aggregate crushing process on crushing strength index, Sustainability (Switzerland) 13 (2021) art. 8353. https://doi.org/10.3390/su13158353

[33] W. Przybył et al., Microwave absorption properties of carbonyl iron-based paint coatings for military applications, Defence Technology 22 (2023) 1-9. https://doi.org/10.1016/j.dt.2022.06.013

[34] A. Szczotok, S. Roskosz, New possibilities of light microscopy research resulting from digital recording of images, Materials Science-Poland 23 (2005) 559-565.

[35] J. Korzekwa et al., The influence of sample preparation on SEM measurements of anodic oxide layers; Practical Metallography 53 (2016) 36-49. https://doi.org/10.3139/147.110367

[36] A. Gądek-Moszczak et al., Nano X-ray Tomography Application for Quantitative Surface Layer Geometry Analysis after Laser Beam Modification, Materials 15 (2022) art. 5935. https://doi.org/10.3390/ma15175935

Terotechnology XIV Materials Research Forum LLC
Materials Research Proceedings 62 (2026) 343-351 https://doi.org/10.21741/9781644904015-44

Paint Coating Technologies for Railway Axles-Current State and Development Directions

Wojciech PAULI[1,a]* and Norbert RADEK[2,b]

[1]Barwa Kielce, ul. Warkocz 3-5, 25-253 Kielce, Poland

[2]Kielce University of Technology, Faculty of Mechatronics and Mechanical Engineering, Al. 1000-lecia Państwa Polskiego 7, 25-314 Kielce, Poland

[a]wpauli@barwa.kielce.pl, [b]norrad@tu.kielce.pl

Keywords: Railway Axles, Protective Coatings, Corrosion Resistance, Non-Destructive Testing (NDT), Epoxy–Polyurethane Systems, Thermo-Indicating Coatings, Impact Resistance (Gritting Resistance), Smart Maintenance, Defence Transport Applications

Abstract. This paper presents a review of current paint coating systems used for protecting railway axles. It discusses normative requirements (EN 13261 [1], DB TL 918 300 [2], ISO 9466 [3], IK 001/2024 [4]) and the test methods used to assess protective and adhesive properties of coatings. One- and two-layer systems are compared, with their advantages and limitations indicated. The final part describes development directions, including thermo-sensitive coatings and systems with reduced or increased thickness, as a response to growing operational and environmental requirements.

Introduction

The running gear of rail vehicles consists of bogies and wheelsets. A wheelset consists of wheels and an axle and may also include drive and braking system components. The wheels are the only elements in contact with the track, bearing the entire vehicle weight transmitted through the axle, which is mounted in the rolling bearings of the bogie. Given this function, the wheelset is a critical safety component of the entire rail vehicle. Damage to a single wheelset can result in derailment, material losses, and even pose a risk to human life.

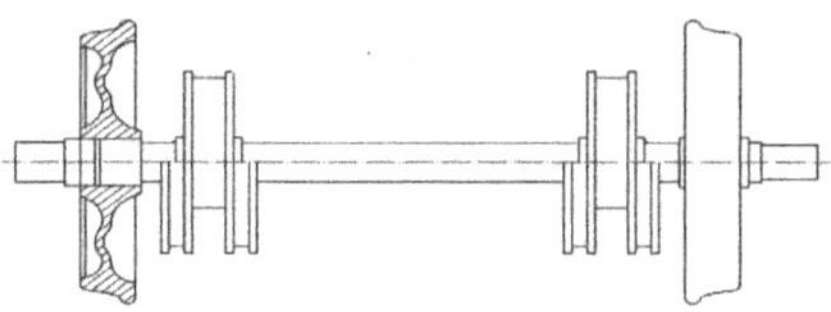

Fig. 1. Schematic of a wheelset according to EN 13260 [5].

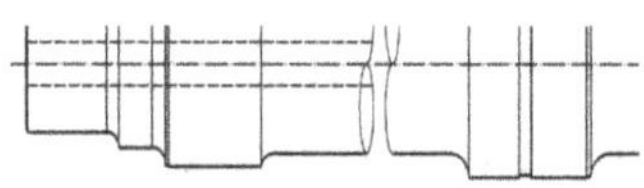

Fig. 2. Schematic of an axle according to EN 13261 [1].

From the moment the vehicle is placed on the bogie, the wheelset is exposed to stresses resulting from the vehicle's weight. During operation, as a rotating component, it is subjected to cyclic fatigue loads. Additionally, as part of the undercarriage, it may be damaged by stones or other solid bodies dislodged from the railway embankment. During its service life, it is also exposed to elevated temperatures, such as those caused by braking or failures, including bearing seizure or loss of lubrication. Being an external component, it is continuously subjected to a corrosive environment depending on the surrounding atmosphere. These factors determine the requirements concerning the material and design of the wheelset, as well as the means of protecting these elements using adhesive coatings.

The purpose of this article is to discuss the requirements, standards, and current state of knowledge regarding protective coatings for railway wheelsets. Finally, key challenges and development paths for future generations of protective coatings are presented, which will contribute to enhancing the safety of railway transport.

Normative Requirements and Test Methods

The requirements for wheelsets, axles, and wheels are defined in the standards PN-EN 13260 [5], PN-EN 13261 [1], and PN-EN 13262 [6]. However, only the standard concerning railway axles, PN-EN 13261 [1], addresses the subject of adhesive coatings. It distinguishes four classes of axle protection:

- Class 1: Axle sections exposed to atmospheric corrosion and mechanical impact.
- Class 2: Axle sections exposed to the effects of specific corrosion products.
- Class 3: Axle sections exposed to atmospheric corrosion.
- Class 4: Unpainted axle sections exposed to atmospheric corrosion.
- Different classes may be applied to a single axle.

For each protection class, the standard specifies the corresponding test methods (Table 1).

Table 1. Tests required for each protection class according to EN 13261[1]

	Class 1	Class 2	Class 3	Class 4
Coating thickness	X	X	X	-
Coating adhesion	X	X	X	-
Impact resistance	X	-	-	-
Chipping resistance	X	X	X	-
Salt spray resistance	X	X	X	-
Resistance to specific corrosive products	-	X	-	-
Resistance of coating to cyclic mechanical stress	X	X	X	-

These requirements are complemented by other normative documents, such as ISO 9466 [3], which defines requirements for protective coatings applied to passenger rolling stock. Individual countries or operators also define their own coating specifications. In Poland, such requirements have been introduced by the Railway Research Institute (IK 001/2024 [4]) and adopted by national railway operators. Deutsche Bahn developed another specification in document DB TL 918 300 Blatt 11 [2], which has become the basis for many systems used in Germany and neighboring countries (Table 2). Other major European countries also maintain their own standards.

Table 2. Selected documents defining requirements for adhesive coatings on railway axles

Standard / Document	Scope of Application
ISO 9466:2025 [3]	Railway applications – coatings for passenger rail vehicles (including bogies and wheelsets).
IK 001/2024 [4]	National document – tests for thickness, adhesion, thermal resistance, and color stability (ΔE*).
DB TL 918 300 Blatt 11 [2]	Deutsche Bahn specification for coatings of freight wagon axles (Protection Group 3).
EN 13261 [1]	Requirements for wheelset axles, including corrosion protection and NDT compatibility.

From these documents arise the necessary tests to be carried out for coatings on railway axles, summarized in Table 3.

Table 3. Tests for adhesive coatings on railway axles

Type of Test	Standard	Purpose of the Test
Film thickness measurement	ISO 2808 [7]	Verification of compliance with the required thickness range.
Adhesion	ISO 2409 [8] or ISO 4624 [9]	Evaluation of coating adhesion to the substrate – required class ≤ 1 (no flaking).
Corrosion resistance (salt spray test)	ISO 9227 [10] / ISO 16773 [11]	Evaluation of anti-corrosion effectiveness in a salt spray chamber.
Resistance to gravel impact (stone-chip test)	ISO 20567-1 [12] / ISO 13261 internal method [1]	Evaluation of mechanical resistance to impacts by solid particles (stones, sand).
Chemical resistance	ISO 2812-1[13]/ ISO 2812-2 [14]	Evaluation of resistance to oils, greases, brake fluids, and cleaning agents.
Mechanical resistance (impact)	ISO 6272 [15]	Impact testing – no cracking or delamination to the substrate.
Thermal resistance / thermo-sensitivity (ΔE*)	IK 001/2024 internal method [4]	Evaluation of color and structure stability after overheating; identification of color change ΔE*.
Heat-humidity ageing	ISO 6270-2 [16]	Simulation of coating behavior under condensation and cyclic temperature changes.

Technologies used in civil rolling stock are also applied in military transport platforms, where requirements concerning corrosion resistance and NDT diagnostics are analogous.

Applied Coating Systems

In industrial practice, two types of coating systems are used: single-layer and two-layer systems (Fig. 3).

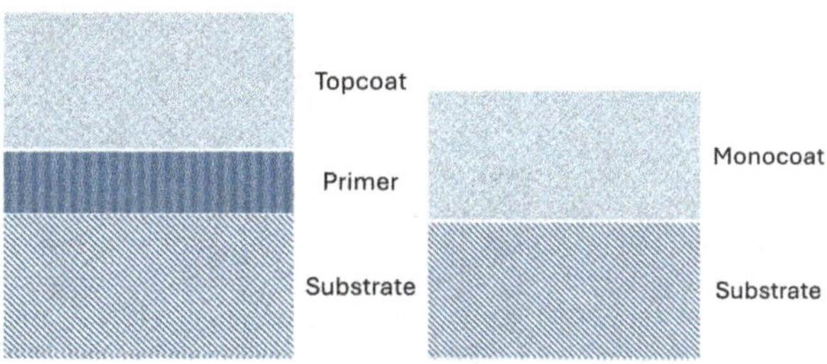

Fig. 3. Schematic of coating systems

Two-layer systems are mainly applied in passenger rolling stock. They are characterized by high mechanical and anti-corrosion resistance, but this increases the cost of painting due to the material costs, as well as the time and energy required to apply two coats.

Single-layer systems are used for freight rolling stock or when fast coating renovation is required. The advantage of such systems lies in their low cost and simplicity, although they exhibit lower mechanical and corrosion resistance compared with multilayer coatings.

Innovations and Development Directions

Adhesive coatings can play a significant role in diagnosing the condition of the running gear. By using thermo-sensitive pigments, the coating can indicate exposure to excessive temperature [17]. Thanks to this property, during standard visual inspections, overheated areas can be identified, and corrective actions can be taken depending on the cause. It is also important that the coatings do not interfere with non-destructive diagnostic methods, such as magnetic or ultrasonic testing.

Another topic addressed in research works is the mechanical resistance of coatings. Two development directions are possible:

- Reduction of coating thickness while maintaining mechanical properties [18].
- Increased resistance by adding an intermediate layer that absorbs and dissipates impact energy [19].

Both approaches offer certain benefits but also introduce risks. The first would reduce costs and simplify repairs, yet could negatively affect protective performance. The second would increase corrosion and mechanical resistance, but would complicate the application process and increase vehicle weight. Selecting the most promising development path requires further analysis and optimization.

Technological, Economic, Environmental, and Social Impact

The implementation of advanced protective coating technologies for wheel axles yields multidimensional benefits, affecting both technological and economic aspects, as well as environmental and social considerations.

A key development direction is the optimization of coatings' operational properties, such as hardness, adhesion, and resistance to mechanical damage. The use of composite coatings and nanolayers, such as Al_2O_3/WS_2 [20] or DLC (Diamond-Like Carbon) coatings [21], directly extends axle service life by improving their tribological behavior [22] and reducing friction [23]. Research on the effect of laser processing on the properties of WC-Co coatings [24] and electro-spark deposition (ESD) coatings [25] demonstrates the potential for precisely adapting the surface structure to the required operating parameters [26]. These technologies are closely linked to the concept of Smart Maintenance, which requires continuous monitoring and diagnostics. Modern coating systems must be compatible with NDT (Non-Destructive Testing), which requires continuous improvement of measurement techniques. An example is the improvement of NDT processes on critical components, such as bearing rings [27] or TOFD-welded joints [28]. The use of nano-X-ray tomography enables the quantitative analysis of surface layer geometry [29]. At the same time, machine learning (ML) techniques facilitate the analysis of surface quality and the early detection of damage in rolling stock components [30]. In the context of safety, the use of special coatings, including fire-retardant (intumescent) systems [31], is also important.

Advanced statistical and mathematical tools, such as Design of Experiments (DOE) analysis, including heuristic [32] and fuzzy [33] approaches, are essential for optimizing technological processes and selecting the optimal parameters. These models enable the quantification of uncertainty in experimental modeling [34] and serve as a response to the challenges of Industry 4.0 [35]. For example, optimizing corrosion conditions may involve selecting inhibitors [36] and adjusting laser power when modifying WC-Cu coatings to enhance corrosion resistance [37]. Studies of heat treatment processes, such as normalizing and tempering, also impact material properties and corrosion resistance [38]. Technology development is also supported by computational tools tailored to various levels of engineering expertise [39]. In the context of Smart Maintenance and technical condition monitoring, vibration mechanics and multiphysics modeling are gaining importance, as seen in the analysis of induction motors [40], inverse problems [41], and the impact of ground vibrations on precision devices [42]. Mechanism reliability and durability can also be studied using hybrid models [43] and static mechanical force amplification analysis [44]. Economic benefits result primarily from extended maintenance intervals (thanks to increased coating durability), reduced material consumption, and optimization of production processes. Process modeling using fuzzy logic allows for optimization under conditions of uncertainty [45].

The environmental aspect is addressed through the pursuit of sustainable development and a circular economy [46]. A key tool in this context is life cycle analysis (LCA), which enables the assessment of the environmental impact of materials and processes, including those for steel wheels [47] and other industrial products [48]. Research focuses on quality-oriented methods for product improvement, taking into account environmental criteria [49] and quality-environmental criteria [46]. These efforts also include minimizing environmental impact through emission control [50] and exploring methods for separating and recycling polymers used in coatings [51]. Recycling and utilizing waste, such as fly ash, for the production of construction materials [52], also align with this trend. In a social context, new coating technologies improve operational safety (through increased axle reliability and the use of thermosensitive coatings to detect overheating) and working conditions. Achieving high quality and occupational safety, for example, in metal foundries [53], is closely linked to the production process of railway components. Improving the quality of casting inspection, for example, silumin castings [54], directly impacts the safety of the final product. Attention to quality and logistics [55], as well as the elimination of production irregularities [56], minimizes the risk of failure. Furthermore, striving to improve the indoor environment (IEQ) in intelligent buildings [57] is analogous to improving operating conditions in railway rolling stock. Advanced statistical methods and data analysis used in the optimization of technological processes [58, 59] have wide industrial applications, which in turn place demands on engineering education [39].

Summary

Paint coating systems applied to railway axles constitute a crucial element of rolling stock operational safety. Their primary purpose is to protect against corrosion, mechanical damage, and overheating in areas subjected to high dynamic loads. According to the requirements of EN 13261 [1], DB TL 918 300 Blatt 11 [2], ISO 9466 [3], and the national document IK 001/2024 [4], coatings must ensure:

- high adhesion and corrosion resistance,
- durability under operational conditions,
- compatibility with non-destructive testing methods (UT, MT, ET).

In practice, two dominant directions in coating technology development can be observed:

- Reduction of system thickness – an economical, faster-to-apply, and low-emission solution, mainly dedicated to freight rolling stock.
- Increased thickness with an elastic interlayer – multilayer systems primarily intended for passenger and high-speed rolling stock, providing enhanced resistance to gravel impact and dynamic loads.

A new and rapidly developing area involves thermosensitive coatings, which enable the visual detection of material overheating. Combining protective and diagnostic functions aligns with the concept of Smart Maintenance, which is the intelligent management of a vehicle's technical condition. A properly selected coating system, tailored to the type of operation (cargo/passenger/high-speed), can significantly extend the axle overhaul interval, reduce corrosion, and improve operational safety.

References

[1] CEN. EN 13261:2020 – Railway applications – Axles – Product requirements. CEN, Brussels, 2020.

Materials Research Proceedings 62 (2026) 343-351 https://doi.org/10.21741/9781644904015-44

[2] Deutsche Bahn AG Systemtechnik. DB TL 918 300 Blatt 11 – Anstrichstoffe für Eisenbahnfahrzeuge – Beschichtungsstoff für Güterwagenradsätze (Korrosionsschutzgruppe 3 nach EN 13261). Deutsche Bahn AG, Berlin, 2023.

[3] ISO. ISO 9466:2025 – Railway applications – Coating of passenger rail vehicles. International Organization for Standardization, Geneva, 2025.

[4] Instytut Kolejnictwa. IK 001/2024 – Requirements and test methods for paint coatings applied to railway axles and bogies. Warsaw, 2024.

[5] CEN. EN 13260:2021 – Railway applications – Wheelsets and bogies – Wheelsets – Product requirements. CEN, Brussels, 2021.

[6] CEN. EN 13262:2021 – Railway applications – Wheelsets and bogies – Wheels – Product requirements. CEN, Brussels, 2021.

[7] CEN. EN ISO 2808:2020 – Paints and varnishes – Determination of film thickness. CEN, Brussels, 2020.

[8] CEN. EN ISO 2409:2021 – Paints and varnishes – Cross-cut test for adhesion. CEN, Brussels, 2021.

[9] CEN. EN ISO 4624:2016 – Paints and varnishes – Pull-off test for adhesion. CEN, Brussels, 2016.

[10] CEN. EN ISO 9227:2017 – Corrosion tests in artificial atmospheres – Salt spray tests. CEN, Brussels, 2017.

[11] ISO. ISO 16773:2016 – Electrochemical impedance spectroscopy for coated metals. ISO, Geneva, 2016.

[12] CEN. EN ISO 20567-1:2017 – Paints and varnishes – Determination of stone-chip resistance – Part 1: Multi-impact testing. CEN, Brussels, 2017.

[13] CEN. EN ISO 2812-1:2023 – Paints and varnishes – Determination of resistance to liquids – Part 1: Immersion in liquids other than water. CEN, Brussels, 2023.

[14] CEN. EN ISO 2812-2:2013 – Paints and varnishes – Determination of resistance to liquids – Part 2: Effects of immersion in water. CEN, Brussels, 2013.

[15] CEN. EN ISO 6272-1:2012 – Paints and varnishes – Rapid deformation (impact resistance) tests – Falling weight method. CEN, Brussels, 2012.

[16] CEN. EN ISO 6270-2:2017 – Paints and varnishes – Determination of resistance to humidity (condensation) – Part 2: Continuous condensation. CEN, Brussels, 2017.

[17] Lucchini RS Group. Wheelset Protection and Coating Systems – Technical Brochure. Lucchini RS, Lovere, 2013.

[18] Crebonit Group. CGRP-M-RAIL-2-EN, Creborail Transportation Coating Systems. Crebonit, Graz, 2021.

[19] Deutsche Bahn Systemtechnik. DB-qualifizierte Nasslacke nach DBS 918 300. Deutsche Bahn Systemtechnik, Berlin, 2023. [Online] Available: https://www.db-systemtechnik.de/dbst-de/assets/unser-technisches-know-how/oberflaechentechnik-12935210

[20] J. Korzekwa et al., Tribological behaviour of Al_2O_3/inorganic fullerene-like WS_2 composite layer sliding against plastic, International Journal of Surface Science and Engineering 10 (2016) 570-584. https://doi.org/10.1504/IJSURFSE.2016.081035

[21] N. Radek et al., Operational properties of DLC coatings and their potential application, METAL 2022 – 31st Int. Conf. Metall. Mater., (2022) 531-536. https://doi.org/10.37904/metal.2022.4491

[22] N. Radek et al., Influence of laser texturing on tribological properties of DLC coatings, Production Engineering Archives 27 (2021) 119-123. https://doi.org/10.30657/pea.2021.27.15

[23] W. Zórawski et al., Plasma-sprayed composite coatings with reduced friction coefficient, Surface and Coatings Technology 202 (2008) 4578-4582. https://doi.org/10.1016/j.surfcoat.2008.04.026

[24] N. Radek et al., The effect of laser beam processing on the properties of WC-Co coatings deposited on steel, Materials 14 (2021) art. 538. https://doi.org/10.3390/ma14030538

[25] N. Radek et al., The effect of laser treatment on operational properties of ESD coatings, METAL 2021 – 30th Int. Conf. Metall. Mater. (2021) 876-882. https://doi.org/10.37904/metal.2021.4212

[26] N. Radek, Determining the operational properties of steel beaters after electrospark deposition, Eksploatacja i Niezawodnosc 44 (2009) 10-16.

[27] D. Siwiec et al., Improving the Non-Destructive Testing Process of the Outer Bearing Ring, Materials Research Proceedings 24 (2022) 174-180. https://doi.org/10.21741/9781644902059-26

[28] M. Patek et al., Non-destructive testing of split sleeve welds by the ultrasonic TOFD method, Manufacturing Technology 14 (2014) 403-407.

[29] A. Gądek-Moszczak et al., Nano X-ray Tomography Application for Quantitative Surface Layer Geometry Analysis after Laser Beam Modification, Materials 15 (2022) art. 5935. https://doi.org/10.3390/ma15175935

[30] D. Nowakowski et al., Application of machine learning in the analysis of surface quality - the detection the surface layer damage of the vehicle body, METAL 2021 – 30th Int. Conf. Metall. Mater. (2021) 864-869. https://doi.org/10.37904/metal.2021.4210

[31] N. Radek, R. Dwornicka, Fire properties of intumescent coating systems for the rolling stock, Communications - Scientific Letters of the University of Žilina 22 (2020) 90-96. https://doi.org/10.26552/com.C.2020.4.90-96

[32] J. Pietraszek, A. Goroshko, The heuristic approach to the selection of experimental design, model and valid pre-processing transformation of DoE outcome, Advanced Materials Research 874 (2014) 145-149. https://doi.org/10.4028/www.scientific.net/AMR.874.145

[33] R. Dwornicka et al., Fuzzy Statistics-Aided Inference in Experimental Design, World Congr. Comp. Mechanics ECCOMAS Congr. (2024). https://doi.org/10.23967/wccm.2024.131

[34] J. Pietraszek et al., The non-parametric approach to the quantification of the uncertainty in the design of experiments modelling, UNCECOMP 2017 – Proc. 2nd International Conference on Uncertainty Quantification in Computational Sciences and Engineering (2017) 598-604. https://doi.org/10.7712/120217.5395.17225

[35] J. Pietraszek et al., Challenges for the DOE methodology related to the introduction of Industry 4.0, Production Engineering Archives 26 (2020) 190-194. https://doi.org/10.30657/pea.2020.26.33

[36] M. Scendo et al., Purine as an effective corrosion inhibitor for stainless steel in chloride acid solutions, Corrosion Reviews 30 (2012) 33-45. https://doi.org/10.1515/CORRREV-2011-0039

[37] M. Scendo et al., Influence of laser power on the corrosive resistance of WC-Cu coating, Surf. Coat. Technol. 259 (2014) 401-407. https://doi.org/10.1016/j.surfcoat.2014.10.062

[38] T. Lipiński, J. Pietraszek, Corrosion of the S235JR Carbon Steel after Normalizing and Overheating Annealing in 2.5% Sulphuric Acid at Room Temperature, Materials Research Proceedings 24 (2022) 102-108. https://doi.org/10.21741/9781644902059-16

[39] M. Radek et al., Matching Computational Tools to User Competence Levels in Education of Engineering Data Processing, Materials Research Proceedings 34 (2023) 453-459. https://doi.org/10.21741/9781644902691-52

[40] A. Goroshko et al., Multiphysics Modeling of Direct-Start Induction Motor Considering Rotor Unbalance, Materials Research Proceedings 45 (2024) 267-276. https://doi.org/10.21741/9781644903315-31

[41] V. Royzman, A. Goroshko, Multiple inverse problem, Journal of Vibroengineering 14 (2012) 1417-1424.

[42] Ł. Łacny et al., Selected overview of the impact of ground motion on the vibrations of particle accelerators, AIP Conf. Proc. 2239 (2020) art. 020025. https://doi.org/10.1063/5.0008950

[43] A. Goroshko et al., Construction and practical application of hybrid statistically-determined models of multistage mechanical systems, Mechanika 20 (2014) 489-493. https://doi.org/10.5755/j01.mech.20.5.8221

[44] W. Jasiński et al., Static Mechanical Force Amplifier on the Example of a Fastener with an Electromagnetic Bolt, Materials Research Proceedings 34 (2023) 246-251. https://doi.org/10.21741/9781644902691-29

[45] J. Pietraszek, The modified sequential-binary approach for fuzzy operations on correlated assessments, Lecture Notes in Computer Science 7894 LNAI (2013) 353-364. https://doi.org/10.1007/978-3-642-38658-9_32

[46] A. Pacana et al., Pro-quality method improving industrial products toward sustainable development with criteria of circular economy, METAL 2023 - Int. Conf. Metall. Mater. (2024) 697-702. https://doi.org/10.37904/metal.2023.4743

[47] A. Pacana et al., Life Cycle Assessment (LCA) of Truck Steel Wheels, Materials Research Proceedings 45 (2024) 213-222. https://doi.org/10.21741/9781644903315-25

[48] D. Siwiec et al., Analysis of aluminum can by eco-costs and life cycle assessments (LCA), METAL 2024 – Int. Conf. Metall. Mater. (2024) 627-632. https://doi.org/10.37904/metal.2024.4975

[49] A. Pacana et al., Life Cycle Assessment (LCA) of Truck Steel Wheels, Materials Research Proceedings 45 (2024) 213-222. https://doi.org/10.21741/9781644903315-25

[50] R. Dwornicka, The impact of the power plant unit start-up scheme on the pollution load, Advanced Materials Research 874 (2014) 63-69. https://doi.org/10.4028/www.scientific.net/AMR.874.63

[51] M. Zenkiewicz et al., Electrostatic separation of binary mixtures of some biodegradable polymers and poly(vinyl chloride) or poly(ethylene terephthalate), Polimery/Polymers 61 (2016) 835-843. https://doi.org/10.14314/polimery.2016.835

[52] A. Pietrzak et al., Application of a Mixture of Fly Ash and Solid Waste from Gas Treatment from Municipal Solid Waste Incineration in Cement Mortar, Materials 18 (2025) art. 481. https://doi.org/10.3390/ma18030481

[53] R. Ulewicz et al., Quality and work safety in metal foundry, METAL 2020 – 29th Int. Conf. Metall. Mater. (2020) 1287-1293. https://doi.org/10.37904/metal.2020.3649

[54] K. Czerwinska et al., Improving quality control of siluminial castings used in the automotive industry, METAL 2020 – 29th Int. Conf. Metall. Mater. (2020) 1382-1387. https://doi.org/10.37904/metal.2020.3661

[55] R. Ulewicz et al., Logistic controlling processes and quality issues in a cast iron foundry, Materials Research Proceedings 17 (2020) 65-71. https://doi.org/10.21741/9781644901038-10

[56] R. Dwornicka et al., Development of a Hybrid Method for Identifying the Causes of Product Incompatibility in Metallurgical Manufacturing, Materials Research Proceedings 34 (2023) 199-206. https://doi.org/10.21741/9781644902691-24

[57] Ł.J. Orman et al., Comparative Analysis of Indoor Environmental Quality and Self-Reported Productivity in Intelligent and Traditional Buildings, Energies 16 (2023) art. 6663. https://doi.org/10.3390/en16186663

[58] E. Skrzypczak-Pietraszek, A. Hensel, E. Skrzypczak-Pietraszek, J. Pietraszek, Chemical profile and seasonal variation of phenolic acid content in bastard balm (Melittis melissophyllum L., Lamiaceae), Journal of Pharmaceutical and Biomedical Analysis 66 (2012) 154-161. https://doi.org/10.1016/j.jpba.2012.03.037, Pharmazie 55 (2000) 768-771.

[59] E. Skrzypczak-Pietraszek, Phytochemistry and biotechnology approaches of the genus Exacum, The Gentianaceae - Volume 2: Biotechnology and Applications (2015) 383-401. https://doi.org/10.1007/978-3-642-54102-5_16

Innovative Technical Aspects of Logistics in International Defense and Civil Cooperation

Ewa KOZIEŃ[1,a *], Paola KOŁODZIEJ[2,b]

[1]Krakow University of Economics, Cracow, Poland

[2]Military University of Technology, Warsaw, Poland

[a]koziene@uek.krakow.pl, [b]pkolodziej10@st.swps.edu.pl

Keywords: Logistics Innovation, Defense Cooperation, Interoperability, Digital Transformation, Supply Chain

Abstract. The article examines the innovative technological aspects of logistics within the framework of international cooperation between the civil and defense sectors. It examines key emerging technologies shaping contemporary logistics systems. The analysis also considers and addresses the challenges associated with system interoperability. The study emphasizes that the ongoing digital transformation of logistics, coupled with the increasing integration of civil and defense domains, fosters the development of resilient, adaptive, and sustainable supply chains, which constitute a cornerstone of security and stability in international economic and humanitarian operations.

Introduction

Modern supply chains are characterized by increasing complexity, an international scale, and a strong reliance on digital technologies. Logistics constitutes not only an element of economic infrastructure but also one of the pillars of global security. The ability to deliver resources quickly and accurately is crucial for both defense operations and crisis management in the civilian sector [1]. Progress in information technology, automation, and predictive systems has given rise to a new paradigm in logistics: one that is data-oriented, integrated, and responsive in real-time [2]. The objective of this paper is to present a review and synthesis of innovative technological solutions that facilitate international collaboration in civil and defense logistics.

The Role of Technological Innovations

Modern logistics increasingly relies on advanced digital technologies that streamline both the flow of materials and the exchange of information. The digitization of logistics processes today encompasses integrated ERP (Enterprise Resource Planning) and SCM (Supply Chain Management) systems, which are supported by predictive analytics and automated supply planning. These solutions enable flexible responses to changing market conditions, reduce the risk of downtime, and optimize resource utilization [3].

Equally important is artificial intelligence (AI), which is used for demand forecasting, risk analysis, and transport route optimization. Thanks to it, faster decision-making, reduced fuel costs, and better alignment of deliveries with customer needs are possible [4]. Another significant advancement is 3D printing, which enables the production of spare parts directly at the place of use. This solution shortens downtime and reduces inventory levels [1].

The next step toward intelligent logistics is the use of Digital Twins, virtual models of real-world processes and objects that enable the simulation of warehouse operations, transport networks, or distribution centers. They make it easier to optimize activities without causing disruptions in real-world operations [5,6]. Blockchain technology is also being increasingly used in logistics, ensuring the security and transparency of data and transportation documentation. It allows business partners to be confident about the authenticity of information and the history of transactions [7,8]. The implementation of modern technologies makes logistics more flexible, precise, and resilient to disruptions. Companies can respond more quickly to market changes, plan operations more effectively, and better meet customer needs. The development of these technologies also plays a crucial role in integrating defense and civilian logistics.

Integration of Civilian and Defense Logistics

The contemporary security environment is characterized by a strong interconnection between civilian and military structures. Both the defense and civilian sectors operate within shared transportation, energy, and information networks, which means that the effectiveness of logistical operations in one area often depends on the ability to cooperate with the other. As a result, the concept of civil-defense logistics integration has emerged, aiming to enhance the efficiency, resilience, and flexibility of the entire supply ecosystem [9]

From a supply chain management perspective, both defense and civilian logistics increasingly rely on the same routes and transport hubs. Seaports, airports, highways, and transshipment centers serve dual-use functions, being utilized both for economic activities and in crisis situations or defense operations [10]. The shared use of infrastructure reduces investment and operating costs and enables better spatial planning.

One of the key challenges of such integration is coordinating planning and maintaining capacity. Reports by the European Commission [11] emphasize that for transport infrastructure to also serve security purposes, it must be designed to allow rapid traffic rerouting, flexible slot allocation, and compliance with the technical requirements of military and emergency systems.

The second dimension of integration involves the transfer of technology and knowledge sharing across sectors. In recent years, numerous technologies originally developed for defense purposes, such as satellite monitoring, real-time data analytics, and AI-based predictive systems, have been successfully applied in civilian logistics [12]. At the same time, civilian innovations, including ERP/SCM systems, cloud-based solutions, and the Internet of Things (IoT), have been adapted for use in operational settings, enhancing resource management and improving the transparency of supply flows [1]. A notable example of dual-use technology is 3D printing, which enables the production of spare parts directly at the site of operations. In the defense environment, it reduces equipment repair time and increases the autonomy of units, while in the civilian sector, it supports the continuity of supply chains during crises. As noted by Sahoo [7], the use of additive technologies in logistics enables flexible responses to disruptions and reduces the need for large inventories, thereby promoting both efficiency and sustainability in transportation processes.

This integration has not only a technical but also an organizational dimension. Joint training sessions, simulations, and operational centers help build trust and interoperability between institutions that have previously operated within different decision-making structures. As noted by García-Pérez [13], data exchange and the standardization of logistical processes contribute to reducing the risk of errors in crisis situations, where the speed of decision-making is crucial.

The integration of civilian and defense logistics is a complex, multifaceted process that entails the joint use of infrastructure, technology, data, and human resources. Its effectiveness relies on adopting unified technical and organizational standards, building mutual trust among partners, and pursuing a long-term strategy of cross-sector collaboration. The future of international logistics lies in cooperative models where the distinction between civilian and defense sectors becomes increasingly blurred, yet transparency and process security remain fully preserved.

Challenges and Benefits of Integration

Despite its numerous advantages, the integration of civilian and defense logistics faces several technical, organizational, and legal challenges. A key challenge lies in the differences between technical and procedural standards applied by entities from different sectors and countries. Complex supply chains often rely on varying interoperability frameworks, which makes it difficult to achieve seamless integration of data and operational processes. Another significant obstacle concerns the protection of information, particularly in systems where civilian and military data streams intersect. Limitations on information access and differing classification can further impede the real-time exchange of data [14].

The third challenge lies in the absence, in many countries, of a unified and secure platform for data exchange among public, private, and defense institutions. This gap results in duplicated efforts, making coordination during crises more difficult. At the same time, striving for both economic efficiency and a high level of security calls for carefully balanced organizational and technical solutions. As Wasielewska-Marszałkowska [6] points out, effective public–private partnership (PPP) in logistics can only be achieved when transparent models of cooperation are established that reconcile economic objectives with the state's security priorities.

Despite the aforementioned barriers, integrating civilian and defense logistics brings numerous economic and strategic benefits. It enhances the resilience of supply chains through mutual support of resources and infrastructure, as well as more efficient use of human and technical potential. In crisis situations, integrated logistics systems enable faster responses to humanitarian, climate, and energy-related disasters, ensuring the continuity of transport and real-time delivery. On an international scale, shared digital and procedural solutions contribute to strengthening coalition capabilities and improving the coordination of stabilization missions [5].

Integration also facilitates the adoption of sustainable logistics models built on digital data exchange platforms, enabling the reduction of environmental impact and enhancing the transparency of supply chain flows [15]. According to Longo [5], integrated logistics networks powered by digital twins can serve as a cornerstone for developing resilient, adaptive, and sustainable logistics systems of the future, where the lines between the civil and defense sectors increasingly fade in favor of a shared objective, ensuring the security and stability of global supply chains.

Social, Technological, and Environmental Impact

The development of logistics has a significant impact on the organization of work in various industries [16,17], improving not only supply chains but also influencing the organization of production streams in such advanced technologies as fire protection coatings [18], special anti-wear coatings [19-21], and anti-corrosion protection [22]. These technologies are an area of effective application of modern research methods in the field of surface engineering [23,24], process optimization [25,26], and non-parametric data analysis methods [27-29], including the particularly important yet challenging issue of estimating forecast uncertainty [30].

The complexity of supply chains and their sensitivity to communication disruptions motivate the need to increase equipment reliability, both from a purely structural perspective [31-33] and from a material perspective [34,35]. This provides an impetus for the development of new tools to support engineers [36] and related methods, including fuzzy description of incomplete data [37,38]. This also inevitably leads to changes in related areas, such as foundry [39], welding [40], and energy [41-43]. This reduces industrial pressure on the environment, enabling wider use of remediation to rehabilitate damaged areas [44] and the use of natural sources of valuable substances, among others, for the pharmaceutical industry [45,46].

Conclusion

Modern logistics is no longer just a support function; it has become a strategic capability. Digital technologies now enable faster response times, greater efficiency, and stronger collaboration between defense and civilian sectors. At the same time, interoperability, cybersecurity, and sustainability form the foundation of tomorrow's logistics systems.

In the years ahead, the integration of automation, data analytics, and eco-friendly technologies will shape the next era of logistics. These innovations will not only enhance efficiency but also strengthen the world's ability to respond to crises, ensuring that logistics truly serve both security and humanity.

References

[1] N. Altay et al. Innovation in humanitarian logistics and supply chain management: A systematic review. Ann. Oper. Res. 335(3) (2024) 965-987. https://doi.org/10.1007/s10479-023-05208-6

[2] I. Menache et al. How generative AI improves supply chain management. Harvard Business Review (2025) https://hbr.org/2025/01/how-generative-ai-improves-supply-chain-management

[3] J. Ahn, E. Huh, S. Choi. Oliot EPCIS: An open-source EPCIS 2.0 system for supply chain traceability. J. Ind. Inf. Integration 32 (2023) art.100451. https://doi.org/10.1016/j.jii.2023.100451

[4] M. Queiroz et al. Blockchain and AI for sustainable logistics. International Journal of Production Economics 247 (2022) art. 108400. https://doi.org/10.1016/j.ijpe.2021.108400

[5] F. Longo et al. The digital supply chain twin paradigm for enhancing resilience and sustainability against COVID-like crises. Procedia Computer Science 217 (2023) 1940-1947. https://doi.org/10.1016/j.procs.2022.12.394

[6] S. Werbińska-Wojciechowska, J. Żurek, K. Nadolny. Digital Twin for operation and maintenance of transport systems. App. Sci. 14 (2024) 3271. https://doi.org/10.3390/app14073271

[7] S. Sahoo, R.K. Jena, S.K. Patel. Blockchain for sustainable supply chain management. Electronic Commerce Research 24 (2024) 33-58. https://doi.org/10.1007/s10660-023-09657-6

[8] P.S. Gandhi Maniam, J.J. Hew, A.W. Kassim (2024). Determinants of Blockchain Adoption in the Australian Agricultural Supply Chain: A Systematic Literature Review. Sustainability 16 (2024) art. 2338. https://doi.org/10.3390/su16052338

[9] B. Kucukaltan, Z. Irani, E. Aktas. Business model canvas for humanitarian operations of logistics service providers. Production Planning & Control 32(11) (2020) 917-932. https://doi.org/10.1080/09537287.2020.1821679

[10] European Climate, Infrastructure and Environment Executive Agency. (2024). Connecting Europe Facility (CEF): Progress report 2024. Publications Office of the European Union. https://cinea.ec.europa.eu/programmes/connecting-europe-facility_en

[11] European Commission, Directorate-General for Mobility and Transport. (2024). Management plan 2024: Directorate-General for Mobility and Transport (DG MOVE). Publications Office of the European Union. https://commission.europa.eu/document/download/9184454a-73fd-4f00 -b9ea-a925a9f532ce_en

[12] T. Lackner et al. Review of autonomous mobile robots in intralogistics. Journal of Manufacturing Systems 70 (2024) 450-464. https://doi.org/10.1016/j.jmsy.2023.09.004

[13] A. García-Pérez et al. Transforming the European transport ecosystem through data-driven operations: interoperability challenges and requirements. Lecture Notes in Intell. Transp. Infrastr. 417 (2025) 417-431. https://doi.org/10.1007/978-3-031-82818-8_31

[14] ENISA. (2023). Good Practices for Supply Chain Cybersecurity. European Union Agency for Cybersecurity. https://www.enisa.europa.eu/publications

[15] R. Dubey, A. Gunasekaran, D.J. Bryde. AI and sustainability in logistics networks. Journal of Cleaner Production 374 (2022) 133-144. https://doi.org/10.1016/j.jclepro.2022.133144

[16] R. Ulewicz, Quality management system operation in the woodworking industry, The Path Forward for Wood Products: A Global Perspective – Proc. Sci. Papers (2016) 51-56.

[17] D. Klimecka-Tatar, R. Dwornicka, The assembly process stability assessment based on the strength parameters statistical control of complex metal products, METAL 2019 – 28th Int. Conf. Metall. Mater. (2019) 709-714.

[18] N. Radek, R. Dwornicka, Fire properties of intumescent coating systems for the rolling stock, Communications – Scientific Letters of the University of Žilina 22 (2020) 90-96. https://doi.org/10.26552/com.C.2020.4.90-96

[19] N. Radek, R. Dwornicka and D. Gontarski, The impact of laser processing on the performance properties of electro-spark coatings, WCCM 2021 and ECCOMAS Congress 1000 (2021) 1-10. https://doi.org/10.23967/wccm-eccomas.2020.336

[20] N. Radek et al., Operational properties of DLC coatings and their potential application, METAL 2022 – 31st Int. Conf. Metall. Mater., (2022) 531-536. https://doi.org/10.37904/metal.2022.4491

[21] A. Kalinowski et al., Laser surface texturing: characteristics and applications, System Safety: Human-Technical Facility-Environment 5 (2023) 240-248. https://doi.org/10.2478/czoto-2023-0026

[22] T. Lipiński, J. Pietraszek, Corrosion of the S235JR Carbon Steel after Normalizing and Overheating Annealing in 2.5% Sulphuric Acid at Room Temperature, Materials Research Proceedings 24 (2022) 102-108. https://doi.org/10.21741/9781644902059-16

[23] D. Nowakowski, A. Gądek-Moszczak and P. Lempa, Application of machine learning in the analysis of surface quality – the detection the surface layer damage of the vehicle body, METAL 2021 – 30th Int. Conf. Metall. Mater. (2021) 864-869. https://doi.org/10.37904/metal.2021.4210

[24] A. Gądek-Moszczak et al., Nano X-ray Tomography Application for Quantitative Surface Layer Geometry Analysis after Laser Beam Modification, Materials 15 (2022) art. 5935. https://doi.org/10.3390/ma15175935

[25] J. Pietraszek, A. Goroshko, The heuristic approach to the selection of experimental design, model and valid pre-processing transformation of DoE outcome, Advanced Materials Research 874 (2014) 145-149. https://doi.org/10.4028/www.scientific.net/AMR.874.145

[26] J. Pietraszek, N. Radek and A.V. Goroshko, Challenges for the DOE methodology related to the introduction of Industry 4.0, Production Engineering Archives 26 (2020) 190-194. https://doi.org/10.30657/pea.2020.26.33

[27] J. Pietraszek, A. Gadek-Moszczak, The smooth bootstrap approach to the distribution of a shape in the ferritic stainless steel AISI 434l powders, Solid State Phenomena 197 (2013) 162-167. https://doi.org/10.4028/www.scientific.net/SSP.197.162

[28] A. Szczotok et al., The bootstrap analysis of one-way ANOVA stability in the case of the ceramic shell mould of airfoil blade casting, Solid State Phenomena 235 (2015) 24-30. https://doi.org/10.4028/www.scientific.net/SSP.235.24

[29] R. Dwornicka, A. Goroshko and J. Pietraszek, The smoothed bootstrap fine-tuning, System Safety: Human-Technical Facility-Environment 1 (2019) 716-723. https://doi.org/10.2478/czoto-2019-0091

[30] J. Pietraszek et al., The non-parametric approach to the quantification of the uncertainty in the design of experiments modelling, UNCECOMP 2017 – Proc. 2nd Int. Conf. on Uncertainty Quantification in Comp. Sci. Eng. (2017) 598-604. https://doi.org/10.7712/120217.5395.17225

[31] A. Goroshko, V. Royzman and J. Pietraszek, Construction and practical application of hybrid statistically-determined models of multistage mechanical systems, Mechanika 20 (2014) 489-493. https://doi.org/10.5755/j01.mech.20.5.8221

[32] M. Krynke, R. Ulewicz, Analysis of the influence of slewing bearing mounting on their static load capacity, Transp. Res. Proc. 40 (2019) 745-750. https://doi.org/10.1016/j.trpro.2019.07.105

[33] A. Goroshko, R. Dwornicka and M. Zembytska, Multiphysics Modeling of Direct-Start Induction Motor Considering Rotor Unbalance, Materials Research Proceedings 45 (2024) 267-276. https://doi.org/10.21741/9781644903315-31

[34] N. Radek et al., The effect of laser treatment on operational properties of ESD coatings, METAL 2021 – 30th Int. Conf. Metall. Mater. (2021) 876-882. https://doi.org/10.37904/metal.2021.4212

[35] N. Radek et al., Technology and applications of ESD coatings before and after laser processing, METAL 2023 – Int. Conf. Metall. Mater. (2024) 500-505. https://doi.org/10.37904/metal.2023.4727

[36] R. Dwornicka, J. Pietraszek, The outline of the expert system for the design of experiment, Production Engineering Archives 20 (2018) 43-48. https://doi.org/10.30657/pea.2018.20.09

[37] J. Pietraszek, Fuzzy regression compared to classical experimental design in the case of flywheel assembly, Lecture Notes in Computer Science 7267 LNAI (2012) 310-317. https://doi.org/10.1007/978-3-642-29347-4_36

[38] J. Pietraszek, The modified sequential-binary approach for fuzzy operations on correlated assessments, Lecture Notes in Computer Science 7894 LNAI (2013) 353-364. https://doi.org/10.1007/978-3-642-38658-9_32

[39] R. Ulewicz, K. Czerwińska and A. Pacana, A Rank Model of Casting Non-Conformity Detection Methods in the Context of Industry 4.0, Materials 16 (2023) art. 723. https://doi.org/10.3390/ma16020723

[40] N. Radek et al., Properties of Steel Welded with CO_2 Laser, Lecture Notes in Mechanical Engineering (2020) 571-580. https://doi.org/10.1007/978-3-030-33146-7_65

[41] R. Ulewicz, B. Krstić and M. Ingaldi, Mining Industry 4.0 – Opportunities and Barriers, Acta Montanistica Slovaca 27 (2022) 291-305. https://doi.org/10.46544/AMS.v2i2.02

[42] J.M. Djoković et al., Selection of the Optimal Window Type and Orientation for the Two Cities in Serbia and One in Slovakia, Energies 15 (2022) art. 323. https://doi.org/10.3390/en15010323

[43] Ł.J. Orman et al., Application of laser treatment technology for boiling heat transfer augmentation, Production Engineering Archives 30 (2024) 259-265. https://doi.org/10.30657/pea.2024.30.25

[44] E. Skrzypczak-Pietraszek, A.G. Pietraszek, Phytoremediation and Allelopathy as an Element of Sustainable Circular Economy to Prevent Environmental Pollution, Materials Research Proceedings 45 (2024) 140-145. https://doi.org/10.21741/9781644903315-17

[45] E. Skrzypczak-Pietraszek, J. Pietraszek, Chemical profile and seasonal variation of phenolic acid content in bastard balm (Melittis melissophyllum L., Lamiaceae), Journal of Pharmaceutical and Biomedical Analysis 66 (2012) 154-161. https://doi.org/10.1016/j.jpba.2012.03.037

[46] E. Skrzypczak-Pietraszek, J. Pietraszek, Seasonal changes of flavonoid content in Melittis melissophyllum L. (Lamiaceae), Chemistry and Biodiversity 11 (2014) 562-570. https://doi.org/10.1002/cbdv.201300148

Materials Research Proceedings 62 (2026) 359-366 https://doi.org/10.21741/9781644904015-46

Human Factors, Technical Culture, and Training for Reliable Machinery Operation in Industry 5.0: An Integrative Framework

Manuela INGALDI[1,a *]

[1]Department of Production Engineering and Safety, Faculty of Management, Czestochowa University of Technology, Al. Armii Krajowej 19b, 42-218 Czestochowa, Poland

[a]manuela.ingaldi@pcz.pl

Keywords: Human Factor, Machinery Reliability, Operator Training, Industry 5.0, Maintenance

Abstract. Contemporary production systems are characterized by a high degree of automation and increasing complexity, which makes machine reliability dependent not only on technology but also on the competencies and behaviors of personnel. This article aims to analyze the role of the human factor in machine operation, as well as the significance of technical culture and training systems in shaping appropriate operator behaviors. The concept of a socio-technical system is presented, illustrating the interaction between humans, the organization, and the work environment, and its impact on machine reliability. Attention is given to the development of technical competencies through both traditional and modern training methods, including digital tools and VR/AR technologies, and their role in reducing human errors and enhancing operational safety. The novelty of the article lies in its integration of humanistic and technical approaches, highlighting system reliability as a result of the synergy between humans, technology, and organization within the context of Industry 5.0.

Introduction

Contemporary production systems are characterized by increasing complexity, a high degree of automation, and extensive use of digital tools. The ongoing technological development is expected to further enhance this diversity, which is typical of the Industry 4.0 and 5.0 paradigms. Companies can invest in increasingly advanced technical solutions to improve the efficiency of both their machinery and processes [1-3]. Recent perspectives emphasize that it is precisely AI that is redefining leadership competencies in 4.0/5.0 environments (e.g., co-decision-making with algorithms, agile communication, ethical oversight), which in turn contributes to operational effectiveness [4].

It should be noted that machine reliability and the safety of operational processes, including the most advanced ones, depend not only on the quality of technology but also on the competencies and behaviors of personnel [5,6]. Human factors, including knowledge, experience, risk awareness, and safety attitudes, remain a key element of machine operation systems, however operator errors continue to be a significant cause of failures and downtime. The literature emphasizes the importance of technical culture and training as tools to mitigate the impact of human errors and enhance operational efficiency [7].

However, in the era of Industry 5.0, where humans act as active co-creators of intelligent systems, an integrative perspective combining technical and humanistic aspects is required. Although automation and robotics have shifted most operations to machines, humans remain responsible for setting machine parameters, coordinating operations, and maintaining continuous functionality, which requires specific skills and competencies that have evolved significantly over time.

This study aims to present the significance of technical culture and the development of technical competencies in the context of machine operation. This is particularly important in the Industry

Materials Research Proceedings 62 (2026) 359-366 https://doi.org/10.21741/9781644904015-46

5.0 era, where humans become co-creators rather than mere users of technical systems. The article aims to demonstrate how a properly cultivated technical culture and training system influence employee behavior, reduce human errors, and enhance machine reliability. The novelty of this work lies in integrating the humanistic approach, including attitudes, responsibility, and the learning process of employees, with the technical aspect of machine operation. The proposed framework enables the assessment of technical system reliability, resulting from the synergy between humans, technology, and organization, while taking into account contemporary challenges related to digitalization, automation, and intelligent production systems characteristic of Industry 5.0.

The Role of the Human Factor in Machine Operation

The human factor remains one of the key elements determining the reliability of technical systems. Despite the increasing level of automation and the use of advanced digital technologies, most operational processes still require the involvement of operators, technicians, and maintenance engineers. Their competencies, experience, risk awareness, and safety attitudes directly influence machine performance and the occurrence of failures.

The literature indicates that as much as 60–80% of operational problems result from human errors, including, among others [8,9]:

- improper equipment handling,
- neglecting maintenance procedures,
- inadequate reporting of anomalies,
- routine actions leading to accelerated wear of technical components.

James Reason's "Swiss Cheese" model [10-12] effectively illustrates the phenomenon of human errors in technical systems. Failures occur only when multiple layers of defense (procedures, instructions, supervision, training) fail to function effectively. In the context of machine operation, this means that properly implemented procedures and training systems can significantly mitigate the impact of operator errors on equipment reliability.

The human factor should be considered within a socio-technical system, where the interaction between humans, machines, the organization, and the work environment determines operational efficiency. Appropriate employee behaviors, stemming from a strong technical culture and continuous competency development, are crucial to maintaining machine uptime, minimizing failures, and enhancing the safety of technological processes.

In this context, activities aimed at developing competencies, standardizing work, and fostering operators' technical awareness are of fundamental importance for the efficient and safe operation of technical systems in modern industry.

Technical Culture in Maintenance

Technical culture is a key element of machine operation systems, influencing the attitudes, behaviors, and decisions of personnel responsible for equipment operation, maintenance, and supervision [4, 13, 14]. It encompasses not only technical knowledge and experience but also adherence to procedures, risk awareness, and responsibility for the proper functioning of the technical system.

As illustrated in Fig.1, machine operation should be viewed as a socio-technical system, in which humans, the organization, and the work environment interact to shape operational processes and influence machine reliability. The human factor is central in this system, linking organizational procedures with machine technology, and its behaviors determine the effectiveness of the entire system.

Materials Research Proceedings 62 (2026) 359-366 https://doi.org/10.21741/9781644904015-46

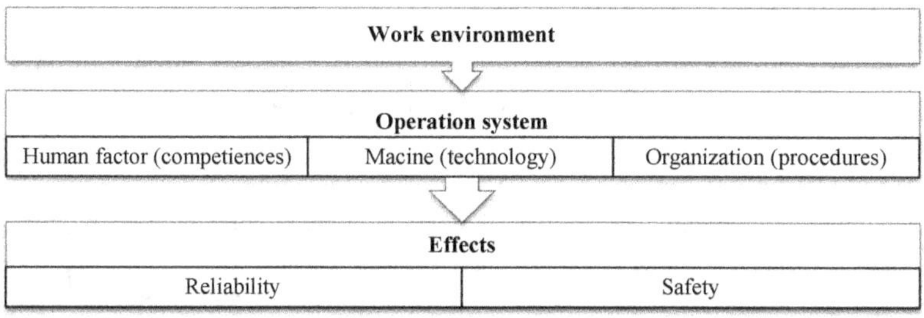

Fig. 1. The Socio-Technical System of Machine Operation [own study]

In practice, a strong technical culture translates into appropriate employee behaviors, as illustrated in Figure 2. The figure shows a direct relationship: the stronger the technical culture within an organization, the more conscious and responsible the personnel's actions, ultimately enhancing the reliability of technical systems. The implementation of work-standardizing procedures, systematic training, and practices that reinforce technical culture thus represents an investment in both the safety and efficiency of machine operation. Incorporating the 'voice of the customer' into procedures (e.g., through requirement classification using the Kano method) promotes alignment between product design and operational practices, reducing the risk of user errors and complaints [15].

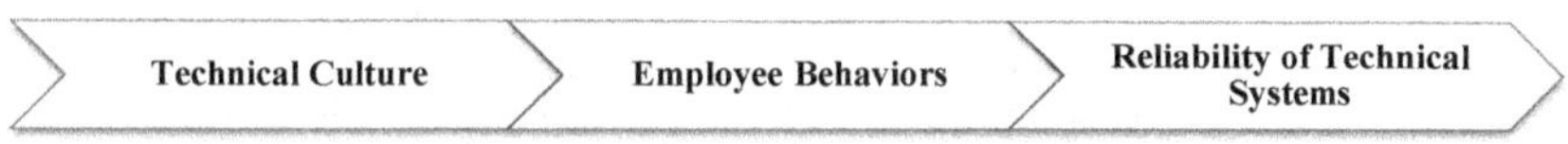

Fig. 2. The Relationship between Technical Culture, Employee Behaviors, and the Reliability of Technical Systems [own study]

In modern industry, particularly within the context of Industry 5.0, technical culture extends beyond mere adherence to procedures. It also encompasses collaboration with intelligent production systems, data-driven decision-making, and active participation in the improvement of maintenance processes. This approach enables synergistic integration of human and technological potential to enhance machine reliability and process safety. A strong technical culture, which encompasses quality awareness, responsibility, and care for the technical environment, fosters appropriate employee behaviors, such as adherence to procedures, timely reporting of anomalies, and proactive preventive maintenance. Consequently, the reliability of the entire machine operation system is improved.

Training and Development of Technical Competencies
The systematic development of employees' technical competencies is one of the key elements that strengthen technical culture and enhance the reliability of the machinery operated. Training provides both practical and theoretical knowledge necessary for the proper operation of equipment, adherence to maintenance procedures, and effective response to unforeseen situations in operational processes.

The contemporary approach to industrial training has evolved from traditional forms of instruction, including lectures, paper-based manuals, and periodic practical exercises, towards modern methods that utilize digital and immersive technologies, such as simulators, virtual or augmented reality (VR/AR), and e-learning platforms. These solutions enable more intensive and safer acquisition of practical experience, as well as real-time monitoring of progress and the adaptation of training paths to the individual needs of employees.

Table 1 presents a comparison of traditional and modern training methods, illustrating the shift in approaches to developing technical competencies in the context of contemporary industry.

Table 1. Comparison of Traditional and Modern Training Methods for Technical Personnel [16-21]

Criterion	Traditional Training	Modern Training (Industry 4.0/5.0)
Delivery Method	Lectures, presentations, paper-based manuals	Simulators, VR/AR, e-learning, interactive applications
Location	Training room or workplace	Any location with access to digital systems
Frequency	Periodic (e.g., once a year)	Continuous, needs-based (on-demand learning)
Participant Engagement	Low to moderate – predominance of one-way instruction	High – learning through experience, simulations, gamification
Knowledge Retention Effectiveness	Limited, dependent on the instructor	High, due to visualization and immediate feedback
Performance Assessment	Knowledge tests, observation	Real-time data analysis, individualized development paths

Modern training methods also facilitate the development of soft skills, such as communication, teamwork, and decision-making in high-risk situations, which are essential in highly automated operational systems. Integrating technical education with the development of interpersonal skills fosters synergy between humans, technology, and the organization, thereby enhancing machine reliability and workplace safety.

Economic, Technological, Social, and Environmental Impact

Coupling human work with a supporting robot [22,23], i.e., a cobot, offers great opportunities to increase efficiency and improve quality [24-26], but also introduces enormous requirements in terms of resource management [27-29] and occupational health and safety [30-32]. It will undoubtedly increase efficiency in areas such as foundries [33], special DLC [34] and ESD [35,36] coatings modified by laser micromachining [37], and welding [38,39], thereby reducing the risk of workplace accidents. Such advanced technologies require the intensive implementation of process optimization methods [40], supported by resampling [41] and incorporating incomplete information [42, 43]. At the same time, it will significantly reduce the volume of industrial waste [44,45], enabling the implementation of biological waste remediation techniques [46] and the extraction of natural resources [47]. This aligns with the guidelines of the circular economy [48-50] and actions for social well-being [51].

Conclusions

Human factors remain critical for the reliability and safety of machine operation. An analysis of the socio-technical system demonstrates that employees' attitudes, knowledge, and behaviors have a substantial impact on the efficiency of technical processes.

A strong technical culture and systematic training form the foundation for shaping appropriate employee behaviors, reducing errors, and facilitating the integration of personnel with modern production systems characteristic of Industry 5.0.

Conclusions drawn from this study:

1. Human factors significantly influence the reliability of technical systems and should be a central element of maintenance strategies.
2. Technical culture and training are crucial for the proper behavior of operators and technicians.
3. Integrating human competencies with modern technologies enhances machine efficiency, safety, and reliability.

In summary, effective machine operation necessitates a comprehensive approach that integrates technical, organizational, and human elements. The development of technical culture and employee competencies represents an investment in the safety and performance of production systems.

References

[1] I. Balahurovska, The Role of Leadership in Industry 4.0. System Safety: Human - Technical Facility – Environment 5 (2023) 66-74. https://doi.org/10.2478/czoto-2023-0008

[2] G. Fragapane et al., Increasing flexibility and productivity in Industry 4.0 production networks with autonomous. mobile robots and smart intralogistics, Annals of Operations 308 (2022) 125-143. https://doi.org/10.1007/s10479-020-03526-7

[3] B. Chidozie et al., The Importance of Digital Transformation (5.0) in Supply Chain Optimization: An Empirical Study, Production Engineering Archives 30 (2024) 127-135. https://doi.org/10.30657/pea.2024.30.12

[4] M. Ingaldi, R. Ulewicz, The Role of AI in Digital Leadership – New Competencies of Leaders. Polish J. Manag. Stud. 31 (2025) 106-122. https://doi.org/10.17512/pjms.2025.31.2.07

[5] K. Czerwińska, A. Pacana, Model of Human Resource Management in Manufacturing Companies Using Key Performance Indicators (KPIs), System Safety: Human - Technical Facility - Environment 6 (2024) 69-78. https://doi.org/10.2478/czoto-2024-0009

[6] A. Kuzior et al., Sustainable consumption and production systems during wartime: needs, challenges, and adaptation strategies, Production Engineering Archives 31 (2025) 212-221. https://doi.org/10.30657/pea.2025.31.21

[7] A. Nirmala et al., Efficiency of artificial intelligence usage on personalized and adaptivity training method for employees at workplace, Polish J. Manag. Stud. 31 (2025) 213-231. https://doi.org/ 10.17512/pjms.2025.31.1.12

[8] M. Krynke, A. Vaško, Management the Safety of Exploitation of the Rotation Mechanism in a Self-Propelled Crane, System Safety: Human - Technical Facility - Environment, 1(1) (2019) 624-631. https://doi.org/10.2478/czoto-2019-0079

[9] K. Knop, Managing Constraints in a Machinery Company Using Theory of Constraints' Thinking Process Tools. Case Study, Management Systems in Production Engineering, 33(3), (2025) 401-413. https://doi.org/10.2478/mspe-2025-0039

[10] T. Diller et al., The human factors analysis classification system (HFACS) applied to health care, American J. Med. Qual. 29 (2013) 181-190. https://doi.org/10.1177/1062860613491623

[11] J. Reason, Human error: Models and management, The BMJ 320 (2000) 768-770. https://doi.org/10.1136/bmj.320.7237.768

[12] T.V. Perneger, The Swiss cheese model of safety incidents: are there holes in the metaphor?, BMC Health Serv Res 5 (2005) art.71. https://doi.org/10.1186/1472-6963-5-71

[13] M. Quintanilla, Tolerance and Technological Culture, Philosophica 66 (2000) 65-71. https://doi.org/10.21825/philosophica.82270

[14] B. Godin, Y. Gingras, What is scientific and technological culture and how is it measured? A multidimensional model, Public Understanding of Science 9 (2000) 43-58. https://doi.org/10.1088/0963-6625/9/1/303

[15] A. Szczyrba, M. Szataniak, Decoding Consumer Preferences in Food Packaging with the Kano Model, System Safety: Human - Technical Facility - Environment 5 (2023) 83-92. https://doi.org/10.2478/czoto-2023-0010

[16] J. Wołowczyk, D. Klimecka-Tatar, Safety of Mountaineering Systems in the Industrial Area, Materials Research Proceedings 34 (2023) 53-61. https://doi.org/10.21741/9781644902691-7

[17] M. Grebski, M. Mazur, Management Strategies To Avoid Professional Burnout, Polish J. Manag. Stud. 26 (2022) 61-75. https://doi.org/10.17512/pjms.2022.26.1.04

[18] R. Prusak, E. Kardas, The Analysis of Mutual Relations and Dependencies of Intellectual Capital Components in Polish Enterprises in the Context of Sustainability Goals, Sustainability 16 (2024) art.7994. https://doi.org/10.3390/su16187994

[19] R.F. Bernardes et al., Millennial Attitudes Toward Online and Traditional Training Methods: The Role of Training Utility and Satisfaction, Cornell Hospitality Quarterly 60 (2019) 320-334. https://doi.org/10.1177/1938965519843400

[20] M.R. Chaudhur et al., Emergence of training and development in organisations - A historical perspective, New Applied Studies in Management, Economics & Accounting, 5(1(17)) (2022) 7-15. https://doi.org/10.22034/nasmea.2022.176185

[21] A.B. Samuels, A.-M. Pelser, Transitioning from Industry 4.0 to 5.0: Sustainable supply chain management and talent management insights, SA Journal of Human Resource Management 23 (2025) art. a2874. https://doi.org/10.4102/sajhrm.v23i0.2874

[22] J.E. Takosoglu, P.A. Laski, Intelligent positioning system of electro-pneumatic servo-drive, Annals of DAAAM and Proceedings of the International DAAAM Symposium (2011) 1641-1642.

[23] R. Ulewicz, M. Mazur, Economic aspects of robotization of production processes by example of a car semi-trailers manufacturer, Manufacturing Technology 19 (2019) 1054-1059. https://doi.org/10.21062/ujep/408.2019/a/1213-2489/MT/19/6/1054

[24] R. Ulewicz, Quality management system operation in the woodworking industry, The Path Forward for Wood Products: A Global Perspective - Proceedings of Scientific Papers (2016) 51-56.

[25] K. Czerwinska et al., Improving quality control of siluminial castings used in the automotive industry, METAL 2020 - 29th Int. Conf. Metall. Mater. (2020) 1382-1387. https://doi.org/10.37904/metal.2020.3661

[26] D. Siwiec et al., Concept of a model to predict the qualitative-cost level considering customers' expectations, Polish Journal of Management Studies 26 (2022) 330-340. https://doi.org/10.17512/pjms.2022.26.2.20

[27] K. Kowalik, D. Klimecka-Tatar, The process approach to service quality management, Production Engineering Archives 18 (2018) 31-34. https://doi.org/10.30657/pea.2018.18.05

[28] A. Kozień, Efficient management of cultural heritage by local government bodies, Wiadomosci Konserwatorskie 2020 (2020) 7-16. https://doi.org/10.48234/WK64HERITAGE

[29] N. Baryshnikova, O. Kiriliuk and D. Klimecka-Tatar, Management approach on food export expansion in the conditions of limited internal demand, Polish Journal of Management Studies 21 (2020) 101-114. https://doi.org/10.17512/pjms.2020.21.2.08

[30] E. Kozień, A. Kozień, Institutional aspects of health and safety at work in poland, System Safety: Human - Technical Facility - Environment 1 (2019) 240-247. https://doi.org/10.2478/czoto-2019-0031

[31] R. Ulewicz et al., Quality and work safety in metal foundry, METAL 2020 – 29th Int. Conf. Metall. Mater. (2020) 1287-1293. https://doi.org/10.37904/metal.2020.3649

[32] R. Caban et al., Work safety during the injection molding process of polymer materials, System Safety: Human - Technical Facility - Environment 6 (2024) 320-329. https://doi.org/10.2478/czoto-2024-0034

[33] R. Ulewicz et al., A Rank Model of Casting Non-Conformity Detection Methods in the Context of Industry 4.0, Materials 16 (2023) art. 723. https://doi.org/10.3390/ma16020723

[34] N. Radek et al., Operational properties of DLC coatings and their potential application, METAL 2022 – 31st Int. Conf. Metall. Mater., (2022) 531-536. https://doi.org/10.37904/metal.2022.4491

[35] N. Radek et al., The effect of laser treatment on operational properties of ESD coatings, METAL 2021 – 30th Int. Conf. Metall. Mater. (2021) 876-882. https://doi.org/10.37904/metal.2021.4212

[36] N. Radek et al., Technology and applications of ESD coatings before and after laser processing, METAL 2023 – 32nd Int. Conf. Metall. Mater. (2024) 500-505. https://doi.org/10.37904/metal.2023.4727

[37] A. Kalinowski et al., The effect of laser infrared radiation on the surface modification of DLC coatings, AIP Conf. Proc. 3130 (2024) art. 020014. https://doi.org/10.1063/5.0204985

[38] M. Patek et al., Non-destructive testing of split sleeve welds by the ultrasonic TOFD method, Manufacturing Technology 14 (2014) 403-407.

[39] N. Radek et al., Properties of Steel Welded with CO_2 Laser, Lecture Notes in Mechanical Engineering (2020) 571-580. https://doi.org/10.1007/978-3-030-33146-7_65

[40] J. Pietraszek et al., Challenges for the DOE methodology related to the introduction of Industry 4.0, Production Engineering Archives 26 (2020) 190-194. https://doi.org/10.30657/pea.2020.26.33

[41] J. Pietraszek, A. Gadek-Moszczak, The smooth bootstrap approach to the distribution of a shape in the ferritic stainless steel AISI 434L powders, Solid State Phenomena 197 (2013) 162-167. https://doi.org/10.4028/www.scientific.net/SSP.197.162

[42] J. Pietraszek, Fuzzy regression compared to classical experimental design in the case of flywheel assembly, Lecture Notes in Computer Science 7267 LNAI (2012) 310-317. https://doi.org/10.1007/978-3-642-29347-4_36

[43] J. Pietraszek, The modified sequential-binary approach for fuzzy operations on correlated assessments, Lecture Notes in Computer Science 7894 LNAI (2013) 353-364. https://doi.org/10.1007/978-3-642-38658-9_32

[44] M. Ulewicz, A. Pietrzak, Properties and structure of concretes doped with production waste of thermoplastic elastomers from the production of car floor mats, Materials 14 (2021) art. 872. https://doi.org/10.3390/ma14040872

[45] A. Pietrzak et al., Application of a Mixture of Fly Ash and Solid Waste from Gas Treatment from Municipal Solid Waste Incineration in Cement Mortar, Materials 18 (2025) art. 481. https://doi.org/10.3390/ma18030481

[46] E. Skrzypczak-Pietraszek, A.G. Pietraszek, Phytoremediation and Allelopathy as an Element of Sustainable Circular Economy to Prevent Environmental Pollution, Materials Research Proceedings 45 (2024) 140-145. https://doi.org/10.21741/9781644903315-17

[47] E. Skrzypczak-Pietraszek, J. Pietraszek, Chemical profile and seasonal variation of phenolic acid content in bastard balm (Melittis melissophyllum L., Lamiaceae), Journal of Pharmaceutical and Biomedical Analysis 66 (2012) 154-161. https://doi.org/10.1016/j.jpba.2012.03.037

[48] A. Pacana, D. Siwiec, Method of Fuzzy Analysis of Qualitative-Environmental Threat in Improving Products and Processes (Fuzzy QE-FMEA), Materials 16 (2023) art. 1651. https://doi.org/10.3390/ma16041651

[49] A. Pacana et al., Pro-quality method improving industrial products toward sustainable development with criteria of circular economy, METAL 2023 – 32nd Int. Conf. Metall. Mater. 2023-May (2024) 697-702. https://doi.org/10.37904/metal.2023.4743

[50] B. Gajdzik et al., Approaching open innovation in customization frameworks for product prototypes with emphasis on quality and life cycle assessment (QLCA), Journal of Open Innovation: Technology, Market, and Complexity 10 (2024) art. 100268. https://doi.org/10.1016/j.joitmc.2024.100268

[51] Ł.J. Orman et al., Comparative Analysis of Indoor Environmental Quality and Self-Reported Productivity in Intelligent and Traditional Buildings, Energies 16 (2023) art. 6663. https://doi.org/10.3390/en16186663

Terotechnology XIV | Materials Research Forum LLC
Materials Research Proceedings 62 (2026) 367-375 | https://doi.org/10.21741/9781644904015-47

The Legal Aspects of a Secret Invention

Mateusz RADEK[1,2,a*] and Krzysztof COLBERT[3,b]

[1]Jagiellonian University, Faculty of Law and Administration, ul. Gołębia 24, 31-007 Cracow, Poland

[2]Cracow University of Economics, ul. Rakowicka 27, 31-510 Cracow, Poland

[3]SWPS University of Social Sciences and Humanities, ul. Chodakowska 19/31, 03-815 Warsaw, Poland

[a]radekmatheus@gmail.com, [b]kcolbert@st.swps.edu.pl

Keywords: Secret Invention, Patent Law, National Defence, Security

Abstract. The institution of a secret invention does not solely stem from patent law by its nature; rather, it constitutes a unique institution at the intersection of patent law and militarization, specifically, national security law [1]. The regulation of the institution of a secret invention differs significantly from the regulation of ordinary inventions, which is evident, among other things, in the special procedure for recognizing an invention as secret, the specific entities eligible to apply for a patent on a secret invention, and the authorities involved in this procedure.

Introduction

When beginning considerations regarding a secret invention, one must first focus on the foundation from which this institution originates, namely, the answer to the question: what is an invention? Polish patent law from 2000 [5] states that patents are granted - across all fields of technology - for inventions that meet three fundamental conditions: 1) they are new, 2) they possess an inventive step, 3) they are capable of industrial application. If even one of these conditions is not met, the granting of a patent for the invention is impossible.

The first condition, novelty, in Polish industrial property law [5] has its definition - an invention is considered new if it is not part of the state of the art. As for the definition of the state of the art itself, the state of the art is understood to encompass everything that, before the date which determines the priority for obtaining a patent, was made available to the public in the form of a written or oral description, by use, exhibition, or disclosure in any other way. Furthermore, information contained in applications for inventions or utility models benefiting from prior priority, not made available to the public, is also considered part of the state of the art, provided they are disclosed in a manner specified by law. Moving on to the second condition, which entails the invention possessing an inventive step, Polish legislation also introduces a definition of it. An invention is considered to have an inventive step if it is not obvious to a person skilled in the art from the state of the art. Regarding the person skilled in the art, this is understood to mean an average practitioner possessing ordinary, generally accessible knowledge in the given field at the appropriate time, equipped with typical means and possibilities for conducting work and experiments. When considering whether a particular solution characteristically evident for an invention will be recognized, it boils down to whether, prior to the priority date and considering the state of the art at that time (no later than the date of filing), an average practitioner in this area, i.e., a person skilled in the relevant technology, would have arrived at it [2].

Finally, the last condition - that the invention is capable of industrial application - according to the definition introduced in patent law, relates to the purpose of employing the invention, hence according to the invention, a product can be obtained, or a method can be utilized, in technical terms, in any industrial activity, not excluding agriculture. Thus, this specific technical solution must be complete, adequately disclosed, socially useful, and guarantee repeatability of the result [3].

Therefore, taking into account the above three constitutive conditions, if a given solution does not collectively meet them, there can be no question of an invention in the normative sense.

Regulations Regarding Secret Inventions in Polish Law

Transitioning to the institution of a secret invention, which is regulated under Polish patent law [5], it is important to emphasize that this particular type of invention must inherently possess solutions that meet all of the three constitutive features of an invention outlined in the previous section, along with other characteristics specific to national security law. Firstly, it should be noted that a secret invention is an exception to the statutory principle of the right to a patent belonging to the inventor or the principle that inventions are patented, rather than covered by a secrecy clause. Therefore, the provisions regarding secret inventions should not be interpreted broadly. Moreover, the mere potential submission for protection of an invention that deserves to be recognized as secret is treated as an incident to which the State should react by classifying such an invention as secret for compensation [1].

The purpose of this institution is to protect values such as national defense and security. Therefore, one of the conditions for recognizing an invention as secret, according to Article 56 of Polish patent law [5], is that the given solution must pertain to the defense and security of the State (material criterion). These are values as precious as "property and other proprietary rights," including industrial property rights, which the legislator expresses by allowing limitations on the exercise of constitutional freedoms and rights due to "security and public order," as stated in Article 31(3) of the Constitution of the Republic of Poland [1, 4]. Moreover, from the State's obligation to protect these national interests derives another condition - the decision to recognize such an invention as secret may only pertain to an invention made by a Polish citizen. In subsequent points of Article 56 of Polish patent law, the legislator outlines the types of solutions that may concern inventions related to the defense and security of the State. To this end, the Polish legislator has utilized open catalogs in both cases, indicated by the formulation "in particular." Therefore, concerning inventions related to the State's defense, these include, in particular, types of weapons or military equipment and methods of warfare, while inventions related to the State's security involve, in particular, technical means used by state services authorized to conduct operational-reconnaissance activities, as well as new types of equipment and methods for their use by these services. These inventions have been detailed in the regulation of the Council of Ministers dated July 23, 2002, concerning inventions and utility models related to the defense or security of the State [6]. In paragraph 2 of the mentioned regulation, the legislator included an open catalog of solutions that may constitute inventions related to the defense or security of the State. Another feature of a secret invention is the confidential information that constitutes the given solution, marked by the following clauses: "strictly confidential," "confidential," "secret," or "classified." These are pieces of information whose disclosure would or could cause harm to the Republic of Poland or be disadvantageous from the perspective of its interests, also during their development and regardless of the form and manner of expression [7].

The Patent Office, if deemed appropriate, directs the submitted invention through the procedure provided for secret inventions and informs the applicant that "steps have been taken to determine the potential usefulness of the solution for the defense or security of the State," and that as a result of these steps, it may issue the applicant with "information on recognizing the solution as secret" [8].

The granting of secrecy status to the invention concerning the defense or security of the State is decided by the following "authorized bodies," namely: the Minister of National Defense, the minister responsible for internal affairs, or the Head of the Internal Security Agency. The person entitled to obtain the exclusive right may notify one of the aforementioned authorized bodies "appropriately according to the subject of the invention," and the body operates "in accordance with its jurisdiction." The invention becomes a secret invention by virtue of the appropriate decision, once this decision becomes final [8].

Evolution of Secrecy of an Invention in Polish Jurisprudence
Polish Patent Law distinguishes two types of inventions: one that can be called 'common' and secret. The quality of secrecy of an invention was first introduced in Article 6 of the Decree on Patents and Inventions from 1919 [9] which listed the requirements for an invention to be deemed secret. Such invention needed to be of defensive or military nature. The Patent Office was obliged to inform the military administration of an application filed for an invention which fulfils the legal requirements of secrecy. The military administration was equipped in the right to demand from the Patent Office deeming an invention as secret. If such demand was filed, the Patent Office was granting patent for the invention deeming it secret and informing the military administration. This was the moment in which the military administration could decide whether it wants to make a purchase of such invention. If so, a compensation had to be determined and paid to the inventor. In case of inventor's denial to compensation's sum the executive power had the right to expropriation. It is important to notice that such proceeding was conducted without making a public announcement. Inventions, Designes and Trademarks Protection Act from 1924 [10] changed the way military administration, then Ministry of Military Affairs, could demand deeming an invention as secret, namely a motion for such action could be filed to the Patent Office regarding inventions that had already been in the military administration's property, before they were patented. That means the Ministry of Military Affairs had the right to purchase the rights to inventions before they were patented and demand from the Patent Office issuing a patent with secrecy clause for them. The model of compensation was regulated more thoroughly than in 1919 Decree in a separate chapter titled 'Expropriation of a patent'. Responsibility for setting the amount of compensation became obligation of the Patent Office and Ministry of Treasury that were supposed to cooperate in this matter. The model was based on negotiations between the State and the inventor. In case of a fiasco of negotiations the State set the amount at its discretion. The inventor had the right to a judicial review of the sum of compensation.

The Ordinance of the President of the Republic of Poland on Designes and Trademarks Protection from 1928 [11] regulated in Article 43 Section 2 the obligation of suspending the patent proceeding done by the Patent Office in case of a dispute between the Ministry of Military Affairs and the applicant. The Invention Law Act form 1962 [12] adopted in the People's Republic of Poland repeated the essential foundations of the secret invention outlined in pre-war legislation. Minimal changes have been made in terms of the patent proceeding that became non-public. As M. Ożóg notices, the fact that the idea of secret invention is applied to the process of its creation is worth highlighting [13].

The Inventions Act from 1972 [14] repeated the majority of 1962 Invention Law Act putting stress on granting the public treasury a patent for a secret invention.

Legal Status of a Secret Invention and Its Nuances
As we can see the core in patenting an invention that falls under the scope of secrecy remained unchanged for decades in form of a formal proceeding conducted by the Patent Office of the Republic of Poland during pre-war era and the Patent Office of the People's Republic of Poland after World War II. It is important to notice however that in current Law on Industrial Property Act from 2000 [5] the lawmaker chose to differ from this tradition by deeming an invention as secret in the form of an administrative decision of the competent authority [15]. This course of action has serious implications and creates theoretical-legal questions. Another thing that needs attention is the fact, that there is no legal definition of an invention [16]. The first question regards the patent issuance. In previous legislation the patent on a secret invention was granted regardless of its

status, that is secrecy or disclosure. The Polish State held a patent for a secret invention and had the right to it. Nowadays, every invention that is deemed secret becomes a subject of unregulated legal status since according to Article 59 Section 1 of the Law on Industrial Property Act the state earns the right to a patent. Meanwhile there is no provision that would regulate patent proceeding of a secret invention in this situation. Therefore, it can be argued that such secret invention's only legal traits are secrecy and the category of an invention. On these grounds it can be argued that secret invention can be viewed as subject of unregulated nature existing outside the legal system if not patented by the state. Not a single civil nor military provision exists that would apply to a secret invention in terms of its use or any other activity. It appears that secrecy excludes any form of legal intervention and regulation simply on this ground. Article 56 of the Law on Industrial Property Act lists two criteria to determine whether an invention should be deemed secret or not; they are defence and national security. Those criteria due to their critical nature might be seen as a sufficient rationale for such manner of regulation, nevertheless, it creates legal ambiguity and hinders legal clarity. The second question that requires exploration is what happens to the secret invention after its disclosure. Article 60 of the Law on Industrial Property Act provides partial answer; secret invention after disclosure can be a subject of a patent proceeding for the period of time not exceeding 20 years from the moment of disclosure. Such wording means that an invention that is no longer secret can start functioning in the legal system as a patented technical solution or tool and fall under legal protection. In previous legislation Polish lawmaker designed a double patent solution; first for the secret invention and second for the disclosed invention. The legal status of a disclosed invention under current legislation is ambiguous since during the period of secrecy it remains in full possession of the state but without any legal basis for being a subject of any legal action. This leads to a conclusion that such invention can be a subject of only factual action like application in combat per instance. It does not require any legal action for this in terms of patent and industrial property law, however, it can be a subject of e.g. a military order for use. This leads to another area of consideration which is the question of the possibility of a secret invention to be a subject of an administrative decision, but this aspect requires separate discussion. In light of these findings, it can be argued that secret invention has its beginning in existence as a legal subject in pleno only in case of a patent issuance after disclosure, since only then it benefits from protection provided by the patent and protection of possession. Naturally, such situation can occur if state files for a patent issuance on an invention it deemed secret however taking into account scarcity of public information in this regard and lack of regulation of the proceeding, such situation is unlikely. Having this concluded, the previous, double-patent solution appears more legally plausible.

Technological, Economic, Environmental, and Social Impact

The legal institution of secret inventions, positioned at the intersection of patent law and national security, primarily governs technologies of strategic importance. These are often dual-use technologies or innovations critical for state defense, whose managed release (or suppression) significantly influences the global and national technological landscape, justifying the necessity for legal secrecy. Consequently, assessing the full impact of these inventions requires a comprehensive look at their technological sophistication, economic implications, environmental footprint, and societal role.

Secret inventions inherently belong to the domain of high-end, advanced materials and processes, with a strong emphasis on reliability and effectiveness, often tied to defense applications. Research focuses on achieving superior performance through sophisticated surface engineering techniques, where the resulting material properties may be considered strategic knowledge. For instance, advanced coatings, such as Diamond-Like Carbon (DLC), are studied for their crucial operational properties and wide potential applications [17]. Further technological advancements involve tailoring the performance of electro-spark deposition (ESD) coatings

Materials Research Proceedings 62 (2026) 367-375 https://doi.org/10.21741/9781644904015-47

through subsequent laser treatment [18], a process critical for improving wear resistance and overall durability in highly demanding applications [19]. Specific research explores tribological behavior using advanced composites, such as Al_2O_3 and WS_2, to reduce friction and enhance machine longevity and energy efficiency [20]. The restricted knowledge surrounding these superior materials creates strategic technological asymmetries between nations.

Crucially, some technologies are directly developed for military use. This includes research into specialized paint coatings based on carbonyl iron, designed for their specific microwave absorption properties, which is essential for signature management in military technology [21]. Another key military application involves analyzing and improving the parameters of effective camouflage systems exposed to operational conditions [22]. Such innovations directly influence national security and military deterrence, making them prime candidates for classification as secret inventions, thereby limiting their dissemination and commercial application. The development of these advanced materials often relies on highly precise diagnostic tools, such as nano-X-ray tomography, for quantitative analysis of the geometry of surface layers modified by laser beams [23]. Furthermore, using neural network approximation is employed to accurately estimate the properties of advanced materials, such as sintered metals [24], demonstrating the integration of artificial intelligence into materials engineering. The economic success of any invention, especially a strategic one, hinges on optimizing production processes and ensuring product quality and reliability. However, inventions under secrecy face unique commercialization challenges, where the primary economic benefit is often realized through exclusive governmental procurement or strictly managed public sector contracts, delaying or preventing traditional market competition. The introduction of Industry 4.0 paradigms presents new challenges for traditional methodologies, such as Design of Experiment (DOE) [25]. The necessity for absolute quality and reliability in strategic, often secret, production lines elevates the importance of rigorous quality management [26]. Advanced statistical and computational methods are vital for achieving process optimization, ensuring that classified production meets stringent standards. This includes the development of expert systems dedicated to streamlining the design of experiments [27]. Optimization itself can be achieved using sophisticated techniques, such as fuzzy regression, which offers robust optimization of technological processes [28] and assembly operations, including flywheel assembly [29]. The fuzzy statistics approach also aids inference in experimental design, effectively managing uncertainty in complex strategic systems [30]. In manufacturing, quality control is paramount. Machine learning is increasingly applied to analyze surface quality and detect damage in manufactured components, improving defect detection efficiency [31]. Furthermore, the quality of critical components, such as the outer bearing ring, is ensured through continuous improvement of non-destructive testing (NDT) processes [32]. Even the quality of specific components, such as cast alloys, can be enhanced using tools like Grey Relational Analysis (GRA) [33]. Optimizing these processes reduces waste, lowers costs, and increases the reliability of the final product, which is particularly relevant for strategic technologies. The economic value is further supported by innovations that extend product life, for example, by improving the corrosion resistance of coatings through precise laser power control [34, 35] or applying advanced laser alloying techniques to materials such as sintered stainless steel [19]. The overall efficiency of quality control procedures in high-volume production, such as the automotive industry, has a significant ripple effect on economic productivity [36]. Beyond the direct technological and economic benefits, secret or strategic inventions and the industries they serve have substantial environmental and social consequences. While the immediate focus of secret technologies is security, their eventual or parallel civilian applications are bound by modern sustainability imperatives. Modern industrial practice must reconcile innovation with sustainability. One critical framework for assessing this is the Life Cycle Assessment (LCA), which is used to quantify the environmental burden of products, such as truck steel wheels [37] or aluminum cans, often in conjunction with an analysis of eco-costs [38]. This pro-quality approach aims to improve products in line with sustainable development, adhering to the criteria of a circular economy [39].

The environmental impact also extends to industrial waste management. Techniques for the electrostatic separation of biodegradable polymers from common plastics, such as PVC or PET, are crucial for improving recycling efficiency [40]. Furthermore, inventive solutions address industrial byproducts, such as the application of fly ash and solid waste from incinerators in cement mortar, promoting resource reuse [41]. The operation of conventional energy sources also carries an environmental load, which can be mitigated by understanding the impact of power plant unit start-up schemes on pollution levels [42]. The transition to new energy paradigms is also represented by the development of hybrid energy storage systems, such as acid batteries [43]. From a social perspective, innovation has a significant impact on public safety and quality of life. An example of this is the development of protective anti-graffiti coating systems for rolling stock, which affects public aesthetics and maintenance costs [44]. Finally, the impact of technological innovation on the working environment itself is significant, as seen in the comparative analysis of indoor environmental quality in intelligent versus traditional buildings and its effect on self-reported productivity [45]. Thus, even inventions originating from national security concerns often transition (following declassification or through dual-use application) into technologies that profoundly shape economic, environmental, and social well-being.

Summary

In conclusion, secret inventions represent a unique legal entity balancing innovation and national security. While historical and current Polish legislation acknowledges the significance of protecting such inventions, legal clarity and regulation consistency remain areas for potential refinement. The evolving notion of secret inventions, particularly their transition from secrecy to disclosure, underscores the need for robust frameworks that protect state interests while fostering transparent legal processes. The double-patent solution previously employed might offer a pragmatic approach, ensuring comprehensive legal protection throughout the invention's lifecycle.

References

[1] M. Ożóg, in: Prawo własności przemysłowej. Komentarz do wybranych przepisów, LEX/el. 2025, art. 56.

[2] M. Jezierska-Zięba, in: Prawo własności przemysłowej. Komentarz, M. Kondrat (Ed.), LEX/el. 2025, art. 26.

[3] M. Jezierska-Zięba, in: Prawo własności przemysłowej. Komentarz, M. Kondrat (Ed.), LEX/el. 2025, art. 27.

[4] Konstytucja Rzeczypospolitej Polskiej z dnia 2 kwietnia 1997 r. (Dz. U. Nr 78, poz. 483 z późn. zm.)

[5] Ustawa z dnia 30 czerwca 2000 r. Prawo własności przemysłowej (t.j. Dz. U. z 2023 r. poz. 1170).

[6] Rozporządzenie Rady Ministrów z dnia 23 lipca 2002 r. w sprawie wynalazków i wzorów użytkowych dotyczących obronności lub bezpieczeństwa Państwa (Dz. U. Nr 123, poz. 1056)

[7] M. Jezierska-Zięba, in: Prawo własności przemysłowej. Komentarz, M. Kondrat (Ed.), LEX/el. 2025, art. 57.

[8] M. Ożóg, in: Prawo własności przemysłowej. Komentarz do wybranych przepisów, LEX/el. 2025, art. 57.

[9] Dekret o patentach na wynalazki (Dz. Pr. P. P. z 1919 r., Nr 13, poz. 137)

[10] Ustawa z dnia 5 lutego 1924 r. o ochronie wynalazków, wzorów i znaków towarowych (Dz. U. z 1924 r., Nr 31, poz. 306)

[11] Rozporządzenie Prezydenta Rzeczypospolitej z dnia 22 marca 1928 r. o ochronie wynalazków, wzorów i znaków towarowych (Dz. U. z 1928 r., Nr 39, poz. 384 z późn. zm.)

[12] Ustawa z dnia 31 maja 1962 r. Prawo wynalazcze (Dz. U. z 1962 r. Nr 33, poz. 156 z późn. zm.)

[13] M. Ożóg, Wynalazek tajny: prawo i praktyka polskie na tle doświadczeń światowych, in: R. Kopeć (Ed.) Przemysł zbrojeniowy. Tendencje, perspektywy, uwarunkowania, innowacje, Kraków 2016, p. 198

[14] Ustawa z dnia 19 października 1972 r. o wynalazczości (t.j. Dz. U. z 1993 r. Nr 26 poz. 117 z późn. zm.)

[15] Article 57 Section 2 of the Law on Industrial Property Act lists three competent authorities equipped in power to deem an invention as secret: the Minister of National Defence, the Minister of Internal Affairs and the Chief of Internal Security Agency.

[16] A. Hołda-Wydrzyńska, Prawo własności przemysłowej, in: R. Blicharz (Ed.), Przedsiębiorca: Zagadnienia wybrane, Katowice 2017, p. 275.

[17] N. Radek et al., Operational properties of DLC coatings and their potential application, METAL 2022 – 31st Int. Conf. Metall. Mater. (2022) 531-536. https://doi.org/10.37904/metal.2022.4491

[18] N. Radek et al., The impact of laser processing on the performance properties of electro-spark coatings, World Congress in Computational Mechanics and ECCOMAS Congress 1000 (2021). https://doi.org/10.23967/wccm-eccomas.2020.336

[19] A. Dudek et al., Laser Surface Alloying of Sintered Stainless Steel, Materials 15 (2022) art. 6061. https://doi.org/10.3390/ma15176061

[20] J. Korzekwa et al., Tribological behaviour of Al2O3/inorganic fullerene-like WS2 composite layer sliding against plastic, International Journal of Surface Science and Engineering 10 (2016) 570-584. https://doi.org/10.1504/IJSURFSE.2016.081035

[21] W. Przybył et al., Microwave absorption properties of carbonyl iron-based paint coatings for military applications, Defence Technology 22 (2023) 1-9. https://doi.org/10.1016/j.dt.2022.06.013

[22] M. Michalski et al., The Influence of Operational Exposure on Changes in Parameters of Effective Camouflage of Coatings Used in Military Technology, Advances in Science and Technology Research Journal 17 (2023) 182-196. https://doi.org/10.12913/22998624/156940

[23] A. Gądek-Moszczak et al., Nano X-ray Tomography Application for Quantitative Surface Layer Geometry Analysis after Laser Beam Modification, Materials 15 (2022) art. 5935. https://doi.org/10.3390/ma15175935

[24] J. Pietraszek et al., The estimation of accuracy for the neural network approximation in the case of sintered metal properties, Studies in Computational Intelligence 513 (2014) 125-134. https://doi.org/10.1007/978-3-319-01787-7_12

[25] J. Pietraszek et al., Challenges for the DOE methodology related to the introduction of Industry 4.0, Production Engineering Archives 26 (2020) 190-194. https://doi.org/10.30657/pea.2020.26.33

[26] R. Dwornicka et al., Development of a Hybrid Method for Identifying the Causes of Product Incompatibility in Metallurgical Manufacturing, Materials Research Proceedings 34 (2023) 199-206. https://doi.org/10.21741/9781644902691-24

[27] R. Dwornicka, J. Pietraszek, The outline of the expert system for the design of experiment, Production Engineering Archives 20 (2018) 43-48. https://doi.org/10.30657/pea.2018.20.09

[28] J. Pietraszek, E. Skrzypczak-Pietraszek, The optimization of the technological process with the fuzzy regression, Adv. Mater. Res. 874 (2014) 151-155. https://doi.org/10.4028/www.scientific.net/AMR.874.151

[29] J. Pietraszek, Fuzzy regression compared to classical experimental design in the case of flywheel assembly, Lecture Notes in Computer Science 7267 LNAI (2012) 310-317. https://doi.org/10.1007/978-3-642-29347-4_36

[30] R. Dwornicka et al., Fuzzy Statistics-Aided Inference in Experimental Design, World Congress in Computational Mechanics and ECCOMAS Congress (2024). https://doi.org/10.23967/wccm.2024.131

[31] D. Nowakowski et al., Application of machine learning in the analysis of surface quality - the detection the surface layer damage of the vehicle body, METAL 2021 – 30th Int. Conf. Metall. Mater. (2021) 864-869. https://doi.org/10.37904/metal.2021.4210

[32] D. Siwiec et al., Improving the Non-Destructive Testing Process of the Outer Bearing Ring, Materials Research Proceedings 24 (2022) 174-180. https://doi.org/10.21741/9781644902059-26

[33] D. Siwiec et al., Improving the Quality of Products from Cast Alloy with the use of Grey Relational Analysis (GRA), Materials Research Proceedings 24 (2022) 204-211. https://doi.org/10.21741/9781644902059-30

[34] M. Scendo et al., Influence of laser power on the corrosive resistance of WC-Cu coating, Surface and Coatings Technology 259 (2014) 401-407. https://doi.org/10.1016/j.surfcoat.2014.10.062

[35] N. Radek et al., The effect of laser beam processing on the properties of WC-Co coatings deposited on steel, Materials 14 (2021) art. 538. https://doi.org/10.3390/ma14030538

[36] A. Pacana et al., Analysis of quality control efficiency in the automotive industry, Transportation Research Procedia 55 (2021) 691-698. https://doi.org/10.1016/j.trpro.2021.07.037

[37] A. Pacana et al., Life Cycle Assessment (LCA) of Truck Steel Wheels, Materials Research Proceedings 45 (2024) 213-222. https://doi.org/10.21741/9781644903315-25

[38] D. Siwiec et al., Analysis of aluminum can by eco-costs and life cycle assessments (LCA), METAL 2024 – Int. Conf. Metall. Mater. (2024) 627-632. https://doi.org/10.37904/metal.2024.4975

[39] A. Pacana et al., Pro-quality method improving industrial products toward sustainable development with criteria of circular economy, METAL 2023 – Int. Conf. Metall. Mater. (2024) 697-702. https://doi.org/10.37904/metal.2023.4743

[40] M. Zenkiewicz et al., Electrostatic separation of binary mixtures of some biodegradable polymers and poly(vinyl chloride) or poly(ethylene terephthalate), Polimery/Polymers 61 (2016) 835-843. https://doi.org/10.14314/polimery.2016.835

[41] A. Pietrzak et al., Application of a Mixture of Fly Ash and Solid Waste from Gas Treatment from Municipal Solid Waste Incineration in Cement Mortar, Materials 18 (2025) art. 481. https://doi.org/10.3390/ma18030481

[42] R. Dwornicka, The impact of the power plant unit start-up scheme on the pollution load, Adv. Mater. Res. 874 (2014) 63-69. https://doi.org/10.4028/www.scientific.net/AMR.874.63

[43] S. Maleczek et al., Tests of Acid Batteries for Hybrid Energy Storage and Buffering System—A Technical Approach, Energies 15 (2022) art. 3514. https://doi.org/10.3390/en15103514

[44] N. Radek et al., Technology and application of anti-graffiti coating systems for rolling stock, METAL 2019 – 28th Int. Conf. Metall. Mater. (2019) 1127-1132.

[45] Ł.J. Orman et al., Comparative Analysis of Indoor Environmental Quality and Self-Reported Productivity in Intelligent and Traditional Buildings, Energies 16 (2023) art. 6663. https://doi.org/10.3390/en16186663

Materials Research Proceedings 62 (2026) 376-384 https://doi.org/10.21741/9781644904015-48

Material Fatigue as a Cause of Construction Drilling Rig Failure

KURP Piotr[1,a *] and GONTARSKI Dariusz[1,b]

[1]Kielce University of Technology, Faculty of Mechatronics and Mechanical Engineering, al. 1000-lecia Państwa Polskiego 7, 25-314 Kielce, Poland

[a]pkurp@tu.kielce.pl, [b]gontar@tu.kielce.pl

Keywords: Fatigue, Construction Machine, Pin Damage, Drilling Rig Failure

Abstract. This study aims to analyze the failure of the BG 1080MGF construction drilling rig, which occurred due to the shearing of the bolt securing the drill bit to the motor shaft. Typical causes of wear in construction machinery components are discussed first, including abrasion, corrosion, material fatigue, overheating, operational errors, and material defects. Particular attention is paid to the phenomenon of material fatigue, which plays a key role in the development of cracks and the weakening of structural components. The analyzed mounting bolt, 20 mm in diameter and with a strength class of 10.9, failed due to the action of variable shear stresses and the presence of a notch in the transition zone between the threaded and smooth parts. Observations of the fracture confirmed the fatigue nature of the failure. Although the bolt used met the strength requirements, the impact of cyclic loading led to the formation of a fatigue gap, resulting in the shearing of the component and the slippage of the drill bit from the drill shaft. The necessity of properly selecting mounting bolts and considering the impact of material fatigue in the design and operation of construction machinery is emphasized.

Introduction

Construction machinery, such as excavators, loaders, bulldozers, rollers, and drilling rigs, operates in very harsh operating conditions. They are exposed to high mechanical loads, vibrations, changing weather conditions, dust, and contact with aggressive substances (e.g., mud, water, gasoline). All of these factors accelerate the wear process, leading to various types of component damage. The main types of wear in construction machinery include [1-3]:

- abrasion – abrasion of surfaces of mating components (e.g., bearings, bushings, bucket teeth, pins),
- corrosion – the effects of water, salt, and other chemicals lead to the degradation of metal surfaces,
- material fatigue – repeated load cycles cause cracking or breaking of structural components,
- overheating – excessively high temperatures (e.g., in hydraulic systems or engines) can lead to deformation, loss of hardness, or damage to seals,
- improper operation – overloading the machine, operator errors, lack of maintenance, and inadequate lubrication,
- material and manufacturing defects – microcracks, material inhomogeneities, welding or casting errors.

Damage can occur to components such as the drive system (shaft cracks, gear wear, seal leaks), the hydraulic system (leaks, pump wear, piston seizure, and hose damage), the chassis and suspension (frame cracks, pin and bushing wear, track deformation), the engine (valve burnout, piston ring wear, bearing seizure), and working components (e.g., buckets, blades, drill bits – edge abrasion, cracking, and deformation).

In this paper, the authors analyze damage to a construction drilling rig caused by the breakage of a component that secures the drill bit to the engine shaft.

Material Fatigue

The fatigue strength of a material is its resistance to varying stresses over time. It is the maximum stress σ_{max} at which the element will not fail after a conventional limiting number of cycles of this stress. Fatigue strength is determined empirically for each element, and the result is illustrated using a Wohler diagram. This strength depends not only on the type of material (its mechanical properties) from which the element is made, but also on the shape of the element subjected to varying stresses [4]. A common fatigue-weakening factor in a component is a notch, which causes local stress concentration. A notch can be, for example, a groove, a recess, a sharp-edged hole, or a thread. An example of a notch is shown in Figure 1.

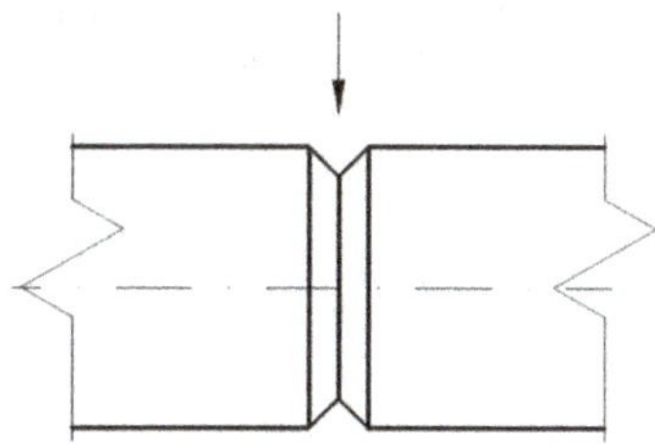

Fig. 1. Schematic showing a notch on a shaft (marked with an arrow) (own study).

Information About the Drilling Rig

The BG 1080MGF drilling rig failed due to the pin securing the drill bit to the engine shaft breaking. The assembly method is shown in Fig.2.

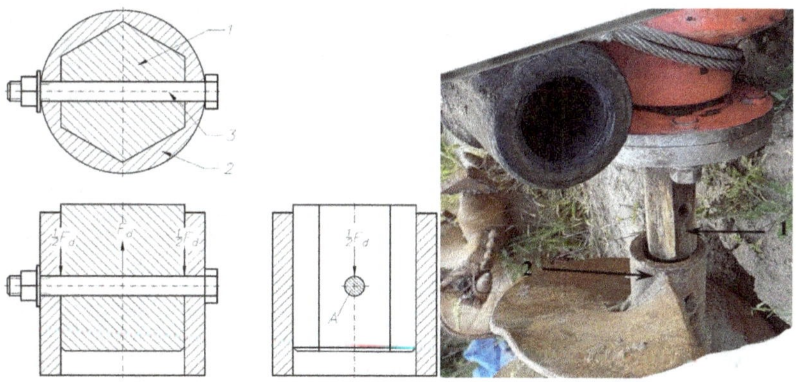

Fig. 2. Diagram of shear force loads on the drill-engine connection. 1 - hexagonal shaft with a hole from the drill engine, 2 - hexagonal drill socket with a hole, 3 - mounting bolt, $\frac{1}{2}F_d$ - value of the shear force acting on the shear cross-section A. (own study).

As can be easily seen, the fastening bolt (3) is mainly exposed to shear - as shown in Fig. 2. The element is considered to be resistant to shear if the actual shear stresses τ in the shear section A are not greater than the permissible stress k_t, which can be described by the relationship:

$$\tau = \frac{F_d}{A} \leq k_t. \quad (1)$$

where:

τ - shear stress, MPa
F_d - shear force, N
A - shear cross-sectional area, mm^2
k_t - permissible shear stress, MPa.

The permissible shear stresses are assumed to be $k_t=(0,5\div0,8)\cdot R_m$ (R_m - maximum tensile stress, MPa). Since, as shown in Fig. 2, the number of sheared sections in the considered design solution of the drilling rig mounting is two, each of the sections is acted on by 1/2 of the shearing force F_d. The sheared section A is the cross-section of the mounting bolt. According to the drilling rig manufacturer's manual [5]:

> When applying tensile and compressive forces to the drilling equipment, care must be taken to avoid bending stress. The manufacturer is not responsible for negligent operations. The risk rests solely with the user.
> Permissible static and dynamic pressure force: F_d=130 kN,
> Permissible static tensile force: $F_{z\,stat}$=46 kN,
> Permissible dynamic tensile force: $F_{z\,dyn}$=22 kN,
> Maximum torque: 1080 daNm,
> Maximum rotational speed: 130 rpm,
> Maximum power: 50 kW.

The element fastening the drill to the engine shaft was a bolt with a diameter of ϕ20 mm, made toan accuracy class 10.9. According to the PN-EN ISO 898 standard [6], bolts are classified into ten strength classes, respectively: 3.6, 4.6, 4.8, 5.6, 5.8, 6.8, 8.8, 9.8, 10.9, and 12.9. The bolt accuracy class marking is stamped on the bolt head. The first value indicates the tensile strength limit, Rm, in MPa, divided by 100. The second value indicates the ratio of the yield strength R_e in MPa (or the conventional yield strength $R_{0.2}$ in MPa) to the tensile strength R_m divided by 10.

Drilling Rig Failure

During the operation of the drilling rig, the drill bit dislodged from the engine, posing a risk of injury to people and property located near the drilling rig. The operator noticed a component falling from the drilling rig's upper mounting. The drill bit then slid out and struck the ground. The cause of this incident turned out to be a loose mounting element – a bolt. The view of the dislodged mounting bolt is shown in Figures 3 through 6.

The bolt received for inspection is manufactured to a precision class of 10.9 (Fig. 4), which indicates that the tensile strength limit of the bolt material is Rm = 1000 MPa. The bolt was made with a partial thread. It is visibly plastically deformed along its entire length in the form of a deflection (Fig. 3). The bolt diameter is ϕ20 mm, while the bolt length is approximately 130 mm (excluding the head). The bolt fracture reveals clear fatigue damage, which weakened the bolt cross-section, and then the remaining cross-section sheared off. The bolt failed at the transition between the threaded and unthreaded lengths (Figures 5 and 6).

Fig. 3. View of the bolt along its length.

Fig. 4. View of the bolt head with strength class markings.

Fig. 5. View of the bolt fracture from the damaged side. 1 - visible burr from the last thread turn, 2 - visible fatigue zone, 3 - visible shear zone.

Fig. 6. View of the end of the bolt. 1 - visible burr from the last thread turn and the fatigue gap behind it.

Technological, Economic, Environmental, and Social Impact

The failure of a critical component of a construction drilling rig due to material fatigue in the mounting bolt has far-reaching consequences that extend beyond the immediate machine failure. These impacts must be analyzed in the context of four key areas: technological, economic, social, and environmental impacts.

The technological implications of failure emphasize the need for advanced diagnostics, improved design, and precise material selection. Observations of fatigue cracking confirm the crucial role of variable shear stresses and notch stresses in the thread area. To prevent such events, it is essential to implement systems that monitor stress changes in real time, for example, by pre-filtering signals in acoustic systems [7]. The process of designing and modeling machines subjected to complex loads necessitates the use of advanced statistical-deterministic models for multi-stage mechanical systems [8] and the accurate modeling of physical phenomena, such as torsional vibrations of beams [9], the dynamic behavior of motors [10], and particle accelerators [11]. A critical element is the precision of manufacturing and assembly. Identifying the root causes of problems enables the development of improved quality control methods. The use of regression modeling (e.g., fuzzy regression modeling in the context of classical experimental design) [12], experimental expert systems (DoE) [13], and heuristic methods [14] is crucial for design validation, thereby minimizing uncertainty [15]. Particular attention should be paid to mechanical strengthening [16] and heat treatment processes, such as induction hardening [17], which affect the microstructure and strength of steel. In the context of materials, the failure underscores the need for a more comprehensive analysis of material properties. The influence of impurities, such as oxygen, on the fatigue strength of carbon steel should be investigated [18]. Even topics as distant as impedance testing of carbon materials [19] or analysis of the statistical shape stability of stainless steel powders [20] provide methodologies for assessing the quality of raw materials. The use of Rapid Prototyping techniques, for example, for pneumatic components [21], influences the speed and quality of implementing design modifications. New technologies also require modern modeling tools. The properties of sintered metals can be estimated using neural networks [22], and plastic forming processes, such as forging, can be predicted using advanced models [23]. Stress analysis, for example, in the rolling process [24] or the study of deformation in composites with prestressing [25], facilitates a deeper understanding of crack formation mechanisms.

The economic consequences of failure are immediate (repair costs, downtime) and long-term (reputation loss, increased insurance costs). The need for proper selection of fasteners and consideration of material fatigue in the design is fundamental to risk management. Failure is a manifestation of technical risk, which should be considered in ex-ante assessment [26]. The increasing complexity of machines and the introduction of Industry 4.0 standards pose new challenges for design of experiments (DoE) methodologies [27], which must account for the instability of production processes (e.g., modeling errors in PCB soldering) [28].

Socially, a construction drilling rig failure poses a risk to occupational safety. Although the cited literature often focuses on advanced statistics in the chemical and biotechnological context [29, 30], nonparametric analysis methods [31] are valuable tools for assessing component reliability data. Safety also has a legal and economic context. In a broader context, one can consider the impact of regulations on economic activity, analogous to discussions on ordoliberalism [32] or tax harmonization [33], which shape the environment for investing in safe equipment. Issues related to biomechanics [34], although from a different field, emphasize the importance of accurate load simulation to protect health and life.

The environmental impact of a drilling rig failure focuses on waste management (damaged machinery, screw) and preventing pollution during construction. The use of inappropriate materials or their premature degradation increases raw material consumption. Research on the corrosion resistance of materials, such as steel in sulfuric acid [35], highlights the importance of selecting materials that are resistant to environmental operating conditions. Component longevity is an element of sustainable development. Although a drilling rig is not a direct renewable energy system, modeling rotating systems such as wind turbines [36] provides engineering knowledge that can be applied to increase efficiency and reduce the environmental footprint of construction operations. The implementation of composites with microstructural analysis [37, 38], also aligns with the trend of seeking lighter and more robust structural solutions.

Summary

The drilling rig's instruction manual states that the tool (drill bit) should be secured to the adapter (drilling rig) with a bolt. Furthermore, the manual states that the maximum clamping force cannot exceed 130 kN. The manual does not specify the type of bolt (in terms of material and diameter) to be used. A ϕ23 mm diameter hole for a securing bolt suggests an M22 bolt (ϕ22 mm diameter). Furthermore, such a bolt should be made of a material that can withstand a force of 130 kN. The ϕ20 mm diameter bolt used is smaller in diameter than the hole suggests. It is made of a material with a tensile strength of R_m = 1000 MPa. By rearranging expression (1) and considering the conditions under which forces act on the bolt, we can calculate the maximum force that this type of bolt can withstand. Assuming the maximum shear stress is $0.5R_m$ and the cross-sectional area of a 20 mm diameter bolt is A=314 mm^2, the maximum pressure force will be F_d=314 kN. This is approximately 2.4 times the permissible force, as specified in the operating manual (130 kN). However, we cannot determine the actual forces acting on the drill bit during operation. It should be assumed that the bolt used was sufficient and appropriate to secure the drill bit to the drilling rig shaft.

It can be concluded that the damage to the mounting bolt, which resulted in the drill bit slipping off the drill shaft, was caused by the action of variable forces on the material. During drilling operation, when the drill bit is being pulled out or pressed in, the mounting bolt is subjected to variable stress. This stress can be caused by forces lower than the allowable limit – the maximum clamping force does not have to be exceeded. Furthermore, the bolt thread acts as a fatigue notch, which increases with successive cycles of variable force. During operation, the fatigue gap from the notch gradually widened, weakening the bolt (its diameter, and therefore its effective cross-section, decreased). At one point, the effective cross-section of the bolt was so small that the forces acting on the bolt caused it to shear, and the threaded end secured by the nut to fall off. As a result, the bolt slipped out of the hole, and the drill bit fell to the ground.

References

[1] G. Mukhopadhyay, Chapter 5 - Failures of construction equipment and accessories: metallurgical root-cause analysis, Handbook of Materials Failure Analysis, Butterworth-Heinemann, 2018, pp. 83-106. https://doi.org/10.1016/B978-0-08-101928-3.00005-7.

[2] Z. Huimin, Construction Machinery Fault Detection Technology and Maintenance Measures, Journal of Engineering Mechanics and Machinery 6 (2021) 37-41. https://doi.org/10.23977/jemm.2021.060106

[3] K. Petroutsatou, I. Ladopoulos, Integrated Prescriptive Maintenance System (PREMSYS) for Construction Equipment Based on Productivity, 2023

[4] C. Bathias, A. Pineau, Fatigue of Materials and Structures: Application to Design and Damage, 2013. https://doi.org/10.1002/9781118616994

[5] Manual: BG 1080 MGF Geotec GmbH Germany

[6] PN-EN ISO 898-1:2013-06 Mechanical properties of fasteners made of carbon steel and alloy steel

[7] I. Dominik et al., Preliminary signal filtering in Self-Excited Acoustical System for stress change measurement, Chinese Control Conference, CCC (2013) art. 6640759.

[8] A. Goroshko et al., Construction and practical application of hybrid statistically-determined models of multistage mechanical systems, Mechanika 20 (2014) 489-493. https://doi.org/10.5755/j01.mech.20.5.8221

[9] E. Augustyn, M.S. Kozien, Analytical solution of excited torsional vibrations of prismatic thin-walled beams, Journal of Theoretical and Applied Mechanics (Poland) 53 (2015) 991-1004. https://doi.org/10.15632/jtam-pl.53.4.991

[10] A. Goroshko et al., Multiphysics Modeling of Direct-Start Induction Motor Considering Rotor Unbalance, Materials Research Proceedings 45 (2024) 267-276. https://doi.org/10.21741/9781644903315-31

[11] Ł. Łacny et al., Selected overview of the impact of ground motion on the vibrations of particle accelerators, AIP Conf. Proc. 2239 (2020) art. 020025. https://doi.org/10.1063/5.0008950

[12] J. Pietraszek, Fuzzy regression compared to classical experimental design in the case of flywheel assembly, Lecture Notes in Computer Science 7267 LNAI (2012) 310-317. https://doi.org/10.1007/978-3-642-29347-4_36

[13] R. Dwornicka, J. Pietraszek, The outline of the expert system for the design of experiment, Prod. Eng. Arch. 20 (2018) 43-48. https://doi.org/10.30657/pea.2018.20.09

[14] J. Pietraszek, A. Goroshko, The heuristic approach to the selection of experimental design, model and valid pre-processing transformation of DoE outcome, Advanced Materials Research 874 (2014) 145-149. https://doi.org/10.4028/www.scientific.net/AMR.874.145

[15] J. Pietraszek et al., The non-parametric approach to the quantification of the uncertainty in the design of experiments modelling, UNCECOMP 2017 – Proc. 2nd Int. Conf. Uncertainty Quantification in Computational Sciences and Engineering (2017) 598-604. https://doi.org/10.7712/120217.5395.17225

[16] W. Jasiński et al., Static Mechanical Force Amplifier on the Example of a Fastener with an Electromagnetic Bolt, Materials Research Proceedings 34 (2023) 246-251. https://doi.org/10.21741/9781644902691-29

[17] A. Szczotok et al., Effect of the induction hardening on microstructures of the selected steels, METAL 2018 – 27th Int. Conf. Metall. Mater. (2018) 1264-1269.

[18] T. Lipinski et al., Influence of oxygen content in medium carbon steel on bending fatigue strength, Engineering for Rural Development 21 (2022) 351-356. https://doi.org/10.22616/ERDev.2022.21.TF116

[19] V. Ptashnyk et al., Impedance investigation of activated carbon material modified by ultrasound treatment, Przeglad Elektrotechniczny 94 (2018) 186-189. https://doi.org/10.15199/48.2018.05.33

[20] J. Pietraszek, A. Gadek-Moszczak, The smooth bootstrap approach to the distribution of a shape in the ferritic stainless steel AISI 434L powders, Solid State Phenomena 197 (2013) 162-167. https://doi.org/10.4028/www.scientific.net/SSP.197.162

[21] S. Blasiak et al., Rapid prototyping of pneumatic directional control valves, Polymers 13 (2021) art. 1458. https://doi.org/10.3390/polym13091458

[22] J. Pietraszek et al., The estimation of accuracy for the neural network approximation in the case of sintered metal properties, Studies in Computational Intelligence 513 (2014) 125-134. https://doi.org/10.1007/978-3-319-01787-7_12

[23] R. Dindorf et al., Prediction of the parameters and the hot open die elongation forging process on an 80 MN hydraulic press, Open Engineering 11 (2021) 528-534. https://doi.org/10.1515/eng-2021-0056

[24] P. Sygut et al., Theoretical analysis of the influence of longitudinal stress changes on band dimensions during continuous rolling process, Archives of Metallurgy and Materials 61 (2016) 183-188. https://doi.org/10.1515/amm-2016-0032

[25] P. Krysiak et al., Comparative Analysis of the Strain in Composite Rings Fabricated with Fibre Prestressing, Materials Research Proceedings 45 (2024) 91-100. https://doi.org/10.21741/9781644903315-12

[26] E. Kozien, M.S. Kozien, The EX ante risk assessment in the project in interval analysis description, Materials Research Proceedings 17 (2020) 31-35. https://doi.org/10.21741/9781644901038-5

[27] J. Pietraszek et al., Challenges for the DOE methodology related to the introduction of Industry 4.0, Prod. Eng. Arch. 26 (2020) 190-194. https://doi.org/10.30657/pea.2020.26.33

[28] J. Pietraszek et al., Modeling of errors counting system for PCB soldered in the wave soldering technology, Advanced Materials Research 874 (2014) 139-143. https://doi.org/10.4028/www.scientific.net/AMR.874.139

[29] E. Skrzypczak-Pietraszek, J. Pietraszek, Chemical profile and seasonal variation of phenolic acid content in bastard balm (Melittis melissophyllum L., Lamiaceae), Journal of Pharmaceutical and Biomedical Analysis 66 (2012) 154-161. https://doi.org/10.1016/j.jpba.2012.03.037

[30] E. Skrzypczak-Pietraszek, Phytochemistry and biotechnology approaches of the genus exacum, The Gentianaceae - Volume 2: Biotechnology and Applications (2015) 383-401. https://doi.org/10.1007/978-3-642-54102-5_16

[31] E. Skrzypczak-Pietraszek, J. Pietraszek, Seasonal changes of flavonoid content in Melittis melissophyllum L. (Lamiaceae), Chemistry and Biodiversity 11 (2014) 562-570. https://doi.org/10.1002/cbdv.201300148

[32] E. Kozień, A. Kozień, Is ordoliberalism democratic? – Legal approach, System Safety: Human - Technical Facility - Environment 2 (2020) 33-38. https://doi.org/10.2478/czoto-2020-0005

[33] A. Kozień, N. Kozłowska, Harmonization and Deharmonization of Excise Duty in the European Union as Contemporary Challenges of the EU Tax Law, WSEAS Transactions on Business and Economics 19 (2022) 816-824. https://doi.org/10.37394/23207.2022.19.71

[34] J. Lorkowski et al., Effects of breast and prostate cancer metastases on lumbar spine biomechanics: Rapid in silico evaluation, Advances in Experimental Medicine and Biology 1096 (2018) 31-39. https://doi.org/10.1007/5584_2018_192

[35] T. Lipiński, J. Pietraszek, Corrosion of the S235JR Carbon Steel after Normalizing and Overheating Annealing in 2.5% Sulphuric Acid at Room Temperature, Materials Research Proceedings 24 (2022) 102-108. https://doi.org/10.21741/9781644902059-16

[36] J. Zwierzchowski et al., Model tests of wind turbine with a vertical axis of rotation type Lenz 2, EPJ Web of Conf. 143 (2017) art. 02150. https://doi.org/10.1051/epjconf/201714302150

[37] A. Wronska et al., Effect of tool pin length on microstructure and mechanical strength of the FSW joints of Al 7075 metal sheets, Communications – Scientific Letters of the University of Žilina 21 (2019) 40-47

[38] P. Krysiak, Microstructural Analysis of Wound Composites with Considerations on the Fiber Winding Force, Materials Research Proceedings 34 (2023) 43-52. https://doi.org/10.21741/9781644902691-6

Terotechnology XIV | Materials Research Forum LLC
Materials Research Proceedings 62 (2026) 385-392 | https://doi.org/10.21741/9781644904015-49

Innovative Coatings Produced Using Technologies that Use a Concentrated Energy Stream for Military Applications

Norbert RADEK[1,2,a*], Jacek PIETRASZEK[3,b], Łukasz J. ORMAN[4,c], Dariusz GONTARSKI[1,d], Damian PRZESTACKI[5,e], Jozef BRONČEK[6,f], Olga PARASKA[7,g], Maria RADEK[8,h], Radosław BIELAWSKI[9,i], and Paweł TWARDOWSKI[5,j]

[1]Kielce University of Technology, Faculty of Mechatronics and Mechanical Engineering, Al. 1000-lecia Państwa Polskiego 7, 25-314 Kielce, Poland

[2]Barwa Kielce, ul. Warkocz 3-5, 25-253 Kielce, Poland

[3]Cracow University of Technology, Faculty of Mechanical Engineering, Al. Jana Pawła II 37, 31-864 Cracow, Poland

[4]Kielce University of Technology, Faculty of Environmental Engineering, Geodesy and Renewable Energy, Al. 1000-lecia P.P. 7, 25-314 Kielce, Poland

[5]Poznan University of Technology, Faculty of Mechanical Engineering, Poznan University of Technology, ul. Piotrowo 3, 61-138 Poznan, Poland

[6]University of Zilina, Univerzitna 1, 01026 Zilina, Slovakia

[7]Khmelnitskiy National University, Department of Chemical Technology, Institutska 11, 29016 Khmelnickiy, Ukraine

[8]Cracow University of Technology, Faculty of Chemical Engineering and Technology, ul. Warszawska 24, 31-155 Cracow, Poland

[9]Science & Technology Organization, Collaboration Support Office, BP 25, F-92201 Neuilly-sur-Seine, France

[a]norrad@tu.kielce.pl, [b]jacek.pietraszek@pk.edu.pl, [c]orman@tu.kielce.pl, [d]gontar@tu.kielce.pl, [e]damian.przestacki@put.poznan.pl, [f]jozef.broncek@fstroj.utc.sk, [g]olgaparaska@gmail.com, [h]mariaradek@wp.pl, [i]radoslaw.bielawski@cso.nato.int, [j]pawel.twardowski@put.poznan.pl

Keywords: Electro-Spark Deposition, Laser Beam Machining, Coating, Military Technology

Abstract. Currently, the role of concentrated energy processing is growing. Processing using concentrated energy sources is primarily used for machining components made of difficult-to-machine engineering materials, as well as for fabricating components with highly complex shapes that would be labor-intensive and time-consuming using traditional methods. This paper analyzes two processing methods used to produce anti-wear coatings: electro-spark deposition (ESD) and laser beam machining (LBM). By controlling electrode erosion parameters, such as current intensity or capacitor capacitance, we can shape coatings with different functional properties. By controlling laser parameters, such as power, scanning speed, and pulse duration, it is possible to create coatings with different surface properties. This paper analyzes these processes in the context of producing coatings for military applications.

Introduction

In the era of globalization, companies achieve product competitiveness by enhancing their manufacturing processes and investing in new technologies. The introduction of new products is often accompanied by the use of new, high-strength materials, leading to a significant increase in demand for unconventional, high-performance machining processes. In addition to optimizing existing and well-established processes, the consistent implementation of new technologies yields results. These include beam technologies using concentrated energy flow [1].

The role of beam technologies is growing in shaping the desired surface features [2,3]. These technologies are primarily used for machining components made of difficult-to-process engineering materials, as well as for manufacturing components with very complex shapes, which would be labor- and time-consuming to produce using traditional methods.

These technologies, which utilize concentrated energy flow, include electro-spark deposition and laser beam machining [4-6]. ESD has already undergone dynamic development and currently holds a well-established position among surface treatments. LBM continues to experience dynamic development in the field of surface engineering. These technologies meet industry requirements for the quality and repeatability of manufactured and refurbished machine components.

A characteristic feature of beam technologies is that materials are processed through the action of a concentrated energy beam. By controlling the energy beam, we can process selected areas of the workpiece. The diameter of the energy beam's impact spot varies from a fraction of a nanometer to several dozen centimeters. This depends on the underlying physical phenomenon occurring in each technology. The smallest impact field is characterized by methods involving a beam of electrons or a beam of laser radiation. The energy beams occurring here can be focused on and controlled by systems of optical or magnetic lenses or mirrors, and the size of the spot at the focus depends only on the wavelength of the electron or laser radiation.

Coatings Produced Using ESD Technology

An example of a potential military application is the use of Stellite coatings applied via electro-spark deposition. Surface layers of this type can be used, among other things, on weapons and military equipment that are subject to abrasive, corrosive, erosive, and fatigue wear. Examples of Stellite coatings applied via ESD to components of U.S. Army military equipment are shown in Figs. 1 and 2.

Fig. 1. M1A1 main battle tank cannon cradle [7].

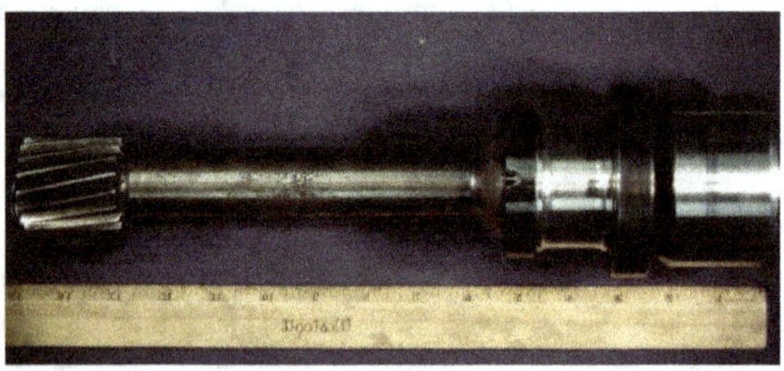

Fig. 2. Photomacrograph of an M1A1 helical gear shaft [7].

The work involved testing the performance of stellite coatings produced by electro-spark alloying. The coating properties were assessed based on microstructure observations, corrosion resistance tests, and microhardness measurements. The surface layers were electro-spark deposited using a stellite electrode (3 mm x 4 mm cross-section, 40 mm long) on a C45 carbon steel substrate. The chemical composition of the stellite electrode is presented in Table 1. The electro-spark coating deposition parameters using the EIL-8A device were as follows:

- voltage U = 230 V,
- capacitor capacitance C = 300 μF,

- current I = 2.1 A,
- exposure time τ = 2 min/cm^2.

Table 1. Chemical composition of the stellite electrode

Elements	C	Cr	Mn	Fe	Co	W
Contents, %	7.0	56.0	1.0	0.6	25.0	10.4

SEM microstructure observations revealed that the resulting layer had a thickness of 10-15 μm, while the heat-affected zone (HAZ) extended approximately 15-20 μm into the substrate material. The microstructure photograph (Fig. 3a) shows a clear boundary between the coating and the substrate. Furthermore, the photograph (Fig. 3a) shows that the coating exhibits high continuity in layer thickness and a lack of pores and microcracks, which may indicate an optimal selection of process parameters.

Linear analysis (Fig. 3b) of the stellite coating revealed an uneven distribution of elements within the coating. Zones containing significant amounts of W, Co, and Cr can be distinguished. Furthermore, the distribution patterns of Fe and Co are typical of a diffusion bond. The coating under study exhibits a uniform distribution of the components.

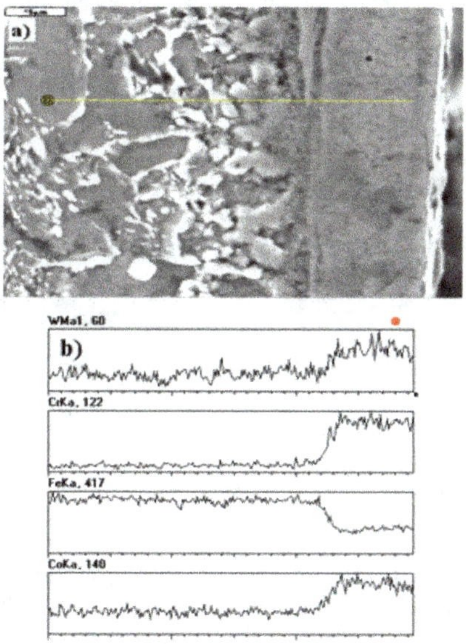

Fig. 3. Microstructure (a) and linear distribution of elements (b) in the stellite coating.

Corrosion resistance tests showed that the stellite coatings were nine times more resistant than C45 steel. The corrosion current density of the stellite coating was I_{cor}=12.7 μA/cm^2, while that of the substrate material was I_{cor}=112 μA/cm^2. This is due to the high chromium content (56%) in the electrode and the absence of microcracks and pores in the stellite coating.

Hardness measurements were performed using the Vickers method, applying a load of 0.4 N. Penetrator indentations were made on perpendicular microsections in three zones: the coating

Materials Research Proceedings 62 (2026) 385-392 https://doi.org/10.21741/9781644904015-49

(white layer), the heat-affected zone (HAZ), and the base material. The hardness of the stellite coatings was almost three times greater than that of the base material (C45 steel). The stellite coating had an average hardness of 492 HV0.4, while the average hardness of the base material was 178 HV0.4.

Laser Treatment of ESD Coatings

The research aimed to assess the effect of laser modification on the properties of WC-Cu coatings produced by ESD. The coatings were deposited on C45 carbon steel samples using an EIL-8A device. Electrodes produced by powder metallurgy through hot pressing from copper and tungsten carbide powders, with a composition of 50% WC and 50% Cu, were used for the coating application. Based on the instructions and our own experience, the following ESD parameters were adopted: voltage U = 230 V, capacitor capacitance C = 150 μF, current I = 0.7 A, and exposure time τ = 2 min/cm^2. Argon was used as the shielding gas.

Laser modification of the surface layers was performed using a Nd:YAG laser (pulsed operating mode), model BLS 720. Samples with electrospark coatings were irradiated with a laser using the following processing parameters: laser spot diameter d = 0.7 mm; laser power P = 50 W, 60 W, 70 W; beam speed v = 250 mm/min; nozzle-to-workpiece distance Δf = 6 mm; pulse duration t_i = 0.45 ms; repetition frequency f = 50 Hz; beam stroke S = 0.4 mm; nitrogen shielding gas Q = 25 l/min.

The research focused on analyzing the morphology of electrospark coatings before and after laser processing. Analyzing the morphology of ESD coatings revealed that they contained microcracks and pores (Fig. 4). Laser beam machining of the electrospark coatings resulted in a homogenization of the chemical composition, a refinement of the structure, and the elimination of microcracks (Fig. 5).

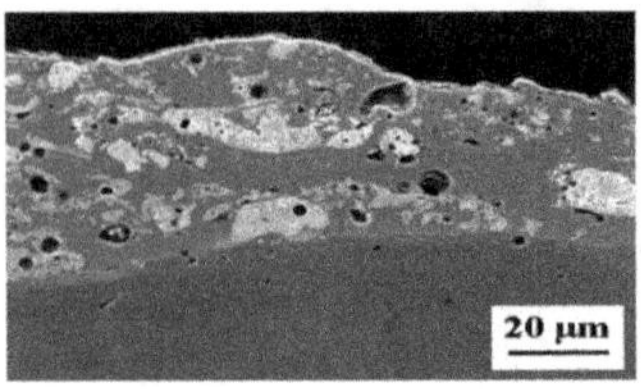

Fig. 4. Microstructure of the WC-Cu coating.

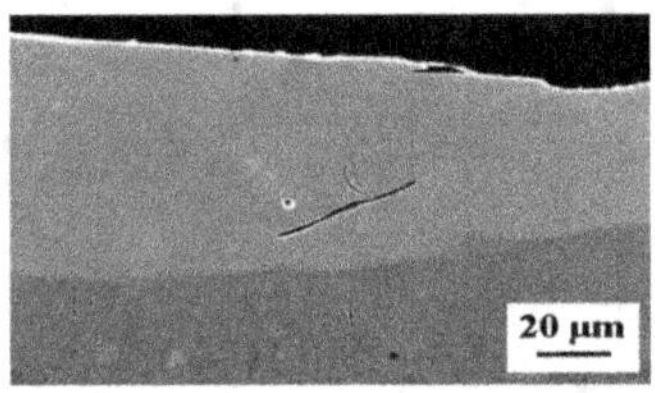

Fig. 5. Microstructure of the WC-Cu coating after laser treatment, P=60 W.

Additionally, measurements of the surface geometric structure (SGS), porosity, hardness, and adhesion of the WC-Cu coatings were performed. LBM positively influenced the adhesion of the coatings and reduced their porosity (Table 2), and to a small extent, their microhardness. The surface geometric structure parameters of the ESD coatings were lower compared to the SGP parameters of the laser-treated coatings. Laser modification resulted in increased corrosion resistance and improved tribological properties (including a reduction in the coefficient of friction by approximately 22%) of the formed surface layers.

Materials Research Proceedings 62 (2026) 385-392
https://doi.org/10.21741/9781644904015-49

Table 2. Porosity test results of WC-Cu coatings before and after laser treatment

Coating	Porosity, %			Average value, %
	Measurement number			
	1	2	3	
WC-Cu	5.7	6.3	3.5	5.2
WC-Cu +laser (P=50 W)	0.6	0.4	0.4	0.5
WC-Cu +laser (P=60 W)	0.4	0.2	0.1	0.2
WC-Cu +laser (P=70 W)	0.1	0.1	0.1	0.1

Laser-modified WC-Cu coatings can perform anti-wear and protective functions, enabling their potential use on weapons and military equipment. Surface layers produced by additive electrical discharge machining, both before and after laser processing, can be used on components of the standard "Grot" rifle (e.g., the gas piston face) and the "Beryl" assault rifle (e.g., the bolt carrier surfaces, gas piston), as well as on components of an infantry fighting vehicle (e.g., track link connectors, drive sprocket ring). Figures 6 and 7 illustrate potential applications of surface layers on weapons and military equipment.

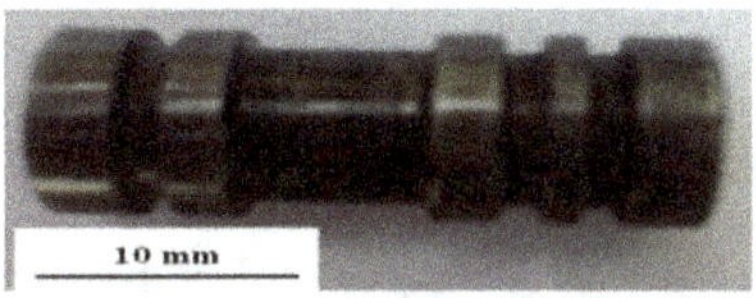

Fig. 6. Gas piston of the standard "Grot" rifle.

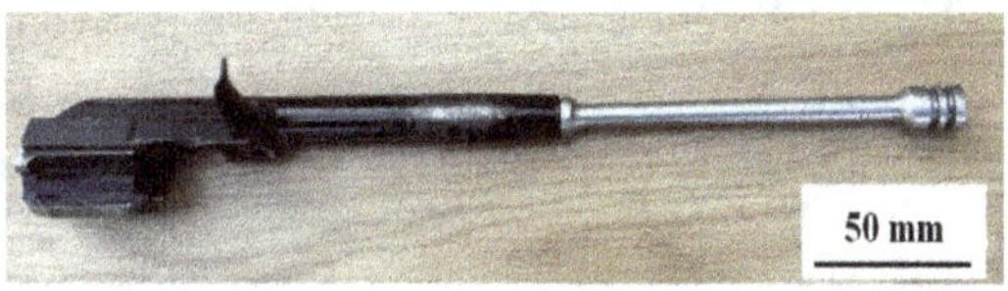

Fig. 7. Bolt with gas piston of the "Beryl" assault rifle.

Technological, Economic, Environmental, and Social Impact

The use of focused energy beam processing techniques, such as electrospark deposition (ESD) and laser beam machining (LBM), is revolutionizing the production of advanced coatings. These technologies enable innovative surface modifications [8] and the creation of composite coatings [9]. ESD is a cost-effective coating method [10] that enables the local deposition of layers, from very thin (micrometers) to thick (millimeters), from virtually any metallic material, ceramic, or their composites.

In technological terms, laser surface alloying [8] and laser texturing [11] significantly alter the performance properties of materials. Modification of ESD coatings through laser remelting results in beneficial changes, including the elimination of porosity and homogenization of the chemical composition, which enhances functional properties [5], including corrosion resistance [12], and overall operational properties [13]. Detailed analysis of the surface layer using advanced imaging techniques, such as X-ray nanotomography [14] and SEM microscopy [15], allows for precise verification of the coating geometry and microstructure. Achieved properties, such as high hardness and wear resistance, are crucial, especially when combined with control of the geometric and microstructural parameters of the oxide coatings [16]. The process precision is supported by advanced statistical methods such as DOE and Fuzzy Statistics [17], which are essential for optimizing technological processes and maintaining stability in the era of Industry 4.0 [18, 19]. Even advanced techniques such as Machine Learning are used for surface quality analysis and early damage detection [20].

ESD is considered a cost-effective technology due to the simplicity of the equipment and the possibility of rapid and local coating deposition [10]. Achieving repeatable and high-quality coatings is crucial from an economic perspective. In this context, statistical and qualitative methods, including logistics systems and quality control [21, 22], have a direct impact on production efficiency, minimizing scrap, and reducing costs. Research on electro-spark alloying of carbon steel, for example, with WC-Co-Al_2O_3 [10], provides data for optimizing cost and quality. The use of advanced statistical methods, including DOE and regression analysis [17], allows for more effective process design and reduction of final product variance. Pro-quality measures are crucial for sustainable development and the circular economy [23]. One of the key advantages of ESD and LBM is their relatively low environmental impact, especially compared to traditional chemical surface treatment methods [10]. The article indicates that ESD coatings are characterized by a lack of harmful environmental impact [6]. An environmental perspective is a key element of modern technologies, which is why quality and environmental assessment methods [24] and full life cycle assessment (LCA) [25], for example, for industrial products, are becoming standard. These analyses, including eco-cost studies [26], are essential for verifying the sustainability of production processes. Also important is research on the use of industrial waste, such as fly ash, in building materials [27], which aligns with the concept of closed circulation and minimizing environmental impact, including pollution [28].

The direct application of these innovative coatings in military technology is the primary focus of this article, with a focus on enhancing the performance of critical components. Coatings developed using LBM and ESD are designed to meet the stringent requirements of military environments [29]. Specific applications include camouflage and anti-wear coatings:

– anti-radar/camouflage coatings: research on the microwave absorption properties of carbonyl iron-based coatings [29] and analysis of the impact of exposure on the camouflage performance of military coatings [30] are crucial for stealth and protection technologies,

– improving operation and safety: improving the performance of coatings, including corrosion resistance [31], directly impacts the durability and reliability of military equipment, reducing downtime and maintenance costs.

Indirect social impact is related to occupational safety in manufacturing processes [22] and overall efficiency and productivity, which can be measured even in the context of environmental quality within industrial buildings [32], resulting in improved working conditions.

Summary

Electro-spark alloying coating is a cost-effective surface treatment technology. This method, along with numerous modifications, has established a permanent place in the field of surface treatment. This is due to the method's features, in particular: the possibility of localized action, the ability to apply thin (from 1 μm) and thick (up to 10 mm) tight coatings from any metallic material, and the simplicity and low cost of the coating application equipment. Another advantage of ESD coating is its minimal environmental impact. Considering technical, technological, and ecological considerations, it can be concluded that the described method meets the criteria defined for modern technologies.

Using laser processing to modify electro-spark coatings results in beneficial changes to the surface layer. By appropriately controlling laser parameters, layers with various functional properties can be shaped, including surface geometry, microhardness, stress state, corrosion resistance, and more. Using laser processing to modify electro-spark coatings enhances their performance by increasing corrosion resistance and introducing a favorable distribution of residual stresses in the surface layer. Laser remelting completely eliminates the coating's porosity and homogenizes the coating's chemical composition.

References

[1] T. Burakowski, T. Wierzchoń, Surface engineering of metals – principle, equipment, technology, CRC Press, Boca Raton, 1999.

[2] M. Madej et al., Properties of diamond-like carbon coatings deposited on CoCrMo alloys, Transactions of FAMENA 39/1 (2015) 79-88.

[3] S. Tofil, The micro machining of polypropylene by UV laser-the influence of laser operating parameters, Materials Research Proceedings 17 (2020) 126-131. https://10.21741/9781644901038-19

[4] N. Radek et al., The impact of laser processing parameters on the properties of electro-spark deposited coatings, Archives of Metallurgy and Materials 2 (2018) 809-816. https://10.24425/122407

[5] N. Radek et al., The WC-Co electrospark alloying coatings modified by laser treatment, Powder Metallurgy and Metal Ceramics 47, 3-4 (2008) 197-201. https://doi.org/10.1007/s11106-008-9005-7

[6] N. Radek et al., Formation of coatings with technologies using concentrated energy stream, Production Engineering Archives 28 (2022) 117-122. https://10.30657/pea.2022.28.13

[7] Information on https://apps.dtic.mil/sti/citations/ADA453366

[8] A. Dudek et al., Laser Surface Alloying of Sintered Stainless Steel, Materials 15 (2022) art. 6061. https://doi.org/10.3390/ma15176061

[9] J. Korzekwa et al., Tribological behaviour of Al_2O_3/inorganic fullerene-like WS_2 composite layer sliding against plastic, International Journal of Surface Science and Engineering 10 (2016) 570-584. https://doi.org/10.1504/IJSURFSE.2016.081035

[10] N. Radek et al., Electrospark alloying of carbon steel with WC-Co-Al_2O_3: Deposition technique and coating properties, Advanced Materials Research 874 (2014) 101-106. https://doi.org/10.4028/www.scientific.net/AMR.874.101

[11] A. Kalinowski et al., Laser surface texturing: characteristics and applications, System Safety: Human - Technical Facility - Environment 5 (2023) 240-248. https://doi.org/10.2478/czoto-2023-0026

[12] M. Scendo et al., Influence of laser power on the corrosive resistance of WC-Cu coating, Surface and Coatings Technology 259 (2014) 401-407. https://doi.org/10.1016/j.surfcoat.2014.10.062

[13] N. Radek, Determining the operational properties of steel beaters after electrospark deposition, Eksploatacja i Niezawodnosc 44 (2009) 10-16.

[14] A. Gądek-Moszczak et al., Nano X-ray Tomography Application for Quantitative Surface Layer Geometry Analysis after Laser Beam Modification, Materials 15 (2022) art. 5935. https://doi.org/10.3390/ma15175935

[15] J. Korzekwa et al., The influence of sample preparation on SEM measurements of anodic oxide layers, Practical Metallography 53 (2016) 36-49. https://doi.org/10.3139/147.110367

[16] J. Korzekwa et al., Influence of the Size of Nanoparticles on the Microstructure of Oxide Coatings, Materials Science 53 (2018) 709-716. https://doi.org/10.1007/s11003-018-0127-x

[17] J. Pietraszek et al., Advanced statistical refinement of surface layer's discretization in the case of electro-spark deposited carbide-ceramic coatings modified by a laser beam, Solid State Phenomena 197 (2013) 198-202. https://doi.org/10.4028/www.scientific.net/SSP.197.198

[18] J. Pietraszek et al., Challenges for the DOE methodology related to the introduction of Industry 4.0, Production Engineering Archives 26 (2020) 190-194. https://doi.org/10.30657/pea.2020.26.33

[19] R. Dwornicka et al., Fuzzy Statistics-Aided Inference in Experimental Design, World Congress in Computational Mechanics and ECCOMAS Congress (2024). https://doi.org/10.23967/wccm.2024.131

[20] D. Nowakowski et al., Application of machine learning in the analysis of surface quality - the detection the surface layer damage of the vehicle body, METAL 2021 – 30th Int. Conf. Metall. Mater. (2021) 864-869. https://doi.org/10.37904/metal.2021.4210

[21] R. Ulewicz et al., Logistic controlling processes and quality issues in a cast iron foundry, Materials Research Proceedings 17 (2020) 65-71. https://doi.org/10.21741/9781644901038-10

[22] R. Ulewicz et al., Quality and work safety in metal foundry, METAL 2020 – 29th Int. Conf. Metall. Mater. (2020) 1287-1293. https://doi.org/10.37904/metal.2020.3649

[23] A. Pacana et al., Pro-quality method improving industrial products toward sustainable development with criteria of circular economy, METAL 2023 – Int. Conf. Metall. Mater. (2024) 697-702. https://doi.org/10.37904/metal.2023.4743

[24] A. Pacana et al., Method Supporting Improving Products in Terms of Qualitative-Environmental, Materials Research Proceedings 34 (2023) 226-236. https://doi.org/10.21741/9781644902691-27

[25] A. Pacana et al., Life Cycle Assessment (LCA) of Truck Steel Wheels, Materials Research Proceedings 45 (2024) 213-222. https://doi.org/10.21741/9781644903315-25

[26] D. Siwiec et al., Analysis of aluminum can by eco-costs and life cycle assessments (LCA), METAL 2024 – Int. Conf. Metall. Mater. (2024) 627-632. https://doi.org/10.37904/metal.2024.4975

[27] A. Pietrzak et al., Application of a Mixture of Fly Ash and Solid Waste from Gas Treatment from Municipal Solid Waste Incineration in Cement Mortar, Materials 18 (2025) art. 481. https://doi.org/10.3390/ma18030481

[28] R. Dwornicka, The impact of the power plant unit start-up scheme on the pollution load, Advanced Materials Research 874 (2014) 63-69. https://doi.org/10.4028/www.scientific.net/AMR.874.63

[29] W. Przybył et al., Microwave absorption properties of carbonyl iron-based paint coatings for military applications, Defence Technology 22 (2023) 1-9. https://doi.org/10.1016/j.dt.2022.06.013

[30] M. Michalski et al., The Influence of Operational Exposure on Changes in Parameters of Effective Camouflage of Coatings Used in Military Technology, Advances in Science and Technology Research Journal 17 (2023) 182-196. https://doi.org/10.12913/22998624/156940

[31] M. Scendo et al., Influence of laser treatment on the corrosive resistance of WC-Cu coating produced by electrospark deposition, International Journal of Electrochemical Science 8 (2013) 9264-9277.

[32] Ł.J. Orman et al., Comparative Analysis of Indoor Environmental Quality and Self-Reported Productivity in Intelligent and Traditional Buildings, Energies 16 (2023) art. 6663. https://doi.org/10.3390/en16186663

Materials Research Proceedings 62 (2026) 393-398 | https://doi.org/10.21741/9781644904015-50

Manufacture and Applications of Electro-Spark Coatings Deposited

Norbert RADEK[1,a*], Monika MADEJ[1,b] and Szymon DRABIK[1,c]

[1]Kielce University of Technology, Faculty of Mechatronics and Mechanical Engineering, Al. 1000-lecia Państwa Polskiego 7, 25-314 Kielce, Poland

[a]norrad@tu.kielce.pl, [b]mmadej055@gmail.com, [c]sdrabik@tu.kielce.pl

Keywords: Electro-Spark Deposition, Coating, Properties, Applications

Abstract. The processes of coating formation on metal parts including electro-spark deposition involve mass and energy transport accompanied by chemical, electrochemical and electrothermal reactions. The paper is concerned with the performance properties of electro-spark deposited coatings. The properties were assessed by analyzing the coating microstructure, microgeometry, microhardness, corrosion resistance and application tests. The studies were conducted using WC-Co-Al_2O_3 electrodes produced by sintering nanostructural powders. The anti-wear coatings were electro-spark deposited over C45 carbon steel by means of an EIL-8A. These coatings are likely to be applied to increase the abrasive wear resistance of tools and machine parts.

Introduction

Sintered carbides are cermets consisting of 70÷96% refractory metal (e.g. tungsten, tantalum, niobium) carbides and a binding matrix that is usually cobalt, sometimes molybdenum or nickel, and occasionally iron. Presently, sintered carbides are a very popular material for making cutting tool edges, especially in turning and milling operations [1].

Sintered carbides are classified into categories depending on their chemical composition, or on WC grain size. WC-Co type carbides are divided, on the basis of WC grain size, into the following groups [2]: coarse-grained - with 3÷30 μm average diameter, standard - with 1.5÷3 μm average diameter, fine-grained - with 0.5÷1.5 μm average diameter, ultrafine-grained - with an average diameter smaller than 0.5 μm.

An interesting option is provided by the manufacture of anti-wear carbide-ceramic coatings that are electro-spark deposited (ESD) using electrodes obtained by powder metallurgy [3,4]. Super-hard coatings can be deposited on the edges of cutting tools, such as, e.g. lathe tools, milling cutters, pinion cutters or taps. It is supposed that the coatings of concern can be successfully applied to those machine components that operate under extreme conditions, including high abrasive wear or impact loads.

Another advantage offered by super-hard electro-spark deposited coatings concerns their low environmental impact. Electro-spark machining does not produce harmful ecological effect. Analysis of properties of coatings obtained from ESD requires many methods. Electro-spark deposition process is characterised by pulse discharges generated between the electrode and the substratum. In electro-spark machining, air is ionised onto the area of high-temperature and high-pressure fields, which results in alloy formation. Electro-spark alloying is a technology used for surface machining. It is characterised by high heat delivery and a very narrow heat affected zone (HAZ), which also refers to the deposition of refractory material coatings. There are alternative methods for applying anti-wear coatings to the ESD method [5-7].

Materials and Treatment Parameters

Coatings were deposited on the C45 grade plain-carbon steel by the ESD method using a portable EIL-8A electro-spark deposition facility (TRIZ, Ukraine). Electrodes containing 85% WC, 10% Co and 5% Al_2O_3, were produced using the powder metallurgy hot pressing route.

The powders were mixed for 30 minutes in the chaotic motion *Turbula T2C* mixer. The mixture was then poured into rectangular cavities of a graphite mould, each 6 × 40 mm in cross section, and consolidated by passing an electric current through the mould under uniaxial compressive load. A 3 minute hold at 950°C and under a pressure of 40 MPa permitted obtaining electrodes of porosity <10% and strength sufficient to maintain integrity when installed in the electrode holder.

The equipment used for electro-spark alloying was an EIL-8A model. Basing on the results of previous research as well as instructions given by the producer, the following parameters were assumed to be optimal for ESA:

- voltage, U = 230 V;
- capacitor volume, C = 150 µF;
- current intensity, I = 2.4 A.

Microstructure Analysis

A characteristic feature of any electro-spark deposited coating is that the new layer has a difficult-to-etch structure, i.e. it remains white. Similar layers are produced by grinding and lapping. What the processes have in common is high temperature and high loads applied locally. Electro-spark deposition differs from grinding and lapping in the process intensity: the pressure of the shock wave from an electric spark discharge is (2-7) $\cdot 10^6$ N/mm^2 and the temperature reaches values of the order of (5-40) $\cdot 10^3$ °C (in grinding it does not exceed 1000°C).

The temperature during electro-spark discharge increases locally and it is much higher than the boiling point of the materials the electrodes are made of. A high heat transfer rate causes the temperature within the layer to fall rapidly to the solidifying point, the thickness of the coating being of the order of several micrometers. The processes of crystallization, phase transition and chemical interaction occur in the solid phase. Electro-spark deposited coatings are fine-grain non-equilibrium structures, which are heterogeneous in composition, microstructure and properties. They are characterized by very high adhesion to the underlying substrate as a result of diffusion or reactive diffusion processes.

A microstructure analysis was conducted for WC-Co-Al_2O_3 coatings using the Joel JSM-5400 scanning electron microscope and the Neophot 2 light microscope. In Fig.1 the microstructure of an ESD WC-Co-Al_2O_3 coating is illustrated. It is clear that the thickness of the obtained layers was from 60 to 70 µm, whereas the heat affected zone (HAZ) ranged approximately from 30 to 40 µm into the substrate. Figure 1 also reveals a clear boundary between the coating and the substrate, pores and microcracks.

A Philips PW 1830 X-ray diffractometer with CuKα radiation, operating at 40 kV and 30 mA, was used for phase(s) identification. As shown in Figure 2, the superficial layer of the coating consists of WC and W_2C as well as a small amount of Co_2C and Al_2O_3. W_2C is known to appear as an intermediate during the formation and dissolution of WC. Besides, it has been found that peaks from the W_2C phase are most intense.

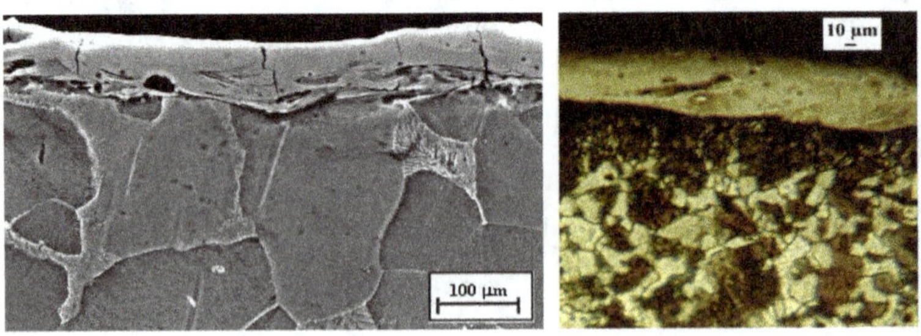

Fig. 1. SEM (left) and LM (right) micrographs of the polished cross section throug a WC-Co-Al_2O_3 ESD coating on C45 steel substrate.

Microgeometry Measurements

One of the main disadvantages of the coatings produced by electro-spark alloying is high surface roughness. By reviewing the literature and analyzing the latest developments in this technology, one can notice that the surface generation process involves erosion of the base material and formation of microcraters and ridges by particles leaving the electrode. The surface is regular with rounded microroughness peaks. The effect of the process parameters on the formation of surface roughness has been described in numerous publications. By controlling these parameters, it is possible to obtain surfaces with pre-determined microgeometry. Electro-spark alloying allows producing surfaces with enhanced roughness called surface relief.

The roughness of the WC-Co-Al_2O_3 coatings was measured at the Laboratory for Measurement of Geometric Quantities of the Kielce University of Technology using a TALYSURF CCI equipment.

The roughness was measured in two directions perpendicular to each other. Then, the average value was calculated: Ra = 4.99÷5.66 μm. The without coatings steel specimens (C45) had a roughness from 0.46 to 0.53 μm. Figure 2 presents an example three dimensional surface microgeometry measurement of the WC-Co-Al_2O_3 coatings.

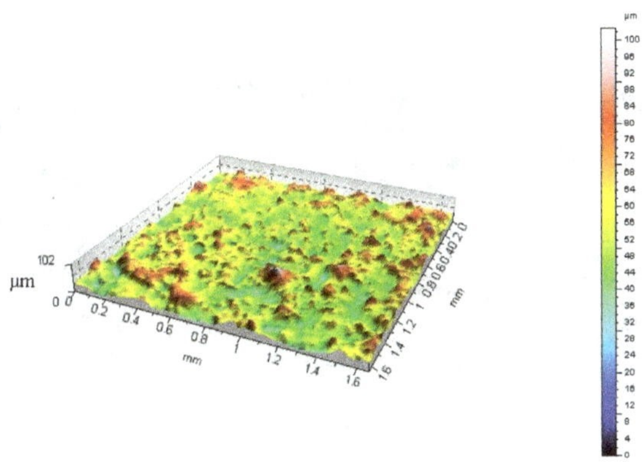

Fig. 2. Three dimensional surface microgeometry of the WC-Co-Al_2O_3 coating deposited.

Microhardness and Corrosion Resistance Tests

The microhardness of the specimens with WC-Co-Al_2O_3 coatings was analyzed applying a load of 0.4 N and using the Vickers method. The indentations were made consecutively in three zones: the coating, the heat affected zone (HAZ) and the base material. The average microhardness of the base material after ESA was 279 HV0.4. The value was the same as that at the initial state. The average microhardness of the WC-Co-Al_2O_3 coating was 906 HV0.4. Thus, there was a 225 percent increase compared to that of the base material. The microhardness of the heat affected zone after electro-spark alloying was 38 % higher in relation to that of the base material.

The corrosion resistance of the WC-Co-Al_2O_3 coatings and the underlying substrate was analyzed using a computerized system for electrochemical tests, Atlas'99, produced by Atlas-Sollich. The potentiodynamic method was applied, because it is reported to be one of the most effective methods of electrochemical testing. The cathode polarization curve and the anode polarization curve were determined by polarizing the samples with a potential shift rate of 0.2 mV/s in the range of ±200

Materials Research Proceedings 62 (2026) 385-392 https://doi.org/10.21741/9781644904015-49

mV of the corrosive potential, and with 0.4 mV/s in the range of higher potentials. Samples with a marked area of 10 mm in diameter were polarized up to a potential of 800 mV. The polarization curves were drawn for samples exposed for 24 hours to a 3.5% NaCl solution so that the corrosive potential could be established. The tests were performed at 21±1°C. The WC-Co-Al_2O_3 coating was reported to have the highest corrosion resistance. The corrosion current density of the coating was 16.8 μA/cm^2, while that of the C45 steel substrate was 35.4 μA/cm^2. Applying the WC-Co-Al_2O coating improved the sample corrosion resistance by approx 100%. The fusion of the coating and the substrate resulted in a considerable heterogeneity of electrochemical potentials on the coating surface. The microcracks in the surface layer also contributed to intensification of the corrosion processes.

Carbide-Ceramic Coating Applications

Tap life testing. Taps made of high-speed cobalt steel (HSSC) were selected for the tests. Three groups of taps were used to test their life. Those included: the tap with TiN coatings imposed with the PVD method, the tap with electro-spark imposed WC-Co-Al_2O_3 coating and the tap without any coating. The coatings were imposed on the tap working parts.

The durability criterion for the tap working part was assumed to be the form of abrasive wear or chipping of the material.

Tap life testing was performed on the Computerised Numerical Control (CNC) lathe, model YDPM BML- 280 (Fig. 3). The items tested were joints made from S235JR steel, manufactured at the facility and used in central heating stoves (Fig. 4).

Fig. 3. Computerised Numerical Control lathe with feeder. Fig. 4. Joint with a threading made on it.

The results of tap life testing are presented in Fig. 5. The in-service testing showed that the tap with the TiN coating had the longest life as it was capable of producing threading in 2,000 joints. The tap with the anti-wear WC-Co-Al_2O_3 coating could make threads in 500 joints.

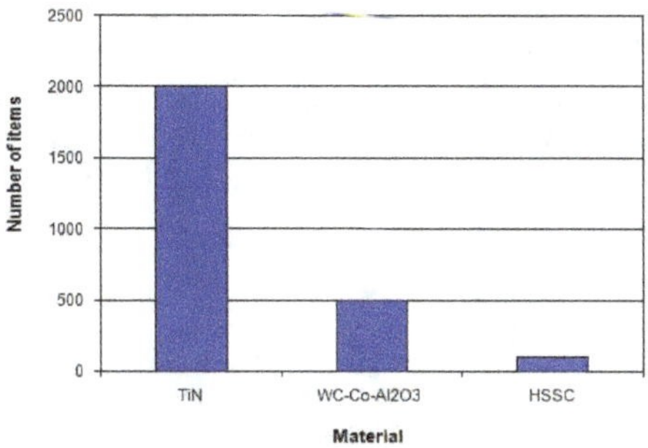

Fig. 5. Dependence of a number of threaded joints on the tap type.

The lowest number of threads, namely 100, was made using HSSC steel tap without a coating. Analysis of the working parts of the taps, which followed the tests, demonstrated the dominant effect of abrasive wear. The material chipping or nicking occurred to a far lesser extent. To sum up, the experimental results show that the tap with the electro-spark imposed WC-Co-Al_2O_3 coating can be used to machine precision threading with CNC lathes, though the number of threads will be smaller when compared with the TiN coated tap (PVD method).

Summary

The following conclusions can be drawn from the analysis and test results.

1. The microstructure analysis revealed that the coating thickness was 60-70 μm, whereas the heat affected zone ranged approximately 30-40 μm. The coatings possessed microcracks and pores.
2. A significant increase in roughness Ra was reported for specimens with WC-Co-Al_2O_3 coatings. Higher roughness, however, is not always considered a disadvantage. Under certain circumstances, valleys in the roughness profile act as lubricant reservoirs, which increases the rate of heat transfer and that of catalysis.
3. The microhardness of the WC-Co-Al_2O_3 coating produced by electro-spark alloying was 906 HV0.4, while that of the base material – C45 steel – was 279 HV0.4.
4. The obtained I_k values indicate over 100% increase in corrosion resistance of the ESD coated sample compared to uncoated C45 steel substrate.
5. Durability tests of taps with WC-Co-Al_2O_3 layers proved their higher durability (about 5 times) in comparison to taps without the layer.

References

[1] Information on https://www.fanar.pl (in Polish)

[2] M. Wysiecki, Nowoczesne materiały narzędziowe, WNT, Warszawa, 1997.

[3] N. Radek, Determining the operational properties of steel beaters after electrospark deposition, Eksploat. Niezawodn. 4 (2009) 10-16.

[4] W. Tarelnik, Combined Electro-Spark Alloying Technologies, Technika: Kiev, 1997. (in Ukrainian)

[5] M. Madej et al., Properties of diamond-like carbon coatings deposited on CoCrMo alloys, Transactions of FAMENA 39/1 (2015) 79-88.

[6] V. Sreenivasulu, et al., Development of protective coating for X8CrNiMoVNb16-13 alloy in high-temperature molten salt environment through high-velocity oxy-fuel sprayed NiCrMoNb

and $C_{r3}C_2$-25NiCr powder coating, J. Proc. Mech. Eng. 237 (2023) 1429-1441. https://doi.org/10.1177/09544089221115481

[7] S. Tofil et al., Surface micromachining of titanium alloys with picosecond laser to increase the adhesion force in glued joints, Proceedings of SPIE – The International Society for Optical Engineering 109740Q (2018) 10974. https://doi.org/10.1117/12.2516646

Keyword Index

e

www.ingramcontent.com/pod-product-compliance
Lightning Source LLC
LaVergne TN
LVHW052221200926
848481LV00001B/170
* 9 7 8 1 6 4 4 9 0 4 0 0 8 *